THE TELECONFERENCING RESOURCE BOOK:
A Guide to Applications and Planning

THE TELECONFERENCING RESOURCE BOOK:
A Guide to Applications and Planning

edited by

Lorne A. PARKER
Christine H. OLGREN

Center for Interactive Programs
University of Wisconsin-Extension
Madison, Wisconsin

1984

NORTH-HOLLAND – AMSTERDAM · NEW YORK · OXFORD

© 1984 by the Center for Interactive Programs, University of Wisconsin-Extension, Lowell Hall, 610 Langdon Street, Madison, Wisconsin 53703, U.S.A.

ISBN: 0 444 86887 9

Published by:
ELSEVIER SCIENCE PUBLISHERS B.V.
P.O. Box 1991
1000 BZ Amsterdam
The Netherlands

Sole distributors for the U.S.A. and Canada:
ELSEVIER SCIENCE PUBLISHING COMPANY, INC.
52 Vanderbilt Avenue
New York, N.Y. 10017
U.S.A.

PRINTED IN THE NETHERLANDS

PREFACE

This publication contains 54 papers that provide a range of information about developments and trends in using teleconferencing for meetings, training and marketing functions in a number of sectors — business, education, health care and government.

The papers address topics of key importance to teleconferencing, which involves the use of interactive audio, audiographic, video or computer systems for electronic group communications. Included are a variety of case study applications that examine teleconferencing from a user's perspective. Other papers describe developments in technology or discuss factors critical to successful implementation, training and usage. Except for the introductory paper, they were selected from three collections of papers compiled by the Center for Interactive Programs, University of Wisconsin-Extension, and printed in conjunction with conferences held in 1980, 1982 and 1983. Altogether, those three collections contain nearly 200 papers on teleconferencing and electronic communications.

The papers are grouped into seven sections:

    Audio Teleconferencing Applications
    Audiographic and Freeze-Frame Video Applications
    Videoconferencing Applications
    Teleconferencing Technologies
    System Planning, Evaluation and Research
    Training and Program Design
    Computer Conferencing

We hope these divisions are useful to readers. We recognize, however, that some papers could be put into multiple categories.

Although the decade of the 80's has not yet reached mid-point, it has brought dramatic changes to teleconferencing. Today's industry is in a transition period marked by new technologies, new applications, and a growing number of product or service suppliers.

On the technology side, we are seeing the trend toward digital systems on many fronts. New teleconferencing bridges, for example, provide many automated features as well as integrated voice and data communication. Audiographic products show the growing development of computerized multi-function workstations for voice, data and visual displays.

Digital technology is also affecting video teleconferencing. Compressed video codecs, or processors, are being used to reduce bandwidth requirements and transmission costs. Their impact on teleconferencing promises to grow as new products are introduced with more efficient compression techniques.

We are also seeing an increase in digital communication networks and a movement toward more international teleconferencing that brings people together worldwide via satellite or other channels. With the growing emergence of international teleconferencing, we need to develop more compatible technologies and increase our understanding about how to effectively use systems that span national boundaries and cultures.

In looking at forces that promote teleconferencing — advances in technology, the socioeconomic climate and its benefits as a communication tool — it appears that teleconferencing is on the threshold of dramatic growth. There are, however, other factors that come into play when one attempts to actually apply the technology. These factors will ultimately contribute to the success or failure of teleconferencing.

Not all teleconferencing applications have been successful, and we have gradually come to realize that widespread acceptance depends on a number of considerations. They range from the design of the equipment to user attitudes and the organizational infrastructure that supports applications. By placing technology within a user context, we can begin to understand the interaction between the technical system and the broader issue of human factors.

The book includes over 20 case study applications. They not only point out the great diversity in the ways teleconferencing is used, but they also underscore the point that teleconferencing is more than technology. The authors' experience with teleconferencing contributes to our understanding of the importance of such factors as user-oriented system designs, advocacy and marketing, implementation strategies, program planning, user training, support services and impact evaluation.

Other papers in the book examine some of these issues separately to provide additional insight into their relationships to successful teleconferencing. These papers are contained in the sections on Training and Program Design as well as System Planning, Evaluation and Research.

As the scope of this book indicates, teleconferencing has many dimensions. It is only through an understanding of these multiple facets that teleconferencing can truly progress to realize its potential as a tool for communications and information management.

This publication is based on the contributions of many people. Special acknowledgment must go to the authors for sharing their experience, knowledge and perspectives and for adding to the information about teleconferencing. We dedicate this book to them and other teleconferencing professionals for their contributions to the field.

Lorne A. Parker
Director

Christine H. Olgren
Associate Director

Center for Interactive Programs
University of Wisconsin-Extension
Madison, Wisconsin

CONTENTS

AUDIOGRAPHIC AND FREEZE-FRAME VIDEO APPLICATIONS

VIDEOCONFERENCING APPLICATIONS

TRAINING AND PROGRAM DESIGN

Contents

COMPUTER CONFERENCING

The Teleconferencing Resource Book: A Guide to Applications and Planning
Lorne A. Parker and Christine H. Olgren (eds.)
Elsevier Science Publishers B.V. (North-Holland)
© Center for Interactive Programs, University of Wisconsin-Extension, 1984

TELECONFERENCING IN THE UNITED STATES:
TODAY'S STATUS AND TRENDS

Lorne A. Parker
Director
and
Christine H. Olgren
Associate Director
Center for Interactive Programs
University of Wisconsin-Extension
Madison, Wisconsin

Introduction

To provide an overview to teleconferencing, particularly recent and projected developments in the United States, this paper focuses on three areas: market trends, system options, and technological trends.

Part 1, market trends, presents some of the data collected in a 1983 survey of teleconferencing users conducted by the Center for Interactive Programs. It includes information about current teleconferencing applications as well as certain factors affecting the marketplace and future demand.

Part 2, system options for teleconferencing, presents background information about the major types of teleconferencing technologies available to users. It is intended as a primer for people who may not be too familiar with teleconferencing and as a summary of the basic types of systems used today.

Part 3, technological trends, discusses the major developments in teleconferencing technologies as well as specific products and services. It examines the current U.S. teleconferencing industry and trends in the emergence of new technological choices.

Part 1. Market Trends

Years ago, the question "Does teleconferencing work?" was commonly heard in telecommunications circles. Thanks to technical improvements, many people now are asking "Which teleconferencing system is best for my group's communication needs?" Few doubt that the technology needed for reliable and efficient teleconferencing is here.

However, trying to determine teleconferencing's current status from existing evidence -- much of which is anecdotal -- can be confusing. While some people herald the technology as an efficient way to cut travel costs and increase productivity, others claim teleconferencing equipment is too complex and

does not meet user expectations. This confusion -- along with the lack of hard data on past, current and future use of teleconferencing -- prompted the Center For Interactive Programs (CIP) to conduct a broad base marketing study of North American teleconferencing users. Results from the study help delineate today's market conditions, describe current users and predict who will next purchase teleconferencing systems.

The Study

The information presented here comes from one of several research projects in this study which is called Teleconferencing Applications and Markets 1984-1990. The CIP team will complete the majority of the study in 1984, but they conducted an initial survey of 149 North American teleconferencing users in early 1983. About 40 percent of the sample were from businesses, 45 percent from educational organizations and 15 percent from government agencies. The CIP staff used computer-assisted telephone interviews to contact those respondents most knowledgeable about the organization's teleconferencing involvement.

Diffusion Of Innovations

The survey's conclusions support Everett Rogers' diffusion of innovation theory. Rogers claims an innovation will diffuse through a user group most easily if it: (1) has relative advantage over what it is replacing; (2) is compatible with the user's current methods and organization; (3) is easy to use; (4) can be examined on a trial basis; and (5) is easy to explain the product's nature and operating instructions.

Rogers' theory holds that the first group to adopt new technology -- the innovators -- is just 2.5 percent of the total market. The next group to adopt is known as the early adopters, who comprise 13.5 percent of the market.

These two groups are followed by the early majority, who make up 34 percent of the market, rely on their peers' opinions to obtain product information, and generally take a "wait and see" approach to new technology. The remaining groups are the late majority (34 percent) and the laggards (16 percent).

At this point, the teleconferencing industry probably has sold the technology to the innovators and to half the early adopters. But the early and late majority groups -- comprising 68 percent of the total market -- still are virtually untouched.

To market to this group, vendors must determine where teleconferencing fits into the organization's communications network and how the technology can solve problems -- many of which the users are not aware exist.

Company Background

Companies responding to the survey represented North America's "leading edge" organizations, which (according to results) tend to adopt numerous telecommunications innovations.

For example, 80 percent or more used audio/visual production facilities or word processing; 60 percent or more regularly sent and received computerized data between sites, had WATS lines or facsimile capabilities or engaged in teletraining; and 40 percent or more used electronic mail, had access to off-site data banks such as news or stock reports, used electronic funds transfers or had 800 numbers.

About 40 percent of the upper-level managers and more than 50 percent of the middle-level managers had computer terminals at their work stations.

First System

Some organizations had been teleconferencing since 1962. Three-quarters of the sample selected audio-only as their first system. About 10 percent of the group chose either audio with audiographics or two-way full-motion video. The remaining respondents installed freeze frame or one-way full-motion video.

These findings suggest users approach teleconferencing cautiously. Teleconferencing growth seems to be evolutionary, from simple to more complex systems. Since audio-only is an inexpensive and gradual change, this system is readily adopted as an organi-

zations first system.

About 60 percent of the sample said their first six months with teleconferencing were successful. Twenty percent said success was mixed, and the remaining group said their first experience was not successful. This indicates substantial variability in users' first experience. Marketers must determine if more training, promotional effort or "hand-holding" will reduce this variability.

The "Shoehorn"

Interpersonal sources had the greatest impact on the organization's decision to adopt teleconferencing. The most influential sources, in order, were vendors, non-organization people (such as a friend or associate), consultants and trade publications.

This indicates that marketing through traditional advertising channels is not as effective as interpersonal influence and direct contact with someone inside the organization-- a person we refer to as a "shoehorn" -- who can help explain the benefits and applications of teleconferencing.

Historically, "high tech" companies have been formed by engineers and technical personnel who are skilled at product design and manufacturing but not marketing. This is generally true of today's teleconferencing industry. Many of the vendors are small companies that tend not to have the financial resources or internal marketing expertise needed to develop an appropriate market strategy or aggressively sell their product or service. The results of the user survey showed that about 90 percent of the organizations contacted the vendor first to acquire information about teleconferencing, suggesting that vendors are not aggressive enough or are mistargeting their marketing effort.

Vendors should also focus their sales message by understanding which teleconferencing benefits potential users expect. According to the survey, vendors most commonly tried to sell teleconferencing by stressing the specific features and technical capabilities of their products. Secondly, the vendors emphasized that teleconferencing, by substituting for travel, could save the company money. These points contrast sharply with the more extensive user expectations, which include the following in order of importance: improved communications between sites, reduced travel time, reduced travel costs, improved information exchange, increased productivity, better meeting efficiency, and other applications-oriented benefits.

Current System

About 70 percent of the respondents reported having an audio-only system currently in operation. Thirty percent said they had audiographics, 19 percent had 2-way full-motion video, 18 percent had freeze frame and 17 percent used 1-way video with 2-way audio (figures exceed 100 percent because many users had more than one system).

Respondents used audio and 2-way full-motion video systems an average of about once per day, audiographics two or three times per week and freeze frame and 1-way full-motion video about once per week.

Survey questions also asked users how satisfied they were with their teleconferencing equipment, in order to discover perceived differences among system types. Although responses indicated little variability, 2-way video was slightly more satisfying than (in order) 1-way video, audio-only, audiographics and freeze frame.

Video's popularity may be due to its similarity to face-to-face meetings, the "yardstick" by which executives often measure teleconferencing. Executives also like audio equipment, perhaps because it is not much different or unfamiliar than a telephone. Audiographics and freeze frame systems often are installed in inappropriate environments and generally require more training and expertise to operate, which perhaps explains their lower satisfaction rating.

Current System Success

About 85 percent of respondents said their organization's teleconferencing experiences have been successful or very successful. Eight percent reported mixed success and the remaining group said their experiences have not been successful.

This represented much less success-rating variability than in the first-system ratings. One-quarter of those reporting successful teleconferencing experiences indicated users accepted the technology and made it part of their work routine. Other reasons given to explain a system's success included cost effectiveness, increased use and travel savings.

The "Hurdle Effect"

The above information suggests the presence of a "hurdle effect," where telecon-

ferencing users must "clear a hurdle" to get from their first teleconferencing experience to a familiarity and acceptance of current teleconferencing technology.

That is, users must lose the awe and fear they feel for teleconferencing before they regard the technology as a transparent and productive communications tool. Those users who do not get across the hurdle tend to drop their teleconferencing commitment and become poor ambassadors to others contemplating adoption of the technology.

Demand Forecasts

The CIP team designed some survey questions to determine the sample's teleconferencing plans, in order to better predict the technology's future. Increased teleconferencing involvement can be interpreted as a "vote of confidence" in the technology and may be a reasonable indication of what other organizations would do after using teleconferencing for several years.

Comparing teleconferencing's past use with its projected demand proved useful. From 1962 to 1976, teleconferencing experienced a slow but definite growth phase as 40 percent of the sample adopted the technology.

Beginning in 1977, a strong surge is clear, with about 60 percent of the sample installing teleconferencing systems. While it took 15 years for 58 organizations to adopt the technology, only five more years passed before an additional 91 users came on line. Thus, 1977 can be regarded as the year teleconferencing began to enter mainstream telecommunications.

Teleconferencing Hours, Investment, Rooms

Comparing 1982 to 1981, the sample reported a net increase of 53 percent in the number of hours spent teleconferencing. The organizations also estimated increasing by an average 33 percent their teleconferencing hours in 1983 compared to 1982.

This near-doubling of time spent teleconferencing in two years is dramatic evidence of the industry's growth. The data also suggest that current and future teleconferencing organizations may double their hourly use of the technology every couple of years.

The sample planned to increase their current teleconferencing investment by 74 percent from 1983-1985. This sizable increase

indicates the sample feels teleconferencing
is worth the monetary investment.

Finally, about 40 percent of the organizations planned to increase the number of
teleconferencing rooms in their companies during 1983. Number of rooms is an important demand indicator, performing the same function
as "housing starts" do to indicate economic
health. In the same way analysts use housing
starts to predict demand for furniture, ap-
pliances, flooring and electrical fixtures, so
too can analysts use room numbers to project
demand for teleconferencing hardware and software sales.

If the organizations follow through on
their plans to increase numbers of rooms, they
will add a total of 655 teleconferencing rooms
by the end of 1983. About 40 percent of the
organizations plan to increase numbers of rooms
in 1984 and 1985, resulting in an additional
854 teleconferencing rooms.

Conclusions: The "Seeding Effect"

The CIP staff believes that survey results indicate teleconferencing growth is inevitable. In industry, most organizations
monitor a few key corporations -- known as
channel leaders to marketers -- for signs of
price fluctuations, new product introductions
and production innovations. Traditionally,
followers become convinced that the channel
leaders' methods are superior and quickly copy
the change.

The CIP team examined major U.S. industries and found that organizations commonly
recognized as channel leaders were using teleconferencing. This finding suggests that,
within the next few years, other organizations
will adopt the technology. The channel
leaders are "planting seeds" of teleconferencing growth within their industry.

In order to facilitate this growth, vendors must determine what they can do to enhance user acceptance. Vendors must understand the client's perception of teleconferencing benefits and adjust their sales tactics
accordingly.

For example, user expectations include
cutting travel time, improving communication
and exchanging information faster during the
adoption phase. However, after teleconferencing is in place, users want the technology to
help them make decisions faster, hold shorter
meetings and improve productivity. The marketing process has to address these changing
needs. Marketers must reach the right people
with the right message -- those key executives who can act as "shoehorns" for the tech-

nology -- in each organization.

Vendors must "see" teleconferencing applications for their clients, who often are unaware of beneficial applications themselves.
By having an in-depth understanding of the
user's communication needs, vendors can create
transparent systems and show clients how teleconferencing can better serve the organization's needs. In this way, the early and late
majority groups will be able to adopt teleconf-
ferencing more easily, thus ensuring the technology's future.

Part 2. System Options

Major Forms of Teleconferencing

In general, there are four major forms of
teleconferencing: audio, audiographics, video,
and computer conferencing. These four methods
differ in complexity and cost but have several
factors in common:

1. They use some type of telecommunications channel and technology;

2. They link individuals or groups of
   people at multiple locations;

3. They are interactive, providing two-
   way communication;

4. They are dynamic and live, involving
   the active participation of people.

Beyond these factors, which contribute to a
shared definition of teleconferencing, the
methods begin to diverge and take on unique
qualities. This section of the paper presents
an overview of the major system options available to users.

Audio Teleconferencing

Audio teleconferencing is voice-only communication. It links people in remote locations via ordinary telephone lines. Audio
systems include telephone conference calls as
well as more sophisticated systems that connect multiple locations--from a handful to
over 200--via a central bridge that ties all
the lines together.

Although audio teleconferencing lacks a
visual dimension, it has some major strengths
in its favor:

1. It uses a readily available and familiar technology -- the telephone;

2. There are 400,000,000 telephones in
   the world that can be used for audio
   conferences;

3. Conferences can be set up on short notice and with less planning than other forms of teleconferencing;

4. It is relatively inexpensive to use;

5. It can interconnect large numbers of locations for a meeting.

There are many system options for audio tele-conferencing, but the most common forms are:

User-initiated conference calls (some-times called "ad lib" teleconferencing);

Operator-initiated or dial-up telecon-ferencing;

Dial-in or meet-me teleconferencing;

Dedicated audio networks.

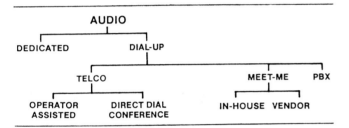

User-initiated conference calls are the simplest and most frequently used type of teleconferencing. A telephone handset, for example, can readily connect parties via three-way calling. Private Branch Exchange (PBX) systems, key telephones, and tele-patchers also provide three-way calling with add-on features.

Telephone or PBX conference calls, how-ever, have limitations. They usually can link only three or six parties and there is little or no control over audio quality. Some PBX systems have the capability to connect 12 or more lines in multi-station conferencing, but the voice quality decreases as lines are added.

For larger multipoint teleconferences, the systems most often used are dial-out bridges, dial-in bridges, and dedicated net-works.

Major Types of Audio Teleconferencing

User-initiated conference call

Easy to use
Links from 3 to 6 sites

Little or no control of audio quality
Uses public telephone network

Dial-up teleconferencing

Calls made to participants
Usually handled by operator
Usually links up to 10 sites
Takes time to call and link parties
Perhaps some control over quality
Uses public telephone network

Meet-me teleconferencing

Participants call into bridge
Links up to 200 or more sites
Control over audio quality
Uses public network or WATS lines

Dedicated networks

Uses leased telephone channels
Private network of fixed locations
Control over audio quality

The main purpose of a bridge is to link channels together so the parties at all loca-tions can hear and talk to each other. While there are many types of bridges, they may be divided simply into dial-out and dial-in bridges.

Dial-out (or dial-up) teleconferencing from a central bridge is generally handled by an operator who calls each location to link it into the conference. It has the advantage of linking more sites than a handset or PBX conference call. It may also provide audio control adjustment for better voice quality, although this feature may be minimal in some bridges. Dial-up conferences primarily use the public telephone network, which is per-vasive, but may not be the most cost-effec-tive channel. The chief disadvantage of dial-out conferences is the amount of time needed for the operator to call each location and link it to the bridge. This becomes es-pecially cumbersome for teleconferences that involve more than 10 sites.

First available in about 1978, the meet-me bridge provides the added flexibility of dial-in teleconferencing. This feature has made it an increasingly popular choice. For dial-in conferences via a meet-me bridge, the parties involved dial a pre-determined num-ber at a designated time. Each call is an-swered either automatically or by an atten-dant operator at the bridge and placed into the teleconference along with other incoming calls. A meet-me bridge usually has features to control audio quality either automatically or by an operator. Individual bridges link from 10 to 25 locations. Used in tandem, they make possible large teleconferences of, for example, 200 sites. Dial-in conferences may use the public network or other channels, such as WATS (Wide Area Telephone Service)for lower long distance rates.

The introduction of meet-me bridges has also opened the door to a new type of teleconferencing service. A number of companies now offer audio bridging services on a per event or contractual basis. Rather than purchase or lease its own on-premise bridge, an organization may opt to buy time on a bridge operated by a commercial company. Related teleconferencing services are also available to clients, such as participant notification of a meeting, tape recording, speaker-microphone equipment rental, and consulting. They may also provide materials--booklets, tip sheets, training sessions -- on how to best use audio teleconferencing.

Commercial bridging services are available from such companies as the Darome Connection, Connex International, and Kellogg Telephone Conferencing Service. Some organizations that have purchased a meet-me bridge for their own teleconferencing applications also sell time to outside organizations, such as the University of South Dakota and Eastern Montana College.

Dedicated networks are another major form of audio teleconferencing. In this case, an organization leases transmission circuits for its private use and forms an established network of fixed locations. Leased channels are generally full-duplex lines that provide a more natural conversation among participants and better voice quality than the public telephone network. They may also be more cost-effective for high-use teleconferencing applications, and may be used for other communications, such as data transmission.

Because dedicated networks connect fixed locations, they lose some of the flexibility of public telephone lines in linking any site into a teleconference. This problem, however, can be overcome by tying public lines into the dedicated network when needed.

Using audio teleconferencing for group-to-group communications involves different equipment considerations than communications among individuals. It usually necessitates that some type of amplified telephone equipment with loudspeaker and microphones be used to accommodate a group. The equipment may be portable and set up in an ordinary room, or the room may be designed specifically for teleconferencing with a built-in audio system and acoustical treatment.

Audiographics

Information other than speech can be sent over telephone lines. This includes an array of audiographics systems that can be used in combination with audio teleconferencing to provide written and

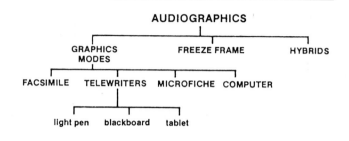

graphic illustrations. As enhancements to voice communication, they offer the following advantages:

1. The flexibility and economy of using telephone lines;

2. Transmission of written, graphic, and print information;

3. Visual illustrations of more complex material.

Audiographics systems offer a number of options to users. The Gemini 100^R Electronic Blackboard, for example, looks like an ordinary blackboard but is really an electronic device that sends writing and hand-drawn graphics over a phone line to remote locations, where it appears on a television screen. Instead of a blackboard, one could use an electronic tablet and pen that sits on a tabletop, or write directly on the TV screen with a light pen. Facsimile machines exchange print documents over phone lines, and computers provide graphics or data that appear on TV monitors at receive sites.

The common element of all audiographics systems is that they transmit visual information over narrowband channels. Beyond that, they differ in technology and capability.

The most common audiographics systems are:
Facsimile;
Electronic blackboards or tablets;
Computer systems.

Facsimile machines are the most frequently used audiographics device. This may be because there are a variety of units available and they can be used for other inter-office communications. They can also transmit a full-page printed document. Like any paper-copying equipment, facsimile reproduces information on paper. However, there is an

AUDIOGRAPHIC OPTIONS

Device	Capability	Examples
Facsimile	Paper reproductions	Typewritten pages
		Documents
	Delayed transmission	Prepared graphics
	Permanent paper copies	Pictures
Telewriters	Hand-drawn graphics	Writing
		Drawings
Electro-mechanical pens	Instant transmission	Outlines
	Shown on TV monitors, projection screens or paper and acetate film	Equations
Electronic blackboard		Graphs
Electronic tablets		
Light pens	Stored on audio tape	
Computer Systems	Text and computer-drawn graphics	Alphanumerics
	Instant or delayed transmission	Computer drawn symbols for diagrams, graphs, schematics, charts
	Shown on TV monitors	Graphics functions (zoom, reduce, label)
	Stored on tape, discs, or in memory	
	Reproduced on paper by printers, plotters and copies	
Random access microfiche and slide projectors	Microfiche images	
	35mm slides	
	Remote random access via telephone lines	
Hybrid systems	Combinations of two or more of the above devices	

important difference. When a document is loaded into a facsimile machine, the information is converted into electronic pulses and sent over a telephone line or other channel to a facsimile machine at a remote location. As the pulses are received, they are reconverted and the information is reproduced on paper. The time it takes to transmit and reproduce the information typically ranges from two to six minutes, but digital machines are available to send a document in less than a minute.

Electronic blackboards and tablets are used by some organizations to generate freehand writing and graphics, such as equations, line drawings, and outlines. The information is electronically produced by using a special pen or surface. It is then transmitted to remote locations and usually shown on a television monitor. It can also be recorded on an audio tape cassette. These devices are especially appropriate to situations requiring real-time graphics support and flexibility in sending planned or spontaneous information. Although electronic blackboards and tablets would appear to be applicable to a variety of teleconferences, not many products are on the market. The Bell System's Gemini 100^R is the only electronic blackboard available. Electronic pens and tablets directly suited to telewriting applications are manufactured by only some five companies. (Other companies offer digitizing tablets for computer systems, but they usually require modifications to operate legibly as telewriters.)

Computer systems as adjuncts to audio teleconferencing provide data and/or graphics illustrations. The storage and retrieval functions of a computer allow data, such as financial information, to be accessed and transmitted to remote locations for a teleconference. With graphics software a computer can also be used to generate and transmit computer-drawn diagrams, schematics, presentation "slides," and other graphics that can be prepared in advance and stored. Some systems may be easy enough to operate so that graphics could be drawn and sent in real-time during a teleconference. The information can include text in different sizes and fonts, symbols for constructing diagrams or charts, and functions for automatically drawn lines, circles, squares, flashing pointers, logos, and so forth. The computer graphics may be selected and positioned by using a keyboard, keypad, tablet, joystick, paddle, or light pen. Printers, plotters or video copiers can reproduce the information on paper.

Few organizations currently use computer systems as adjuncts to audio teleconferencing. This is probably due to the fact that most computer systems are designed for data processing and require keyboard operations that are not particularly simple to execute. Another is that the display format is not always suitable for teleconferencing, especially group-to-group communications. This situation is changing, however, as new computer technologies and software make it easier to access data and generate graphics. The introduction of microcomputers and business graphics software, for example, provide displays of text and graphics that are similar in appearance to 35mm slides commonly used for presentations. Some of the systems incorporate electronic tablets or pens to produce graphics by simply touching a menu selection.

Another recent development is the introduction of the voice/data terminal, a product that combines both voice and data communications in a small desktop unit.

As computers become easier to use and less expensive to purchase, they may find their way into more teleconferencing applications.

A different type of audiographics systems is a random access microfiche or slide projector. At least one manufacturer now markets systems that allow microfiche or slide projectors to be randomly accessed and controlled via a telephone line to display information at remote locations. Using a telecommunications interface, the fiche or slide machine responds to tone signals sent over the telephone line. When a signal is received and translated, the machine selects the desired fiche or slide and projects it onto a screen.

Video Teleconferencing

Not only do video teleconferencing systems include voice communication and graphics, they also provide images of people. For many users, pictures of people create a "social presence" that resemble face-to-face meetings. Although video systems are more expensive, they have two major advantages:

1. They show pictures of people so you can see and hear the speaker at the same time;

2. They show three-dimensional objects as well as graphics.

Many people think of video teleconferencing as full-motion color television using broadband channels. That is one type of system, but there are two other options. One of the newest technologies is the video compression codec. It reduces the amount of information in a full-motion color picture so it can be sent over a narrower transmission channel at much less cost. Although the compressed picture looks very much like the original, the technology is not yet perfect and there may be some occasional blurring. Many industry experts, however, predict that it will have a major impact on video teleconferencing, especially with the improvements scheduled for the next generation of codecs.

Another option is freeze-frame video. It transforms a moving picture into a still, or "frozen," image which can be sent over a regular telephone line. While it can be used for pictures of people, its chief advantage is for showing three-dimensional objects and graphics at a fraction of the cost of full-motion or compressed video systems. Freeze-frame is sometimes called an audiographics technology, but its ability to show pictures of people gives it an ingredient common to video teleconferencing.

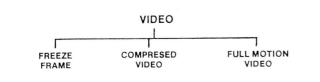

Videoconferencing systems can be divided into three basic types:

    Freeze-frame video;
    Compressed video (near-motion);
    Full-motion video.

These technologies differ essentially in how information is sent and how it appears on the TV screen.

A freeze-frame system uses telephone channels--usually a normal voice-grade line or occasionally a data-grade circuit--to transmit the video information. Because of the narrow bandwidth of a voice-grade telephone line, the image takes a number of seconds to reach the receive location. It then appears on the TV monitor as a still picture. The image is usually scrolled onto the screen from top to bottom much like a window shade being lowered. The transmission time varies from a few seconds to over a minute, depending on the amount of information sent and the bandwidth capacity. An average time would be 30 seconds for a black and white image. Because the freeze-frame

picture is still, or "frozen," showing no motion, these systems are most appropriately used for objects, graphics, or slides. The chief advantages are lower equipment costs, lower transmission costs, and flexibility in linking multiple sites. Slow-scan systems are very similar to freeze-frame, and the terms are often used synonymously. Common generic terms include still frame and captured frame.

Major Types of Videoconferencing

Freeze-frame video

Uses a regular telephone line
Still image
Transmitted in a number of seconds
Black and white or color

Compressed video

Uses a Tl data circuit
Rapid movements may be blurred or jerky
Transmitted instantly
Usually in color

Full-motion video

Uses a wideband channel
Full, continuous motion
Transmitted instantly
Usually in color

A compressed video system uses a telephone data circuit--a Tl carrier of 1.5 or 3 megabits--to transmit video, voice, and data. It reduces the full video information by using a compression technique to eliminate some of the redundant information. The result is that the original video signal of 100 million bits is reduced to 1.5 or 3 million bits. The compression is performed by a picture processor, or codec. The video picture appears in real-time (instantly) at the receive location, but there may be some jerkiness or blurring of fast movements. Video codecs, which became available in 1981, have made near-motion the newest form of videoconferencing. Its main advantage is the significant reduction in the amount of bandwidth needed to transmit a moving image. This greatly decreases transmission costs, but the price of a codec itself is still high. Most codecs presently cost over $100,000, but the prices are falling as the technology progresses. Advances in technology are also improving the image quality and reducing transmission rates. Codecs are now capable or rates ranging from 512 kilobits per second to 2.048 megabits, and codecs operating at 56 kilobits are being introduced.

A full motion video system uses wideband

channels--6 megaHertz for color analog--to
send video, voice, and data. Because of the
large channel capacity, it transmits a full
video picture with continuous motion in real-
time. The picture quality is most readily
compared to broadcast television, and it is
natural in appearance. If video codecs improve in cost and performance, however, wideband videoconferencing may no longer be a
popular alternative to long-distance communications, mainly because of the high cost of
wideband channels. Rather, future applications may be used primarily for special event
conferences or local communication between
sites within a metropolitan region where microwave and cable channels can be leased at a
lower cost.

Computer Conferencing

While voice communication is a shared
element in audio, audiographics and video
systems, it is absent in computer conferencing. It is a means by which many people can
communicate with each other through computer
terminals that are linked together. These
meetings are not bounded by space or time.
Users can control the pace of their communications, choosing when and to what degree they
wish to participate in a conference. It
includes these major advantages:

1. The flexibility and economy of using
    telephone lines.

2. People need not be present at the
    same time but can join at their con-
    venience.

3. A conference can be on-going over days
    weeks, months, or longer.

4. People can gather new information in-
    between participating.

Computer conferencing requires keyboard
operations, but systems are becoming easier
to use. Users typically type their messages
to other conference participants on a sta-
dard computer terminal that is linked to a network by a telephone line. Messages from other
participants are received at the terminal
each time one joins the conference. Today's
computer conferencing systems may link networks of microcomputers--small and relatively
inexpensive terminals that are becoming very
popular. They may also connect larger mini-
computers of those commonly employed for design and drafting functions. And, they may
involve a network of remote computer terminals linked to a host main frame computer
that stores large data bases accessible to
all participants.

Computer conferencing systems can also

differ in network and control arrangements.
Most systems that are used for long-distance
communications employ a "star" network. In
this arrangement, one computer--a micro, mini,
or main frame--serves as a hub or central
point. Each remote terminal is linked to the
hub, and information from the terminal is
sent to the central computer and then relayed
to other terminals on the network. Another
design that is commonly used for local communications (e.g., intra-building) is a "ring"
network, in which the terminals are linked
directly to each other as points in a circle.

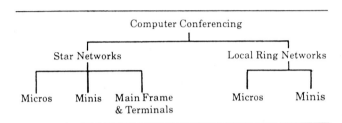

Admittedly, we have just barely skimmed
the surface in discussing teleconferencing
options. Many specific types of products and
services are available for audio, audiographic, video, and computer conferencing. In addition, users may opt for private facilities
with on-premise equipment; off-premise services such as public videoconferencing rooms;
or they may combine resources in shared networks or consortia.

Part 3. Technological Trends

In looking at the trends in teleconferencing technologies, some major factors stand
out: (1) steady progress in audio teleconferencing equipment over three decades,
(2) rapid advances in videoconferencing after
the mid-1970s, (3) the effect of communications satellites, (4) the impact of digital
technology on equipment and transmission services, and (5) the increased availability
of product options and customer services.
The last three factors in particular distinguish the 1970s from the 1980s and deserve to be examined in greater detail because of the continuing impact they will
have on teleconferencing.

Communications Satellites

From the first satellite communications
in the early 1960s through a multitude of
experiments in the 1970s, satellites have

progressed technologically to the point where they will play a major role in communications in this decade.

According to a recent study by Frost & Sullivan, the U.S. production of communications satellites will more than double by 1991 to reach $630 million, from $300 million in 1980. Sales of earth stations in North America, which was a $522 million market in 1980, will quadruple to $2.1 billion in the next 10 years. Another study, by International Resource Development, predicts that by 1986 there may be 35,000 earth stations installed in the U.S.

Among the major forces in satellite growth are the integration of satellites into long-distance networking by common carriers and the expected proliferation of private business networks for data, electronic mail, video and voice. Some sources expect digital earth terminals to be on the rooftops and grounds of almost every major shopping center and office complex in the country, providing network services for intercity communications (Expanding Satellite Systems, 1982).

As of March 1982, there were 15 operational U.S. and Canadian communications satellites, with an additional 23 planned for launching. And, the amount of traffic through U.S. domestic satellites is doubling every three years. Those satellites scheduled for launching represent an investment of millions of dollars each by such companies as GTE Satellite, AT&T, SBS, Southern Pacific, Hughes Communications, Western Union Space Corporation, and RCA (Morgan, 1981).

The next generation of satellites also promises to be more powerful, make more efficient use of transponder capacity, and provide on-board switching. For example, the present Canadian Anik and SBS satellites operate in the 14/12 gigaHertz frequency range that allows for smaller earth stations. Joining them in the next few years will be the Advanced Westar, GTE's GStar, and Southern Pacific satellites. Another important development is TDMA technology, or time division multiple access (Lombard, 1981). With TDMA, which is used on the SBS satellites for example, a transmitting earth station can send a digital bit stream burst in an allocated time slot, permitting more channels to be sent through a transponder and more efficient use of assigned channel capacity. It also reduces signal interference and includes on-board channel switching to send the signals to the appropriate receive stations.

Satellites are just beginning to have an impact on teleconferencing, especially in the use of videoconferencing and private networks for digital communications.

Digital Communications Services

In the trend toward the integrated services digital network, which will be a major development in the 1980s, satellite carriers are making important advances in providing intercity services for voice, data, and video. This includes Satellite Business Systems (SBS), American Satellite Company, RCA, Western Union, and M/A-Com.

Among its satellite services, SBS provides private wideband networks to clients. Early SBS customers like Allstate Insurance, Hercules, and Boeing already have private networks of earth stations on-line for communications that include videoconferencing. Other early customers of private networks are Control Data Corporation, J.C. Penny, Westinghouse, Crocker National Bank, Travelers Insurance, Bank of America, and a number of others.

At least two of the SBS customers are planning to serve as resale carriers--Allstate and Isacomm. Allstate plans to provide video communications to the business and public sector via a network linking seven cities initially. Isacomm (Insurance Systems of America Communications Services) resells service to the insurance industry and plans to have public videoconference rooms in 40 cities in 1984. By reselling channel capacity, companies like Allstate and Isacomm make it feasible for smaller organizations to use satellite networks on a shared basis. Otherwise, the cost and capacity of wideband private networks have so far made them feasible for only large corporations.

American Satellite (ASC) through its satellite data exchange (SDX) service provides voice, data, and video communications via small earth stations located on the company's premises. They now have over 50 earth stations operating and over 35 under contract. Operating since 1974, ASC pioneered the use of digital transmission for private networks via satellite and small earth stations, and the first such installation was at Sperry-Univac in 1978.

RCA Communications provides a service called "56 Plus" for voice, data, facsimile and freeze-frame videoconferencing. Western Union offers a similar service. Macomnet, a part of the M/A-Com companies, also provides digital network services for voice, data, facsimile, and videoconferencing.

Another company with digital services is, of course, AT&T. It introduced digital communications with pulse code modulation (PCM) via Tl data carriers in 1962. Providing the equivalent of about 24 digitized voice channels, the lines were initially installed for short hauls within large cities. AT&T presently has about 1 million circuit miles of Tl carriers installed in the U.S. With the advent of compressed video teleconferencing systems which use the channel's capacity of 1.544 megabits, the availability of Tl circuits is critical to future applications. Related to Tl channels are two recent moves by AT&T. Its Picturephone Meeting Service uses a High-Speed Switched Digital Service to link public rooms throughout the U.S. with two Tl channels for videoconferencing at 3 megabits. The network will also be extended to private rooms on customer's premises. AT&T recently filed for another service called Terrestrial Data Service to provide single Tl channels on a point-to-point basis, which could be used to link videoconferencing rooms at two locations. Among its other digital services, AT&T provides Dataphone Digital Service in several hundred cities with speeds of up to 56 kilobits.

Among other developments in telecommunications services for intercity links are the value-added carriers. Their major thrust is to interface incompatible equipment so they can communicate with each other. Included are GTE's Telenet, Tymnet, Uninet, Autonet, Graphnet, and Faxpax. Other carriers, like MCI, provide voice channels as an alternative to AT&T's public network.

Other developments in communications that will impact on the automated office and teleconferencing are digital local area networks, digital termination services, and fiber optics. Progress is being made on local area networks to link incompatible workstations--data computers, word processors, copiers, digital PBXs--by a single intra-building network that can be connected to intercity service and worldwide systems. Xerox, for example, is trying to make its coaxial cable Ethernet the standard for local networking. New PBX systems with "super switches" that act as central controllers for integrated voice and data are also battling for the local networking market. A third option is Datapoint's ARC System, which is a combination of Ethernet and the super PBX. AT&T has announced the Advanced Information Systems/Net I for local networks to link multiple devices in the automated office.

Digital termination service for local access to earth stations are also emerging as a major development in the 1980s. A number of companies were recently approved by the Federal Communications Commission to provide digital termination systems or digital electronic message service. One such service is available from Local Digital Distribution Company, a partnership of M/A-Com and Aetna. It offers local data distribution via small microwave terminals (RAPAC) or over cable networks (CAPAC) for access to an earth station. Interestingly, it is very similar to the X-Ten service proposed in 1980 by Xerox but then withdrawn. Like X-Ten, it uses local packet-switched microwave at the newer and less used frequency of 10 gigaHertz.

In summary, telecommunications services--and teleconferencing--are indeed feeling the impact of digital technology. Although digital services are just beginning, they will see dramatic growth through this decade. This includes wideband services for integrated digital networks of voice, data, and video via satellite for intercity links. Local networks will also be digital, providing local access to earth stations via microwave or cable channels. Within the buildings themselves, a variety of equipment will be made compatible through network management systems like Ethernet or via super PBXs. Digital channel capacities of 1.5 megabits, 448 kilobits, or 56 kilobits will be used for voice, graphics, data, and video for teleconferencing purposes. As more services and new companies move into telecommunications, more channels will be available to users and costs will fall. Fiber optics will also expand to other cities, and they will be increasingly used by private companies for short distances, especially between buildings. We are moving toward the Electronic Age, with networks moving information among offices, between buildings, from city to city, and out onto worldwide "highways."

Digital Equipment

Although the application of digital technology has been relatively recent, it has had a dramatic impact on teleconferencing equipment. It has led to a number of new options in improved designs. It has also opened the door to sophisticated signal processing techniques that are only just beginning to be applied. Witness the digital codecs for compressed video. When first introduced a short time ago, they reduced channel bandwidth to 3 or 6 megabits. No sooner had the ink dried on the marketing brochures than 1.5 megabit compression became available. In 1983 companies introduced compression techniques to send a moving picture over a 56 kilobit channel. Other developments include transmission rates ranging from 512 kilobits per second to 2.048 megabits per second as well as compatibility between North American and European standards.

Digital technology is also being used in new audio equipment. Digital bridges, for example, provide automatic adjustment of line quality and, in one model, a voice recognition system to respond to the person on-line. A new all-digital bridge being developed by Bell Laboratories will link up to 60 locations for audio or audiographics conferencing. Called the Network Services Complex, it has separate bridge modules for audio or graphics and is designed primarily for AT&T's operator-assisted conference calls. The bridge will be able to automatically call up the phone numbers that have been previously keyed in via a Touch-Tone telephone. The bridge is also designed to transmit graphic information at speeds up to 56 kilobits. A similar bridge for private use is also being developed by Bell Labs.

In addition to bridges that can make graphics transmission more efficient on digital phone lines, important progress is being made in other areas of audiographics. Although developments in audiographics have lagged behind audio and video through the 1970s, we are now seeing the introduction of a number of new products. They include high-resolution graphics, electronic tablets, light pens, and hybrid devices that combine graphics and video. New integrated voice/data terminals with interfaces to hard copiers and videodiscs are also emerging.

These products are the forerunners of even more sophisticated systems that are due to appear by the mid-1980s, if not sooner. They include fully integrated workstations for digital voice, graphics, data, and video with access to hard copiers and data storage. They will employ various technologies: laser document scanning, touch sensitive screens, compression techniques for fast transmission, multi-site conferencing features, computer graphics, and so forth. The most difficult problem to overcome is not the technology itself, but the design of the human-machine interface.

In fact, the human-machine interface and the whole area of ergonomics and human factors will probably be the overriding issue of the decade. With the advent of digital equipment and transmission services, teleconferencing technologies will expand and improve in performance at a rapid rate. The primary question is, how will people use them? Related to this issue is the area of customer services.

Customer Services

One of the barriers to teleconferencing through the 1970s was a lack of turnkey systems, integration services, and training in product applications. It was also difficult to find the vendors, as marketing was not especially aggressive nor widespread. Most potential users had to design their own systems, buying components from a number of specialized vendors. Once the system was installed, little training was available on how to use it. Rather, it necessitated a trial-and-error approach in learning to manipulate the technology and employ presentation techniques appropriate to the medium.

Fortunately, part of the growth in teleconferencing is due to new companies moving into the turnkey, integration, and consulting areas. They will design the system and rooms, integrate the components for a turnkey package, work with acoustic engineers and others to install the system, and may provide some training on how to operate the equipment.

As the technologies and system integration improve, it focuses attention on what is still the weakest link in teleconferencing applications: human factors. It includes ergonomics and the design of better user-oriented systems. However, it also includes broader issues such as: how will the system be integrated into the organization, who will use it, how will it be employed, how is effective software designed, how should it be promoted to gain acceptance? All of these questions fall within the realm of client services--needs assessment, implementation planning, training--and need to be addressed by people in the teleconferencing industry if systems are to be truly effective and successful.

Summary

In looking at the status of today's technologies and applications, teleconferencing seems poised for growth. A number of factors are contributing to a readiness posture. They include advances in technology and a maturing of the industry that are meshing with organizational concerns about productivity, communications, operating costs, competition. Teleconferencing has also proven its effectiveness in many situations, and it has demonstrated its ability to meet diverse needs.

There are, however, warning signs. Most involve human factors and those "soft" areas that relate to users and applications. Emerging as primary issues are the design of user-oriented systems, implementation strategies, training, software, and support infrastructures that promote successful applications and the integration of teleconferencing into an organization. There has been a tendency for people to put technology first, rather than focusing on applications and end-users. The growth of teleconferencing depends on the recogition of human factors, as well.

The Teleconferencing Resource Book: A Guide to Applications and Planning
Lorne A. Parker and Christine H. Olgren (eds.)
Elsevier Science Publishers B.V. (North-Holland)
© Center for Interactive Programs, University of Wisconsin-Extension, 1984

14

AN EXPERIENCE IN APPLYING TELECONFERENCING
TO THE HOLDING AND COORDINATION OF
A NATIONAL SALES MEETING

Thomas W. Hoff
Director
Communications Development
Roche Laboratories
Division of Hoffmann-La Roche Inc.
Nutley, New Jersey 07110

The Department of Communications Development of the Roche Laboratories Division of Hoffmann-La Roche, Inc. was formed about three years ago to identify, test and integrate new communications techniques, technologies, media and concepts into our marketing and internal communications activities. Shortly after the establishment of the group, one of its members prototyped an acoustically coupled speakerphone capable of remotely triggering a carousel projector. The existence of this device led to a rather intensive and extensive search of the world literature on research in teleconferencing and to the making of a number of personal contacts with individuals actively involved in practicing the art.

During the ensuing several months, a variety of scenarios were developed in which teleconferencing could be productively employed to help:

1.   Market a new product.

2.   Provide continuing professional education to remote groups of health care professionals.

3.   Train district sales managers and salespeople.

4.   Hold and coordinate sales meetings.

5.   Provide individual salespeople with home office technical support.

In order to gain management funding and support to develop and test these scenarios, it was decided to make a presentation to management via teleconferencing. In other words, use the media to present/inform/promote the media.

On December 14, 1976, using the prototype speakerphone and a carousel projector, a presentation was made which clearly developed the cost versus travel versus time-saving potentials of portable audio teleconferencing, coupled with pre-mailed graphics (slides, videotape, flip charts, etc.). A scenario was painted which showed the potentials for application by members of each of the departments represented by the management group. These included field sales management, sales promotion, distribution, professional services, marketing research, product management, and sales training.

The presenters were located in a room on another floor. They were able to openly dialog with the management group and controlled the slides by push button. The net effect was not very different from the typical business presentation in which one or several individuals appear before a group and present a fairly well thought out presentation using slides, flip charts or overheads.

The net result of the presentation was a set of specific agreements to proceed in the development of a variety of pilot tests. Worthy of note here is the technique by which these specific agreements were obtained. The presenters had previously placed a flip chart and felt pen in the conference room occupied by the management group. At the conclusion of the presentation, they asked one of the group to act as scribe and jot down the names of individuals to be assigned to work on a teleconference pilot with Communications Development. The presenters then polled the management group -- one at a time -- to solicit a person's name from each management area -- a great technique that would not have worked any better face-to-face and which clearly demonstrated one of the potentials of the media.

During the two years following this presentation, specific pilot projects were initiated, completed and assessed in each area. Teleconferencing has become a regularly considered/utilized communications alternative within our organization.

For purposes of this case study, one of our applications will be reviewed.

Our sales organization is composed of approximately 60 divisions of 10 to 12 people (for each division) including a local division manager. The divisions are organized in eight regions, four in a north area and four in a south area. In May 1978, sales management found it necessary to rapidly organize sales meetings for the purpose of communicating some important new strategic and follow-up information on a newly launched product and on another major product. The information they wished to communicate required the direct involvement of the two involved product marketing teams. Opportunity for direct and open dialog between the salespeople and product and top management was deemed essential.

The only way to facilitate this personal, direct dialog in the past was by arranging either large regional or national meetings. In this instance, the time pressures and budget mitigated against such a strategy. The lead times were far too short to secure adequate hotel space for June meetings. At this point, sales management asked if and how teleconferencing could help. The goal was to let each of the 60 individual divisions meet in a local hotel in their base city, which would reduce hotel and travel costs and reduce lost field time.

A proposal was pulled together in a few days and accepted. The objectives and content are summarized as follows:

- to maximize communication between field staff and home office/marketing personnel

- to maximize inter-regional communication

- to minimize dollar and other costs of extensive travel and work time lost for both home office and field people

- to eliminate the arduous task of separate repetitive presentations by home office management to each of the eight sales regions

- to decrease the tremendous expenses associated with hosting large assemblages of sales representatives and management at traditional national or regional meetings

The meeting was to begin with opening messages beamed from the home office from the general sales manager to the entire field staff and from the area and regional sales managers addressing their respective people. The national sessions focused on two product areas. Major audio presentations would be made from the home office for these two products, supplemented by synchronized on-site videotapes and slides. The product teams at the home office would present sales objectives and strategies and would be available on the teleconferencing "bridge" to respond to questions from field staff members. Each division would be able to hook up on a rotating basis with the product team panels to get immediate answers, thereby enabling the various divisions to hear answers to problems similar to their own as manifest in other sales areas. The national teleconference sessions were then followed by divisional workshops held at the various meeting sites to address specific areas of concern to each division and by a final Q & A teleconference with the respective product team at the scheduled conclusion of the workshops.

Approximately six weeks later, our national teleconferencing media was up and running.

Prior to the national meeting, there were meetings of division sales managers to discuss meeting plans. These meetings made use of teleconferencing and provided the mechanism for introducing the concept and for the training of each division manager in the fine points of hooking up and using the teleconferencing equipment. At each site, a member of the home office staff hooked up a Darome acoustic convener and dialed their "meet me" bridge at the appointed hour. One of the members of our department then remotely presented an interactive slide presentation to all of the division managers assembled demonstrating equipment hookup, etc. During these meetings (or shortly after in the mail), all of the division managers received the slides, videotapes, conveners, telephone credit cards, bridge numbers, meeting flow outline, schedule and other materials they would need for the national teleconference meeting.

One of the most complicated and most critical aspects of the preparation for the teleconference was the creation of the master teleconference meeting schedule. Since the bridge could handle only 40 lines, the meeting was designed to allow the teleconferencing portion to shift between the 30 northern area divisions and the 30 southern area divisions. See example schedule attached for June 12 and 13. Note that the key was to have the north sales area discuss and work on products A and B on June 12 while the southern sales area worked on other issues. On June 13, they reversed.

The time zone situation created another scheduling concern. The divisions on the West Coast had to be on line at 7:00 a.m. local time while those in the East did not have to start until 10:00 a.m. The Westerners were through early, the Easterners worked into the evening.

As mentioned previously, we utilized the Darome acoustic convener, an instrument with which we had had previous satisfactory experience. This unit made it possible to avoid all of the hassles of arranging 60 special lines to be put into 60 different hotels serviced by various Bell companies as well as several independents. The only local requirement was to be sure the meeting room selected had a phone from which outside calls could be placed. A further necessity was to review with the local hotel management and its switchboard operator that the line would be in rather constant use and to ask the operator not to disconnect it.

All of our managers have video playback equipment. Most have access to slide projectors and screens and most hotels can supply any of these items with little advance notice. Therefore, all AV equipment was supplied locally.

We utilized the Darome teleconference services "meet-me" bridge. The operators served as timekeepers and meeting expediters to assure that the schedule was maintained. They, therefore, stayed "on line" at all times so they could step in to problem solve, if necessary. They were given the responsibility of interjecting to advise a speaker that his equipment was not functioning properly, his line was bad, or his speaking voice was too weak. They also could pull people out of the bridge in order to troubleshoot any technical problems with them.

Since this was both Darome's and our first experience with a conference

of this size, once a scheduled conference was set up, each site was generally left on line after the scheduled presentation/discussion until the next use. This was done to make things simpler for the sites and because of our uncertainty as to whether all sites could be dropped and reset reliably several times each day. Simply put, we felt, once we've got 'em, let's hold onto 'em. This long-term hook up, while much more expensive, did allow a number of impromptu conferences between various divisions to be set up for idea sharing, problem solving, etc. All a site had to do was advise the bridge.

As can be seen from the master schedule, all sites knew what visual element they would be using for each part of the teleconference. In the case of the slides, all slides were numbered in the transparency portion so as to be visible on the projection screen. The presenter could thus merely ask for the slide by number and be assured that "everyone out there" was looking at the same thing. Videotapes also bore large numbers, i.e., Tape 1, etc.

Every attempt was made to include Dr. Lorne Parker's basic teleconferencing guidelines. All speakers were introduced with a picture to help establish presence. Long speeches without dialog were avoided. Humanizing was fairly easily accomplished as all participants were well known to each other.

Two "sending" sites were established in the home office -- one for each product team. These sites used hard wired conveners and were fitted with all of the same audiovisual equipment that was at the remote hotel sites. Backup acoustic conveners on a separate line were also in place.

In addition, a "listening post" identical to a remote hotel site was set up and any and all home office staff was invited to drop in and listen in at their leisure.

The other presenters, such as the general sales manager, area sales managers, and regional sales managers, made their various presentations or served on Q & A panels, etc., directly from the remote sites they were visiting. This tended to add a "between us" to the national scope aspect of the meeting and reduced the top down from the home office impact.

Technically, the national teleconference sales meeting worked quite well. There were occasional sites that improperly hooked up their acoustic couplers but they were usually quickly remediated. Transmission from the home office was lost for a couple of minutes once.

The single major problem occurred on the second day. Just after all sites called in and the operator was about to turn the conference over to one of the product teams, a local power failure hit the Bell ESS station, and all lines were dropped. Within 10 minutes, every site recognized (via busy signals) that they had been dropped and redialed -- a great testimonial to the training they had received ·and to the flexibility of the "meet-me" system.

Our assessment of the national conference as a substitute for a major (central or regional site) meeting is positive. A rather extensive post-meeting survey was conducted among division managers and sales representatives. The results indicate that:

- The participants did not feel there were serious technical problems, i.e., couldn't hear, couldn't get on line, although 40 percent reported observing some sort of technical problem.

- The participants enjoyed hearing from the top management people who addressed them. They liked getting immediate authoritative answers to questions and hearing positions, etc. of colleagues at other sites.

- 70 percent thought that the meeting was equal to or better than previously attended large meetings without teleconferencing.

- Some participants, especially the division managers, felt that the rigid schedule imposed by teleconferencing created an undesirable amount of inflexibility and clock watching pressure.

- Some felt that the Q & A portions were too large to encourage maximum interaction.

Some of the additional observations made are worth reporting here. This media can prove to expedite the feedback/decision process. The need to make a major strategic promotional change was uncovered during several sessions and confirmed by linking all sites together. A decision to effect this change was literally made on the spot by adding into the national conference the appropriate key management personnel. The normal time frame for such feedback to be confirmed, a decision made and then communicated could be measured in weeks.

We also discovered that this media can minimize some of the psychological barriers of face-to-face meetings. When a question was asked by one of the sites, it was in effect being asked by the group of peers in that room, with the group's full support. The home office person's answers were judged by all sites. This occasionally resulted in bold demands for straight, decisive answers as the asking group had the compunction to challenge the respondent.

While some of the participants found the tight scheduling a problem, some of the home office personnel felt that it tended to force coverage of all the material and assured all agenda items were addressed.

From a cost point of view, considerable economies were indeed realized based on the savings realized on travel and hotel space alone. These savings -- well in excess of $100,000 -- do not include the lost field/office time that would have occurred by transporting 600 plus salespeople to regional, area or national type meetings and 15 to 20 executives to those same sites.

There are no specific plans for using teleconferencing on this scale in the near future. It will, however, be a viable alternative as each subsequent sales meeting is considered. The teleconference technique is, however, regularly utilized by field management for weekly/monthly communications, feedback and problem-solving purposes.

***

EXAMPLE

NATIONAL TELECONFERENCED SALES MEETING
North Area
Monday, June 12 and Tuesday, June 13

* Denotes on-line teleconference

Pacific	Mountain	Central	Eastern
6:30 am*	7:30 am*	8:30 am*	9:30 am*

Call Darome bridge using credit card
and <u>assigned</u> <u>regional</u> <u>bridge</u> <u>telephone</u>
<u>number</u>.

7:00*   8:00*   9:00*   10:00*

(on-line) Meeting begins - Field Sales
Manager introduces General Sales Manager and welcomes group. (Have slide
tray #1 on carousel and ready to go.)

7:05*   8:05*   9:05*   10:05*

(on-line) General Sales Manager's talk.
Slies 1-30

7:20*   8:20*   9:20*   10:20*

(on-line) Field Sales Manager/Regional
Sales Manager provide meeting overview.
Slides 30-41

7:40*   8:40*   9:40*   10:40*

(on-line) Product A overview -- Product
Director introduced by Field Sales Manager. Use videotape #1 product A.

7:45   8:45   9:45   10:45

(off-line) Product A workshop conducted
by Division Sales Managers - Use moderator's guide as suggested flow, utilize
resources specified.

> Note: During the six-hour work-
> shop, regional Sales Planning
> Manager will tap into each divi-
> sion -- to answer questions, pro-
> vide input, etc. <u>DO</u> <u>NOT</u> <u>HANG</u> <u>UP</u>
> <u>YOUR</u> <u>PHONE</u> <u>DURING</u> <u>THIS</u> <u>PERIOD</u>!
> (Lunch to be taken during this
> six-hour period.)

1:45*   2:45*   3:45*   4:45*

(on-line) Area Q & A with teams, Field
Sales Manager, General Sales Manager.

2:45   3:45   4:45   5:45

(off-line) Product B - Bridge operator
will instruct when to play product B
videotape #1.

2:47   3:47   4:47   5:47

(on-line) Status Update - Product Director for product B will instruct when to
play product B videotape #2.

> The next half hour consists of
> various presentations by members
> of the marketing group for product
> B.

3:20   4:20   5:20   6:20

(off-line) Each Division Sales Manager
writes down questions (maximum of three)
his division would like to ask the product B team. (Darome operator will explain process before meeting segment.)

3:30*   4:30*   5:30*   6:30*

(on-line) Product B team will randomly
poll divisions and answer questions.
Ample time will be allowed to divisions
with unanswered questions to ask the
team.

4:00   5:00   6:00   7:00

(off-line) Product B workshop (1 hour).

5:00*   6:00*   7:00*   8:00*

Regional Sales Manager wrap-up (10
minutes) to his respective region.

                    Day 2

No teleconferencing - local divisional
business meetings.

Day sequence reversed for 30 divisions
in southern area.

The Teleconferencing Resource Book: A Guide to Applications and Planning
Lorne A. Parker and Christine H. Olgren (eds.)
Elsevier Science Publishers B.V. (North-Holland)
© Center for Interactive Programs, University of Wisconsin-Extension, 1984

19

PUBLIC INVOLVEMENT IN THE LEGISLATIVE PROCESS THROUGH TELECONFERENCING
THE ALASKA EXPERIENCE

Sioux Plummer
Teleconference Coordinator
Legislative Affairs Agency
Division of Public Services
State of Alaska
Juneau, Alaska

It is impossible to talk about the effects of teleconferencing on the legislative process without first describing the stage and the players. Environment, geography, cultural and educational differences, accessibility to comforts, transportation and means of subsistance are all factors of impact to the technology of a teleconference system such as the Alaska State Legislative Network.

There have been previous reports to this one, describing in detail the origination of Alaska's system and technical detail regarding its components.[1] This paper will touch briefly on the history and development of the network; emphasis will be given to impact and results since its inception.

The Legislative Teleconference Network, herein referred to as the LTN, is now in its third year of operation. Most significant at this time is the impact teleconferencing has had on the whole legislative process. By legislative process we mean the passage of bills and generally expanding information and communication of all kinds relating to the workings of state government. We in Alaska have found the effects of teleconferencing to not only be an interesting phenomenon, but one which provides very legitimate and profound results.

Setting the Scene

Some people are very aware of the "uniqueness" of the State of Alaska, more are not. For the sake of clarity, and to best understand the causes of some effects of teleconferencing on the Alaskan legislative process, it is important to set the stage and make note of the total environment.

Picture, if you will, a state twice the size of Texas with a total population less than that of Omaha, Nebraska. It is a state of vast extremes: the largest in land area and the smallest in population; it is first in income per person, yet has the least amount of electric power available than any other state.

Alaska is a land of extreme weather conditions, some not uncommon to other states with its limited access to the comforts that lessen the effects of those extremes. Alaska has impassable mountain ranges, plus it boasts the nation's highest mountain, Mt. McKinley. It has desert-like areas where nothing grows; it has more coastline than all the other states combined and fewer roads and highways than probably all of Madison, Wisconsin.

Varieties of people make Alaska culturally interesting and cause communication concerns. We are a state of mostly Eskimo, Aleut and Indian natives, interspersed by Caucasions throughout. Our people are aboriginal types to university-educated; some reside in small villages and fish camps while others live in sophisticated urban centers much like cities in the "lower 48". Transportation for some is dog sled and more recently, snow machine. Many people have 4-wheel drive jeeps or cars, while others find the standard, modern automobile perfectly suitable.

Highways are limited due to great distances between urban centers and difficult terrain. Juneau, the capitol, completely lacks a highway out of the city to anywhere--it is land-locked and relies on the Alaska Marine Highway (the ferry system) and daily commercial jet flights. Roughly 75% of the state's residents are concentrated in a dozen or so communities, many of which are primarily, or solely, accessible by air. There is a small railroad, but light aircraft serves the bulk of the state's transportation needs, and often, on a sporadic basis.

The scene and environment for legislative teleconferencing is now set. We can see then, the need to communicate with varieties of peoples, over vast distances, often with detrimental weather conditions, and regarding subject matter that can vary greatly in importance from region to region.

A Need is Defined

Obviously, legislators need to talk to their constituents and the citizenry needs access to their elected representatives. There are definitely some basic constraints involved with trying to do just this. A classic example is Mr. Citizen in Bethel, Alaska wishing to "testify" at a public hearing to be held by a legislator and his/her committee in Juneau. Without a teleconference system, Mr. Citizen has two options in order to be heard by his legislator--he can write a letter or get on an airplane and fly to Juneau. Mr. Citizen is a fisherman by trade and the committee hearing is on legislation greatly effecting his region, business and life. He WANTS to be heard! Previous to the accessibility of teleconferencing, he probably would not have been heard. Either lack of time or money for travel to the Capitol would no doubt have prevented his presence at the hearing. And, not being one to write letters, much less to an "important" official, he probably won't get his concerns down on paper and in the mail to his legislator. Anyway, he fears that his letter will be lost in the "bureaucratic shuffle", so what's the use?

So we have, in simple terms, defined the need. In 1977, the Alaska State Legislature made note of this need and decided to tackle the problem. Legislative Resolve 93 established the LTN Task Force, assigned to research, develop and recommend a suitable system, one which would be administered by the Legislative Affairs Agency.

This was not a simple task. Had this group represented the legislature in Washington, for example, (the nearest state to them) the story would be quite different. Keep in mind that Alaska only became a state in 1959. Modern telecommunications, let alone the relatively new technology of teleconferencing, was not exactly the by-word in those days. Great strides have taken place in twenty years, however, and the past decade has shown change in the basic MTS telephone network from a military owned and operated communications system to modern satellite and earth station use.

The development of a teleconference network was a very new thought though, and by the time the deed was done, those involved--consultants, legislators, coordinators, and the one and only long lines carrier in the state, RCA (now known as Alascom, Inc.), all became well-versed in the factors of teleconferencing, but more particularly, in just trying to communicate with each other. Previous reports and history of the development of the network clearly indicates there were difficulties in the beginning and throughout the project.[2] RCA was asked to provide technology and service they really had not dealt with before, and the task force was essentially a group of interested experts, but none of whom had ever been directly involved in such a project either.

System Description

The task force did indeed accomplish their assignment, and by December 1977 the order for a dedicated, 4-wire leased circuit was given to RCA. The circuit was to be full-time and linked to six communities: Anchorage, Fairbanks, Juneau, Nome, Ketchikan and Bethel. (See map attached) The circuit was also to provide facsimile transmission capability.

By February of 1978 the system was in place as ordered. The sites became "teleconference centers", with the coordinator's office located in Juneau. Scheduling and direction came from Juneau, as did the origination of most hearings or meetings that were teleconferenced. The following September, Washington, D.C. was included on the circuit. The Washington offices of our two U.S. Senators were added, as well as the Office of the Governor. This afforded tremendous scope to the network and has helped bring certain issues and priorities of state government closer to federal government.

The RCA/Alascom-provided circuit terminates in local telephone company facilities in each community, none of which, incidentally, are the same company. As the long lines carrier, Alascom is required to coordinate and be responsible for the inter-connection to each of these separate independent telephone companies. (None are Bell affiliated, some are small cooperatives and others are municipal utilities.) The telephone companies extend the circuit through their own cable facilities from Alascom's earth stations or toll centers to the office where the teleconference equipment is located, terminating into a standard, modular jack. At that point, we take modular cord in hand, and conveniently plug our portable equipment into the circuit.

The conference equipment, simply operated, is a WE50 A1 conference set, which comes equipped with two small mikes and an enclosed set mike. It looks much like a speaker telephone. Since the circuit is dedicated, the hand-set on the conference equipment is not utilized for dialing--ours is not a "dial-up" network. It is "on" 24 hours a day; the only way one can "delete" a site from access to the system is to merely turn the conference set off.

The set is compatible to other types of microphones, in particular, push-to-talk Darome mikes are conveniently used, as are Shures in conjunction with mixers for use with many participants, such as for a large committee in a Capitol hearing room.

The System Today

Since its inception, the network has grown, both in terms of where the circuit goes and by its increase in use. Now, it extends beyond the six original communities and presently includes Sitka, Dillingham, Wasilla, Valdez, Kotzebue, Kenai Peninsula and Barrow. (see map attached)

Valdez is the only site that is commonly referred to as a "teleconference center", because it strictly offers access for teleconference participation. All other sites have evolved into full-fledged Legislative Information Offices, offering not only telecopy services via the LTN (sometimes to all sites at once!), but other information services and communications devices such as computer-based data terminals providing instant access to bill history and the convenience of electronic mail. Three of the original sites were already providing legislative information to their communities, but it is fair to say that the development of the LTN contributed to the entire progression of the other locations.

Needless to say, with direct access to the Capitol from thirteen Alaskan communities plus Washington, D.C. and visa-versa, communication has increased dramatically state-wide. There are now few remote locations in the state that cannot readily reach a teleconference facility and have a voice in state government.

Today we conduct three to five teleconferences almost daily, sometimes back to back. Most are public hearings, and quite often, are conducted to include all sites on the network, excepting Washington, D.C. Some are more regionalized, or are scheduled to include only those sites that share the same time zone. (Besides the vastness of the network, we deal with four time zones in the state. Scheduling is indeed a challenge!)

Since the most frequent use of the network is legislative and for the primary purpose of teleconferencing public hearings, the capitol complex in Juneau has many drops, that is, modular jacks, connected to the circuit. These drops are located in fourteen committee rooms in the capitol and two adjoining buildings thus affording committees the convenience of conducting hearings comfortably from their own turf. This is appreciated, as it prevents legislators and staff from having to carry their prepared materials and files to a central teleconference room. Instead, the teleconference staff wheels the portable equipment on a cart up elevators, down halls and across the street, much to the delight of each committee.

Additional drops are located in the Legislative Affairs Executive Director's office and in the Teleconference Coordinator's and Legislative Information offices, all in the Capitol building. Occasionally, small teleconferences are held right in the coordinator's office where equipment is permanently connected to the circuit. Each field office has one to three drops and there are drops in locations specifically used for video conferencing which will be discussed later in this paper.

Procedures and Protocols

One of the Task Force's original concerns was to prevent the "technology" of teleconferencing from interrupting or interferring with the formal public hearing. Definite procedural protocols and understandings had to be established and adhered to by the users from the very beginning.

This has been quite successfully achieved by basically following the routine "Robert's Rules of Order" format, as done in face to face meetings. Each teleconference site is staffed by a "moderator", who is in charge of equipment, plus acts as an extension to the chairman of the teleconferenced meeting. Probably the most effective protocol is the fact that the chairman does indeed "control" his/her own meeting or hearing. The moderator assists by polling sites, preparing those on the network for the chairman's call to order, and sharing or receiving any pertinent pre-meeting information with the other moderators. The moderator remains present throughout the entire teleconference, adjusting equipment occasionally, assisting with witnesses present and responding to needs of the chairman. At the close of each teleconference, the moderators "sign-off" from their locations and remove the equipment from the committee room in the capitol and make necessary rearrangements, if any, in the field offices. In case of line noise or any technical disturbance, it is the duty of the Juneau-based moderator to report the trouble to Alascom repair. The moderator is also responsible for familiarizing witnesses and observers with equiment and protocols before each teleconference.

Moderators often have dual roles in their offices, as they also act as legislative information officers. They have on hand copies of bills and other documentation as back-up materials for given teleconferences. On short notice, a committee may telecopy additional information just prior to a teleconference, which is an added advantage to all concerned.

Alaska's teleconferencing and information falls under the auspices of the Legislative

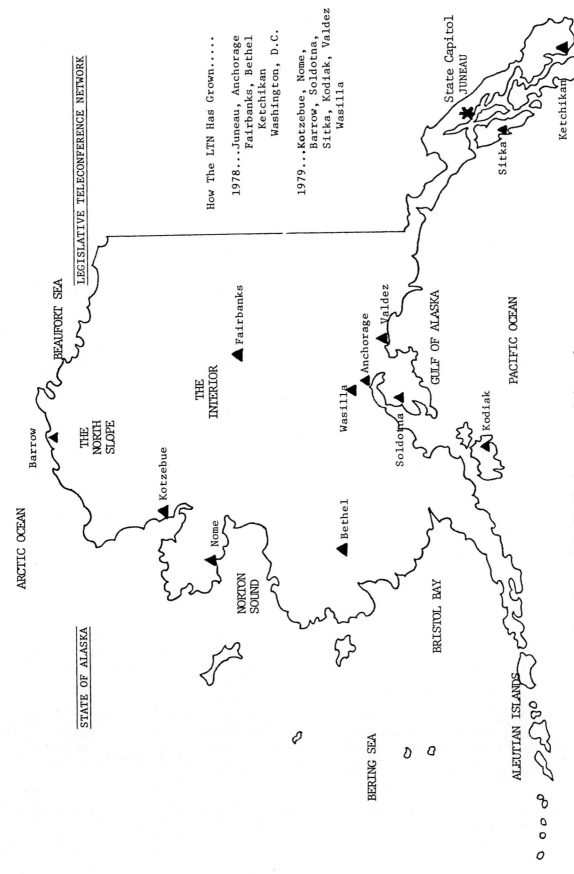

STATE OF ALASKA

LEGISLATIVE TELECONFERENCE NETWORK

ARCTIC OCEAN

BEAUFORT SEA

Barrow

THE NORTH SLOPE

Kotzebue

Nome

NORTON SOUND

THE INTERIOR

Fairbanks

Bethel

BERING SEA

BRISTOL BAY

Wasilla

Soldotna

Anchorage

Valdez

GULF OF ALASKA

Kodiak

PACIFIC OCEAN

ALEUTIAN ISLANDS

Sitka

Ketchikan

State Capitol
JUNEAU

How The LTN Has Grown.....

1978...Juneau, Anchorage
        Fairbanks, Bethel
        Ketchikan
        Washington, D.C.

1979...Kotzebue, Nome,
        Barrow, Soldotna,
        Sitka, Kodiak, Valdez
        Wasilla

***** The LTN is located in 13 Alaskan communities, plus *****
      Washington, D. C.

Affairs Agency's Division of Public Services. That Division provides employee orientation either in Juneau or Anchorage prior to each legislative session, which begins January of each year and runs approximately five months. During this orientation, equipment and protocol training is provided, plus the very positive factor of moderators meeting and sharing problems and experiences.

A moderator's "how to" handbook and a teleconference "user's guide" was developed this year to supplement training and to answer on-going user queries.

A very important facet of involving the public in state government through teleconferencing is publicity. Access to the media varies from community to community, as does success at reaching constituents. Publicity protocol requires that legislative committees advertise and promote their own public hearings. However, we again act as extensions of the committee whenever possible, by our moderators displaying posters in their offices, having up-to-date schedules on hand, and by following up with public service announcements to local radio, T.V., or newspaper sources in their own communities.

Other Users

The LTN is not solely used by the Legislature. The original legislative mandate included offering the network to other users on a priority basis, and only when it is not in use by the Legislature. The eligible users, in order of priority are:

> Legislative Affairs Agency
> Executive and Judicial Branches
> Other State Agencies
> Non-profit Organizations

During the legislative session, most teleconferences are legislative. However, when space is available, other state agencies, in particular, find space on the network, as do non-profit organizations such as health agencies. One particularly successful teleconference was presented by the Governor's Commission on the Status of Women; post-hearing reports indicate topics and individuals were represented for the first time that could not have participated without access to the LTN. Another successful non-legislative program is an all-day workshop on Sudden Infant Death Syndrome, presented by the State Department of Health and Social Services.

During the interim, there is more time available for non-legislative use, although permanent legislative committees do continue to utilize the LTN. Because the information offices are not staffed year around, we do on occassion charge a small fee to non-legislative users. This helps to cover costs for keeping a moderator "on-call" at those locations. Otherwise, regular funding continues to cover LTN costs.

Non-legislative requests are accepted at the user's "risk"; that is, a legislator can pre-empt something previously scheduled on 24 hour's notice. This rarely occurs; with careful tracking and constant updating of schedules, we seem to accommodate most everyone.

With frequent teleconferences, reliable personnel becomes essential. Most locations need at least one part-time back-up to the regular employee. During session, Juneau requires two full-time moderators plus a part-time person who primarily handles evening teleconferences. Anchorage has a similar setup.

A regular and vital use of the network is by the Division of Public Services for administrative purposes. We conduct daily "chatter" to all offices whenever possible for schedule updates, advice on problems (such as mail service and faulty data terminals, etc.), and exchange of bill and other legislative-related information. "Conference calls" are regularly held by the teleconference coordinator or another administrator to gather all sites together for pertinent news, comments, and quite often, just for the sake of good morale for all. The LTN is a distinct asset in keeping such an extensive chain of offices in touch with one another and with the agency's administration.

Video Teleconferencing

In 1978, funds were set aside to provide the opportunity for twelve two-way video conferences, in addition to the already established audio-only system.[3] Video conferencing is considered part of the LTN, but with some different aspects that require it to be presented as a separate subject.

Whereas audio-only teleconferencing is provided via an in-place circuit at permanent locations, video conferencing differs greatly. It is not a "system" that can be quickly utilized and easily scheduled.

To develop full-motion, two-way video capability, the task force went to considerable effort. A great deal of time and understanding of the concept, technology and requirements were spent by consultants and again, RCA/Alascom. In another innovative move to advance telecommunications throughout the state, once again the Alaska legislature's requirements resulted in expansion of the state's telephone utility, particularly in the area of satellite and microwave development.

The final outcome was an agreement with Alascom that allows use of their earth stations up to twelve times in a twelve-month period for $123,000 per year, and stretched over a five-year period. This fee is to pay Alascom for installing the necessary equipment at their terminals to allow two-way video. The thirteen and subsequent conferences in a given twelve-month period will cost only $1,400 per. The first "year" of the agreement began June 18, 1979.

Nine locations can participate in video productions; seven are two-way and two are receive only. The state's leased Satellite T.V. Demonstration Project transponder is utilized, plus a video "window" is regularly scheduled for two hours each week with the Alaska Public Broadcast Commission, which administers satellite programming.

The Last Mile

While the agreement with Alascom established "long line" video transmission, establishing termination facilities in each community was not quite as simple. Small information offices are not T.V. studios! As it stands, local T.V. facilities in private cable companies, college studios, and in two cases, the offices themselves, are secured for each video conference, and at expense beyond the basic $123,000. These "last mile" arrangements were at first cumbersome and inefficient. We are finding now, however, after five video presentations thus far in 1980, a more definitive procedure has resulted. This is in part due to coordination and production assistance by the Legislative Affairs Agency's new Media Center and also because video teleconferencing just became a "comfortable medium" this year. For a variety of reasons, some technical, most people-related, video teleconferencing did not get fully tested nor utilized in 1978/79. We do feel now that we have demonstrated video teleconferencing as a viable entity and fully expect to continue with more video productions through the remainder of the 1980 session.

Different From Audio

Video differs from audio-only in many respects, but most obviously by the addition of faces and graphics. While audioconferencing is probably unequaled as an inexpensive way to hear extensive testimony, video conferences speak to a different need.[4] Video adds more personality

and "color"--it gives an intimacy not always established by voice-only.

Video conferences were fairly well-suited to the medium this year. One was a demonstration of equipment where a picture was essential to understanding, another was a presentation on telecommunications by a reknown guest speaker and the others were on the emotional and popular issue of Alaska's Permanent Fund (the state's oil reserve monies).

Assessments

To date, how do we feel about the LTN and our video endeavors? For the audio-only system, we are comfortable and pleased with a network that works well, is showing results by its increased popularity and improved access, and by and large, is not technically troublesome or difficult to maintain and operate. 1979 Session use of the LTN nearly doubled since 1978 despite the fact that the 1979 session was considerably shorter than the previous one. Three to four teleconferences were held most days, that has frequently increased to five. This increase can be attributed primarily to greater public awareness of the LTN's effectiveness, growing acceptance by legislative users, larger number of communities served and its suitability in dealing with the unique problems presented by the necessity of considering "sunset" legislation.

There is no way to truly evaluate this impact we speak of or measure the effect of citizen participation in the legislative process, we can only vouch for its existance. Judy B. Rosener, from the University of California, states, "...there is little agreement among and between citizens and administrators as to their goals and objectives". She and others agree there is little known about the "effectiveness" of citizen participation.

We base our positive feelings on the mere fact that more people than ever before are dealing direct with the legislative process; they are indeed becoming part of it. And, Alaskan legislators can represent their districts more realistically through teleconferencing. Weekly and monthly "constituent" meetings are now regularly held by legislators in Juneau, where they conduct evening talks about local problems with their constituents back home. This positive impact is further reinforced by the fact that all this is cost-effective; the process is expanding while costs for travel and time spent are lessened considerably.

Because video conferencing seems yet to be completely demonstrated, we are hesitant to make absolute judgement. And, whether or not the expense justifies the means over audio teleconferencing has not yet been fully reached and evaluated. With present staffing and facilities, plus the popularity of audio conferencing, suffice it to say we are generally pleased that our agreement limits us to twelve videos per year. On the other hand, the rate of success thus far has been good, and we have proven video conferencing can indeed be done, and with professional integrity.

Future Plans

Our assessment indicates we are generally satisfied with the LTN. Naturally, nothing is ever perfect, and even great ideas can be improved upon. With that in mind, and due to user comments and requests, we continue to research improved equipment, options and ideas.

One feature the original system is lacking is signalling. Since we do not have a dial-up network, there is no means to "call" a site that has the conference set turned off. Our one failure of note was an attempt to contract the construction of signalling devices, which was done with poor results. Not all the "decoders" worked, and those that did, failed rapidly.

Due to their own preference, the Washington, D.C. offices keep their sets turned off most of the time, which is understandable due to the time zone difference and no real need to "chatter" with Juneau on a daily basis. But, on occasion, there is need to reach those offices, and placing a long-distance call to Washington, D.C. to ask for the set to be turned on does seem ludicrous.

We also have infrequent need for "off-net" capability, that would allow access to 2-wire telephones via our 4-wire system. In particular, certain issues or legislation make including a remote village desirable, where otherwise a permanent LTN drop is unnecessary.

Certain problems will always exist in any telecommunications system, no matter how simple or complex. Our basic system is technically a simple design. It becomes more complicated to maintain and trouble-shoot when you consider it connects via satellite, earth station and terrestrial carrier, and terminates into various local telephone company facilities. Because this system grew rapidly beyond the original expectations some "make-do" practices have resulted here and there, thus resulting in a particular problem which was lack of enough battery power for five sites to send (only) telecopy via the network.[7]

Alascom originally provided this "talk battery" at their toll centers, but no longer does this. So, we are required to supply this to our customer-owned equipment. This becomes complicated due to our also owning separate 500-style (single line) telephone sets, modified to 4-wire, for the explicit purpose of telecopying over the network. The reason for the extra phones is that the "Trimline"-shaped handset on the WE50A1 conference set does not fit into the acoustical coupler in the Xerox telecopiers! Thus, we have separate phones that also plug into the circuit. A resolve to this Rube Goldberg set-up is desirable, and we have been working with our Division of Communications on designing a small power device that will supply the power boost that is needed.

There have also been inquiries regarding simultaneous teleconferencing, that is provision for two or three teleconferences to be conducted from Juneau to various sites--all on the same day, same time. This came to light particularly during scheduling of evening constituent meetings; it required some juggling and cooperation by a number of people. There just didn't seem to be enough week nights to facilitate the majority of the 60-person state legislature!

Primarily due to these three concerns, we have investigated the new SS-4 bridging system developed by Bell Labs/AT&T, plus the Darome Meet-Me Conference bridge. We anticipate adding this or similar equipment to the existing network configuration by year end 1980. Alascom is strongly recommending we re-configure in some manner to alleviate what they call an over-load problem. They contend that the existing circuit has so many drops on it (approximately 50) that we are degrading its entire integrity plus making it more susceptible to various noise problems. We are hoping our present investigations will confirm and/or resolve that problem.

Future plans also include expansion of the network into more communities, probably also by year end. Funding is being looked into now for the addition of Seward and Homer, located along the Gulf of Alaska, and Sand Point, which is on the Aleutian chain.

User Reaction

What do our users say? Evaluation forms have been provided at most teleconferences, giving opportunity for good and bad comments. Those forms that come to Juneau from the field sites almost unanimously mention that the participant would not have given testimony had there not been access to the LTN. Many comments refer to time away from jobs, not to mention money saved by not having to go to Juneau. Legislators and staff remind us now and again of

how glad they are that committees are traveling
less, spending less, and plus how helpful it is
to have instant communication with the information offices in their districts. A recent
comment made by a state representative who is a
frequent LTN-user, "It sure beats the mail!"
Another legislator, who well typifies the
frequent and happy LTN user, said, "The system
works better than was ever expected; it really
has become a tool for the legislature. The best
feeling of all is to talk to a constituent about
a matter only amended the same day--that is real
efficiency."

BIBLIOGRAPHY

1.   Goldin, Lawrence. <u>Legislative Teleconference Task Force Report</u>, December 1977.

2.   Goldin, Lawrence. Letter to Representative
Mike Miller, December 23, 1977.

3.   Fromuth, Peter. <u>Final Report on Telecon-
ferencing</u>, July 1978.

4.   Fromuth, Peter. <u>Final Report on Telecon-
ferencing</u>, July 1978.

5.   Plummer, Sioux. <u>Report to Legislative
Council</u>, July 14, 1979.

6.   Rosener, Judy B. "Citizen Pariticipation:
Can We Measure Its Effectiveness?" <u>Public
Administration Review</u>, September/October
1978.

7.   Plummer, Sioux. <u>Report to Legislative
Council</u>, July 14, 1978.

The Teleconferencing Resource Book: A Guide to Applications and Planning
Lorne A. Parker and Christine H. Olgren (eds.)
Elsevier Science Publishers B.V. (North-Holland)
©Center for Interactive Programs, University of Wisconsin-Extension, 1984

THE USE OF TELECONFERENCING FOR TRAINING BANK PERSONNEL

Ellis P. Waller
Marketing Director
Financial Education and Development
Division of Don Jones and Associates
Middleton, Wisconsin

Training within the banking industry has traditionally been conducted through in-house training, by supervisors or the bank's personnel department. As the banking industry became more complex, banks relied upon more outside sources for training and education. The American Bankers Association and the American Institute of Banking, provided courses and other training materials for bankers. Individual state banking associations offered courses, seminars, workshops and schools to meet the training needs of their members. Associations such as the Bank Marketing Association and the Bank Administration Institute provided training in their particular areas.

At the present time, bankers can choose from a wide variety of training sources to fill virtually any training need. However, current economic conditions have reduced bank profits and caused many banks to reduce their training budgets. Fewer employees are being sent to outside training programs as the expense of travel increases. As a result, the attendance at conferences, seminars, and workshops is declining. State Banking Associations have not been immune to this trend. A reduction in the attendance at their programs affects the quality and quantity of the educational programs they can offer their members. The small and medium sized banks making up a state banking association's membership are affected most since their internal training capabilities are often limited.

Teleconference Demonstration

The Illinois Bankers Association, in addition to looking for ways to increase attendance at its seminars and conferences, wanted a way to take training programs to its members throughout the state. When the firm of Financial Education and Development of Middleton, Wisconsin demonstrated teleconferencing to the Illinois Bankers Association in the summer of 1981, I.B.A. officials realized how it could meet both of their objectives.

The brief presentation used two Darome 610B convener units, one located at Financial Education & Development in Middleton, Wisconsin and the other one at the I.B.A.'s Chicago office. Slides were advanced by a Financial Ed staff member in Chicago, as the speaker made the presentation from Middleton, Wisconsin. This brief demonstration was enthusiastically received by the I.B.A.'s Director of Education and was followed up with a proposal to conduct a series of Banking Professionalism Seminars throughout the State of Illinois.

Financial Education and Development of Middleton, Wisconsin has conducted similar "live" programs throughout Illinois for a number of years and teleconferencing appeared to be an excellent way to augment these programs by taking the training to the bankers rather than requiring bankers to travel long distances for their training. The Banking Professionalism Seminars were scheduled for five consecutive Tuesday nights in January and February of 1982.

The Illinois Bankers Association was responsible for selecting the cities and training facilities, as well as enlisting the assistance of volunteer coordinators to handle the details of setting up the teleconference equipment, advancing the slides and serving as moderators for the groups. The seven Illinois cities selected were located throughout the State with 270 miles separating the northern and southern-most cities. Marion, the southern-most city was 375 miles from Madison, Wisconsin where the conference bridge was located. Populations of the seven locations ranged from 400 in Carlock to 23,600 in Macomb.

To keep the cost of the conference series to a minimum, the I.B.A. decided to hold them in bank conference and community rooms. Classroom style seating arrangements were used when the availability of tables and the size of the audience permitted.

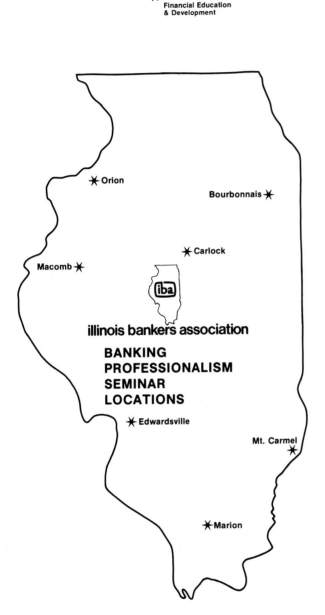

Madison, WI
Financial Education
& Development

Orion

Bourbonnais

Carlock

Macomb

(iba)

illinois bankers association

BANKING
PROFESSIONALISM
SEMINAR
LOCATIONS

Edwardsville

Mt. Carmel

Marion

Teleconference Coordinators

At each location, the Illinois Bankers
Association enlisted the assistance of a bank
employee to act as a coordinator for each teleconference. In all cases, the coordinator was
employed at the location selected. The
coordinators were all bank officers including
two presidents. In several cases, the
coordinators shared their responsibilities
with one of two other individuals who handled
details at the weekly conferences.

Coordinator Training

The coordinator's role was a vital one
for the success of the entire teleconference
series. The coordinators had never used
teleconferencing equipment nor were they
familiar with the content of the program.
They were relied upon to set up the teleconference equipment, dial into the conference
bridge in Madison, Wisconsin and handle any
problems that might occur during the conference series.

The teleconference coordinators handled
registration and advanced the slides while
the speakers made their presentations. Since
it was not possible to assemble the coordinators prior to the first teleconference for a
training session, it was decided to conduct a
"dry run" teleconference a week prior to the
first teleconference. The "dry run" provided
the coordinators with a "hands on" opportunity
to set up and connect the teleconference
equipment and to use it while discussing their
concerns with the other coordinators. A
representative from the Illinois Bankers
Association participated in the "dry run"
teleconference along with a representative
from Financial Education and Development.

Communication With Coordinators

It was imperative that every potential
problem be addressed to minimize any problems
once the teleconferences were underway. From
the day the coordinators were selected by the
I.B.A., a steady flow of letters and telephone
calls was directed to each coordinator and
the I.B.A. representatives. This began with
a check list of conference room requirements
as well as audiovisual equipment needed. It
ended with a shipment of teleconference
equipment prior to the "dry run". Phone calls
were made to each location to determine if
there were any problems in connecting the
teleconference equipment to the phone lines.

In several cases, the banks elected to install a telephone line in the conference room where the conferences were held rather than using extension cords to nearby phones. An excellent rapport developed between Financial Education and Development's staff and the conference coordinators. All technical matters were taken care of before the "dry run" teleconference. To make sure no questions remained unanswered, a 23 page Coordinators Manual was prepared covering every item from a description of the conference bridge to the shipment of the teleconference equipment and slides back to Financial Education and Development. The Coordinators Manual described the teleconference equipment in detail and a color-coded reproduction of the Darome convener unit was included to assist the coordinators or their assistants in connecting the equipment. The manual also outlined the general nature of the teleconference series, the use of the equipment and directions for each of the five teleconference sessions.

"Dry Run" Teleconference

The "dry run" teleconference, held a week before the first conference, was intended to duplicate the actual conditions of the evening teleconferences. During the "dry run" teleconference, individual coordinators had an opportunity to use the equipment in the same setting as in an actual teleconference. The "dry run" was held at 10:00 a.m. and all locations were bridged into the network within a ten minute period. The "dry run" was quite uneventful since the manual had answered most of the participant's questions. Once everyone felt comfortable with the equipment and had no further questions, the teleconference was concluded. The "dry run" lasted approximately 40 minutes.

A Darome 610B convener unit was shipped to each conference location prior to the "dry run". Four Darome 413B microphones were also included. At one location the attendance of 72 people made an extra four microphones necessary. Extra telephone and microphone extension cords were included. In one location the conference room was exceptionally large. The coordinator successfully used two hi-fi speakers to augment the convener unit speaker.

The conference leaders conducted the first two conferences using a regular Darome microphone. This proved to be rather confining even though the talk bar was taped down. The speakers are accustomed to presenting conferences in front of large audiences and working directly with three to

six projectors and two screens. For the third, fourth, and fifth conferences, the speakers used an Electro/voice miniature dynamic RE51 microphone and headset. This allowed the speakers to have more freedom of movement and more flexibility when using course materials.

Speaker's Presentation

Because the speakers were conducting conferences in other states during this teleconference series, the presentations were made from a number of different locations. The first conference was conducted before a live audience in Marion, Illinois, one of the seven conference sites. The second was conducted from the Illinois Bankers Association's office in Chicago, Illinois. For the third teleconference, the speaker made the presentation from Financial Education and Development's conference room in Middleton, Wisconsin. For the fourth teleconference the speakers spoke from Helena, Montana and Middleton, Wisconsin. The fifth was conducted from a hotel room in Atlanta, Georgia and Middleton, Wisconsin.

If the speakers had presented the conferences in person at each of the seven Illinois locations, it would have required over 5000 miles of automobile travel during the 5 week series. During the week of the third teleconference, Southern Illinois was hit by a record snow fall which closed the schools for a week and made travel virtually impossible on the nights of the conferences. As a result, two conference locations were rescheduled for the week following the last scheduled conference session. During the fourth conference, travel conditions were equally bad and two locations chose to cancel the conferences. These were also rescheduled. Rescheduling would have been difficult had this been a "live" presentation.

During the third and fourth teleconferences, executives from several state banking associations were also bridged into the teleconference network to experience teleconferencing firsthand. While they did not have the advantage of the course handout material or the visual presentation, they were impressed with the interaction which took place between the groups and the speakers.

Audiovisual Presentation

The use of volunteer coordinators to advance the slides at the seven teleconference

Teleconference session being conducted from the
office of Financial Education and Development, Middleton, WI

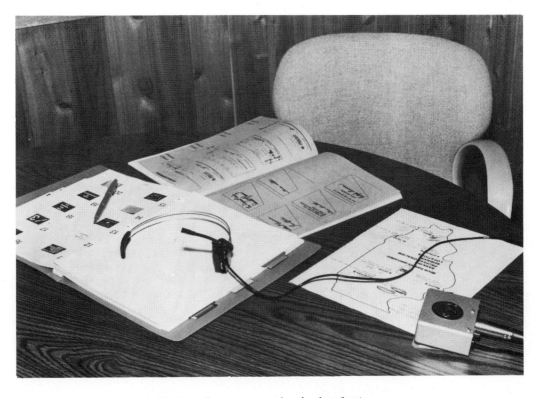

Teleconference speaker's headset,
Coordinator's Manual and Course Materials

locations was a matter of some concern to
Financial Education and Development. The
coordinators were unfamiliar with teleconferencing and the programs. An electronic
solution was thought to be the best way to
ensure that all seven projectors would be
advanced at the proper time. However, the
price was well beyond the budget of the
Illinois Bankers Association. As an alternative, each coordinator was provided with
colored xerox reproductions of all slides to
be used in the presentation. These slides
were numbered for each of the five conferences.
Occasionally it was necessary for the speakers
to refer to a specific slide number, but in
most cases the conference leaders gave cues
such as "Next we will see..." or "Now we
have..." As it turned out, the advancing of
slides was not a problem for the coordinators
and it gave them an opportunity to actively
participate in the program.

Coordinators Manual showing reproductions of teleconference slides

Pre-Conference Promotion

The Illinois Bankers Association made two
mailings of a promotional brochure to Illinois
banks. The I.B.A. also handled all registration details directly through the coordinators.
Since the conferences were held in bank conference rooms, there was a maximum capacity
of 300 people. The I.B.A. anticipated a maximum attendance of 210. The day before the
first seminar, registration reached 270 or
28% more than the I.B.A.'s optimistic estimate.
A number of walk-ins increased the attendance
to 291, 38% more than the I.B.A.'s original
estimate.

Teleconference Agenda and Content

Each of the five conference sessions was
designed to discuss two banking professionalism topics. These ranged from counterfeit
currency to time management. The conferences
were attended by tellers, bookkeepers, new
account personnel, customer service representatives, customer contact employees, and
bank officers. The first teleconference
began 15 minutes early to familiarize the
audience with the teleconference equipment.
All subsequent sessions started promptly at
6:55 p.m. and ended at 8:45 p.m. A 10
minute break was included.

The agenda was kept flexible to allow
for interaction between groups but careful
timing made it possible to remain on schedule
throughout the series.

Banking Professionalism Conference Topics

The following topics were discussed
during the five week conferences.

    Currency Facts
    Identifying Counterfeit Currency
    Cash Handling Tips
    Opening Lasting Checking Accounts
    The New Account Interview
    Proper Endorsements
    The Bank's Exposure on Endorsements
    The Federal Reserve System
    Check Preparation
    Becoming a Better Time Manager
    Overcoming Procrastination

Course Materials

All attendees received a 60 page note-

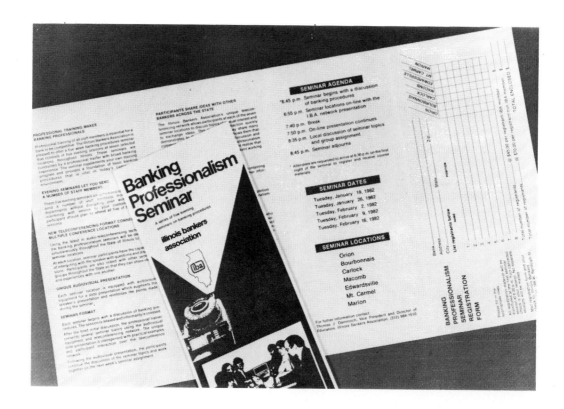

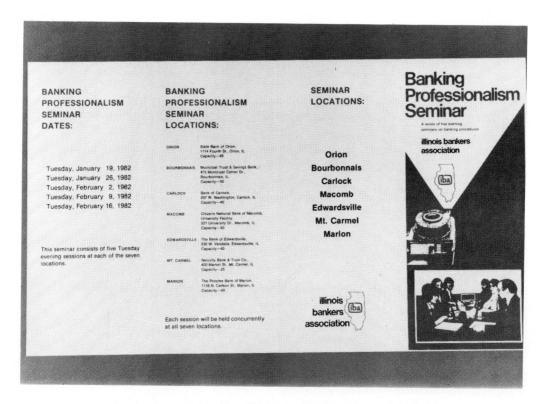

Brochure sent by the Illinois Bankers Association
to promote the Banking Professionalism Teleconference Series

book of course materials containing agendas,
outlines of each evening's conference and
descriptive material to augment the slide
presentation. The materials were referred to
during the teleconference by the speakers
and could be used on the job after each conference.

Each evening's conference included one or
two exercises or quizzes which were used as
a basis for discussion. The teleconference
coordinators reported that there was a great
deal of discussion at each location during the
conference in response to the speaker's
questions.

Completion Certificates

The Illinois Bankers Association provided completion certificates for each
attendee. The certificates were sent to each
attendee's bank for presentation by management at the conclusion of the teleconference
series.

Interaction

The conference bridge made tapes of the
teleconferences and when they were analyzed,
they revealed an unusually high amount of
audience participation and interaction. At
each location, frequent off-line discussions
took place in response to the speakers'
questions and comments. A high level was
reached on the third teleconference when the
speaker asked a total of 35 questions and received 29 replies in addition to the 22
questions that were asked by the audience.
This total of 51 responses, during a two hour
teleconference, was many times greater than
the interaction in a live conference before a
similar sized group.

Conference Evaluations

At the conclusion of the teleconference
series, participants and coordinators were
asked to complete evaluation forms. The
coordinators' evaluations echoed the senti-
ments of the participants as to the success
of the teleconference series.

The teleconference coordinators were
officers of the banks in which the five conferences were conducted and many of their
own employees attended. They were, therefore,
in an excellent position to judge the quality
of the training and the overall quality of
the teleconference series.

Each coordinator and assistant coordinator
was asked to rate aspects of the teleconference and to write comments on a number of
other areas. Their evaluations were as
follows:

1. On the whole, how would you rate the
Banking Professionalism Teleconference?

2. The quality of the training and instructions you received for the equipment and
its operation?

3. The ease with which you were able to set
up and operate the equipment?

4. The ease with which you were bridged into
the conference?

5. The effectiveness of the presenters in
getting the idea across?

6. The ease with which questions could be
asked and answered during the conference?

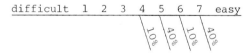

7. How easily did the conference participants
adjust to the teleconference medium?

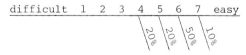

8. In terms of the quality of learning how
would you compare this teleconference
to a "face-to-face" conference?

The one area where the evaluation forms indicated less than the anticipated response, was that of asking and answering questions. The coordinator's evaluations did not correspond to the high amount of interaction between the speakers and the individual locations. The coordinators were asked to provide a diagram of the conference room layout used and it quickly became apparent that the difficulty in asking questions was a function of audience size, layout, and microphone location. Because of the large attendance and limited space, several of the coordinators arranged their rooms theater style with microphones in the front of the room rather than on each table in a classroom style layout. Unless a question was relayed verbally to a coordinator individuals had to leave their seats and walk to the microphone to ask their questions. This effort to reach the microphone was the reason for the response on several of the coordinators evaluations.

Teleconference coordinators were also asked to comment on the conferences. Their responses are as follows:

"I thought the conferences were excellent"

"I enjoyed it and would like to be involved in more of them"

"This is a very good way to reach the most people in the shortest span of time. I believe this is the method we need to develop"

"I think its a very exciting program and I look forward to participating in the future"

"They were super!"

"...very successful. I would hope this type of learning tool will be used in the future. It gave some "smaller banks" an opportunity to participate in a way they couldn't for a large meeting"

"Feel that the slides really help. Thought the notebooks were excellent"

"Had a great time. Met a lot of great people"

"Cost per participant was low enough that several persons from a bank could attend"

"Participants were more apt to ask questions than if a speaker was in front of them"

When asked to describe the audience reaction, the coordinators comments were all very positive, ranging from "very impressive" to "timely", "useful", "well handled" and "very positive". General comments were as follows:

Coordinators felt that attendees learned about as much as with a live speaker.

Coordinators suggested that 21 additional topics be presented using teleconferencing, particularly programs for new employees.

All thought that the $45.00 price was right.

Coordinators were divided on the length of the conference series. Half felt five 2-hour sessions were best, the other half felt three to four sessions would be an improvement.

When comparing the amount of discussion with a "face-to-face" conference, coordinators felt the amount of discussion was the same as or greater than a "face-to-face" presentation.

I.B.A. Involvement

The Illinois Bankers Association's staff members made it a point to visit each location once during the series of teleconferences. This provided the opportunity to evaluate the conferences first hand and to talk with their members in the field. At the first session, the Vice President and Director of Education for the I.B.A., Mr. Thomas J. Dammrich delivered his comments from his suburban Chicago home. At the final conference, Mr. William H. Hocter, Executive Vice President of the I.B.A., addressed the group from a phone booth on the Illinois Tollway. This served to bring the Association closer to its members and was welcomed by the attendees.

Conclusion

In summary, the series of five Banking Professionalism Conferences were especially well received by the Illinois Bankers Association and its members. The conference coordinators did their jobs exceptionally well and the 291 attendees were very positive about the results. There was little indication that the teleconference interfered with the quality of the instruction and according to the evaluations it increased the interaction among the seven groups and within each group. The audience appreciated the fact that they did not have to travel long distances to the conference, although several did travel 30 to 60 miles to attend each session. Holding the conferences in banks made the atmosphere more informal and conducive to learning.

As a result of the success of the I.B.A.'s Banking Professionalism teleconference series, a second series of Bank Security Teleconferences is scheduled for July of 1982. This series will consist of three evening sessions in ten locations. The format will remain the same, but attendance is expected to increase because of the shorter time commitment required of the attendees.

The Illinois Banking Professionalism Teleconferences satisfactorily demonstrated that teleconferencing can be a cost effective and efficient medium for providing quality training for bankers throughout a wide geographic area.

Participants at the Edwardsville, Illinois
Banking Professionalism Teleconference

The Teleconferencing Resource Book: A Guide to Applications and Planning
Lorne A. Parker and Christine H. Olgren (eds.)
Elsevier Science Publishers B.V. (North-Holland)
© Center for Interactive Programs, University of Wisconsin-Extension, 1984

USING AUDIO TELECONFERENCING FOR SALES AND MARKETING:
HOW TO ACCELERATE THE ADOPTION PROCESS

George R. Silverman
President
The Teleconference Network™
Division of Market Navigation, Inc.
Orangeburg, New York

In this paper, I will outline how to use audio teleconferencing to compress and accelerate the marketing cycle of products and services.

Marketing is the practice of profitably creating the maximum number of transactions which benefit both buyer and seller. Sales is a sub-category of marketing, the practice of persuading people to buy your product or service.

Both marketing and sales are extremely difficult and frustrating disciplines. Even if a product is genuinely superior, it takes many years for it to be fully accepted by the marketplace. For instance, it typically takes about seven years from market introduction for a superior new drug to be accepted by physicians. It takes about fifteen years for a new agricultural product to be accepted by farmers. An educational innovation takes about twenty five years for acceptance! So, by this latter figure, it will take until well into the 1990's for teleconferencing to be fully accepted by the educational community.

What's holding things back? In the last few decades, there have been tremendous advances in market research and the entire technology of persuasion: More sophisticated advertising, upgrad-ing of sales forces, better promotional materials, more professional communication and more efficient means of educating prospects about our products. Yet the time it takes a product to reach full market share has not significantly decreased.

In 15 years of consulting with companies on the applications of teleconferencing to marketing problems, I think that I have discovered the reasons why this frustratingly slow adoption process is so resistant to acceleration:

The development of effective marketing and sales campaigns depends on being able to construct a delicately balanced system so that all elements in the marketing mix work together. Salespeople, ads, mailings, brochures, seminars, trade shows, etc. must all complement each other. At the root of developing such a system, is communication. Now, communication these days is fast, with electronic mail, overnight letters, voice messaging, phone calls, etc. However, much of the essential communication in marketing and sales is interactive communication, and it is this that has not been appreciably accelerated yet. Let me repeat: The inability to deliver certain types of interactive communication is the bottleneck to selling products faster.

Specifically, there are three types of interactive communication which can make a dramatic impact on product adoption if only they were used more: I. group idea stimulation, II. peer learning, and, most importantly, III. word of mouth. All of them are very difficult to deliver and control, which is why they are not used more. However, they are deliverable relatively easily via teleconferencing.

What I'd like to do is walk you through the marketing of a product, from before it is even developed, until it has reached full market acceptance, showing you how its adoption can be accelerated by delivering these types of inter-active communication via various teleconferencing programs.

But, before I do that, I'd like to explain a little more about these different types of interactive communications.

I. Group Idea Stimulation

The first type of interactive communication is group idea stimulation. It is based on the finding that most people generate more and better ideas and information when they are interacting in groups rather than working as individuals. Frequently, many people have pieces of the puzzle, and by putting them together, the whole picture emerges.

An example is problem solving sessions, where people should be included who are geographically dispersed. By using teleconferencing, these people can be brought together quickly and easily. These sessions can be used to generate the ideas for new products, or solve problems relating to already existing products.

Another kind of idea stimulation group is the focus group. It is the market research technique whereby groups of people are brought together to discuss a product. Through their group interaction, more can be brought out than in individual interviews. This is a very commonly used technique for consumer groups, since it is easy to get them together face to face. But it is not used as much as it should be with more difficult to reach respondants, such as physicians, executives, managers, salespeople, etc. The busiest of these people are usually too busy to attend face to face meetings. Again, teleconferencing is the answer. Telephone focus groups provide a very powerful way of generating new ideas, and testing old ones. They are severely underutilized. Their use at all steps of the marketing cycle would save many products from costly pitfalls, as well as provide more targeted product positioning.

II. Peer Learning

The second type of interactive communication mode that can accelerate product acceptance is peer learning. People who are at similar levels of attainment learn from each other because they are able to complement each other's knowledge. This is particularly applicable to your sales force, where 20% of your salespeople typically generate 80% of the business. If you could somehow spread the knowhow of the "stars," you would generate a lot more business, faster. But it's costly to take salespeople out of the field. Again, simple audio teleconferencing is the answer. The salespeople participate from ordinary phones at home in the evening, or early morning, not during sales time.

III. Word of Mouth

Before we get to our illustrative case in accelerating product acceptance, I will spend some time going into word of mouth in some detail, because the delivery of word of mouth made possible through teleconferencing is a major marketing breakthrough.

The most important limiting factor in the adoption of a new product is the time that it takes for people to gain enough experience with it to feel comfortable enough to adopt it fully. Yet, there has always been a mechanism for accelerating experience. It's the oldest and most powerful force in the marketplace: Word of Mouth. If there were a way to harness this force and make experience sharing easy and less time-consuming you could reach your full market share faster.

The "Confidence vs. Experience" Dilemma

Let's look, for example, at how physicians adopt new drugs. A physician can't have confidence in a drug without a great deal of experience with it; but he won't get extensive experience without having confidence. However, while "experience is the best teacher," it is also the most risky. What if the drug has unforeseen side effects, or is trickier to use than anticipated? No one wants to be unsuccessful, to say nothing of

unintentional harm to patients. This is why the physician wants to hear from peers before trying a drug, or expanding from trial to full adoption.

Studies show that physicians typically try a new drug on three to five patients over many, many months. These are usually patients refractory to other drugs. Because of the small numbers and the severity or refractory nature of their conditions, there are inconclusive results. One patient seems to improve, another seems to be unresponsive, another deteriorates and one moves to Florida! Consequently, it takes an average of 7 years for a new drug to reach its full market share.

Fortunately, there is a less risky alternative to direct experience: indirect experience through word of mouth. It is, in effect, a way for a physician to try out a new drug on someone else's patients.

Indirect Experience Through Word-of-Mouth

Studies have repeatedly shown that word of mouth from peers is the most powerful triggering mechanism for getting physicians to write their first few prescriptions. This is not to say that advertising, salespeople, sales promotion, publicity, experts' testimonials, conferences, symposia and the other communication channels are not important. These channels provide information. But after information comes confirmation: Are the claims true? Do the differences claimed make a real difference in actual practice? How will the drug work out in my hands, in a typical practice, with patients like mine? What are the negatives that the company may be soft-peddling?

From the marketing point of view, word of mouth has its definite drawbacks. It is unpredictable, slow and doesn't provide an opportunity to correct misconceptions. It can be disconcertingly like its close cousin, rumor -- fast to spread the negatives, slow to spread the positives, if it does so at all. Frequently, when asking a peer for information about a product, the words used are "Had any trouble with -----?" So, the most powerful force in the marketplace, word of mouth, like lightning, hurricanes and tornadoes, is also the most out of control and potentially most destructive.

Peer Word-of-mouth Teleconferences

There is, however, a way to harness the tremendous power of word of mouth. That way is peer word of mouth teleconferences: Hundreds of groups of a dozen or so physicians each are conducted, over a specially designed telephone conference system, by an independent third party in a completely objective manner. During the course of the discussions the physicians share and compare their experiences with a new drug, creating a larger pool of experience from which to draw their conclusions. Most importantly, they are able to notice subtle differences which they probably would have missed acting alone. They also have more confidence in their findings if other people have come to the same conclusions through a process of rational deliberation. (For the issues where the group's experience is likely to be insufficient, excerpts from groups conducted with experts can be played into the session.)

During their conferences the physicians are carefully guided through the decision-making process through the use of the Decision Map™. This is a flowchart of the stages and steps that physicians go through in making a solid adoption decision. It has been my experience that if they skip steps, they bog down and don't adopt the drug, or take an inordinately long time. From an analysis of the Decision Map™, an agenda is developed which gets physicians to look at all of the important issues in considering whether to use the drug. By the end of each session physicians reach firm conclusions about exactly how and where they will be using the drug. Peer word of mouth teleconferences have been found to have a dramatic and immediate effect on physicians' prescription writing.

Illustrative Case: How to accelerate product acceptance through teleconferencing

Now that I have described the various forms of interactive communication that will accelerate the product adoption process, I can show you how to put it all together into a marketing program. Let's take an imaginary product to illustrate the process. You'll have to use your imagination to apply it to your particular products. Not everything that I describe will be applicable in your case.

Let's take a product in the pharmaceutical industry, since that is where I have the most experience. A particular pharmaceutical company has a great deal of expertise in the cardiovascular area.

1. They have an advisory group of expert cardiologists which meets several times a year, by teleconference, with company research and development people to advise them on new developments in the field and to suggest new research and development directions. A need is identified for a longer acting form of one of the company's drugs. This is an idea which might never have occurred to the company's people. This kind of experts advisory group is rarely convened on a regular basis, again because of the logistical difficulties of getting people together face to face. The difficulties are easily overcome through teleconferencing.

2. Making the drug longer acting can be accomplished by a variety of means. One can change the molecule, combine it with other drugs, or change the physical properties into an ointment, a patch, microencapsulated granules, etc. So a series of idea generation groups is convened to get ideas for how to develop this new drug. In this way experts from different locations can be brought in, as well as consultants from around the country. They can meet in several successive meetings as easily as picking up their phones. Various modes of brainstorming, as well as other creative problem solving techniques are used.

3. At the same time, several user advisory groups are convened in order to see which of the directions under development would meet the greatest acceptance in the marketplace. These groups consist of more typical physicians, who would ultimately be prescribing the product. At first they discuss the new directions very generally, but as the research and development group makes more and more headway, the advisory group gives more specific input into what is being developed.

4. Once a new drug is developed, it takes about six to ten years and usually well over ten million dollars to get it approved by the Food and Drug Administration. A limiting factor is the amount of time it takes to coordinate the clinical studies which are conducted by independent clinical investigators. Often, it is important that these clinical investigators follow the same protocol. Here, again, teleconferences are used to coordinate the studies and make sure that they are being conducted properly. If your products are not pharmaceutical or agricultural, you probably do not have clinical investigators. But this kind of teleconferencing is probably applicable to other types of project management situations.

5. Now the product is pretty far along, and looks like it will be launched in a few months. It's time to turn to several groups of star salespeople, who meet regularly to tell the home office ways they have discovered of selling the company's products. Everyone knows that the best salespeople do not do things exactly as they are told by the home office, they do it better. That's why they are star salespeople. This enlightened company does not try to fight this trend, they take advantage of it. They regularly tap the ideas of their best salespeople and use them for advertising ideas, sales "how to" bulletins and other uses. In this case, the home office wants to know how these sales stars would approach the features of the new product, turn them into benefits and how they would express those benefits to real people. Again, the sessions are conducted via audio teleconference, outside of sales time. The advertising agency can listen in live, or hear tapes later.

6. We now enter a market research phase. Several concept testing focus groups are conducted. In these groups, real practicing physicians tell us what they think of the new concept,

express their skepticism, tell us what evidence they would require (and in what form they would require it) to get them to try the drug.

7. These lead into Persuasion Design Groups where we actually try out some of the persuasion strategies developed in the above two kinds of groups to see which of them work best and to develop the persuasion elements even further. They are essentially laboratories in which the actual words which are going to be used can be tried out in nationwide groups before they are committed to ads, salespeople, brochures, etc. As people react, the messages can be improved and tried out on the next group. We then construct a Decision Map™, which is a flowchart of the stages and steps people will have to go through in the decision making process. The Decision Map™ allows us to have a picture of the whole decision making process, so that we can make sure that every part of it is covered properly. This refinement process results in very powerful messages.

8. Ad and Copy Testing Sessions also consist of the kind of practicing physician who will be the best prospect for prescribing the drug. In these, we mail out proofs or mock-ups of ads and brochures and have them react, again tightening the message and avoiding things which rub people the wrong way. Since this is audio teleconferencing, we mail out the materials in envelopes which are labeled with a request not to open them until the time of the conference.

9. Prior to launch, Experts Sessions are conducted, so that the nation's leading experts, many of whom have been involved in the original studies, can comment on the practical use of the drug. They can summarize the studies, talk about the benefits of the drug and recommend how to start patients on the drug. These sessions are then distributed in cassette form, transcripted as a paper and/or played in excerpted form to word of mouth sessions, described later.

10. Sales meetings: Now its time for the pre-launch sales meetings. These will most likely be held face to face. However, I mentioned that this is a enlightened company. The Sales Vice President regularly conducts meetings of regional managers via teleconference. The regional managers regularly conduct teleconferences of their district managers and the district managers in turn conduct teleconferences of their salespeople. So, information passes up and down the line very very quickly in this organization. Changing market conditions are responded to so fast that the competition can't figure out how this company can be so responsive. As a result, after the pre-launch meetings, the company can follow the acceptance of this drug. They will know what difficulties the salespeople are having, what is working well and what support is needed.

11. The salespeople also invite physicians who are known to be innovators and early adopters to Word of Mouth Sessions. Hundreds of these small group sessions will be conducted, in the evenings, with physicians participating from their homes. Physicians will review the case for using the drug and be offered the opportunity to try the drug on a carefully selected group of patients. They will then reconvene in order to pool their experiences and see if the drug lives up to its promises. Instead of trying the drug on five patients over more than a year, the physician can try the drug on five to ten patients in a matter of weeks, return to the follow up session, and have a pool of 50 - 100 patients to evaluate. If the promise of the drug holds up, the physician will be a full adopter after the session. This is the single most powerful method I know of for accelerating the adoption of a difficult-to-evaluate product. As discussed previously, it is the best way of delivering word of mouth, and word of mouth is the most powerful force in the marketplace.

12. About six months after launch, more Word of Mouth Sessions are conducted in which physicians who have now adopted the drug share their experience with the later adopters, who are still skeptical. The adopters can learn tips and refinements in the drug's use, and the skeptics can ask questions. Since it is very hard to argue with success, the skeptics end up wanting to try the drug and come back to a follow up session.

13. While all of this is going on, some of the salespeople are not performing as well as the company would like them to. So, we set up Peer Learning Teleconferences in order to combine people who are successfully selling the product with those who are having difficulty. It is much easier for a salesperson to learn from another salesperson than from anyone else. The discussion is on a very practical, "how to," level, so that everyone goes away with new things to try.

14. Throughout the entire marketing cycle, even as the product comes to maturity, Telephone Focus Groups regularly keep track of the thinking of customers and prospects. We see where they need additional support and information, and develop new themes and uses for the product.

Summary

The old view of selling and marketing is to bombard people with information with enough frequency to get them intrigued enough with the product to try it. This is very inefficient and costly. The more modern view of selling and marketing is to provide a support system to the prospect which will guide him/her through the decision making process. This is much more targeted, efficient, faster and less costly.

But to do this requires the kind of deep understanding that comes out of group, interactive communication with customers, prospects, outside consultants and salespeople at all levels. In the past, this was so difficult that marketing was more of a seat-of-the-pants art. With the simplicity of audio teleconferencing, where everyone participates from ordinary phones in their homes or offices, much more refinement is possible.

Most importantly, and I think it's a fundamental breakthrough in marketing as important as the invention of ads, mailings and salespeople, is the fact that now, for the first time in human history, word of mouth can be produced, predicted and controlled using teleconferencing as its delivery system.

The Teleconferencing Resource Book: A Guide to Applications and Planning
Lorne A. Parker and Christine H. Olgren (eds.)
Elsevier Science Publishers B.V. (North-Holland)
© Center for Interactive Programs, University of Wisconsin-Extension, 1984

APPLICATION OF TELECONFERENCING TO
HEALTH CARE

Susan D. Roeder, M.A.
Producer and Director, Teleconference Workshops
Professional Services Department
Roche Laboratories
Nutley, New Jersey 07110

Medicine, and health care in general, has reached a juncture where the printed word -- although still important -- cannot keep pace with evolving technology and clinical findings. Much of patient care is based on peer consultation and the decisions of the medical team.

Peer consultation and medical teams can take many forms: each element must be able to act and react quickly and accurately; the decisions of each depend on shared knowledge and experience. In most instances this knowledge sharing is verbal and well suited to the application of teleconferencing. The applications of this medium are virtually unlimited if a network is well designed.

The most effective design for a network is based on a multiple usage concept; a system designed for a single purpose is neither cost nor time efficient. A health care network should provide programs for administrative personnel, medical staff development and patient management.

Networks should also be designed for a specific geographic area. The prevailing attitude is that teleconferencing should link national and international centers. While this can be easily accomplished technologically, it may not be effective or desirable. Effective networks can be designed on a much smaller, more specific scale.

Several states, for example, Georgia, Texas and Oklahoma, have networks that link 100 or more medical facilities within their boundaries. These networks have become the model for other states to emulate. In addition, efficient networks could be developed for a single county or for institutions within major metropolitan centers.

Following are six specific uses of teleconferencing in health care that could be considered when designing a network. These examples are intended as a stimulus for the broader application of the medium.

Interdisciplinary Consultation

Periodic meetings can be scheduled for consultation purposes. These meetings can and, perhaps, should be scheduled in advance and on a regular basis, e.g., weekly or monthly, in much the same fashion as grand rounds. The entire medical team -- physician, pharmacist, nurse and consulting specialists -- can be assembled in each of several affiliated centers to discuss specific patients, difficult cases or new clinical findings. The topic or topics of the conference would be determined by current needs.

This type of approach may be of particular benefit to the medical staff of a small rural hospital. Conferences linking rural centers to other community hospitals or to large urban hospitals and teaching institutions could provide specialists not otherwise available.

A specific area of increasing importance to overall patient care is nutrition. Dieticians are an integral part of the medical team with regard to hospitalized patients as well as those about to be discharged or managed on an outpatient basis.

Another evolving specialist not routinely on the staff of smaller hospitals is the clinical pharmacist. Their knowledge of pharmacokinetics, drug interactions and appropriate drug regimens is vital to the medical team decision. Any number of other specialists could be added to this interdisciplinary approach: infectious disease, surgical, obstetric.

These sessions would probably be most effective if run as informal discussions. Conferences would require a moderator or coordinator at each site and an agenda. Appropriate specialists would have to be invited and given patient records or specific objectives to be covered. Major points could be recorded on a blackboard or large lecture pad at each site by the designated moderator.

The equipment for these conferences need not be elaborate. In most cases an audio-only network would be sufficient. This approach could, of course, be expanded depending on the equipment available at each location.

Continuing Education Programs

Perhaps the broadest application of teleconferencing in health care is in the area of continuing education and medical staff development. The mandatory requirements for physicians, pharmacists and nurses to participate in such programs have created a large and growing arena for teleconferencing. Programs can either be developed specifically for teleconferencing or programs can be adapted from other media.

Programs could be developed on specific topics of current interest to a large audience and presented as accredited programming in the same fashion as on-site seminars. The instructor in this case would be conferenced to each site rather than travel to each location.

These programs are much more structured than simple lectures and must include written learning objectives and testing material. These courses, in most cases, must pass an educational review committee and are approved by the director of medical education for each hospital.

One such example would be in-service programs for nurses. Programs on any topic could be made available to groups of nurses on the floor or at their work stations. The daily routine is not disrupted, nor is patient care. Laboratory technicians and pharmacists are other specialized medical staff who benefit greatly by educational programs brought directly to them.

The key to continuing education with regard to this audience is that it be convenient and practical. Medical staffs have day-to-day patients to care for and a rapidly changing case load. The need is for information to enhance diagnosis, improve therapy and expedite the healing process. Information must be current, accurate and immediately applicable to patient care. No other medium offers the potential of teleconferencing for bringing up-to-date information directly to the work place. No other medium offers the potential for a vast educational network of interactive programming.

Teleconferencing has also long been recognized as a vehicle for medical education. Audio-only conferences are effective, as are slow-scan or full-motion video conferences. Computer medicine is a growing area and as this concept evolves, so will the increasing interface with teleconference programs and technology.

Satellite-Facility Staff Meetings

Many large health care facilities have satellite centers at the city, county or state level. Staff meetings could be conducted weekly or monthly via the teleconference system. These meetings could be

run using a simple audio-only link with
an agenda and any print material sent out
beforehand. The audio-link in this case
could be a local conference call; no
special bridging service would be required.

Satellite facilities may be small
neighborhood clinics, out-reach centers or
special screening centers for hypertension
and other disorders. Many of these facilities may have a minimal staff, perhaps only
two or three. Without coverage, staff
members are not free to travel even across
town to attend meetings, yet they need to
be kept abreast of developments and procedures within the hospital or institution at
large.

Budgetary considerations, staff and
procedural changes affect all facilities.
On-going dialogue and current information
enhance the operation of any institution
or business in a way that annual reports,
memos and lengthy procedural manuals
cannot. Dialogue reinforces understanding
of material; live meetings also provide an
environment to easily identify and rectify
any problems.

Once such a network is in place,
special meetings can be easily arranged to
deal with emergency situations or administrative needs that may arise. A benefit
of this type of network is definitely
flexibility. Meetings can be convened when
necessary by appointed staff members and
support personnel, such as accountants or
members of the board of directors, can be
invited to participate. The configuration
of the network is limited only by the
requirements of each meeting.

Guest Lectures

Lectures are part of most hospital
routines. Staff personnel, visiting faculty
and guest lectures are regularly scheduled to
speak at hospitals, institutions and universities. Although these lectures are not
specifically designed for teleconferencing
and are much less structured than continuing
education programs, they could easily be
transmitted to other local, regional or
national centers via an audio-only link.
If lectures are illustrated by slides,
arrangements can be made in advance for
duplicate sets to be sent to participating
sites.

An advantage of this type of network is
the ability of participating sites to record
the lecture for later use by interested
clinicians unable to attend. Audiotapes and
slides (if used) can be packaged and kept in
hospital or university libraries. Extensive
libraries can be built in this way for very
little expense.

This type of resource is as valuable as
medical journals and textbooks. Resident
teaching programs and journal clubs could
access this information for use during regularly scheduled meetings. All medical staff
members -- physicians, pharmacists, nurses
and technicians -- could benefit from such a
network and library.

Psychiatric Consultation

Psychiatric illnesses are among the more
difficult to diagnose and treat. In many
cases, an apparent psychiatric disorder may

be a sign or symptom of a neurological or
other physical disease. In addition,
psychiatric patients often have multiple
disorders/diseases occurring simultaneously
and require complicated drug and therapeutic
regimens. Consultation among numerous
specialists -- psychiatrist, neurologist,
pharmacist, dietician, endocrinologist -- is
essential and referral may be required.

Teleconferencing has obvious application
as a medium for consultation. In this case,
however, the design of the network may be
more difficult; conferences would be on an
"as needed" basis rather than on a regular
schedule. The consulting team may vary
depending on the patient and treatment. In
some cases, written patient records may not
be sufficient; videotaped interviews may have
to be sent to each consultant. If referral
to another institution is necessary, that
medical team should be part of the conference
as well.

An ad hoc consulting network should be
scheduled through a bridging service that
can provide conferences without geographic
limitation. The ability to record these
conferences centrally from the bridge is
also a consideration; follow-up and
complete patient records are essential in
such cases.

After a core consulting team has been
identified, their role could be expanded by
creating teaching conferences for the staff
and residents of the participating hospitals
and institutions. Each conference could

deal with a specific case and patient
management decisions. Once again, the
content is limited only by the needs of
the conference participants. The network
itself, theoretically, has no limitations.

Nursing Home Pharmacists Conferences

Another health care group that would
benefit by peer conferencing are nursing
home pharmacists. The design of the network would have to accomodate multiple
sites, but only a small number of parti-
pants. Most nursing homes are staffed by
only one, possibly two, pharmacists so
conferences format would have to be well
planned to incorporate the advantages of
the medium.

The interactive features of a conference would best be handled by a skilled
moderator. An agenda with specific
discussion points and segments should be
provided to each pharmacist prior to the
conference. Depending on the topic of
the conference, each pharmacist could be
assigned a particular item for presentation. An alternative approach would be
to coordinate topics on a rotating basis,
each pharmacist being responsible for one
conference in the series.

Conclusions

Teleconferencing is an ideal medium
for health care. For all the examples
described here, the networks are flexible
and the equipment needs are minimal. The
critical factor is the coordination and
organization of the networks. However,
once a viable network is in place, users will
recognize additional applications for it and
a natural evolution and growth will occur.
To be effective, the network and the concept
of teleconferencing cannot be restricted to a
few at the administrative level.

Any network should be easily accessed by any number of people at multiple levels within the organizational structure. Health care is no different. The primary intent of teleconferencing in medicine is peer interaction and support with the bottom line of better patient care.

The Teleconferencing Resource Book: A Guide to Applications and Planning
Lorne A. Parker and Christine H. Olgren (eds.)
Elsevier Science Publishers B.V. (North-Holland)

DEVELOPING A MULTIPLE PRODUCTION/MULTIPLE RECEIVING STATION
INTERACTIVE AUDIO TELECONFERENCING NETWORK

Donald L. Cordes, Ph.D. and Steven R. Smith
Continuing Education Center
Veterans Administration Medical Center
Washington, DC

BACKGROUND

The Veterans Administration (VA) Health Care System is one of the largest in the world. Within its Department of Medicine and Surgery (DM&S), there are 172 VA Medical Centers (VAMCs) and various Out-Patient Clinics scattered across the United States, Guam, Philippines, and Puerto Rico.[1] The DM&S has personnel totaling more than 200,000.[2] Recently the VA was designated as the first-line backup to the Department of Defense in the event of a national emergency.

A system as large and complex as the VA has gigantic training needs. Many VAMCs are great distances from population centers and affiliated academic institutions. Their access to continuing education (CE) offerings is often limited. The Office of Academic Affairs (OAA) has the responsibility of providing CE for these people. This job has become increasingly difficult in the face of tight budgets and inflation. The cost of providing CE has dramatically risen without concomitant budget increases. Presently the purchasing power is half that of 1977 levels. The challenge was to develop a format which would meet many of the CE needs of DM&S personnel at a lower cost than is currently being expended.

At first, the task of selecting an alternative educational methodology to explore for the VA system seemed paramount. After examining the need, a review of the literature indicated numerous options. Various self-instructional methods--closed circuit television, satellite television, and teleconferencing were among the potential alternatives examined.

A decision to field test an interactive audio teleconference network within the VA was made on the basis of the teleconferencing literature which indicated that little potential was lost by utilizing audio as opposed to video.[3] Visual support materials need to be avaliable for learners, but a full video link was not a necessary expenditure. This decision to field test interactive audio teleconferencing coincided with the desire to build a network with a relatively simple and inexpensive technical requirement.

Field Test

The primary purpose of the field test was to examine the appropriateness of educational teleconferencing in the VA setting. Teleconferencing as an educational delivery methodology had been shown to be effective in many different settings. Therefore there was an effort to provide information on as many aspects of teleconferencing as possible in order to assist in making a decision to implement this format throughout the VA system.

The field test was further designed to provide the OAA with practical experience in the event of a final decision to implement system-wide.

The overall field test spanned from January 1981 through January 1983. Figure 1 shows the various activities and their relative time frames.

The field test consisted of 32 educational teleconferences. The production of these teleconferences was shared among eight production sites. They included the seven Regional Medical Education Centers (RMECs) geographically distributed throughout the U.S. and the Continuing Education Center (CEC) in Washington, DC.

The 13 receiving sites were selected according to criteria including size, function, and distance from available CE. These VAMCs were primarily located in the Great Plains and Rocky Mountain States but ranged from Altoona, Pennsylvania to Ft. Harrison, Montana. The University of South Dakota was selected as the bridging facility because their ongoing teleconferencing--both process and quality--matched VA requirements (see Figure 2).

For the 32 teleconferences four health disciplines were selected as primary target audiences--physicians, nurses, dietitians and social workers. Eight teleconferences were

Figure 1. TELECONFERENCE FIELD TEST OVERALL TIMETABLE

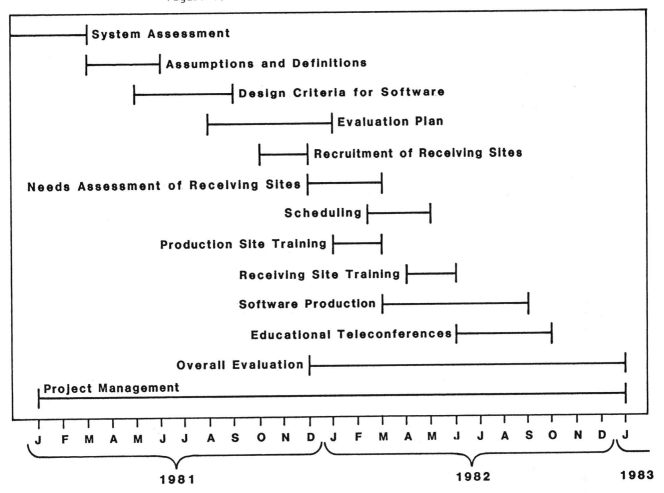

Figure 2. FIELD TEST LOCATIONS

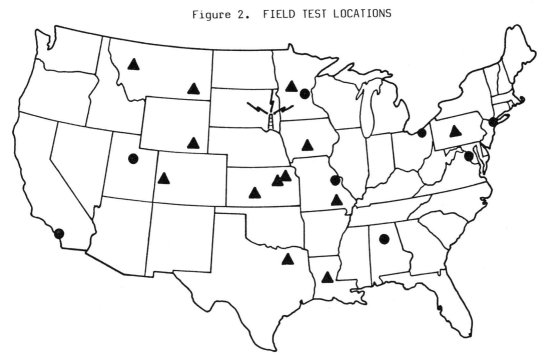

(Figure 2. Field Test Locations) ▲ <u>Receiving</u>
<u>Sites</u> - Alexandria, LA; Altoona, PA; Bonham,
TX; Cheyenne, WY; Fort Harrison, MT; Grand
Junction, CO; Knoxville, IA; Leavenworth,
KS; Miles City, MT; Poplar Bluff, MO; St.
Cloud, MN; Topeka, KS; Wichita, KS.
● <u>Production Sites</u> - Birmingham, AL;
Cleveland, OH; Long Beach, CA; Minneapolis,
MN; Northport, NY; Salt Lake City, UT;
St. Louis, MO; Washington, DC.
<u>Bridge Site</u> - Sioux Falls, SD

presented in each discipline--two different
disciplines per week. This phase of the
field test ran from June through September
1982. Each program was presented in a
standard format--one-hour long, interactive
audio teleconferences supplemented by print
and non-print media. This standardized
approach made it possible for the receiving
sites to receive similar programs from any
one of the eight production sites. Only one
teleconference was transmitted from the
bridge at a time, but all 13 sites received
it simultaneously.

Equipment

The equipment used in the field test
involved:

1. standard telephones with modular
    jacks
2. speakers with microphones
3. one bridge
4. long distance lines

Any standard telephone set which had
modular jack connections (RJ 11) would serve
the purpose. The phone was needed to dial
the bridge. Communications could then be
either via the handset or the Darome Convener
(Model 1610-X). In simplistic terms the
convener consisted of a speaker box and four
microphones. The bridge used was a Darome
Co-Convener Meet-Me-Bridge Teleconference
System (Model 2020). Direct distance dial
commercial lines were used for the field
test. Federal Telecommunications System
(FTS) lines and bridges were tested and used
in the early stages of the project. The FTS
lines were not of sufficient quality to do
educational teleconferencing. All receiving
and producing sites used the same equipment
itemized above. The bridge was located in
Sioux Falls, South Dakota.

Implementation Strategies

A field test of this complexity brings
with it a number of potential pitfalls.
These pitfalls can be clumped into two
major categories--technical and human. An
underlying assumption of the field test was
that both categories are of equal importance
but that the human category would be more
difficult to develop.

The human category is so potentially
problematic because the initiation of a new
learning methodology requires both the
producers and receivers of the learning
experience to undergo change. Change brings
with it fear of the unknown and a host of
other concerns. A Theory of Institutional
Change articulated by Hall provided guidance
in assessing and dealing with the various
levels of concern. [4]

Three primary strategies were used to
address the concerns people may have had.
They were:

1. Assessment
2. Involvement
3. Training

Assessment

Prior to beginning the field test, all
VAMCs were surveyed via questionnaire
regarding teleconferencing and its use for
CE. This system-wide survey focused on
both current teleconferencing uses and the
willingness to participate in pilot testing
of educational teleconferencing.

The production sites were assessed
immediately after the decision to embark on
some type of teleconferencing pilot test.
That assessment sought to identify the
existing teleconferencing expertise at each
of the production sites and to identify
peripheral resources and capabilities
necessary to support teleconferencing. The
assessment was accomplished through the use
of a questionnaire followed by a site survey
by the project director. The receiving
stations were assessed as to their educational needs to be addressed in the pilot
test. This assessment involved the Service
Chiefs in Internal Medicine, Nursing, Social
Work and Dietetics.

Involvement

A decision was made to involve both
the production and receiving sites in as
much of the decision making as possible
regarding the design and implementation of
the field test. It was felt that involvement
in the decision-making process would enhance
the likelihood that those involved would
develop ownership for the project.

Production Sites

The involvement of the production sites was manifest in several ways. Initially a meeting was held with representation from each RMEC and the CEC to specify the assumptions and relevant operational definitions of a VA teleconferencing system. Follow up input was obtained via written reaction to draft documents developed from meeting notes.

A second major area of involvement dealt with the educational software to be used for the field test. This software constituted agenda, objectives, outline, bibliography, visual materials, etc. which were sent to each site in advance. Since ease of operation required that all educational teleconferences follow a standard format, it was necessary to achieve consensus regarding both the format and the design criteria for developing educational teleconferencing software. Following a meeting with representative instructional designers from each production site, an interactive process via the mails refined the initial drafts and allowed for input from everyone at the production sites.

The third major area of involvement was the evaluation plan for the field test. Four production site representatives participated in the evaluation meeting. The resulting evaluation plan was sent to each production site for additional input and/or critique.

A final major area of involvement was the selection of content areas to be developed into educational teleconferences. After completing the assessment of educational needs at the Medical Centers, a list of those topics which fit the teleconferencing format was sent to each production site. The eight production sites' representatives met via teleconference to decide on content assignments. A round-robin selection process enabled the production sites to select topics for development which matched their strong resource areas.

Receiving Sites

The receiving sites were involved in the decision-making process in two ways. Initially, the Medical Centers were given the opportunity to participate on a voluntary basis in the field test. Secondly they were asked for input as to the best times of day and days of week to conduct the teleconferences. A composite of those times and bridge availability ultimately determined the times of the teleconferences.

Training

Learning through the medium of audio teleconferencing requires different behaviors on the part of both the producers and recipients of the learning experiences. To assure that both groups were properly prepared for the field test, special training exercises were developed and conducted.

Production Site Training

The preliminary production site assessment indicated that there were virtually no personnel experienced in teleconferencing in the RMEC system. This necessitated the development of a comprehensive course on how to develop and implement an educational teleconference. It was decided to use the teleconferencing medium to teach teleconferencing. This had a number of advantages:

1. it enabled the production site personnel to experience teleconferencing from the learner perspective,

2. it enabled the learning experience to extend over a several week time period allowing for staff questioning and interaction, and

3. it enabled all relevant production site staff to participate instead of one or two per site who could have attended a workshop.

The training consisted of a seven part training series of interactive audio teleconferences. The conferences consisted of a short lecture supplemented by visuals and handout materials followed by questions and discussions. The teleconferences in the training series followed the software design criteria specified by the teleconferencing task force and thereby served as models of what audio teleconferences are and how they are conducted.

Additionally the project management office at the CEC was always available to answer any questions on the part of production site staff. This openline function was a much used service and provided untold additional training.

Receiving Station Training

Typically little attention is paid to the process of orienting potential participants to the role of teleconferencing learner. Also, assumptions are frequently made regarding the teleconferencing learning environment and the equipment needed for teleconferencing. This project attempted to overcome these potential pitfalls by

providing specific training sessions for
on site coordinators and the learners
themselves. To accomplish this task two
video tapes were developed: one for on site
coordinators and one for learners. Additionally, pamphlets and manuals were
developed and made available as hard copy
reinforcement of the videotape information.
The videotapes and printed information were
presented at each Medical Center by a member
of the project management staff. This on
site visit also allowed for additional input
into such things as room preparation,
equipment arrangement, etc. Prior to
beginning the educational teleconferences,
two dry-run administrative teleconferences
were held to provide the on site coordinator
and the potential learners a chance to set up
and operate the equipment and become familiar
with the format and process.

The primary intended outcome of this
recieving station training was to familiarize
the learner with the medium. In other words,
the learners should be able to concentrate on
the material to be learned not on the fact
that the instructor is in a remote location.

Evaluation Strategy

The field test evaluation was conducted
in accordance with a comprehensive evaluation
plan. The plan specified data to be collected, sources of data, potential collection
methodology and data analysis. It also
called for formative and summative evaluative aspects.

In addition to focusing on the main
purpose of the field test, the evaluation
also gathered information regarding the
resources consumed, unintended outcomes
(both positive and negative) and to determine
teleconferencing's non-educational uses in
the agency.

Data

Several types of data were collected.
They were:

o perceived benefit
o cognitive change
o perceived quality
o cost (total resources)
o attitudes concerning teleconferencing
o technical performance

Data Sources

Data were gathered from a number of
relevant sources. They were:

o participants/learners

o production site personnel
o technical personnel
o receiving station coordinators
o receiving station management
o project management staff

Collection Methodology

Numerous methods of gathering data were
used. All participants at the 32 educational
teleconferences were asked to rate the conferences by responding to a standard set of
questions. These questions dealt with such
things as relevance of the program to their
jobs, technical quality, educational quality,
etc. In 10% of the cases pre-post multiple
choice cognitive exams were administered
in conjunction with the teleconferences.
Structured interviews were conducted with a
random sample of participants, service
chiefs, on site coordinators and hospital
directors at the conclusion of the field
test. Additionally, both perceptions from
production site personnel, and actual
documented cost and time data were collected.
A comprehensive log was kept by bridge
personnel of the technical aspect of the field
test. Finally, each educational teleconference was monitored by the project management
staff with a comprehensive log kept of the
field test.

Data Analysis

Each production site did the tabulation
and analysis for the teleconferences they
implemented. Means were calculated for the
standard data set and written comments were
categorized for each teleconference. The
pre-post multiple choice exams were graded
and gains/losses were calculated. Frequency
distributions were done on all written
comments after they were sorted into categories. A frequency distribution was done
on the standard data set with percentages
calculated for the means of each item.
Finally the interview data was categorized
and integrated with the other data to provide
meaning for the numerical data.

Summary

The implementation of a multiproduction/multi-receiving station teleconferencing network is an extremely complex
process. As contemplated, it was absolutely
essential to have the genuine cooperation of
both production and receiving station
personnel. The ideal situation occurs when a
cooperative spirit develops into ownership for
the project.

The process for developing ownership in
a new entity, like teleconferencing, can best

be described with the word involvement. Those groups and individuals who will be directly affected must be involved as early as possible in goal setting and decision making which will guide and direct the project. As a result of the cooperation displayed, the VA was able to successfully field test a complex audio teleconferencing network and open the door to additional means for meeting the training needs of DM&S personnel.

References

1. Consolidated Address and Territorial Bulletin 1-D, Veterans Administration, Washington, D.C., March 1982.

2. VA Employment by Appropriation in Program, RCS P-38, Nov. 30, 1982.

3. Hammond, Sandy and Williams, Ederyn, "A Brief Review of the Work of the Communications Studies Group, 1969-1977," in Meet-Me Seminar On Teleconferencing (University of Wisconsin-Extension, February, 1981), p. 24).

4. Hall, Gene E., "Facilitating Institutional Change Using the Individual as the Frame of Reference", University of Texas, Austin - Research and Development Center for Teacher Education.

The Teleconferencing Resource Book: A Guide to Applications and Planning
Lorne A. Parker and Christine H. Olgren (eds.)
Elsevier Science Publishers B.V. (North-Holland)
© Center for Interactive Programs, University of Wisconsin-Extension, 1984

52

TELEDUCATION IN TEXAS: CONTINUING EDUCATION FOR
MENTAL HEALTH PROFESSIONALS

Joyce L. Sanders, R.N., M.S.
Assistant Director/Teleducation Coordinator
Office of Continuing Education

Linda J. Webb, Dr.P.H., Director
Office of Continuing Education

Peter Baer
Chief, Media Services

Texas Research Institute of Mental Sciences
Texas Department of Mental Health and Mental Retardation
Houston, Texas

Introduction

COMNET

The Texas Department of Mental Health and Mental Retardation's network, COMNET, was established in December 1981 to meet the varied communication needs of the Department. It is a four wire, audio only, dedicated system, leased from Southwestern Bell, that link the Department's twenty-eight facilities (state hospitals, state schools, and Human Development Centers) (figure 1). Each facility has a designated conference room and identical equipment (leased from Bell) consisting of a Darome 611 convener and a table microphone. Some of the larger facilities have purchased additional conveners and microphones and have had additional jacks installed to allow for more conference rooms. A special feature of the network is a two-wire bridge which allows one additional location to participate. This feature is utilized to bring in outside faculty or other participant groups. The network generally operates between the hours of 8:00 a.m. and 5:00 p.m., however it is possible to do programming for 24 hours of the day. The capacity of the system is for one program at a time. COMNET is presently being used for professional continuing education, administrative meetings and informal contacts.

Many people are involved in the operation of COMNET. At the Department's Central Office in Austin, Texas, there is a COMNET administrator who is assigned responsibility for the overall management, coordination, documentation and evaluation of the system and a COMNET

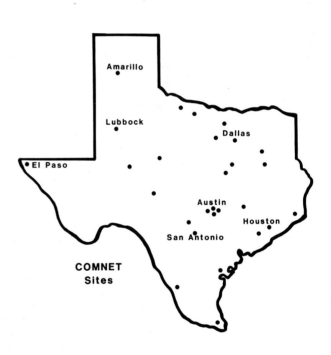

Figure 1

operator who is assigned the technical responsibility for establishing connection of facilities on the network, the activation of equipment and coordination of network scheduling of activities and programs. In addition to the Central Office there is, at each of the twenty-eight facilities, a COMNET administrator. This individual's responsibility is for local management, coordination, documentation and evaluation of the COMNET activities and guiding other staff on network use. Under the administrator at each facility is a COMNET operator, who is responsible for receiving calls and establishing connection on to the

network, activation of conference room equipment, scheduling with the Central Office, assigning log numbers and ensuring documentation
of activities and programs. There is technical support staff responsible for audiovisual
equipment, programs, or other unique requirements related to supplementing COMNET activities and programs. The Office of Continuing
Education works very closely with facility
COMNET administrators and operators to ensure
smooth coordination of programming.

Office of Continuing Education

The Office of Continuing Education of the
Texas Department of Mental Health and Mental
Retardation is located at the Texas Research
Institute of Mental Sciences (TRIMS) in Houston. The mission of the Office is to provide
continuing education for the clinical professionals in facilities of the Department.
Toward that goal, the Office: sponsors continuing education programs using a variety of
methods, i.e. workshops/seminars/ institutes
(at TRIMS, regionally and in-house); accredits
programs for continuing medical education
credit and continuing education units; and
works collaboratively with universities and
other organizations to cosponsor programs.
While all these methods have been and still
are successful, factors such as the size, dispersion and diversity of the target population; increased cost of travel; and the time
constraints of both the participants and the
faculty restrict the Office's ability to fulfill its mission. Subsequently, when COMNET
was established to meet the varied communication needs of the Department, we in the Office
of Continuing Education immediately recognized
the potential of teleconferencing to help us
better reach or serve our statewide clientele.
What follows is a brief description of the evolution of the Teleducation Program developed
by the Office of Continuing Education.

TRIMS Teleducation Program

Several months of researching (literature
and site visits) the state-of-the-art of teleconferencing and the educational application
of the technology, revealed several factors
essential to the success of teleconferencing.
The factors pertinent to the development and
design of the TRIMS' Teleducation Program are:
1) program design; 2) program faculty; 3) marketing and 4) evaluation.

Program Design

Pulling together what we'd learned from
our research and our experience with a teleconference pilot project done a year prior to
the establishment of COMNET, it became clear
that there were significant differences in the
design and implementation of traditional face
to face educational programs and teleconference programs. We took the established concepts and guidelines and modified them to meet
the specific needs of our teleconferencing
efforts. The following represents the program
design that evolved out of those modifications
and which currently exist:

- The length of individual program segments vary from one hour to two hours.
  All programs over two hours must be
  highly interactive and have scheduled
  breaks.

- All programs carry the appropriate continuing education credit - CME (continuing medical education), CEU (continuing education unit) and/or APA (American Psychological Association).

- Program formats vary, but are basically
  fifty percent information-out (lecture,
  panel, case studies, slide/videotape)
  and fifty percent information-in (questions and answers, discussion, brainstorming).

- The program faculty must possess subject matter competency as well as the
  qualities of a good teacher.

- Faculty are strongly encouraged to use
  audiovisual support material - slides,
  videotapes and handouts.

- Question and answer sessions are handled at random from the sites rather
  than by roll call thus conserving time
  for additional questions and discussion.

- Programs are generally presented on
  established days and times (1st Wed-
  nesday and 3rd Thursday from 3:00-4:30
  or 5 p.m.) to facilitate participant
  orientation and acceptability.

- A workbook is developed for each program and includes a photo of the faculty, program overview, objectives,
  program outline and recommended readings (figure 2).

• Program support material (workbooks, evaluation, attendance forms and slide/videotapes) are mailed to the facilities at least two weeks prior to the program.

• Each program is evaluated with the computerized evaluation used for traditional programs.

Figure 2

Program Faculty

In teleconferencing, as well as with other educational methods, the faculty is an essential component in providing high quality continuing education. The faculty is important from the traditional standpoint of being an expert in their content area and equally important, for their qualities as an educator. Program faculty for the teleducation programs are primarily selected from the varied mental health and mental retardation professionals at the Texas Research Institute of Mental Sciences (TRIMS). While the basic subject matter and topic areas are determined by needs asses-

sment, departmental mandates, participant input and the like, the specific content of each program is left to the faculty. However, the preparation of the faculty for teleteaching is the responsibility of the Office of Continuing Education.

The TRIMS faculty in general has considerable experience in the traditional lecture mode and workshop format. However at the inception of our network none of the faculty had any expertise or prior experience in teleteaching. Thus we were confronted with having to educate the faculty about how to both design a teleducation program and how to effectively teleteach. To help accomplish this task, an initial workshop was conducted by Lynn Graham from the Teleconference Network of Texas. Lynn met with the faculty (live in person) and provided a two hour training session prior to our initial nine months of programming. After nine months of programming, another training session was held for faculty who would be involved in programs during the upcoming year. Using the expertise available at the University of Wisconsin, the training was held over the teleconference network.

Both of these training sessions were very valuable in providing faculty with needed training. The initial on site visit helped acquaint the faculty, who were totally unfamiliar with the methods and equipment, with teleconferencing. Having someone who was familiar and comfortable with the equipment was very important. Lynn was able to convey, during this two hour period of time, a certain excitement about the use of teleconferencing that was as important in working with this faculty as training them in the particulars of teleteaching.

With the faculty now familiar with teleconferencing the second training session was provided over the network. The two hour session modeled a good teleducation program as well as covered key elements of teleteaching and designing a teleducation program for teleconferencing.

One of the difficulties in working with the faculty in teleconferencing is that by and large these people are very comfortable with the lecture format. Often times their face to face workshops are not as interactive as they should be for good educational programs. Fortunately, however, because we are using a new medium, we can readily emphasize the importance of an interactive format and the need to use case studies, slides, videotapes, and provide sufficient time for discussion. In fact it is probably easier for us to get them to be more interactive with the new medium, than it would be to try to get them to change their old ways with the traditional workshops.

In addition to attending the training sessions the majority of the faculty have attended a teleducation program, conducted by another faculty prior to their scheduled date on the network. By watching other faculty use the medium and observing the difficulties encountered in not having a live audience, they are better prepared for their own teleconference. Unfortunately, in some instances this is the only training that the faculty may have received. In addition, an educational coordinator from the Office of Continuing Education meets with the faculty in designing the program and to help prepare them to teleteach. Articles and the Wisconsin videotapes on teleteaching are also available to the faculty.

In general our faculty have been very responsive to teleconferencing especially after they have offered one program on the network. It takes a certain amount of encouragement to get them to agree to teach a program; however, once they have offered the course they are very enthusiastic about the large number of people that they are able to reach and some of their reservations about teleteaching are diminished. As with any change and the use of a new method there is an inevitable amount of resistance. However given the mission of the Office of Continuing Education and the size of the state of Texas, learning to use teleconferencing and to teleteach is more efficient and often preferable to traveling to Lubbock, Abilene, and other far reaches of the state.

Just as the faculty have become more familiar with the network, we as program coordinators have also become better educated about faculty needs and components of a good program. The plans for the future are to schedule, at least once a year and preferably twice a year, a teleteaching conference for upcoming faculty who have not previously used the network and a refresher for our experienced faculty.

Marketing

With the target audience already well defined for continuing education programs in the Texas Department of Mental Health and Mental Retardation, the Office of Continuing Education's marketing objective for teleducation is to reach individual mental health professionals as directly and as inexpensively as possible. The primary tools for promotion and publicity are the brochures and individual program flyers which are sent to COMNET facility administrators who distribute them to professionals within their facility.

The Continuing Education calendar is distributed annually and lists all continuing education programs for the coming fiscal year. The more detailed publicity for teleducation is the quarterly brochure. All teleducation programs are listed with a full description, i.e. time, date, faculty, origination site, program summary, and registration information. As a follow-up to the calendar and brochure, one page program announcements are mailed for each individual program approximately one month prior to the teleconference. Again, like the brochure, full description and registration information are listed. The registration information indicates the type of registration needed for a particular teleducation programs. In most cases the COMNET facility administrator will register the facility at least two weeks in advance. Participants will register at the time of the program. However, because of the format for some programs, participants must register in advance through their COMNET administrators. Individual pre-registration is needed when a limited number of participants can attend, i.e. small group sessions which will call for a lot of individual interaction among participants and faculty. In these cases a large audience would make the program ineffective.

Other avenues used to promote teleducation programming include monthly calendars and newsletters which are sent out by the Central Office to all network facilities. Also announcements of upcoming programs are made at the beginning and end of each program and during weekly system checks.

Evaluation

The standard evaluation form, in figures 3a and 3b, used to evaluate all continuing education programs sponsored by the Office of Continuing Education, is also used to evaluate the teleducation programs. The evaluation is designed to discern some level of participant satisfaction in the areas of facilities and accommodations, learning aids, program structure and format, future course projections and quality of individual faculty presentations. Some additional information for office management is collected, including demographic data (education, primary job function), how the participant learned of the program and why the participant registered for the program. The evaluation form is computerized to provide graphic data on 1) the mean rating on each of the 16 program items; 2) mean rating on each of the five faculty evaluation items with a separate report for each faculty member; 3) summary scales for each of the three primary evaluation areas including personal/social arrangements, learning aids and program structure and format. Also provided is a comparative ranking of each program and program faculty with other programs sponsored by the Office of Continuing Education. (See figures 4a and 4b for a sample of an evaluation printout)

The percentile ranking is a comparison of the mean rating of a particular program with the mean ratings on the same questions for all other programs sponsored to date. The farther the mean score is from the overall average, or the 50th percentile, the higher or lower the ranking. For example if the rating is at the 25th percentile, 75 percent of the programs received a higher average rating, 24 percent received a lower ranking. If the ranking is in the 80th percentile, the program was rated higher that 79 percent of the other programs and 20 percent received a higher ranking.

The results for the first two or three teleconferencing programs indicated that the programs were ranked in the lower 50 percentile of all continuing education offerings particularly discouraging in light of the considerable amount of time and energy put into the programs. After carefully reviewing the evaluation results some changes were made in the format to allow for more interaction. After these changes the ratings, in fact, did improve. However, overall rankings for the teleducation programs did not begin to compare with the higher ratings of our face to face workshops.

Table 1 shows the title and dates of the teleducation programs, the program mean and ranking, (question 25 from figure 3a), and the mean ranking for the faculty (question 46 from figure 3b) As you can see, the program ratings have improved over time. The faculty ratings have been consistently higher than the program ratings.

In trying to interpret and use these results, we are attempting to ascertain whether or not the overall satisfaction with the program is compounded by the considerable technical difficulties we have encountered with the network. For example, if the sound is breaking up or it is difficult to hear the speaker or some sites are lost during transmission, the overall satisfaction with the program will probably be influenced. Another

factor which might influence the ratings is the use of a new medium which is certainly different than face to face workshops and it may not be "fair" to compare the rankings of the teleducation programs with the face to face workshops. In the future we have decided to separate out the teleducation programs from other workshops. Then we will have some indication of the rankings of all the teleconferencing programs relative to each other.

In conclusion our evaluation system has been very useful to us in determing whether or not the programs have improved over time, and in assessing the consumer's satisfaction with teleducation. Also these results are provided to the faculty who use them in planning future teleducation programs. In particular, feedback is provided on the quality of the handouts and the audiovisual materials which is used to encourage faculty to increase their use.

Summary

In conclusion it is important to note that the technical difficulties associated with using teleconferencing have created certain problems in attempting to provide teleducation for professionals in the Texas Department of Mental Health and Mental Retardation. We have experienced considerable problems with our network including: individual sites, as well as the system, going "down" in the middle of a conference; the overall quality of the audio reception (static and volume too low); and the two wire bridge, which has been totally inadequate and unreliable. These problems have no doubt influenced our participants' responses and reactions to teleducation programs. For example if it takes 15 to 20 minutes to get the network working and participants have to wait for a program, they are not as receptive and as likely to attend future programs. Likewise faculty who experience considerable difficulties with the network are reluctant to teleteach in the future. Program design, well trained faculty and marketing are certainly all important to the success of our teleducation program. However, without equally dependable technical equipment, all of our program efforts are for naught. We have been working hard to correct the technical problems, however, we have found it far easier to alter design aspects and educate the faculty than it has been to clear up the technical problems. Once we can get the technology and the educational components synchronized we should have a far more successful teleducation program in Texas.

OFFICE OF CONTINUING EDUCATION
Texas Department of Mental Health and Mental Retardation
Texas Research Institute of Mental Sciences

PROGRAM TITLE _____ **CODE**_____ [1-5]

Participant Information

[6-7] **Education (Check One)**
1. MD
2. PhD Psychologist
3. Ma. MS Psychologist
4. EdD. MS. MED Education
5. RN
6. LVN
7. MSW. MSSW Social Work
8. BA. BS Case/Social Work
9. Rehab (OT. PT. etc.)
10. Other _____

[8] **Primary Job Function (Check One)**
1. Administrator/Supervisor
2. Clinician/Therapist
3. Researcher
4. Clerical/Support Worker
5. Training/Education
6. Trainee
7. Other _____

[9] **Facility (Check One)**
1. Comm. MHMR Center
2. Human Dev. Center
3. State Hospital
4. State School
5. Central Office
6. TRIMS
7. Other _____

Instructions

Please circle the number that best represents your reaction to each of the following statements by using the following code: Circle

1.....if you **STRONGLY DISAGREE** (SD)
2.....if you **DISAGREE** (D)
3.....if you **SLIGHTLY AGREE** (SA)

4.....if you **AGREE** (A)
5.....if you **STRONGLY AGREE** (STA)
NA....if statement is **NOT APPLICABLE**

PERSONAL/SOCIAL ARRANGEMENTS

	SD	D	SA	A	STA	NA	[CARD 1]
Hotel facilities excellent	1	2	3	4	5	NA	[10]
Meeting room and other facilities excellent	1	2	3	4	5	NA	[11]
Breaks and/or food arrangements adequate	1	2	3	4	5	NA	[12]

LEARNING AIDS

	SD	D	SA	A	STA	NA	
Physical facilities excellent	1	2	3	4	5	NA	[13]
Quality of printed materials excellent	1	2	3	4	5	NA	[14]
Quality of audio-visual materials excellent	1	2	3	4	5	NA	[15]

PROGRAM STRUCTURE AND FORMAT

	SD	D	SA	A	STA	NA	
Program brochure described program adequately	1	2	3	4	5	NA	[16]
Program accomplished stated objectives	1	2	3	4	5	NA	[17]
Scope of objectives appropriate to program structure	1	2	3	4	5	NA	[18]
Group session appropriate size	1	2	3	4	5	NA	[19]
Question/Discussion periods adequate	1	2	3	4	5	NA	[20]
Size of general session adequate	1	2	3	4	5	NA	[21]
Program content current	1	2	3	4	5	NA	[22]
Program relevant to my needs	1	2	3	4	5	NA	[23]
Expect work to benefit from participation in program	1	2	3	4	5	NA	[24]
General level of satisfaction excellent	1	2	3	4	5	NA	[25]

FUTURE COURSE PROJECTIONS

	Yes	No	
Recommend repeat of course to others in field	1	3	[26]
Consider attending another more basic program	1	3	[27]
Consider attending another more advanced program	1	3	[28]

Figure 3a

REASON COURSE WAS TAKEN

Please check all of the following reasons which best describe why you attended this program.

[29]__Obtaining course credit   __Course Subject matter [30]

[31]__Program faculty   __Location of the program [32]

[33]__Requirement of supervisor/facility   __Other_____[34]

LEARNED OF PROGRAM FROM

Please check all the following which best describe how you learned of the program.

[35]__A brochure mailed to me   __A colleague [36]

[37]__My supervisor   __A professional journal ad [38]

[39]__A newspaper release   __An agency newsletter [40]

[41]__Other_____

FACULTY EVALUATIONS

Please rate each of the following speaker(s) by placing a circle around the best response according to the following code: Circle

1.....if you **STRONGLY DISAGREE** (SD)   4.....if you **AGREE** (A)

2.....if you **DISAGREE** (D)   5.....if you **STRONGLY AGREE** (STA)

3.....if you **SLIGHTLY AGREE** (SA)   NA....if statement is **NOT APPLICABLE**

	SD	D	SA	A	STA	NA	[CARD 1]
Excellent command of subject matter	1	2	3	4	5	NA	[42]
Adequate use of materials	1	2	3	4	5	NA	[43]
Level of organization excellent	1	2	3	4	5	NA	[44]
Response to questions excellent	1	2	3	4	5	NA	[45]
Overall presentation excellent	1	2	3	4	5	NA	[46]

	SD	D	SA	A	STA	NA	
Excellent command of subject matter	1	2	3	4	5	NA	[47]
Adequate use of materials	1	2	3	4	5	NA	[48]
Level of organization excellent	1	2	3	4	5	NA	[49]
Response to questions excellent	1	2	3	4	5	NA	[50]
Overall presentation excellent	1	2	3	4	5	NA	[51]

GENERAL COMMENTS

Figure 3b

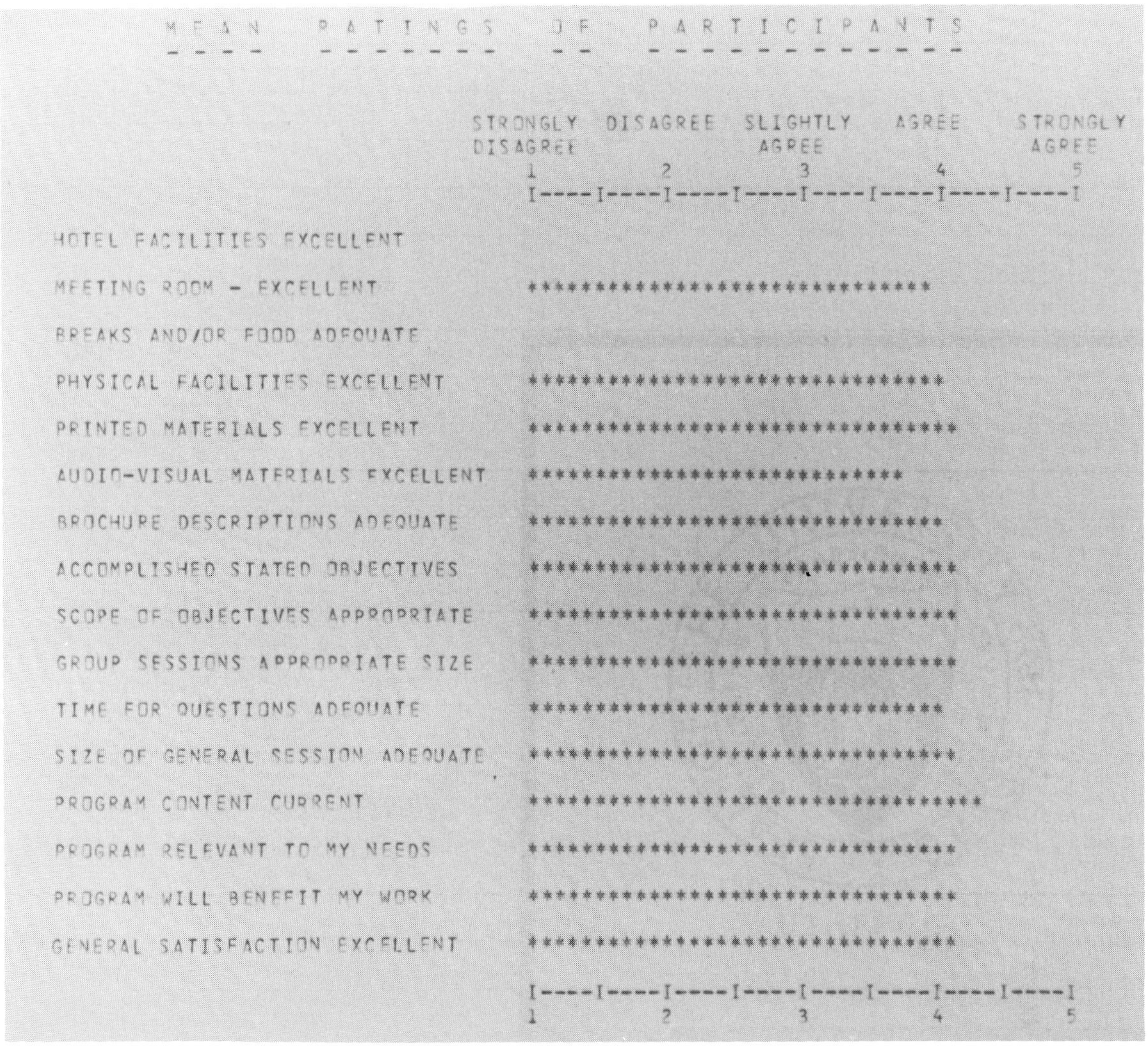

```
M E A N   R A T I N G S   O F   P A R T I C I P A N T S
----   -------   --   ------------

                              STRONGLY  DISAGREE  SLIGHTLY   AGREE    STRONGLY
                              DISAGREE            AGREE               AGREE
                              1         2         3          4        5
                              I----I----I----I----I----I----I----I----I

HOTEL FACILITIES EXCELLENT

MEETING ROOM - EXCELLENT      *******************************

BREAKS AND/OR FOOD ADEQUATE

PHYSICAL FACILITIES EXCELLENT *********************************

PRINTED MATERIALS EXCELLENT   **********************************

AUDIO-VISUAL MATERIALS EXCELLENT ****************************

BROCHURE DESCRIPTIONS ADEQUATE ***********************************

ACCOMPLISHED STATED OBJECTIVES ************************************

SCOPE OF OBJECTIVES APPROPRIATE ************************************

GROUP SESSIONS APPROPRIATE SIZE ***********************************

TIME FOR QUESTIONS ADEQUATE   ***********************************

SIZE OF GENERAL SESSION ADEQUATE ************************************

PROGRAM CONTENT CURRENT       **************************************

PROGRAM RELEVANT TO MY NEEDS  **************************************

PROGRAM WILL BENEFIT MY WORK  *************************************

GENERAL SATISFACTION EXCELLENT *********************************

                              I----I----I----I----I----I----I----I----I
                              1         2         3         4         5
```

Figure 4a

```
M E A N   R A T I N G S   O F   F A C U L T Y
----   -------   --   -------

FACULTY : CARLO C. DI CLEMENTE, PH.D.

                              STRONGLY  DISAGREE  SLIGHTLY   AGREE    STRONGLY
                              DISAGREE            AGREE               AGREE
                              1         2         3          4        5
                              I----I----I----I----I----I----I----I----I

EXCELLENT COMMAND OF SUBJECT  *****************************************

ADEQUATE USE OF MATERIALS     ***************************************

LEVEL OF ORGANIZATION EXCELLENT ****************************************

RESPONSE TO QUESTIONS EXCELLENT ***************************************

OVERALL PRESENTATION EXCELLENT **************************************

                              I----I----I----I----I----I----I----I----I
                              1         2         3         4         5
```

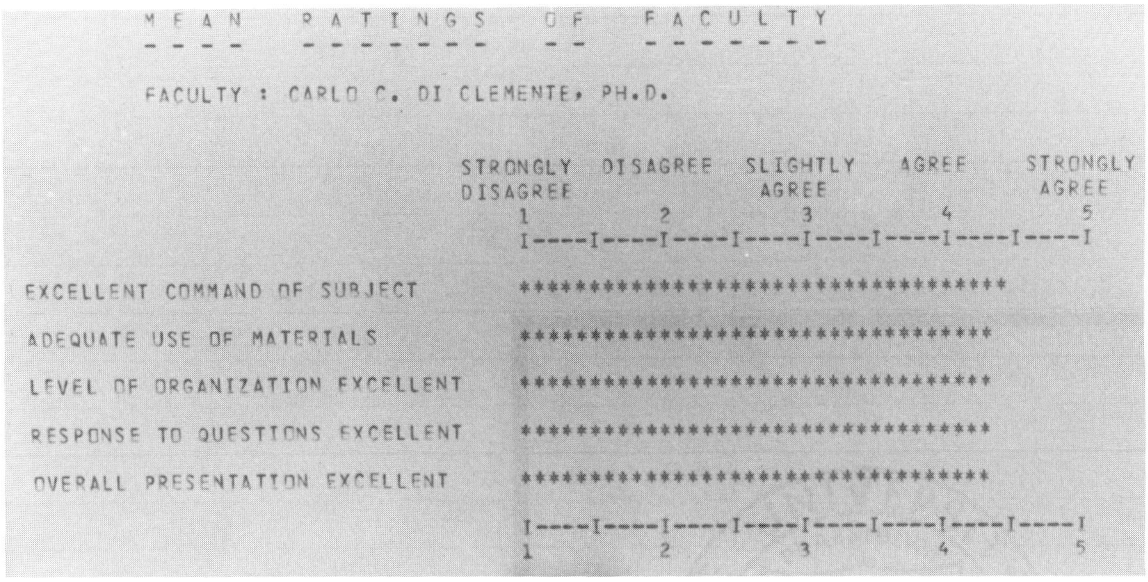

Figure 4b

J.L. Sanders et al.

TABLE I

DATE NAME OF TELEDUCATION PROGRAM	NUMBER PARTICIPANTS	PROGRAM OVERALL MEAN	PROGRAM PERCENTILE RANKING	FACULTY MEAN	FACULTY RANKING
January 6, 1982 DSM-III Overview	417	3.8	27	4.0	64
February 3, 1982 DSM-III Concepts and Use Major Classifications	352	3.7	25	3.9	60
March 3, 1982 Disorders Of Infancy, Childhood And Adolescence & Development Disorders	320	3.8	29	3.9	60
April 7, 1982 Psychosexual, Facitious, Impulse Control And Adjustment Disorders	218	3.9	33	4.0	60
May 5, 1982 Substance Use And Organic Mental Disorders	188	4.0	43	4.1	67
June 2, 1982 Personality Disorders	158	4.2	53	4.4	74
July 7, 1982 Schizophrenic, Psychotic Not Elswhere Calssified And Paranoid Disorders	132	4.0	42	4.2	71
August 4, 1982 Affective Disorders	99	4.0	42	4.1	67
September 1, 1982 Anxiety, Somatoform And Dissociative Disorders	75	4.1	44	4.2	69
October 21, And November 18, 1982 Receptor Theory Mechanism of Action of Drugs-Part I & II	146	3.8	23	4.1	65
December 1, 1982 Art Psychotherapy With Schizophrenic And Psychotic Patients Part-I	123	4.1	47	4.3	72
December 8, 1982 Chronic Broncitis	55	4.2	57	4.5	77

The Teleconferencing Resource Book: A Guide to Applications and Planning
Lorne A. Parker and Christine H. Olgren (eds.)
Elsevier Science Publishers B.V. (North-Holland)
© Center for Interactive Programs, University of Wisconsin-Extension, 1984

TELECONFERENCING –A CASE STUDY FOR A MAJOR PROJECT IN THE U.K.

Ray Winders, B.A. M. ED.
Co-ordinator Education and Educational Technology
Plymouth Polytechnic
Devon, England.
and
John Watts
Technical Developments Officer
Council for Educational Technology for the U.K.
3 Devonshire Street
London, England
On Secondment from British Telecom

Introduction

A Difference in Attitudes

'My department is in possession of full knowledge of the details of the invention (the telephone), and the use of the telephone is limited.'
Engineer-in-Chief, British Post Office 1877

'One day, every town in America will have a telephone.'
Mayor of a small American town, circ 1880

These quotations represent two century old opinions of the potential usefulness of the telephone. In some respects these quotations can still apply today. The lack of a driving force, prejudices or fears of one sort or another, have inhibited the introduction and development of telecommunication services as a useful educational tool. For education and business in the U.K. these opinions have just begun to change.

This change is being wrought because the Council for Educational Technology for the United Kingdom (CET) and other bodies have begun to develop and encourage the use of new Distance Learning techniques. In the past with one important exception the accent has been almost solely on the use of written material. The exception of course, has been the Open University with its innovative development of many new techniques including that of telephone tutoring and counselling.

In conjunction with the conception of these relatively new teaching techniques the economic climate for both education and business has begun to change and in most instances has worsened.

It was this situation which gave a new impetus to developing new approaches to education and in particular emphasised the need for better industrial and vocational training. Thus various new driving forces appeared which have made the use of the telephone as an educational tool a more attractive proposition.

An insight into the relationship between these new educational methods will be given later as part of the description of the study for the project.

A Short Survey of Educational Teleconferencing

in the U.K.

This survey seeks to give the reader an insight into the way in which teleconferencing is currently being used or studied in the U.K. In general terms it will be seen that some of the activity is designed to test the methods and technology for a known application, whilst others are aimed at stimulating interest prior to implementing and testing new applications. The rather involved position regarding the supply and development of equipment will be covered later.

The survey is taken country by country and describes the organisations involved, their applications, problems, and hopes for the future.

Diagram 1

(1) WALES

Wales is a country which is mountainous and sometimes, in winter, relatively inaccessible. The population is concentrated to the north and south in the industrial areas, the centre of the country being devoted almost exclusively to sheep, cattle and mountains. Opportunities for vocational training and further education can be restricted by distance, terrain and weather, thus making this an area where distance education, properly supported, can flourish to good effect.

Welsh National School of Medicine

A tutoring service, based on the University Hospital in Cardiff, is being provided using teleconferencing between five separate hospital sites. The service started in December 1982 and is presently being used to tutor fourth year Medical degree students. Previously prepared and distributed video tapes provide the basis for each tutorial.

The School of Medicine is operating a British Telecom bridge (the Teleconference

Unit 1A) on the Public Switched Telephone Network (PSTN) and has installed modified Loud Speaking Telephones in the lecture rooms and a lecture theatre. It is important to note that this is only the second privately operated bridge to be used for educational purposes, the first being the Open University. It is indicative of the importance attached to the project that steps were taken to find suitable equipment and to obtain the close cooperation of British Telecom.

There were early supply problems with finding suitable loudspeaking telephones and then with working out a modification to the microphones in order to increase the equipments' flexibility. Once the problems were overcome the system was brought quickly into service and has since proved to be very reliable.

Should the project prove the service to be educationally effective the potential then exists to extend the service to the remainder of the 17 hospitals in Wales where medical training takes place. Major savings are expected to be made in time and in the travelling and subsistance budgets for students and tutors alike.

Vocational Training and the potential for

Telephone Tutoring

This idea can only be briefly described because it has not yet reached the approval stage. In order to provide a more effective method of giving vocational training to students and school leavers in the 14-18 age range the Manpower Services Commission, a Government agency, is investigating the possibility of setting up a trial Technical-Vocational Education Centre.

This centre would provide supplementary technical education which would be of direct relevance to industry and commerce, by helping to provide students with suitable skills and a more effective and efficient transition from scholastic to adult life.

The centre in question would serve an area with a 20 mile radius and a population of 1.5 million encompassing 35 schools and 6 other colleges. In order to reduce the movement of staff and pupils between schools and centre to a minimum, and hence keep costs low, study centres would be set up in each location. These coupled with a requirement for tutoring and counselling, would certainly benefit from the support of a teleconferencing system.

(2)   NORTHERN IRELAND

This is an area in which the possible applications for telecommunications in education are being carefully studied. The uses being considered include links, both audio and visual, between Ulster Polytechnic and the University, which are 70 km apart, and the support of Outcentres for both teaching and administration. It is felt that the forthcoming introduction of new teaching techniques and equipment will rapidly stimulate the growth in the number of applications under consideration.

(3)   SCOTLAND

This is another country which is very rugged. It has a large number of mountain ranges and lakes. Travel to and fro from the large population centres is often long and tedious, frequently being completely disrupted in winter. Consequently a number of educational bodies in Scotland have been actively engaged in studying and testing the usefulness of the telephone system as an educational tool. In particular the Scottish Basic Education Unit is currently disseminating information and encouraging the use of teleconferencing as part of Distance Learning and Training courses.

Community Education - Argyll and Bute

The Community Education Office are running Distance Learning courses in school level English and Mathematics for those who did not obtain these basic qualifications before leaving school. This is proving to be a very useful service as it is offering an education with the opportunity for interactive discussion by teleconference for a number of very remote Island communities.

The application uses British Telecom's Public Conference Call Service which is based in Glasgow. Their main problems are due to the lack of automatic gain control on the bridge, and the effects of the weather on the microwave radio links to the islands. The introduction of improved and more versatile equipment would encourage the spread of this embryonic service.

Scottish Highlands

In this area the Public Conference Call Service is being used to support the training of Voluntary Youth Leaders and Community Workers in a variety of subjects. This can be regarded as an ideal example of the effective use of the teleconference as the course population is low and the region large and mountainous. Prospects are good for its future use.

(3)   ENGLAND

Southtek. A potential large scale use for teleconferencing has been identified in the South East of England. This is one of the Manpower Services Commission Open Tech projects.

The Open Tech scheme provides for Open Learning for mature technicians in the engineering industry. In this case it is the focus for the Southtek Consortium, which, in the beginning will comprise of thirteen Technical and Further Education Colleges, three Local Education Authorities and a similar number of employers. In order to support the Open Learning Modules a number of study centres for tutorial and practical work will be established in colleges and on the premises of the larger employers.

Should the potential for teleconferencing be proved for this application then it could form an integral part of the Open Tech developments as they spread to the rest of the U.K.

The Open University. The Open University may be regarded as the "Father" of teleconferencing in the U.K. The very essence of its modus operandi meant that efficient and effective forms of tutoring and counselling had to be found to support its courses, especially the ones with a dispersed and low population.

A great deal has been learnt from their various implementations of the technology and techniques. Their work has been a great inspiration to those involved in designing or implementing projects involving the use of teleconferencing.

Currently they are very much involved with the introduction of the next generation of the interactive audio-visual aid known as Cyclops. Briefly, it is to be implemented as a plug-in cartridge for the BBC microcomputer, greatly extending its range of applications and flexibility.

Council for Educational Technology for the U.K.. Surprisingly, as far as can be ascertained at the time of writing, there are no other current full time users of teleconferencing in education, apart from that of the Open University. There have been a number of experiments, conducted by the Council for Educational Technology, to test the educational and technical aspects of teleconferencing.

Cambridgeshire Primary Schools Trial. In this experiment a group of very small and isolated primary schools were equipped with loudspeaking telephones and were linked using the Public Conference Call Service. The aim of the trial was to test the feasibility and usefulness of a variety of applications, including the sharing of resources, experiences and specialist teaching.

The very long local telephone lines and the lack of automatic gain control on the bridge caused a number of technical problems which were very discouraging for the users. An attempt to use a new experimental meet-me bridge was defeated by the unreliability of the equipment. The patience and understanding of the teachers and pupils cannot be praised highly enough.

Valuable lessons about the type and quality of equipment were learned and will be discussed later. Despite the limited success rate with the conference calls enough was learnt about the applications tested to judge that Primary School education could benefit from its future use.

Training for Trainers. As part of a CET project in the South West of England work was carried out in testing the training methods of the trainers themselves. One of the tools tested was the use of teleconferencing in support of the techniques employed in Distance Learning. From this work and a study of the educational trends in general a commitment was made to test teleconferencing for every possible application. Thus the idea of the Plymouth Polytechnic Project was born.

South West England

The area selected for the 1983-86 project is South West England and will be centered on the city of Plymouth. In USA terms distances in Britain are short. The rail journey from

London to Plymouth is in fact almost equivalent to the distance from Chicago to Madison. What is however significant is the journey time to reach a class in a specified location.

The South West region is characterised by high moorlands such as Exmoor and is dissected by broad river estuaries. Both create long detours on apparently short point to point journeys. Public transport has almost disappeared, particularly in the evenings when most working students can attend classes. The rural areas are served by a medieval road network, along which it is difficult to maintain a speed of even 20 mph. Though the area is relatively free from snow, heavy rains and fogs create problems in the winter, whereas in the summer tourist traffic clogs every road. In the U.K. an evening class would typically run from 7 p.m. to 9 p.m.. I have a student who leaves work on the north coast at 5 p.m. and arrives back home at 11 p.m.. Our telephone conferencing proposal may free him from some of his odysseys!

The UK Educational Context

A brief description of the present UK educational scene is necessary here.

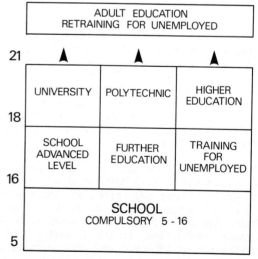

OUTLINE OF EDUCATION IN ENGLAND & WALES

School is compulsory from 5 - 16. There is no grade system. 20% will obtain an academic General Certificate of Education in specified subjects and some of these will continue to Advanced Levels at school. A further 50% will obtain a Certificate of Secondary Education again in specified subjects, the remaining 30% will leave with little evidence that they have attended school at all.

We are particularly concerned here with the post-16 age group who have left school. The Colleges of Further Education which exist in each town of about 10,000 person catchment, offer both full-time and part-time courses mainly in vocational subjects. Electronics, accounting, catering, hairdressing would be typical departmens in a UK college. These colleges also take students who are released by their employers on a one day a week basis or on blocks of a few weeks to obtain relevant technical qualifications. They also offer Advanced level courses in academic subjects as an alternative to the schools, but these are in the main taken by adults in the evenings.

There has been a significant increase in the proportion of students taking courses beyond 16 as a result of very high youth unemployment. Some students are staying on at school or taking full-time courses in Further Education but a new development is special courses for the young unemployed financed by the Department of Industry. Some of these courses are offered in special skill centres, but others are offered in the Colleges of Further Education. There are interesting frictions at the traditional interface between Education and Training. The Department of Employment pays the trainees and specifies training objectives and methods. The traditional Further Education course is more student centred.

Higher Education is provided by Universities, Polytechnics and Colleges of Higher Education. The Universities offer a wide range of full-time graduate and post-graduate courses all on-site. They each have a limited extra-mural function in offering cultural courses, often on an evening class basis at several locations in their region. In my own village the University of Exeter currently offers Oil Painting and 20th Century Political Thought, but nothing else. The Polytechnics are in effect technological Universities. There are 30 in th U.K., each with an average of 6000 students. Most courses are full-time though many of these have up to a year in industry or commerce on what is known as a sandwich basis. Typically, a student taking a Business Studies Degree would spend two years in the Polytechnic, one year in a firm, and come back for a final year in College. The Polytechnics serve as a focus for technical and business education in their regions, sometimes using local further

education colleges as outstations. They also offer advanced level day-release courses leading on from the further education colleges. It is not uncommon for a student to leave school at 16 and to have reached degree level standard in his profession by day release and evening attendance until his late twenties. The Higher Education Colleges have developed from the former Teacher Training Colleges and, as well as offering degrees in Education, now offer broadbased degrees, mainly in the Humanities.

For the student who is unable to attend a scheduled class, there are considerable problems. In almost every village, the school is used on some evenings for educational purposes, but as mentioned above, the courses offered depend upon a local enthusiast. The only alternative is some form of correspondence course. Though there are commercial colleges, the major provider is the Open University. The Open University offers a very wide range of degrees. Students are sent regular study materials and assignments, and these are supported by Radio and Television broadcasts and summer schools. With so wide a range of provision in so small a country, why mount a telephone conferencing project?

Development

In the period 1981 - 82 a Distance Learning Project was financed by the Council for Educational Technology UK. It was centered on the South West region with 2 main aims.

(1) To identify "good practice" in educational and industrial training

(2) To identify problems

The conclusions were clear:

(1) There were several examples of creative and effective work, but almost all were in total isolation from others in the same field.

(2) Common to all distance learning projects was the problem of communication both between teachers and students, and between students themselves.

(3) Telephone conferencing was unknown.

A telephone conferencing system with a centrally located bridge had been available for some years but what little marketing was done was aimed at the business user. The system used standard lines with reasonable access, but with enough problems to limit confidence.

The only significant user of telephone conferencing has been the Open University, where a small number of regional experiments has been taking place. A project in the Scottish Highlands in 1982 used 650 hours on one to one telephone calls, 268 hours of conference calls linking individuals, and 10 hours to study centres, using loudspeaking telephones. Some students were on remote islands and for many, the phrase "my only lifeline" is used of the telephone tutorials. Each student was allowed three hours of free calls. George[1], in her review of the year, classifies use as follows:

1 One to one calls   – problem solving and counselling

2 Conference calls linking individuals at home   – discussion of course assignments and study materials

3 Loudspeaking telephones   – limited use because the student has to travel to the centre.

The Open University is developing an expertise and has published a useful handbook.[2] An interesting aspect of the Scottish experiment is that 1/3 of the students have no home telephones. Some even drove to a public call box in order to participate in telephone conferences!

P.P.P.

A limited experiment was carried out as part of the 1982 South West Project. A small group of students on industrial placement were linked through the central bridge to a tutor at Plymouth Polytechnic for a half-hour tutorial. The evaluation[3] listed positive comments as "reduced the feeling of isolation", "I found out what other students were doing", "enabled me to make a comparison between my job and others", and, perhaps most significantly, "comforting to know other people were having similar problems". The negative comments stemmed from inexperience in using the network, and included "uneven contribution between participants", "too much time spent on a few questions". The most common complaint was of poor technical quality, and it is at this point that discussions began to establish a more extensive telephone conferencing project using a newer bridge located in Plymouth Polytechnic itself.

PROJECT

Proposal

There are five distinct applications of telephone tutoring which have been identified during our initial investigations:

(a) Distance learning at home or in small groups

(b) Conferences, course committees, etc.

(c) Students on industrial placement

(d) Courses to which expert inputs from a distance would be valuable

(e) Courses held at more than one centre

In order to conduct a carefully controlled and evaluated project, it is proposed to operate in two stages:

STAGE 1

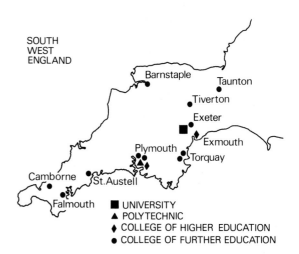

SOUTH WEST ENGLAND

Barnstaple   Taunton
Tiverton
Exeter
Exmouth
Plymouth   Torquay
Camborne   St.Austell
Falmouth
■ UNIVERSITY
▲ POLYTECHNIC
♦ COLLEGE OF HIGHER EDUCATION
● COLLEGE OF FURTHER EDUCATION

a) Bridge at Plymouth Polytechnic Learning Resources Centre

b) Group tuition studios at

    (i)   Plymouth Polytechnic

    (ii)   Plymouth Polytechnic Hoe Centre

    (iii)   Plymouth College of Further Education

    (iv)   Cornwall Technical College

These four locations would allow the full range of activities listed above to be evaluated. The two Polytechnic locations (i & ii) would provide an excellent demonstration and training facility for later national and international dissemination.

STAGE II

At this stage additional colleges could be linked to provide a South West England network.

Colleges would be selected from

(i)   Somerset College of Arts & Technology

(ii)   Exeter College of Further Education

(iii)   North Devon College of Further Education

(iv)   South Devon Technical College

(v)   Seale Hayne College of Agriculture

(vi)   Derriford Hospital Education Centre, Plymouth

In addition, Cornwall Technical College has associate centres at Truro, Falmouth, Penzance and Wadebridge, any or all of which could be usefully linked.

(a)   Distance Learning

A number of people in the South West live more than two hours driving from any college. Those who can only attend in evenings find journey times and driving conditions even more difficult. There are sizeable minorities who are unable to attend colleges at all. Though some colleges have facilities for disabled students, there are many who are unable to attend specific courses. There are shift workers, particularly in the Naval dockyard, whose work schedule varies on a three week rota: 6 am to 2 pm, 2 pm to 10 pm, 10 pm to 6 am. An unusual group are pharmacists. In the UK a drug-store is only allowed to open when a qualified pharmacist is in attendance. This, in effect, means that except for the very largest stores, pharmacists cannot be released for updating. Another group are those who need specialist tuition, but are not sufficient in numbers to form a viable group. An extreme example of this is a Diploma in Acoustics offered in Cornwall. In any one year, only 30 people in the whole country need this qualification.

To meet the needs of the above groups, a

number of distance learning courses have been organised. They are print based with, in the case of the Diploma in Acoustics, cassette tapes, and in a more elaborate Electronics Course, a box of parts, instructions, workbook etc. To date, none of these has used telephone conferencing, though there has been a limited use of the telephone for individual tutorials.

The following courses have already applied for inclusion in the telephone conferencing proposal.

Diploma in Acoustics
   National Course tutored from Cornwall Tech.
                        (32 students)
Social Work Diploma
   Based on Polytechnic with fieldwork tuition

National Diploma in Educational Technology
   Polytechnic to homes

Professional Courses Business School
   Plymouth Polytechnic to homes

T.E.C. Electronics II
   Plymouth C.F.E. to students at home

Flexistudy 'A' level & 'O' Level G.C.E.
   Plymouth C.F.E. to homes
   Cornwall Tech. to homes

T.E.C. Construction Studies
   Plymouth C.F.E. to workplace

T.E.C. British Telecom Technicians
   Plymouth C.F.E. to homes and workplaces
   Cornwall Tech. to homes and workplaces
   Somerset Tech. to homes and workplaces

(b)   Conferences

Each course in the Polytechnic has management and advisory committees, which in turn report to National Bodies. If some of these meetings, which may well, given travel and lost time, cost $300 for an afternoon, could be replaced by teleconferences, there would be considerable saving. The Health Service and the Social Services will also use the Plymouth Bridge as part of the project. The student welfare service of the Polytechnic has developed a service to support unemployed graduates. Mutual group support by telephone would be very beneficial.

(c)   Students on Industrial Placement

Several Polytechnic courses are in a sandwich mode with up to a year in industry. Currently the equivalent of 3 full-time staff ($100,000) are involved in visiting students on Management Courses alone.

The following courses have sandwich placements:

Agriculture B.Sc.

Business Studies B.A.

Civil Engineering B.Sc.

Communication Engineering B.Sc.

Department of Trade Navigation Officers

Electrical and Electronic Engineering B.Sc.

Health Visitors Certificate

Mechanical Engineering B.Sc.

Mechanical Engineering H.N.D.

Social Work Certificate

Social Policy & Administration B.A.

The above are all Polytechnic courses. Use of telephone tutoring would save money, but would also enable students to talk to each other in discussion. There is current national concern about the isolation of students on placement some of whom withdraw from courses during this period.

Though the above are long placements an even more expensive commitment is part of the Unified Vocational Preparation Scheme in which half of an unemployed leaver's time is spent in industry and each student must be visited once a week. There are 200 of these students at Plymouth C.F.E. alone. The use of some telephone group tutoring would have considerable cost savings.

(d)   Expert Contributions from Outside

The proposal is to invite an outside speaker to contribute information or answer questions. Normally costs in cash and time for a busy officer or politician would be prohibitive, but a 15 minute call to the local M.P., with questions prepared by a group, would be effective. The Chief Fire Officer, Chief Education Officer, Head of Police, Famous Footballer were among the first suggestions.

(e)   Courses held at more than one Centre

A number of courses are held at a network of colleges. Students on these would benefit from telephone contact and expertise of lecturers could be made more widely available. By using the loudspeaking telephone we are proposing to link the following:

Agriculture B.Sc.& B. Sc. Hons
   Polytechnic - Seale Hayne

Scientific and Technical Graphics
   Polytechnic - Cornwall Tech.

Diploma in Management Studies
   Polytechnic, Seale Hayne, Exeter

Certificate in Education (FE)
   Polytechnic - Cornwall Tech. - Taunton

Social Work Diploma & Advanced Diploma
   Polytechnic - all F.E. Colleges

B.Sc./HND Electrical Engineering
   Plymouth Poly. - Cornwall Tech.

TRAINING

Our own limited experience together with indications from the British Open University and indeed in previous papers at this conference has emphasised the need for training both teaching staff and students in the use of the telephone. Before the Plymouth project is fully operational all staff will undertake training in technical operation and more importantly in the organisation of telephone tutoring. The two Polytechnic sites and College of Further Education in Plymouth offer 3 situations which are physically separate but within a mile of each other so that training groups can be split between the sites. Plymouth Polytechnic will later in the project be used for demonstration and training on a wider scale beginning with a Council for Educational Technology Conference in January 1984.

Teleconferencing Equipment and Facilities

The Situation in the U.K.

The revolution that is sweeping through British Telecom is beginning to cause some drastic changes to the way in which equipment and services are provided. It involves at present, liberalization or the ending of British Telecoms monopoly in the supply of equipment for connection to the Public Switched Telephone Network. This, by its very nature requires the generation of new British Standards for a whole host of Tele-communications equipment, and is consequently a long and involved business.

Due to quite natural business pressures the standards work is being firstly carried out to cover the equipment which will have the largest potential market.

Typically one of first is the standard type of telephone. Until the time that a specification is agreed and issued it can leave British Telecom with the monopoly for the supply of particular ranges of equipment. This is very much the case with the supply of equipment for teleconferencing applications. Unfortunately this only leaves the education market with standard telephones and loud-speaking telephones and a variety of teleconferencing applications that cannot be effectively served.

Public Conference Call Service

Strictly speaking British Telecom has the monopoly for the provision of operator controlled Public Conference Call Services on the Public Switched Telephone Network. The service is provided by a number of operator centres in different parts of the country. The equipment used gives some amplification but no automatic gain control and consequently only functions well on standard quality lines. Currently two British made bridges are being evaluated for possible replacement purposes.

Experience has shown that when the conference call service is heavily used by educational bodies, the best results are obtained by allowing that body to operate the bridge itself, and thus provide a wide range of backup facilities. British Telecom has recognised this fact and is allowing the Open University and CET to run their own bridges so that these new applications can be properly monitored and evaluated.

Technical Problems - A Summary

Undoubtedly many of the problems that have been identified have been found by our counterparts in other countries. The solutions too are probably similar, except that they may well be arrived at in a different way.

Virtually all of the educational teleconferencing in the U.K. is conducted over the PSTN, this is mainly due to the need for flexibility and the heavy cost of private circuits. The effects of long local lines and variable volume levels were quickly recognised as being due to the lack of agc on the public bridges. Many conferences have failed because one or more participants could not be heard. The solution is obvious, long awaited, but slow in coming and that is to replace the bridge. The solution at the moment lies solely in the hands of British Telecom.

The other major problem to be identified is the practicality of the various types of loudspeaking telephone that are available. In all cases all that we have been able to use are a variety of standard desk top models, the majority of them difficult to use with more than two or three participants. Experience and experimentation has shown that an effective solution can be had by adding additional microphones with push-to-speak buttons and additional loudspeakers. It is felt that one new LST designed with flexibility in mind would serve the majority of applications at a reasonable cost. Once again the solution lies with British Telecom.

Cyclops

Cyclops is the name given to the interactive visual aid that was developed by the Open University, British Telecom and Aregon. It is capable of acting as a stand-alone teaching device or as terminal connected at each location of a teleconference. The system allows each user to input text from a keyboard, to draw freehand on the screen with a light pen or to draw on a bit pad. As information is input it is transmitted to the other participants through a modem and the bridge, each participant requiring a second telephone line. Speech and more complex pictures can be prepared and stored on a stereo cassette recorder and then transmitted as part of the lesson. Cyclops has now been proven to be an important audio-visual aid and has thus secured a future for itself.

An important step forward this year is the implementation by the Open University of Cyclops as a plug in cartridge on the BBC microcomputer. With the addition of a low cost light pen and additional facilities such as a library of graphics routines it then becomes a very powerful low cost audio-visual aid.

The next step forward is to reduce the number of telephone lines required from two to one. This can be achieved by using a speech plus data modem.

Speech plus Data Modem

This development is being considered very carefully by the Open University and British Telecom as its implementation can be regarded as being very useful for business as well as education. When fitted to a telephone, Cyclops or other forms of interactive device, can then penetrate more freely to all single line locations including the home.

The exact specification for the modem is
being worked out at the moment and for the time
being for commercial reasons must remain
confidential.

Interactive Involvement

Because of the lessons learned about the
importance of getting the technology to the
right level of usefulness with sufficient
facilities and reliability, before proceeding
to test the full range of applications the CET
became involved at an early stage in
researching and defining the technical problems
and requirements.

Throughout the past year the CET has
liaised closely with the users and potential
users of teleconferencing, learning of their
experiences, problems and requirements. All
of this information has been passed on to
British Telecom and last November the breakthrough came when a new group was formed within
British Telecom Enterprises. This group is now
studying the market potential of
teleconferencing most carefully to see how it
might best be served.

The following items have been summarised
and put forward as being essential for
development.

(a) Versatile loudspeaking telephone with
     facilities for microphones, loudspeakers,
     tape recorders etc.

(b) Speech plus data modem possibly
     incorporated within the LST

(c) Improved teleconference bridge with multi-
     conference facilities, tape recorder
     access etc.

Should all that is hoped for be achieved
and the equipment be produced the Plymouth
Polytechnic Project will form an important test
bed for the new items.

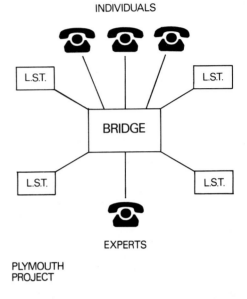

PLYMOUTH
PROJECT

The Proposed System

For the first stage of the project it is
planned to have three study centres and a
lecture theatre equipped with the versatile
loudspeaking telephones and Cyclops terminals.
A twenty line operator controlled bridge will
be installed in the Learning Resources Centre
at Plymouth Polytechnic. All of the locations
will be connected to the PSTN, the bridge
telephone lines being given PBX hunting so that
a meet-me service can be provided when required.

Initially it is proposed that the bridge
is equipped with a switching matrix so that any
of the conferees can be switched into any one
of a number of simultaneous conferences. This
will also allow the setting up of sub
conferences during the main conference. The
bridge will also be equipped with a call cost
accounting system and facilities for tape-
recording. These will be used for course
support, evaluation and normal accounting
procedures.

The project is being structured to have a
high degree of flexibility. This is being done
to ensure that equipment development can follow
the users needs, and will be especially
important as the project enters its second
phase. Because of teleconferencing's novelty at
the moment it is impossible to judge how far
its applications will reach in the South West
of England. It is hoped of course that there
will be an explosion of practical applications,
giving the opportunity to develop more complex

forms of bridge equipment. The second phase will be structured to cater for these developments.

Future Ideas

There is sufficient interest being shown nationally for quite a few interesting ideas to be floated. One of the more interesting is for a computer controlled digital bridge that can have its combining functions controlled remotely by an intelligent terminal, giving a tutor real time control over the conference. This could be done by an extension to the BBC computers Cyclops facilities.

Somewhat further in the future is the transition from an analogue system to one which is totally digital from end to end. The family name for this Telecommunications Service is "System X" and the transmission facilities are provided by the "Integrated Digital Services Network". This service will offer a total data rate of 80 Kbits which is split into 64 Kbits for speech or non-voice, 8 Kbits for non voice and 8 Kbits for signalling. The first two channels are independent and may be used to make simultaneous calls. This will mean a dramatic explosion in the range of facilities available. For teleconferencing one can envisage enhanced audio bandwidths, high speed viewdata, interactive audio-visual aids and high quality freeze-frame T.V.

The Future

The Plymouth Polytechnic Trial will provide a test-bed for future development in the United Kingdom. The Government is already committed to a major development of technician training in a distance mode. Telephone communication particularly when visual data transmission becomes available is crucial to this. In the current recession, all public bodies are endeavouring to cut costs and the development of successful experience in conference and committee techniques will save millions of pounds.

The project, as described, is making more cost effective what already exists. The developments in technology already in hand will, however, create totally new opportunities for learning. Group problem solving activities with access to accurate information and expert advice are now becoming possible. Access via telephone line to major teaching programmes with the possibility of interactive or even creative learning will add a new dimension.

It is easy to be carried away on a wave of generalised enthusiasm, but the project will be rigorously evaluated. The main areas of evaluation will be:

(1) Cost-benefit for each type of usage

(2) Problems encountered in operation

(3) Training, induction, and attitudes of staff and students

(4) Effectiveness for learning and for business transactions

(5) New possibilities for learning created by this facility

During the project a major task will be dissemination of results and applications. Though a final report will be made at the end of the project we are proposing continuous interactive action research. In the UK there has been no focus for telephone conferencing activity. This project bringing together the Council for Educational Technology, British Telecom, and the Department of Industry in a composite regional trial will provide an essential test-bed for future development.

We come to this conference with a great deal to learn. We are fortunate that we can draw on the long experience of participants in this conference to sustain us through this challenging project.

REFERENCES

1 George, J.
"Tutorials at home - a growing trend"
Open University 1979

2 "Tutoring by Telephone - a handbook"
Open University 1982

3 Hutton, D B A.
"The Telephone Conference; planning, preparation, Assessmene and Evaluation"
University of Bath 1983

The Teleconferencing Resource Book: A Guide to Applications and Planning
Lorne A. Parker and Christine H. Olgren (eds.)
Elsevier Science Publishers B.V. (North-Holland)
© Center for Interactive Programs, University of Wisconsin-Extension, 1984

THE AIR FORCE INSTITUTE OF TECHNOLOGY
TELETEACH EXPANDED DELIVERY SYSTEM EXPERIMENT

G. Ronald Christopher
Chief, Plans and Evaluation Division
Air Force Institute of Technology
Wright-Patterson Air Force Base, Ohio 45433

Background

The rapid growth of knowledge and the increasing sophistication in technology offer a significant challenge to educational institutions. This challenge is quite evident within the Air Force Institute of Technology (AFIT), which is responsible for providing undergraduate, graduate, and professional continuing education to the Air Force, and in certain content areas, the Department of Defense (DOD). AFIT accomplishes its mission objectives through resident instruction at Wright-Patterson AFB and various types of nonresident instruction.

In the specific area of professional continuing education (PCE), neither manpower nor facilities have kept pace with the need. In the 1978-79 academic year alone, over 7,000 students received professional continuing education in resident courses - less than 45% of the 15,000 who needed resident PCE that year.

Resident PCE is designed to foster intense concentration while minimizing the time students are absent from their duty stations. Course length ranges from three days to ten weeks. Instruction occurs during a 6-7 hour class day, five days a week. Many blocks of instruction are presented by functional experts who are not assigned to the AFIT faculty. They teach as guest lecturers in courses designed and managed by the permanent faculty.

Nonresident professional continuing education is provided through seminars, workshops, on-site offering at the students' location, and correspondence courses. Approximately

10,000 students are instructed annually in these modes. More requests for nonresident education exist than AFIT can accommodate.

Meeting the education demand poses a dilemma. How can AFIT provide education to more students in existing courses and concurrently develop new courses without increasing either the number of faculty or overall instruction cost?

This dilemma is particularly evident in the School of Systems and Logistics. Numerous courses have 2-3 year student backlogs, while numerous requests are pending for new courses. Faculty are restricted in the amount of time available for course development due to heavy commitments in providing instruction in existing courses.

Limited physical facilities both within AFIT and at WPAFB, limited faculty, and limited budget preclude a solution based upon increased resident student attendance. Expansion of current modes of nonresident instruction is also limited, since an increase would require additional faculty, increased travel, and increased support personnel.

During the search for a resolution of this dilemma, use of the telephone as an educational delivery system was considered. Since 1973 both the Civil Engineering School and the School of Systems and Logistics have routinely used commercial dial-up telephone services to provide limited length (1-2 hrs) instruction to single remote locations. Teleteach or Telelecture was the name given to this delivery mode.

Recent technological advances in tele-communications now offer expanded capabilities. In early1979 American Telephone and Telegraph began commercial marketing of a device which can transmit, through telephone lines, material written upon an electronic blackboard and regenerate the writing at distant locations on a standard TV monitor. The electronic black-board offers a significantly expanded capabil-ity to the use of the telephone for education-al purposes.

Further research into the educational viability of telephonic instruction revealed that over 37 telephone networks now convey instruction to civilian students who are remote from the point of origin. The most common network patterns are within specific states or within limited geographical areas. The acknowledged leader, the University of Wisconsin, serves over 35,000 students annu-ally through its statewide telephonic network.

Telephonic networks generally convey con-tent to serve practicing medical, legal, and agricultural professionals as well as students in agronomy, business, engineering, and mathe-matics. As yet, there has been no indication that any content discipline is unsuitable for telephonic transmission. Some programs offer academic credit, others meet professional continuing education requirements, while others carry no formal credit. Program length varies somewhat, with the majority adhering to the advanced education schedule, i.e., one to two hours a day, one to two days per week.

Research has shown consistently that learning is not significantly affected when telephonic instruction is compared to tradi-tional classroom instruction. An excellent review of the literature is provided in Myrless Hershey's dissertation, A Comparison of the Effectiveness of Telephone Network and Face-to-Face Instruction for the Course "Crea-tive Classroom," Kansas State University, Manhatten, Kansas, 1977.

The success of telephonic networks in the civilian sector and AFIT's previous limited use of the medium strongly suggested that a dedicated telephonic delivery system using the electronic blackboard might offer the solution to our need to educate more people within current resources.

AFIT administrators identified courses within the School of Systems and Logistics where significant student backlogs existed. These courses are designed to meet the require-ments of essentially two major Air Force Commands, the Air Force Logistics Command (AFLC) and the Air Force Systems Command (AFSC). The majority of potential students

are stationed at a limited number of bases.

Following an AFIT proposal, both commands agreed to establish a telephonic network with classrooms at specified bases. Command par-ticipants agreed to share the cost because the system would permit students to receive needed instruction while working part time and remain-ing at the work location. No travel and per diem costs would be involved. A preliminary cost analysis indicated that the cost of the delivery system could be offset if approximate-ly 360 students received instruction without incurring travel and per diem expenses. Addi-tionally, the student backlog could be rapidly eliminated, since instruction provided at AFIT to a regular class of 24 students could be received by approximately 120 additional stu-dents at AFLC sites and 96 additional students at AFSC sites. This could occur without addi-tional faculty or facilities.

Subsequently, two dedicated telephonic networks were installed in September 1979. One network connects AFIT with five Air Logis-tics Centers (ALCs) and a second network links AFIT with four Air Force Systems Command (AFSC) locations.

Sites are geographically dispersed throughout the United States and encompass all time zones. Using the two separate net-works, two courses can be offered simulta-neously: one for the AFLC sites, and one for the AFSC sites. Each course originates from a separate classroom at AFIT. The originating classrooms and each remote site classroom are connected through two pairs of dedicated tele-phone lines. One pair sends and receives ver-bal expressions, while the other pair trans-mits writing generated upon the electronic blackboard. Each site is able to transmit as well as receive. Therefore, presentations may originate from any site. Necessary visu-als, in the 35 mm slide format, are made available to each site and may be controlled by the presenter. All verbal and blackboard written communication during each class is recorded on stereo-audio tape. Replay of classroom sessions is at the discretion of each site monitor.

Consideration of the time zone differences and normal resident class scheduling resulted in an expanded Teleteach schedule of daily four hour system use from 12:00 to 4:00 p.m. EST. Presentations originating at AFIT are made before a student audience employed at Wright-Patterson AFB, Ohio. Ten-minute class breaks occur each hour.

The additional capabilities available in this Teleteach approach -- the electronic blackboard, dedicated lines, and recording of classroom sessions, combined with AFIT's previous Teleteach/Telelecture delivery mode -- suggested labeling this new delivery system Teleteach Expanded Delivery System (TEDS).

Course Description

These courses provide the data used in the evaluation of the Teleteach Expanded Delivery System.

The Materiels Management Course, Log 220

This course is designed to improve the management effectiveness of key personnel assigned to Materiel Management and related AFLC activities which provide support to the Air Force and other DOD agencies. It is intended to familiarize the student with the structure, philosophy, policies, functions, processes and systems of Air Force Logistics, with particular reference to their impact on Materiel Management.

Student eligibility is predicated upon course director approval of a standard form. Students usually range in grade from captain to lieutenant colonel and civilian grades GS 09-14. The Log 220 course contains 96 student contact hours. The instructional format is basically lecture, with allowance for student questions. A simulation exercise, involving teams of five students, serves as a culminating experience. Class size ranges between 22-28. A large number of guest speakers make presentations in this course.

The Fundamentals of Acquisition Management Course, Sys 326

This course is designed to provide an overview of the management process by which USAF systems are acquired and in particular the role and responsibilities of the program office as they relate to the acquisition process. Students receive instruction involving the Air Force Systems Command, the acquisition process, the budget process, the program office, engineering management process, the contracting process, integrated logistics support, program control, the management review process, and the interrelations with other Air Force Commands.

This 60-hour course is intended for persons recently assigned to acquisition management positions. Although the great majority

of students are lieutenants, some higher ranking officers and NCOs also attend. Civilians ranging in grade from GS 05-14 are eligible. This course was given for the first time using TEDS. The instruction format is basically lecture and discussion. Most instruction is provided by AFIT personnel, although several guest speakers are involved.

The Principles of Contract Pricing Course, QMT 170

This course provides a basic understanding of cost and price analysis policies, procedures, techniques, and negotiation strategy.

This course is sponsored by DOD and is mandatory for civil service employee progression in the contracting career field. Persons in the comptroller career field also constitute a significant portion of class enrollment. Eligibility is predicated upon successful completion of a Mathematics Screening Evaluation and acceptance by the course director of a standard form. Military students range in grade from second lieutenant to colonel, while civilian grades range from GS 05 to GS 14. Normal classroom enrollment is 25. This 98-hour, three-week course is presented numerous times annually through resident instruction at WPAFB and onsite where sufficient student numbers warrant. AFIT faculty conduct the QMT 170 course. Instruction occurs within a problem-solving format where students apply mathematical formulae to various pricing situations. An integrating problem involving student teams serves as a culminating experience.

Differences

The manner in which AFIT employs its telephonic delivery system raises several distinctive concerns which have not been addressed by published research.

We must consider the nature of the influence upon students of a four-hour, five-day a week class schedule. Most available research depends upon class schedules of 1-2 hours a day, 1-2 days per week. The effect of numerous speakers vs one or two during a course offers another challenge to available research findings. We need to measure the effect of time of day when instruction is received, and we need to identify the impact of remote classroom supervision by persons who are unfamiliar with course content upon student performance and system acceptance.

These factors, both individually and in combination, insist that a thorough evaluation be conducted and preclude a priori system acceptance based upon currently available research.

Research Questions

A comprehensive evaluation is being conducted in which these research questions are being addressed:

Are student groups (control/experimental) comparable in terms of education level, grade/rank, age, and entry level knowledge?

What effect upon academic achievement did the TEDS have compared to resident and on-site delivery of the same courses?

What differences in academic achievement occurred between resident student groups receiving instruction face-to-face with the presenter and student groups receiving instruction without face-to-face presentations when both groups used the TEDS?

To what extent was the TEDS acceptable to students, their supervisors, presenters, visitors, and site monitors?

To what extent did students and their supervisors consider course value to be significantly different when resident on-site and TEDS instruction occurred?

To what extent did students and their supervisors consider the TEDS schedule acceptable?

Does the cost of the TEDS exceed that of resident instruction on a per capita student basis?

The Evaluation Plan

The evaluation plan permits acquisition of data to determine if the Teleteach Expanded Delivery System is acceptable educationally, affectively, and financially. Data are being collected to reveal (a) learner accomplishment, (b) learner, supervisor, and presenter system acceptance, (c) system functional reliability, (d) visitor impressions, (e) interaction among sites, and (f) system cost. Several uses of the TEDS with several different courses provide data for analysis.

The effectiveness of the delivery system will be determined through these comparisons: (1) regular resident class(es) vs TEDS

classes(es), (2) regular resident class(es) vs resident TEDS class(es), (3) resident TEDS class(es) vs remote TEDS class(es), (4) on-site class(es) vs TEDS class(es), (5) on-site class(es) vs resident TEDS class(es), and (6) TEDS offerings will be compared when repeated. Total evaluation will be complete in approximately one year.

All data collection instruments were designed by AFIT.

Student Learning (Achievement)

Student learning (achievement) is being measured by comparing content pretest performance and subsequent content examination scores.

Perceived student learning is being ascertained through use of a student end-of-course critique and an appraisal by both the student and his/her supervisor six months after course completion. Presenters are asked to assess their own effectiveness after using the TEDS.

Presenter Appraisals

Student teams in each classroom will rate each presenter. Resident team ratings will be compared to remote team ratings.

System Acceptability (Affective)

Students, their supervisors, presenters, site monitors, and visitors are being asked for their opinions of the TEDS. Students provide their impressions using the TEDS version of the end-of-course critique and the six-month post course student questionnaire. Students' supervisors will be asked to express their opinion of the TEDS schedule and the part-time availability during instruction of their employees. A Site Monitor Daily Critique collects site monitors' impressions of the delivery system. Visitors to any TEDS classroom are asked for their reaction to the system.

System Acceptability (Operational)

System operational performance is documented daily at each site by students and site monitors. A daily system performance log is maintained at the central patch panel.

System Acceptability (Cost)

System cost data include all lease and installation changes. Comparisons are made between system costs and the costs which would have been incurred had the students come to Wright-Patterson AFB for resident instruction.

Findings

As of 1 March 1980, three courses had been provided via the TEDS. One course (Sys 326) was given twice. Complete data analysis has only been accomplished on two courses to date. Results reported here are strictly preliminary and serve principally as guidelines for subsequent system use.

Achievement

Data from the two courses (Log 220 and Sys 326) are somewhat contradictory. The Log 220 resident and remote TEDS students revealed essentially the same academic achievement. Resident TEDS students in the Sys 326 course scored significantly higher than remote TEDS students on measures of learning achievement.

System Acceptance

Student acceptance of the TEDS system varied widely from very high acceptance to almost total rejection. No remote site group accepted the schedule although class meeting times ranged from 9 a.m. - 1 p.m. (PCT) to 12 to 4 p.m. (EST). Commanders of AFSC and AFLC have expressed their enthusiasm for the networks.

Cost

Using only these first two courses offered via TEDS and including installation costs, there was a $48,000 cost avoidance realized. Projected cost avoidance for the remainder of the fiscal year using the current course schedule will approximate another $250,000.

System Operation

There were periodic audio transmission problems involving interference at remote sites due to air traffic control overrides. Some instructors' voice volume varied greatly.

The electronic blackboard and the dedicated audio lines presented almost error-free operation.

Discussion

Although there are insufficient data currently available to arrive at definitive conclusions, there are indications that the nature of course content, age/experience of students, and environmental conditions at remote sites play an important role in student achievement and system acceptability. Some tentative conclusions have been made and actions taken.

One course (Sys 326) was reduced from 80 hours to 60 hours after its first offering because students felt there was too much content provided in a short time. One remote site classroom was replaced by a larger one. Local site monitor activities varied and where these monitors remained involved every class hour, the students were more receptive to the delivery system. Instructor microphones were unidirectional which caused volume variation during transmission. Omnidirectional mikes have been ordered. Some remote sites had not received their student microphones which inhibited student interaction. The electronic blackboard was not used to any great extent, yet 77% of the students felt it should remain a system component.

One apparently influencial factor in student negative reaction to the delivery system was the fact that some supervisors expected students, even though attending 4 hours of class per day, to fulfill their 8 hour job responsibilities.

The Future

Action is being taken to remedy the external negative influences upon students, to correct transmission problems, and to improve the remote site classrooms. Some content revisions are being made.

Results of subsequent system use will provide more reliable information concerning the effect upon learning and system acceptance since most initial "unknowns" will have been identified and corrected.

Later studies will provide the definitive data necessary to properly answer the research questions.

Already AFSC is planning to expand the system by adding 11-15 remote sites. The AFLC

has used the system on several occasions to
conduct workshops, seminars, and short courses.
Actual saving of TDY monies has been documented.

Conclusions

The concept of Teleconferencing in the
Air Force is gaining acceptance. We look for
system expansion, greater system use, and
increased diversity of applications.

The Teleconferencing Resource Book: A Guide to Applications and Planning
Lorne A. Parker and Christine H. Olgren (eds.)
Elsevier Science Publishers B.V. (North-Holland)
© Center for Interactive Programs, University of Wisconsin-Extension, 1984

THE AIR FORCE INSTITUTE OF TECHNOLOGY
THE AIR FORCE REACHES OUT THROUGH MEDIA: AN UPDATE

G. Ronald Christopher
Chief, Plans and Evaluation Division
Air Force Institute of Technology
Wright-Patterson Air Force Base, Ohio 45433

In August 1979 the Air Force Institute of Technology (AFIT) "came on line" with the Teleteach Expanded Delivery System (TEDS). This system reaches out from Wright-Patterson Air Force Base (WPAFB), Ohio to ten remote sites in eight states taking needed instruction to students at their places of employment.

During each year of system use, an extensive evaluation was conducted to determine the system's effectiveness, acceptability, and cost. Annual evaluation reports are published as Air Force Technical Reports.[1,2] This paper presents an overview of findings during the first two years of operation. Significant consistent findings as well as noticeable changes are identified.

Evaluation

The same four courses were evaluated during both years. Minor changes were made in the student end-of-course critique data collection instruments, some TEDS sites improved their facilities, and overall management improved. The evaluation focused on six research questions. Data were collected and compared using analytical methods.

The six research questions can be divided into these major areas: comparison in learning achievement between those who took a resident course (nonTEDS) and those who received the course using TEDS and between face-to-face TEDS students and remote TEDS students; student's acceptability of the Teleteach method and scheduling; and cost comparison between providing resident instruction and using the TEDS.

Learning

The comparisons below involve only those in which student groups were from the same population according to demographic and pretest results.

In four instances where regular resident students (nonTEDS) were compared to students taught through TEDS, only one instance revealed a significant difference in learning. This difference favored the TEDS students.

In four instances where regular resident students (nonTEDS) were compared to resident (face-to-face) TEDS students, no statistically significant difference in learning was found.

Three separate comparisons involving TEDS taught students revealed two instances where the resident (face-to-face) TEDS students scored significantly higher than remote students. In the third comparison there was no statistically significant difference in learning.

Acceptance

Data revealed that the TEDS method was marginally accepted (53%) for full length courses (average length 89 hours). The system was highly accepted (83%) for the short courses (average length 3 1/2 hours).

In 1980 students in all time zones expressed dissatisfaction with the TEDS schedule. In 1981 students in the Eastern and Pacific Standard time zones expressed satisfaction with the schedule while Central and Mountain Standard time zone students remained dissatisfied. The reason for this change is not clear since the schedule itself did not change.

Table 1

Time Zone	Class Hours
Eastern Standard	1200-1600
Central Standard	1100-1500
Mountain Standard	1000-1400
Pacific Standard	0900-1300

Costs

Table 2 below reveals the TEDS costs and cost benefits realized in FY 80-FY 81.

Table 2

Costs
FY 80

System Cost		$272,421
Installation	$ 23,730	
Purchase	25,960	
Lease (Aug 79-Sep 79)	10,312	
Lease (Oct 79-Oct 80)	212,419	
Total	$272,421	

Cost Avoidance		$795,829
AFIT	$523,702	
Others	272,127	
Total	$795,829	

Cost Benefit		$523,408
Cost Avoidance	$795,829	
System Cost	- 272,421	
Total	$523,408	

FY 81

System Cost		$224,718
Lease (Oct 80-Sep 81)		

Cost Avoidance		$695,151
AFIT	$611,300	
Others	83,781	
	$695,151	

Cost Benefit		$470,433
Cost Avoidance	$695,151	
System Cost	- 224,718	
	$470,433	

FY 80 - FY 81

Cost Benefit		$993,841
FY 80	$523,408	
FY 81	470,433	
Total	$993,841	

System cost is actual amount paid. Cost avoidance was determined by computing the cost which would have been incurred if remote students/others had come to WPAFB. Cost benefit is the difference between system cost and cost avoidance.

Service

During the two year period, 2982 students and 1761 others were involved in the evaluation. The AFIT cost avoidance was computed only upon 1247 remote students.

Conclusions

Both comprehensive evaluations revealed that TEDS students achieve at least as well as students taught in residence at Wright-Patterson AFB for those courses included in the study. Students marginally accepted TEDS as a delivery method when lengthy courses were presented but were very receptive to the shorter courses. Acceptance of the TEDS schedule was mixed according to time zone. TEDS was shown to be a cost effective educational delivery system which afforded cost benefit to the Air Force of over $993,000 in two years.

The consistency of the findings substaniates the contention that teaching remote students via interactive audio and graphics is as effective as face-to-face instruction.

BIBLIOGRAPHY

1. Christopher, G. R. and Milam, A. L.; Teleteach Expanded Delivery System Evaluation. Air Force Institute of Technology, Wright-Patterson AFB Ohio, January 5, 1981.

2. Christopher, G. R. and Milam, A. L.; Teleteach Expanded Delivery System Evaluation. Air Force Institute of Technology, Wright-Patterson AFB, Ohio, December 15, 1981.

The Teleconferencing Resource Book: A Guide to Applications and Planning
Lorne A. Parker and Christine H. Olgren (eds.)
Elsevier Science Publishers B.V. (North-Holland)
© Center for Interactive Programs, University of Wisconsin-Extension, 1984

TELEMEDICINE IN NORTHWESTERN ONTARIO**

Christopher A. Higgins*

Earl V. Dunn#

David W. Conrath*

*Department of Management Sciences, University of Waterloo

#Department of Family and Community Medicine, University of Toronto

**This study was funded by the doctors of Ontario via PSI Foundation, Ontario

ABSTRACT

In this paper a slow scan telemedicine system designed to assist in remote health care delivery is described and the nature of its use is outlined. The system is operating out of six communities in northern Ontario, and two teaching hospitals in Toronto. At present it is being utilized approximately forty times per month. Usage of the system can be divided into three broad categories; medical consultation, education and social therapeutic contacts. Specific examples are given. After two years of operation the system has been accepted, usage is expanding and the communication mode is now considered an integral part of health care delivery. The equipment is operated by the regular health workers without expert help, and has been felt to be a significant contributor to better and more personal health care.

INTRODUCTION

For the past several decades increasing attention has been given to the problems of health care delivery to remote and isolated populations. Suggested solutions can be placed in one of three general categories: the use of non-physician providers, physician incentive plans and telemedicine. The latter of these, telemedicine, involves the use of telecommunications technology to assist in health care delivery. In this paper an operational telemedicine system is described and various aspects of the system are discussed.

The telephone and radio have been used for many years to assist in providing health care. In 1950 Greshon-Cohen and Cooley described their successful attempts to transmit X-rays via standard telephone lines. The results were satisfactory, but the cost of the equipment was excessive. During the next twenty years technological advances were made, and by the late 1960's several telemedicine projects had been initiated. The best known of these connected the Logan Airport in Boston with the Massachesetts General Hospital by means of a two-way broadband video system (Murphy & Bird, 1974).

Throughout the early 1970's there were many experiments testing telemedicine technologies (Park, 1974; Conrath et al, 1975). Unfortunately, most were experiences rather than experiments in the true sense of the word. Little was done in the way of evaluating the impact of the communications technology, and comparisons among alternatives were ignored. More recently, several studies have compared alternative communication modes (Conrath et al, 1977; Moore et al, 1975; Sanders et al, 1976). Each of these experiments came to the same basic conclusion: Increased sophistication of technology (and thus cost) gave little or no increase in the ability of the technology to assist in health care delivery. In other words, for most purposes, cheap, reliable technology was as good as the more expensive equipment available.

With these results in mind, it was now time to make some comparisons in a situation where the need was real. Consequently, we installed a slow scan television system tying together five isolated communities, a regional hospital and two tertiary hospitals. This system was to be compared with existing telephone and radio communication networks in terms of effect and costs versus benefits.

THE AREA

The Sioux Lookout Health Zone in northwestern Ontario provides health care to over 10,000 people living in an area of 285,000 square kilometers (twice the size of the state of New York). The people live in 27 small communities, none of which is large enough to support or justify a resident physician. For the most part, the only transportation between communities is by small bush aircraft, on floats in the summer and on skis in the winter.

Seven of the larger communities, with populations ranging from 350 to 1,000, have nursing stations with resident nurse practitioners who provide daily health care. These nursing stations have minimal laboratory and X-ray facilities and a small complement of beds so that patients can be kept under observation for short periods of time. However, for most investigations or hospital treatment the patient must be transferred to the Zone Hospital or to the 'south'.

The other smaller communities have indigenous health aides who provide emergency care and some daily health care. Their training is very limited and they work under the remote supervision of a nursing station. Nurses make regular bi-weekly visits to these communities, accompanied by a doctor approximately once a month.

The Zone Hospital, the medical and administrative center of the system, has a complement of four family physicians and the Zone director (traditionally a physician). In addition, specialists from the Universities of Toronto and Manitoba make regular visits to the area. The Zone Hospital physicians attempt to visit each nursing station twice a month and each other community monthly, although this is not always possible, weather being a significant factor. The distances between the health aide stations, the nursing stations and the Zone Hospital are not trivial, averaging about 75 kilometers from health aide to nursing station and about 250 kilometers from nursing stations to the Zone Hospital. Toronto is over 1500 kilometers away.

THE SLOW SCAN SYSTEM

The slow scan telemedicine system operates in eight locations: two in communities with health aides, three in communities with nursing stations, one at the Zone Hospital and two in Toronto teaching hospitals (one pediatric and one adult).

Each telemedicine unit consists of a speaker telephone (Northern Electric Companion II), a slow scan video system (Colorado Video), send and receive monitors (Electrohome), a video camera (Dage) and an X-ray view box. The hospital units are augmented with two store monitors so that several pictures can be viewed simultaneously. The slow scan video system consists of a frame storage device, a transmitter, a receiver and a modulator/demodulator. The camera is equipped with different lenses and has a zoom facility. It operates with simple room illumination although an X-ray view box is required for roentgenograms. The slow scan system is capable of transmitting black and white pictures over the telephone lines to another location (or locations using conference bridging) in 78 seconds. Common transmissions are of X-rays, E.K.G.'s, skin rashes and burns.

The system has been in operation since August, 1977, and by the summer of 1979 had been used on over 500 occasions (approximately 40 cases per month). It is also used for regular weekly educational programs. The equipment is operated for the most part by the regular health workers, that is, by the nurses, doctors and other indigenous people. There are no regular project staff at each location to maintain and operate the equipment. In spite of this lack of trained technicians and operators, there has been infrequent and short 'down time'. This is due to the simplicity of the equipment and the fact that there is a complete set of spare replacement parts to install when needed. Trouble shooting and exchange of equipment are accomplished by nurses and other non-technical people.

SYSTEM UTILIZATION

The system is used for three main purposes; medical consultation, education and social therapeutic contacts. The uses have been divided fairly evenly among the three areas, although recently the educational focus has become more active. With over 500 cases it would be impossible to discuss all aspects of the telemedicine system. However, we will outline the common applications in each area and give some examples.

Medical Consultation

The two most frequent types of consultations have been those between nurses and doctors and those between doctors and subspecialists. The majority of these have been for the transmission and interpretation of X-rays, but there have been consultations regarding E.K.G.'s, plastic surgery (burns), dermatology, physiotherapy, microscopic slides and dentistry. As long as there are no technical problems, the transmitted image has nearly always been satisfactory for the consultant. The transmission of data has facilitated discussion and clarified problems. There has been little concern regarding the quality of the transmission. It does take some time and familiarity with the equipment to be able to get an ideal picture but satisfactory transmissions are easy to obtain. In fact, there have been occasions when a nurse who has never used the equipment, on the instruction of the physician at another location has successfully transmitted X-rays.

The direct costs of these consultations are usually less than $10.00 for those between nurses and doctors, and less than $25.00 for those between doctors and subspecialists. Many of these exchanges have resulted in cost savings. To illustrate, let us go through a sample case. An individual fell from a snow machine and injured his leg. He was taken to the nursing station and X-rayed. There was an obvious fracture but the nurse was not certain whether she should transfer the patient to the Zone Hospital or to a larger center for special orthopaedic care. She contacted the Zone Hospital and arranged with the family physician to transmit the X-rays for his opinion. By agreement, 15 minutes later the nurse telephoned the telemedicine office where the physician had turned on the equipment and was prepared to receive a transmission via slow scan

video. Within 10 minutes the physician had received three views of the ankle and had ascertained that there was a trimalleolar fracture. This fracture could not be handled adequately at the Zone Hospital and so the patient was transferred directly to a larger facility for definitive management.

What were the costs and benefits in this case? The doctor and nurse took 15 minutes extra to make their decision. The telephone and slow scan transmission costs were about $3.50 (excluding capital costs). The patient had one direct airplane transfer rather than two trips. This was much more comfortable for the patient and saved about $150.00 in air fare.

Education

The educational potential of this telemedicine system is being explored and expanded. The first regular use of our system for formal education began in July, 1978, when weekly X-ray rounds for the staff at the Zone Hospital were initiated. Each week a senior radiologist in Toronto reviews X-ray films selected by the Zone physicians. While these rounds are designed for educational purposes they frequently result in changes in patient management. As many as 18 films have been transmitted and discussed within the hour demonstrating the capacity for rapid setup and transmission.

Once a month, using conference calls, a radiologist conducts similar more basic rounds for the northern nurses at the three nursing stations with the video system. Additionally since May, 1979, weekly E.K.G. rounds have been conducted from Toronto. There is more structure in the E.K.G. rounds as the cardiology staff in Toronto are attempting to review common problems in an ongoing way. These rounds which cost about $60.00 per hour are attended by both medical and nursing staff members. The system has also been used for seminars such as psychiatry and plastic surgery. Furthermore, the E.K.G. and X-ray rounds have been accepted by the College of Family Physicians of Canada as hour for hour required study credits for membership in the College.

The system has been used for inservice education. For example, a plastic surgeon in Toronto observed a burn

patient via the system. He was able to
ascertain the degree of the burn and
the need for skin grafting. He outlined
the technique for debridement and management of the burn. The discussion and
the video pictures were recorded and
later replayed for the nurses on the
ward caring for the child. This resulted in better management for the child
and was educational for the nurses.

There have been uses of the system
for patient education. The hospital
physiotherapist has used the system to
instruct the family of a stroke patient
on how to do the appropriate exercises
so that the patient could return to her
home. The nurse in the remote nursing
station observed the instruction and was
able to supervise the family's endeavours. The management was successful
with an earlier return than normal of
the patient to her own family.

Social Therapeutic Contacts

Admissions to the Zone Hospital
are especially difficult for the India-
ns because of their very close community and family ties. Patients and their
families are typically disturbed by the
separation and the distance. The telemedicine system is used on a regular
basis to lessen the isolation and loneliness felt by the hospitalized patients. For example, prenatal patients
who have been transferred to the Zone
Hospital for delivery have weekly sessions with their family back home.
They can both talk and view pictures of
each other. After a baby is born, a
picture of the child is transmitted
back to the family to aid in the early
bonding process.

The system has been used to lessen
isolation and fears related to being
away from one's family. For example, a
woman with several children was flown
from her community to the Sioux Lookout
Zone Hospital for investigation of several symptoms. Within two days the patient stated that all her symptoms were
gone. After talking to and seeing her
family via slow scan television she related that she had been worried about
her children. After seeing them she
was reassured about their well-being.
Because of this reassurance she admitted that her symptoms were still present and agreed to remain in hospital
for investigation.

Since the inception of this telemedicine project data have been collected on diagnosis, management and other
aspects of health care for all clinical
encounters in the Sioux Lookout Zone.
The 1978 encounters are summarized in
Table 1.

Table 1

Summary of 1978 Encounters

	Nurse	Health Aid	Doctor	Total
Patients Seen	33,162	9,536	5,151	47,849
Diagnoses	43,407	12,044	6,781	62,232
Referrals	1,560	472	257	2,289
Communi-cations	691	659	48	1,422
Transfers	603	181	143	927

In 1980, after two years' of data
have been processed, we will compare
those communities which have slow scan
television units with those serviced
by telephone and/or radio. Changes in
health care will be investigated and a
cost/benefit assessment related to the
telemedicine system will be carried
out.

There was initial resistance and
hostility to the system. However, the
results of a recent questionnaire indicate that after two years there have
been considerable attitudinal changes
and the system is now accepted.

CONCLUSIONS

The slow scan video system is feasible to use and provides a valued
service. The costs are not exorbitant.
It can be used for consultations, educational sessions and for reducing the

loneliness of isolation and separation. One of the more welcomed features of the system is the educational aspect which is continually expanding.

The present system is relatively inexpensive, the operating costs are low and the system can be maintained by the medical staff with minimal help from technical experts. There are cost savings but as yet a full cost-benefit analysis has not been done. Down time and equipment problems have been minimal. The quality of transmissions of pictures is adequate and the staff elect to use the equipment more than 40 times per month. Early indications are that the system alters health care and it appears that this is for the better.

BIBLIOGRAPHY

1. Conrath, D.W., Buckingham, P., Dunn, E.V. and Swanson, J.N. (1975) 'An Experimental Evaluation of Alternative Communication Systems as used for Medical Diagnosis', Behavioural Science, 20, 5, 296-305.

2. Conrath, D.W., Dunn, E.V., Bloor, W.G. and Tranquada, B. (1977)- 'A Clinical Evaluation of Four Alternative Telemedicine Systems', Behavioural Science, 22, 1, 12-21.

3. Gershon-Cohen, J. and Cooley, A.G. (1950). 'Telegnosis', Radiology, 55, 582-587

4. Moore, G.T., Willemain, T.R., Bonanno, R., Clark, W.D., Martin, A. R. and Mogielnicki, R.P. (1975). 'Comparison of Television and Telephone for Remote Medical Consultation', New England Journal of Medicine, 292, 729-732.

5. Murphy, R.L.H. and Bird, K.T. (1974). 'Telediagnosis: A New Community Health Resource', American Journal of Public Health, 64, 2, 113-119.

6. Park, B. (1974). An Introduction to Telemedicine: Interactive Television for Delivery of Health Services, New York City, The Alternate Media Center at New York University.

7. Sanders, J.H., Sasmor, L. and Natiello, T.A. (1976). An Evaluation of the impact of Communications Technology and Improved Medical Protocol on Health Care Delivery in Penal Institutions: Final Report. The University of Miami, Coral Gables, Florida.

The Teleconferencing Resource Book: A Guide to Applications and Planning
Lorne A. Parker and Christine H. Olgren (eds.)
Elsevier Science Publishers B.V. (North-Holland)
© Center for Interactive Programs, University of Wisconsin-Extension, 1984

DIVERSIFICATION IN THE USE OF SLOW-SCAN TELEVISION IN MAINE

Judith A. Feinstein
Director
Northern Maine RAISE
Presque Isle, Maine
and
Anne Niemiec
Director
Interactive Telecommunications Systems
Medical Care Development, Inc.
Augusta, Maine

INTRODUCTION

That there is a need for telecommunications in rural areas is becoming obvious. Distance, poor or nonexistent transportation systems, economic influences, and, in northern areas, long winters, are among the factors which contribute both to perceived and real feelings of physical and professional isolation and to the attractiveness of telecommunications as a resource. For a number of hospitals in northern Maine, it is a resource whose value can be demonstrated and which is now being shared with other community organizations and resources.

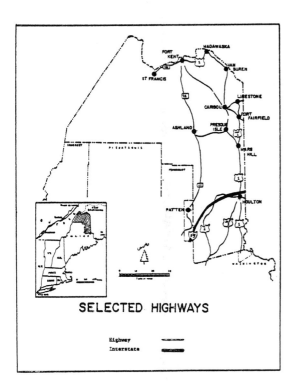

SELECTED HIGHWAYS

Highway   ——————
Interstate   ——————

Aroostook County, a political subdivision the size of the states of Rhode Island and Connecticut, has a population of some 95,000 people. Its largest city is Presque Isle, which has about 11,000 people and the major commercial airport and is geographically in the center of the County. From Presque Isle it is 165 miles to Bangor, the nearest big city, and 240 miles to Augusta, the state capital. Interstate Route 95 was extended relatively recently as far as Houlton, located at the County's southern end.

This relative isolation, as well as some occasional and very real limits on travel and accessibility, has contributed to a strong sense of identity among those who live and work in Aroostook County and, at the same time, has encouraged an equally strong tradition of self-reliance. The five hospitals in the County, all community-based, benefit from and contribute to that tradition.

The purpose of this paper is to critically describe the development of a process that began with hospitals connected by a full-motion video system, was adapted a few years later for hospitals connected by a slow-scan television system, and now is being used by a variety of organizations to meet information and training needs. This system is working to increase the awareness of the possibilities of telecommunications as a resource for development. Hudson et al stated that...

> to a considerable extent the slow development of rural telecommunications arises from an insufficient awareness of the role telecommunications can play in supporting such aspects of economic development as rural programs in agriculture, government administration, health, commerce, and transportation...[1]

The awareness of this role can be enhanced or expanded by demonstrating the benefits potentially available through the technology.

It is important that those benefits be clearly related to the particular environment and the needs of those who will be using the system. The potential benefits should make sense in terms of those needs. Factors we have found to be critical include not just cost-effectiveness but also local perceptions of the appropriateness of the technology to their particular needs.

That the hospitals are community-based and responsive to the needs and interests of their respective communities has been an important factor in the increasing diversification of the slow-scan telecommunications system. People tend to view the hospitals not only as places to receive health and medical care but also as general community resources. Public access to meeting rooms and now the willingness of the hospitals to facilitate use of the system by community groups, represents a clear manifestation of this dynamic.

The hospitals themselves have a long history of cooperation. The chief executive officers and several other administrative groups meet regularly for joint planning and problem-solving purposes. In particular, these hospitals, for the past nine years have been supporting a regional shared educational service whose goal it is to develop and coordinate continuing education opportunities for health care personnel. Through this agency, Northern Maine RAISE (a Regional Approach to Improved Health Services through Education), the value of cooperation through shared services has become a fact to the participating institutions.

BACKGROUND

The Aroostook County Telecommunications System (ACTS) is a slow-scan television network which connects five hospitals in Aroostook County and provides for an interconnect with the Central Maine Interactive Telecommunications System (CMITS). The system allows both audio conferencing and still-image video conferencing. The CMITS is a two-way broad band microwave television system which connects participating health institutions in Lewiston, Augusta, Togus, and Waterville. Merging the slow-scan system with the broad band television system in central Maine demonstrates an economical method for improving educational offerings and medical care as well as saving travel and energy in remote areas.

The Department of Education's (formerly the Department of Health, Education and Welfare) Telecommunications Office granted

Medical Care Development, Inc. (MCD), a non-profit health care research and development organization, monies to initiate the slow-scan telecommunications project in Aroostook County. The equipment was assembled by Lake Systems Corporation in Newton, Massachusetts, using the specifications provided by MCD's telecommunications personnel. The audio equipment consists of a Darome miniconvener and the slow-scan equipment utilizes slightly modified Robot slow-scan transceivers. The entire system is connected using dedicated voice-grade telephone circuits and a telephone company bridging device.

The system provides: (1) talk back and interaction with program originators in the central portion of the state, (2) the ability to utilize the slow-scan equipment for intra-county conferencing and programming, and (3) interconnection with the CMITS so that Aroostook health professionals may lecture and participate more fully in the educational activities outside the County.

The participants are now contributing to the cost of the system. For a description of how this system was established see "A Case Study: An Inexpensive, Easy to Operate, Slow-Scan Telecommunications System in Northern Maine," by Robert Ellis. The system did not become fully operational until the summer of 1981.

DIVERSIFICATION PROCESS

This telecommunications system was constructed for the primary purpose of improving health services by enhancing communication in the rural communities of Aroostook County. Development of the use of the system in the hospitals was hampered by the delays in its becoming fully operational, but both the number of hours and range of use is increasing. Use is now averaging 40 hours a month. This system is being used for already established groups, for the establishment of new groups who would not normally connect with one another, for organized educational uses, for informal peer exchange, for training for ongoing planning, and for single project planning.

Once in place, funds were sought by MCD and received from the Farmers Home Administration to plan for increasing the range of use of this slow-scan television system for community and business development in Aroostook County. The form this takes varies depending on communications needs. It includes meetings between individuals, meetings of groups, formal and informal learning sessions, and consultations.

Hospitals are, along with other organizations, looking to partnerships to help defray rising costs. Planning for diversification of the applications of this telecommunications system allows better utilization of limited resources in a variety of ways and lowers unit costs for use of the system.

Such diversification is a further development of a process of planning by participants which has successfully been utilized by CMITS and now with ACTS. Planning is done by groups composed of one or more representatives of a particular discipline or professional group from each community. These groups meet over the teleconferencing system to discuss possible joint programs based on needs they have identified. The group is then involved in the design, implementation, and evaluation of the program which is facilitated through the system.

The objectives of this diversification process are an extension of the objectives developed for the CMITS and ACTS which center on the cooperative sharing of resources.

The objectives are:

1. To expand intercommunity planning and development among the public and private organizations in rural Aroostook County through increased use of an innovative two-way slow-scan telecommunications system.

2. To actively involve business representatives, public officials, medical and other human services representatives, and economic, environmental, and health planners in discussions of cooperative projects that can be more efficiently managed through the use of the telecommunications system.

3. To introduce developmental initiatives and programs of state and federal agencies to Aroostook representatives through the use of the telecommunications link between Aroostook County and Augusta, Maine, the location of the state capital.

The first step in increasing the range of use outside of the hospitals was to compile lists of potential users and this was done in a variety of ways, including asking people who were already involved with the system. These lists include community groups, businesses, service agencies, and local governments who were then notified of the availability of the telecommunications system for joint planning. Planning committees are being or have been formed and meetings arranged over the telecommunications system involving people with interests in similar problems or developmental ideas for the different activities. Needs are identified that could be met by the cooperative sharing of resources.

As part of the planning process, demonstrations are arranged to aid in considering how the system can meet the specific needs identified by each group. Prior to or during initial meetings, an analysis of current communications patterns takes place. The demonstrations include background on the equipment and a discussion of the different configurations possible, such as site-to-site private communications and multi-site conferences. Ways that the video and audio systems can be used to improve communication are discussed including the use of graphics and other media techniques such as taping and replay. Demonstrations are arranged for specific agencies. Open house demonstrations are held for one hour every other week to which anyone with an interest is invited. These introductory procedures have led to new uses of the system by agencies such as the Department of Education and the Bureau of the Elderly.

Two different types of planning activities in the hospitals, coordinated by RAISE, successfully utilized the System during the past year; these examples illustrate the use of this technology as a planning tool.

The first of these involved the hospital diabetes educators, most of whom are nurses with multiple responsibilities, and one pharmacist. The project actually began in the spring of 1981, when the group met face-to-face to discuss the needs for and content of a regional education workshop geared for staff nurses on delivering diabetes education to inpatients. As discussion progressed, slow-scan was identified as the medium of choice for this sort of presentation, for three reasons: it would eliminate travel and expense; weather would not be a factor; and a tape could be made for later use. A number of the educators volunteered to be involved in the planning of the program, and meetings were held via slow-scan. These subsequent meetings led to the development of a teaching manual. Both the group and RAISE staff felt that use of slow-scan as a planning and coordinative tool was very efficient and effective and that it had been a positive experience. The group decided to plan future meetings on the system on a quarterly basis for sharing new information, updating the manual, and general discussion.

A second user group was developed in response to requests for educational programming for Central Sterile Supply department personnel. The number of people involved is relatively small, but they have unmet continuing education needs that were identified in their first meeting. Reaction was positive to the possibility of using the system to cooperatively meet these needs. What resulted was a schedule of bi-monthly meetings to allow for continuing discussion, information-sharing, and problem solving. The participants have become increasingly comfortable with using the system and reported their last meeting, in early February, as highly interactive and very productive.

An additional example of the use of this technology was planning for the purchase of a mobile CAT scanner that would be used by all the hospitals. Programs are now being planned to enable a hospital in central Maine which has had a unit for some time to share what they have learned from using it.

The system has been used by the Bureau of the Elderly to conduct training in the use of a new assessment form. Sharing among area agencies for the aging is being discussed.

The system has also been used to meet social needs. The mother of one employee at a hospital in northern Maine was transferred to a hospital in central Maine. Because the employee had very young children, she was not able to travel to central Maine to visit her mother. Worry about her mother prompted her to see is she could use the system. It was arranged and she brought her two children and grandmother to the hospital in northern Maine to talk with her mother; they were

pleased at the visual assurance that all was well.

PROCESS AND PROTOCOLS

Slow-scan teleconferencing creates a unique relationship between voice and image. They are not as tightly bound as over a full video communications link or face-to-face meeting and can be used to supplement or counterpoint one another in novel ways. In addition, because one does not have continuous nonverbal feedback as at a face-to-face meeting, one needs to do some things differently. One's skills (or lack thereof) at conducting meetings is more obvious.

More structure is needed in running meetings on a slow-scan system. As a result, meetings tend to be shorter with more getting accomplished in less time. There is still some social chit-chat but less then at a face-to-face meeting. To take advantage of the psychological "presence" factor possible with the technology, people are encouraged to send a picture when they speak.

As with a video link of any type, care must be taken to keep graphics readable. Such care usually results in more impact. While written guidelines about visuals are available, people often need to experience their importance firsthand. Examples of good visuals are an integral part of our demonstrations, so that initial planning with first time users includes discussion of appropriate visuals which can enhance their meeting or presentation.

Disadvantages may include unseen interruptions and side conversations at sites not being heard or seen on the monitor. Careful planning and some basic protocols can help to mitigate against these sorts of disruptions. For example, the presenting speaker or meeting chairperson can "check in" with participants at regular intervals to keep them involved in the meeting. When there are several people at one site who begin discussing the business of the meeting among themselves, it is helpful if someone lets the entire group know what's going on, and that the sense of the discussion will be passed along to them.

At the same time, however, when there are two or more participants at one site (particularly when they represent the same institution), the system affords them the opportunity to check responses privately or plan and refine strategies. Again, silence may be a problem, so that letting others know that a quick conference is taking place is usually appreciated.

IMPLEMENTATION ISSUES

The support and interest of the chief executive officers of the hospitals has been crucial to the success of this system, corroborating research in business which has demonstrated the importance of the visible involvement of top-level management in the introduction of teleconferencing.[2] They use it for regional administrative meetings, particularly in the winter, or for ongoing discussion of specific issues. Their support and interest demonstrate a commitment which needs to be visible to employees of the hospitals.

One strategy for implementation has been to introduce new employees to the system and ask them to make short presentations both as a way of their meeting people and introducing to others the resources they bring to the hospital. One reason for paying particular attention to new people is that there is no preconceived attitude against the system to overcome. Because of the problems and time involved with implementation, an attitude of resistance is not uncommon among some long time employees.

As with many other new technologies, peer influence and involvement has proven to be a factor. Positive experiences help create and maintain an open attitude and enthusiasm about using the system and encourage further exploration of its potential capabilities. Seeing peers achieve a level of comfort and competence also helps the new user gain a sense of his or her own ability to use the system. This effect was clear during the meetings of the Central Sterile Supply personnel group, where initial interaction was among one or two individuals and support from RAISE and MCD; over time, most of the others have become active participants.

It is clear that the first uses are critical. Another example from the hospitals can illustrate: in the fall of 1980, a group of operating room and recovery room nurses were brought together on the system to plan a training workshop. This happened during the time before the system became fully operational. The system performed unevenly through several meetings and a majority of participants were dissatisfied. Although they did continue to meet for some time, the planning for the workshop was accomplished through a number of small, in-person,

on-site meetings, involving considerably more time and travel for the workshop coordinator.

On the other side, we have the experience of the demonstration arranged for Cooperative Extension Service Staff of the University of Maine. It was a particularly interactive session in which there was no awkwardness or perception of the technology as a barrier. Verbal and visual exchanges were constant and there was almost the same feeling as sitting around a table.

It does take time for people to get used to the equipment. If anything goes wrong, especially the first time people use the system, then it takes longer. As with many other things, where the need is greater, the resistance is less. There is a range of responses to problems of technological difficulties. Some will react immediately, with no thought of returning and make statements like, "I knew it wouldn't work anyway." It may well be part of a larger attitude or fear of technology. Many times what we call a "people problem" (someone incorrectly using the equipment) is attributed to the technology. The technology gets blamed for problems, no matter what the cause, because for the casual user, what is perceived is that the system did not work. Others, however, are able to manage the problems, and somehow accept (perhaps on faith) that the difficulties are temporary. A problem with the camera focus control during the demonstration for the Department of Education is an example. People were obviously disappointed with the unfocused, blurry pictures, but did not dwell on the problem and went on with their business.

As part of the training process, people are encouraged to have fun with the equipment. Playing with it decreases the time it takes for people to become comfortable with the equipment and allows them to get over the initial fear of it.

ACCESS TO RESOURCES

Most of the resources shared thus far have been local, but this slow-scan technology allows access to resources in other parts of the country using telephone lines with other locations that have Robot transceivers or to create audio-only links with any location. With this ability we have accessed audio only and audio and video programs from California, North Carolina, Maryland, and Illinois. Gaining access to other rural areas of the country to share

methods of problem solving and information is
being explored currently with Missouri, North
Carolina, and neighboring New Brunswick.

The system is a member of the Association
of Hospital Television Networks, a national
consortium of 30 regional television networks
providing educational services to staff and
patients of over 900 hospitals. Association
activities include development of programming
distributed nationally by satellite,
developing mechanisms for sharing among
member networks, and consultation with
producers on educational design and network
services. In the past year we have received
eight satellite programs, including an
international one from Florence, Italy.

Although our initial focus in northern
Maine has been on slow-scan video, we are
exploring a variety of technological options,
realizing that different options may be more
appropriate for certain needs. We are
concerned with technological fit; we work
with individuals and groups to choose and
then effectively use the technology to meet
their needs.

REFERENCES

1. Hudson, H.E., Goldschmidt, D., Parker,
    E.B., & Hardy, A. The Role of
    Telecommunications in Socio-Economic
    Development: A Review of the Literature
    with Guidelines for Further Investi-
    gations. Keewatin Communications, Palo
    Alto, California, May 1979.

2. Green, D., & Hansell, K. Teleconfer-
    encing: A New Communications Tool.
    Business Communications Review. March-
    April 1981, 10-16.

The Teleconferencing Resource Book: A Guide to Applications and Planning
Lorne A. Parker and Christine H. Olgren (eds.)
Elsevier Science Publishers B.V. (North-Holland)
©Center for Interactive Programs, University of Wisconsin-Extension, 1984

TELECONFERENCING AT HONEYWELL

David C. Prem
Senior Project Manager
and
Susan M. Dray
Principal Human Factors Productivity Engineer

Honeywell Inc.
Minneapolis, Minnesota

Introduction

Two years ago at this conference, a presentation was given detailing the development of an internal teleconferencing program for Honeywell Incorporated. Today's presentation presents an update on the Honeywell program. Specifically, this presentation addresses the following topics:

o Review of the beginnings of the program.

o Expansion of the program.

o Promotion of the program.

o Follow-up activities.

o Development of additional applications.

o Future plans for teleconferencing.

Review of Program's Beginning

Honeywell is a multi-national company, with operations in several dozen countries scattered across the globe. World-wide revenues were $5.5 billion dollars in 1982. Current world-wide employment with Honeywell is approximately 94,000. Honeywell is active in several different market areas. A representative sampling of products Honeywell produces includes information systems, avionics instrumentation, environmental control systems for buildings, process control systems for industry, medical instrumentation, semiconductors, and electronic components.

Honeywell operates on a decentralized basis. Each operating division, there are some 30 plus divisions, operates as an independent business with bottom line profit and loss responsibility to the Corporate Management. Each division provides the bulk of its own G & A support. A relatively small corporate staff provides support only in those areas where it is more effective to

do so on a centralized basis than on a decentralized basis. Examples of such centralized support include legal counsel, telecommunications services, real estate support, and the administration of field sales offices.

In the late 1970's, Honeywell, like other companies, was hit by rapidly increasing travel costs. Costs increased by approximately 20% annually over a period of several years, with projections at that time for similar annual increases over the forseeable future. Teleconferencing was identified as one potential means for regaining a measure of control over these rapidly escalating costs. Management directed the formation of a pilot project team, and charged the team with the objective of developing an implementation plan for a teleconferencing program at Honeywell.

After a detailed study of travel patterns within Honeywell, and of the types of meetings that were held, the team recommended the construction of dedicated teleconference rooms at six major Honeywell facilities. The team further recommended non-dedicated facilities at those locations where projected teleconference volumes were not high enough to justify dedicated rooms. Specific locations for dedicated rooms were determined through the detailed study of traffic patterns, and the possibilities of travel expense reductions between specific destinations.

The six dedicated teleconference rooms are similar in design, construction, and furnishings. The rooms were constructed using guidelines for audio teleconference rooms developed for Honeywell by AT&T. The rooms each have minimum dimensions of 16 x 20 feet and each is capable of comfortably accommodating 10 to 12 active meeting participants.

All rooms are located in areas which are convenient and accessible to users, and in relatively quiet areas within a building.

The rooms are located away from building
noise sources, and away from outer walls.
The rooms are not in areas which could be
considered to be on someone's personal turf
(e.g. executive row). Furnishings in the
rooms are comfortable, but not opulent. The
intent is for the facilities to be
comfortable so that meeting participants
enjoy using the rooms for meetings, but not
so plush that the participants feel
uncomfortable in using the facilities.

All rooms are equipped with high quality
audio systems. All rooms have an overhead
transparency projector and a retractable
screen. A high speed facsimile transceiver
and a transparency maker are available for
use in ante-rooms of the facility.

Program Expansion

During the past two years, the Honeywell
teleconferencing program has been a success,
and has expanded in several areas.

Three additional dedicated
teleconference rooms have been constructed
and five more rooms are currently on the
planning boards. The new and projected
dedicated rooms are or will be construceted
using the specifications developed by AT&T.
Usage of the existing dedicated rooms has
been continually increasing. In 1982,
average monthly usage of the rooms increased
by 35%. Monthly usage of the nine dedicated
rooms currently totals 350 hours. Average
monthly travel cost savings through the use
of the rooms amount to $150,000. Each of
the dedicated rooms has recovered the
initial investment of $25,000 to $30,000
through travel cost savings over period of
six to nine months.

Teleconferencing activities using
non-dedicated facilities have shown
significant increases over the past two
years, both in number of locations using
portable conferencing equipment and in the
frequency with which the equipment is used.
Although the exact number of portable
conference units currently in use has not
been determined, the number of locations
using this equipment is increasing based on
feedback we continually receive from the
field. In addition to increases in domestic
usage, locations in Europe and Mexico
currently use portable teleconferencing
equipment on a regular basis for meetings
with Honeywell locations in the U.S.

The frequency of multi-point
teleconferences has increased dramatically
over the past two years. Multi-point

teleconferencing, as measured by billings
from the vendor for services, increased by
65% during 1982. The demand for multi-point
teleconferencing has increased to such a
level that it is now cost effective to
provide internal facilities for multi-point
conferences. A 20 port Darome conference
bridge has been installed in Honeywell's
Corporate Headquarters and a multi-point
conference bridging service is being offered
internally commencing April, 1983.

Initially, each room was equipped with a
Rapifax 100 model high speed digital
facsimile machine. As a cost saving
measure, these units were replaced with
Group II machines which transmit at a speed
of three minutes per page versus 30 - 40
seconds per page by the Rapifax units.
Usage of the machines plummeted. A
transmission speed of three minutes per page
has been found to be unsatisfactory in the
environment of a live teleconference
meeting. The Group II machines have
recently been replaced by 3M Model 9140
Group III machines.

In 1982, slow Scan TV (SSTV) systems
were installed in four of the dedicated
teleconference rooms. The systems are
produced by US Robot, and each consists of
several cameras, monitors, and a control
unit. One camera is mounted in the ceiling,
and is used for picturing graphic displays,
manufactured piece parts, diagrams, etc.
Other cameras are strategically placed
throughout the room to capture pictures of
meeting attendees, the moderator, and
writing on the flip chart, or writing on the
front writing board. Up to now, usage of
the SSTV systems has not met our original
expectations. User resistance to usage of
SSTV results from the relatively few
locations that have SSTV, the requirement
for preparing special graphics used with the
ceiling camera, the general satisfaction
with audio teleconferencing, and a lack of
expertise in using the slow scan equipment.
We continue to work with current and new
users of teleconferencing to further orient
the users in the operation of the SSTV
equipment and in the identification and
development of teleconferencing applications
for which SSTV can be used.

Teleconference Support and Promotion

One of the reasons the teleconference
program at Honeywell has experienced the
acceptance and enjoyed the success it has,
is because of a comprehensive program of
support and promotion for teleconferencing.
The support program was developed very early

during the implementation of the
teleconferencing program. A comprehensive
support effort for each dedicated room
commences prior to a room becoming
operational. A typical support effort
includes the following:

o An orientation on teleconferencing
  presented on an as requested basis for
  employees new to teleconferencing.
  When a new room becomes operational,
  there is a heavy demand for such
  orientations. This demand decreases
  after a few months.

o Training support in the operation of
  equipment in the teleconference room
  is provided as part of the orientation
  session.

o A room coordinator at each location
  with responsibility for:
  - scheduling the room,
  - insuring that the room is kept
    clean, and
  - charging the users for use of the
    room through an internal
    charge-back system.

o Maintenance of the equipment in the
  room. Proper maintenance of the
  equipment minimizes the probability of
  equipment malfunctions. Equipment
  breakdown during a teleconference
  session is extremely disruptive and
  can result in the cancellation of the
  session.

o Providing operational assistance in
  the event that session participants
  need such assistance. Such help is
  especially important to new or first
  time users of teleconferencing.
  Should new users have bad experiences
  during an initial session, it is
  unlikely they will use
  teleconferencing again Conversely, if
  the participants initial experiences
  are positive, it is likely that they
  will become advocates of
  teleconferencing.

The process of promoting
teleconferencing throughout an organization
the size of Honeywell is significant, and
one that takes a considerable period of
time. When a new service, such as
teleconferencing, is introduced to an
organization, a comprehensive promotion
campaign serves a dual purpose; one, it
makes the using community aware of the
service, and two, it identifies potential
applications for the service.

There are several different methods of
promotion. An effective promotion campaign
will simultaneously employ several of these
methods simultaneously. The various
promotion methods, through a synergistic
process, reinforce each other in getting the
desired message across to the potential
using community. The following methods of
promotion are either currently being used or
have been used:

o A column addressing teleconferencing
  topics appears in the quarterly
  organizational newsletter.

o News items about teleconferencing
  applications appear periodically in
  the company newspaper.

o Flyers addressing a particular aspect
  of teleconferencing are produced on a
  quarterly basis for posting on
  bulletin boards.

o Ticket inserts suggesting the use of
  teleconferencing as an alternative to
  traveling to on-site meeetings are
  placed in ticket folders of employees
  traveling on company business.

o Presentations about teleconferencing
  are given whenever requested. These
  presentations are tailored for the
  audiences addressed.

o An orientation video tape has been
  produced. The ten minute tape is
  available for showing to anyone
  wishing to see it.

o First time users of teleconferencing
  are given free usage of a
  teleconferencing room for their
  initial session.

The teleconference coordinators,
exchange information on specific promotional
activities at their respective locations
during regularly scheduled quarterly
meetings.

Follow-Up Activities

A comprehensive follow-up program has
been in effect since the early stages of the
teleconferencing program. The follow-up
program consists of the following:

o Moderator's check sheet turned in at
  the completion of a conference.

o Monthly reporting by coordinators on
  conference room usage and estimated
  travel cost savings.

o User questionnaires on reactions to effectiveness of teleconferencing, facilities, and support.

We have found the information obtained through the follow-up activities important for the following reasons:

o Usage patterns of the dedicated rooms are accurately monitored.

o Trends in usage patterns are identified.

o Contingency plans based on these trends are developed in a timely fashion.

o Reports to management on the effectiveness of teleconferencing, and the cost savings resulting from usage are prepared and circulated. In 1982, we averaged a monthly cost savings of $150,000 across the nine rooms. Information of this type greatly increases management's awareness of the value of teleconferencing.

o Helps insure that the dedicated rooms are being effectively used, and helps identify where corrective action should be taken in the event that the rooms are improperly used.

A survey of teleconferencing users was conducted during the Fall of 1982. The purpose of the survey was the determination of the reactions of teleconferencing users to both the process and to the facility. 53 users from five different locations returned completed questionnaires. About half of the respondents had used teleconferencing more than ten times. They were classified as "experienced users." The rest of the respondents were equally split between new users (2 - 5 times) and intermediate users (5 - 10 times).

We used the classification based on user experience in order to analyze answers to the first six questions which dealt with ease-of-use issues. Questionnaires were divided and analyzed by location for questions 7 through 12. A copy of the questionnaire is included in Appendix 1.

Most respondents felt that teleconferencing facilities are "easy" to "very easy" to use. As we expected, intermediate and experienced users tended to rate the facilities easier to use than new users, probably due to increased practice and familiarity. Experienced and intermediate users also rated the overall

ease of teleconferencing higher than new users.

Most respondents also felt that teleconferencing works well. In general, the effectiveness of teleconferencing was rated "the same as" the effectiveness of face-to-face meetings for the following:

o staff meeting

o project reviews

o meetings with people from more than one location

o meetings where participants know one another

However, teleconferencing was rated as "worse" than face-to-face meetings for meetings with conflict or those meetings where participants were unknown to each other. Relatively few respondants had participated in training or marketing/sales meetings using teleconferencing.

We analyzed the remainder of the questionnaire by location rather than by user experience level. At least 90% of the respondents rated the rooms as visually pleasing and the furnishings "comfortable." The respondents felt the following would make the room more useful, comfortable, or appealing:

o electronic blackboard (17)*;

o full video (16);

o clocks (13)

o plants (12);

o prints on the walls (10)

*Number of respondents.

Four-fifths of the respondents had participated in multipoint conferences (which necessitated use of a bridge), and half of these had participated in confidential discussions during the course of a multipoint meeting. Only 20% of those participants involved in confidential meetings felt there had been a potential security problem.

The training for users appears to be both widespread and totally adequate. Only one respondant of those who had received training felt that it had not been adequate. However, the location with the largest number and proportion of respondants who reported that they had not received any training, was the location where ease-of-use, efficiency, comfort, and security were all rated low. This suggests that training is critically important not only because it teaches people how to use the equipment, but also because the lack of training negatively influences people's perceptions of the technology.

New Application (Teletraining)

One application of teleconferencing that, until a few months ago, had not been pursued is teletraining. In the Fall of 1982, we arranged for orientation sessions on applications of teleconferencing to training. Training managers from different Honeywell locations were invited to and attended one or more of these sessions.

One of the results of these sessions was the realization that not only was there a high level of interest in teleconferencing on the part of the Honeywell training community, but some training organizations had started planning for the implementation of teletraining. Because Honeywell is decentralized in its operations, it became quickly apparent that unless some form of centralized coordination with respect to teletraining was established, various training organizations would be duplicating efforts and would not likely benefit from the experiences of others. In order to avoid this situation, a steering committee for teletraining has been established, training organization have been advised of its formation, and have been invited to participate. The committee was established for the following purposes:

o Share information on teletraining applications.

o Share information on successes as well as failures.

o Share information on resource availability.

o Share information on course development.

o Identify teletraining applications which cross organizational lines.

o Develop a pool of teletraining expertise.

o Share information relating to costs of teletraining.

One of the larger training organizations within Honeywell has inaugurated a pilot project to evaluate the effectiveness of teletraining in its training operations. Its plans call for the use of the Gemini Electronic Blackboard in two locations and associated TV monitors in four locations. Should this pilot program be successful, similar facilities will be established in additional locations. Both the curriculum and the number of locations participating in teletraining will be expanded.

Future Teleconferencing Plans

Honeywell's teleconference program has thus far concentrated on audio, audio plus graphics, and SSTV conferencing. For the near-term future, continued growth in these forms of teleconferenncing is anticipated.

Like many other organizations, Honeywell has considered full motion video. On two occasions Honeywell conducted ad hoc teleconferences employing broadeast full motion video and two way audio. Both conferences were used to introduce new products. There will likely be additional ad hoc teleconferences of this type in the near future.

The implementation of full motion video in Honeywell has been considered. Implementation of full motion video has for the present been deferred because:

o Costs of full motion video are such that its use within the Honeywell environment would not now be cost effective.

o Potential users of full motion video have not been clearly identified.

o Current and future advancements in technology are likely to lower the costs of full motion video teleconferencing.

One of the technological advancements which appears promising is the use of T-carriers as the transmission facility for full motion video. The T-carrier used in conjunction with appropriate instrumentation is theoretically capable of transmitting what appears to the end user to be live or full motion video. This transmission facility has a considerably narrower band width than what is currently used for full motion video. Plans are currently being formulated for the implementation of this capability, although an implementation timetable has not been set.

Conclusion

The Honeywell teleconferencing program is currently in its third year of existence. The program has a proven track record of success. Both usage and acceptance are increasing rapidly. The company is realizing net monthly cost savings of approximately $125,000 through the use of teleconferencing. Future rates of savings are likely to increase as the use of teleconferencing increases throughout the company. Honeywell management is aware of the teleconferencing program, pleased with the progress of the program, endorses the use of the service, and is encouraging the planning for future enhancement and expansion of the program.

APPENDIX I

TELECONFERENCING QUESTIONNAIRE

We are gathering some information on how people like using the teleconferencing facilities available at Honeywell. We need your help. Please fill out the following questionnaire and return it to your teleconferencing coordinator. We do not need your name; we want this to be anonymous. Thank you!

1. What is your location? _____

2. How often have you used teleconferencing? (Circle one)

    a. Never b. Once c. Twice to 5 times d. 5-10 times e. More than 10 times

3. How easy do you think the equipment in the conference room is to use? (If you haven't used a particular piece of equipment, please put "N/A" on the line.)

    ____ Microphones

    a. Very easy b. Easy c. Neither easy   d. Difficult e. Very difficult
                                  nor difficult

    ____ Phone hookup to activate the room

    a. Very easy b. Easy c. Neither easy   d. Difficult e. Very difficult
                                  nor difficult

    ____ Overhead projector

    a. Very easy b. Easy c. Neither easy   d. Difficult e. Very difficult
                                  nor difficult

    ____ Facsimile machine

    a. Very easy b. Easy c. Neither easy   d. Difficult e. Very difficult
                                  nor difficult

    ____ Meet-me bridge (for multi-point calls)

    a. Very easy b. Easy c. Neither easy   d. Difficult e. Very difficult
                                  nor difficult

    ____ Camera controls (where applicable)

    a. Very easy b. Easy c. Neither easy   d. Difficult e. Very difficult
                                  nor difficult

    ____ Monitors for slow scan (where applicable)

    a. Very easy b. Easy c. Neither easy   d. Difficult e. Very difficult
                                  nor difficult

4. Overall, how easy do you think the teleconference room at your location is to use?

    a. Very easy b. Easy c. Neither easy d. Difficult e. Very difficult
                                  nor difficult

5. Overall, have you found that teleconferencing works well?

    a. Always b. Usually c. Sometimes d. Rarely e. Never

6. For the following types of conferences, does teleconferencing work better than, the same as, or worse than a traditional face-to-face conference? Circle N/A if you haven't participated in such a meeting.

Staff or other regular meeting	Better	Same	Worse	N/A
Project review/status meeting	Better	Same	Worse	N/A
Training meeting	Better	Same	Worse	N/A
Marketing/sales meeting	Better	Same	Worse	N/A
Meeting with people from more than one location	Better	Same	Worse	N/A
Meeting with conflict	Better	Same	Worse	N/A
Meeting with people you know	Better	Same	Worse	N/A
Meeting with people you don't know	Better	Same	Worse	N/A

7. How comfortable is the room?

    a. Very   b. Comfortable c. Neither d. Uncomfortable e. Very
      comfortable   uncomfortable

8. How visually pleasing is the room you usually use?

    a. Very   b. Pleasing   c. Neither   d. Unpleasing e. Very
      pleasing   unpleasing

9. What additions, if any, would make the room more useful, comfortable and appealing? (Check all that apply)

___ Plants

___ Prints or paintings on the wall

___ More comfortable furniture

___ Better lighting

___ Clocks with time in different zones shown

___ Electronic blackboard or writing tablet

___ Slow scan

___ Full video (not slow scan)

10. Have you ever held a multipoint teleconference?
___ Yes   ___ No

If so, did you discuss any confidential or proprietary information?
___ Yes   ___ No

Did you feel security was a problem?
___ Yes   ___ No

11. Did you get any training in how to use your teleconferencing room?

___ Yes   ___ No

If so, did you feel it was adequate?

a. Totally b. Partially c. Not at all.

12. Any comments?

The Teleconferencing Resource Book: A Guide to Applications and Planning
Lorne A. Parker and Christine H. Olgren (eds.)
Elsevier Science Publishers B.V. (North-Holland)
© Center for Interactive Programs, University of Wisconsin-Extension, 1984

TELECONFERENCING IN WISCONSIN:
ADDING FREEZE-FRAME HIGHLIGHTS 18TH YEAR

Marcia A. Baird
Associate Director
Instructional Communications Systems
University of Wisconsin-Extension
Madison, Wisconsin

It's well known that, in Wisconsin, teleconferencing is viewed as an important educational delivery tool. For 18 years the interactive audio and audiographic networks have linked rural and urban areas, fulfilling the Wisconsin Idea by making the University System's research, teaching and public service resources available to all state residents. During the years we've had many onlookers from other educational groups and businesses, and many of these have followed in our teleconferencing footsteps.

In 1982, and now in 1983, teleconferencing in Wisconsin continues to be in the spotlight.

During this time we've had two special challenges: one with a new 26-site freeze-frame network, the other with a Wisconsin Telephone Company proposal that could significantly increase our private line network costs.

It's fair to say that these challenges are not very different from those faced by a growing number of organizations. Increasingly, new products and enhanced teleconferencing systems are being put into place. The timetable to install and troubleshoot the hardware is frequently as incredible as ours was. The training of instructors and meeting leaders on how to operate the equipment and, more important, how to design a successful

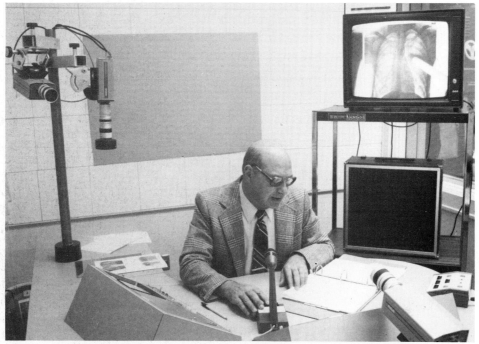

Instructors using Wisconsin's freeze-frame network
can originate programs from any of the 26 statewide
sites. Still black-and-white pictures are transmitted through the dedicated telephone network and
displayed on television monitors at all sites.

Teleconferencing: Wisconsin Style

Today, more than 40,000 Wisconsin citizens--mostly adults--participate annually in teleconference courses near or in their home communities. Below, the teleconferencing systems and audio services operated by the Instructional Communications Systems unit of the University of Wisconsin-Extension.

EDUCATIONAL TELECONFERENCE NETWORK (ETN)
ETN links 200 conference rooms at courthouses, UW campuses, hospitals and libraries with a two-way dedicated audio system. Participants at each site hear the program through a speaker and use tabletop microphones to ask questions or make comments. Everyone can talk with each other as if they were in the same room. Programs can originate from any network site or any telephone in the world.

Continuing education courses are offered in such areas as education, business, agriculture, nursing, music, library science, communications and law. ETN also is used to link administrators, faculty and community-based staff across the state. More than 300 programs are offered annually, totalling approximately 1,900 hours.

STATEWIDE EXTENSION EDUCATION NETWORK (SEEN)
SEEN links 26 statewide sites with two-way audio and freeze-frame video. Installed during 1982, freeze-frame technology allows instructors to send still pictures of themselves, students, charts, slides, models and hand-drawn graphics to all sites. Again, microphones allow easy two-way interaction among all participants. Programs can originate from any network site.

SEEN offers 50 continuing education courses annually in engineering, business, library science, music, history and environmental resources. Undergraduate and graduate credit courses are also offered in engineering, education and music. The system is used approximately 750 hours annually.

ADMINISTRATIVE .ELECONFERENCE NETWORK (ATN)
ATN is the audio portion of SEEN. It links 30 locations, primarly on the UW System four-year and two-year campuses. The network is used for faculty meetings, advisory panels, board meetings, curriculum sessions and field updates. Meetings can originate from any network site or any telephone. ATN is scheduled approximately 250 hours annually.

MEET-ME BRIDGE
The ICS Meet-Me Bridge links people together from any telephone, anywhere in the world. Participants dial a pre-assigned telephone number and are immediately connected for the two-way audio conferencing. Or, participants can be called. Large conferences can be subdivided into small groups and later reconfigured.

Twenty Wisconsin sites are equipped with portable Meet-Me teleconferencing equipment for easy group participation. Or, individuals can use standard telephones. The Wisconsin bridge can join up to 20 locations, but it is also used in tandem with other bridges for conferences linking 40, 60 or more locations.

Meet-Me is used regularly for meetings, credit courses and continuing education programs. Usage totals approximately 700 hours annually.

SPECIALIZED NETWORKS
ICS also provides technical coordination and operation of two teleconferencing networks serving specialized audiences.

The Vocational Teleconference Network (VTN), a two-way dedicated audio network, is used by the State Board of Vocational, Technical, and Adult Education for meetings among statewide staff. VTN links 24 sites. The CESA #11 Teleconference Network links 25 high schools in western Wisconsin. This dedicated audio network is used for language courses, teacher inservice, support staff training and meetings.

AUDIO PRODUCTION AND DUPLICATING
ICS also is a one-stop audio production and duplicating center offering such services as cassette, reel-to-reel and cartridge duplication, bulk (blank) cassette sales, cassette packaging and distribution, and audio studio services. Projects are diverse: recording, duplicating and packaging of nationally distributed cassette learning packages, duplication of public service announcements; mixing music and sound effects, slide/tape pulsing and production, and tape sales. Projects are produced for both Wisconsin and out-of-state clients.

teleconference is always a critical task. And, the upcoming AT&T divestiture of the Bell Operating Companies will impact all of us. Telephone rate increases and private line restructure proposals hit hard when other resources are especially tight.

As this article goes to press, the private line restructure case, originally filed in 1978 and re-filed in 1981, is still awaiting decision by Wisconsin's Public Service Commission. The freeze-frame network has been in daily use for seven months.

Implementation of the freeze-frame network has been particularly rewarding for it has both stretched and strengthened our teleconferencing organization. The remainder of this article focuses on that implementation.

THE DECISION FOR FREEZE-FRAME

Two years ago University of Wisconsin-Extension made a commitment to upgrade the visual component of the Statewide Extension Education Network (SEEN). Instructional Communications Systems (ICS) coordinates SEEN as well as several other teleconferencing and audio production services described on the next page.

Since 1969 SEEN had interconnected 23 sites, combining a two-way dedicated audio system and one-way audiographics system. One dedicated voice grade telephone line was used for audio, and another was used for visuals. The heart of the visual system was an electrowriter transmitter manufactured by Victor Comptometer Corporation, Chicago. Instructors used the electrowriter to transmit hand-drawn graphics, including charts, diagrams, formulae and text. The electrowriter pen position and movement created tone signals which were carried over the network to receivers around the state. The signals were translated by the receiver onto acetate and then projected onto a screen in each SEEN classroom. Tabletop microphones allowed for easy interaction between learners and instructors.

The electrowriter system served UW-Extension classes well for over a decade. Major network users during

that time were engineering and small business departments. Most of the courses were continuing professional education although some credit classes were offered each semester by UW System campuses. The equipment, however, became costly to maintain and repair during the mid to late 1970's. The equipment, developed decades earlier, was becoming increasingly obsolete and we began a serious search for another visual system.

ICS engineers and technicians conducted an extensive survey of commercially available products that could transmit visual information over telephone lines--electro-mechanical pens, the electronic blackboard, light pen video writers, graphics tablets, facsimile and freeze-frame video.

We also queried SEEN users and non-users. What kinds of visuals would faculty like to use to support SEEN classes? Did they need a graphics tablet, a keyboard for alphanumerics, still picture capability? What kind of flexibility did they want in the equipment--instructor-controlled operation, or technician-controlled? Did they want to originate programs outside of Wisconsin?

Our telephone survey and informal discussions revealed that both instructors and their participants wanted more than just a replacement of SEEN's hand-drawn graphic device. Alone, it was too limiting. The concensus was for a more versatile visual system. In addition to the ability to send free-hand graphics, instructors also wanted to illustrate statewide programs with slides, drawings and diagrams from textbooks and magazines, photographs, written material and three-dimensional models.

Ideally, some wanted a full-motion video system reaching approximately 50 statewide sites. The start-up and monthly operating costs associated with a full-motion video teleconferencing system interconnecting the state however, were difficult to justify. If we were going to provide some kind of video, we knew we were looking at slow-scan or freeze-frame video. With our SEEN voice grade audio channel along with the voice grade data network already supporting electrowriter signals, freeze-frame offered us an economical video system to reach geographically dispersed locations. It also was a flexible system--allowing us to make slide, graphics and other visual presentations across the state or across the country. Finally, we felt it could easily fit into a modular "multi-mode" video system and could easily be upgraded to improve performance.

At this same time we also had the opportunity to gain valuable hands-on experience with freeze-frame technology. Over a two-year period, several ICS staff taught telecommunications credit courses--via freeze-frame--from Madison to graduate students in the Interactive Telecommunications Program at New York University. Several other staff were involved in the technical set-up and operation.

This learning experience, combined with our user surveys and hardware investigation, convinced us and other UW-Extension faculty that this technology met our needs and resources. UW-Extension administration agreed and we plunged into design specifications and contract bidding.

PROJECT IMPLEMENTATION

Two major equipment requirements surfaced. We established the need for a resolution of 256 x 256 with 64 grey levels, allowing a single frame to be transmitted in 35 seconds over a voice grade channel. Instructors also wanted a two-frame memory system to facilitate the smooth flow of information. With this feature, an instructor could "program" the visuals: one visual could be sent and stored in memory while participants viewed and discussed another. With the push of one button the stored image could then appear on the screen at all sites with the appearance of instantaneous availability. Without this memory feature all images would gradually wipe onto the screen, either from left to right or from top to bottom.

These requirements, along with numerous others, comprised our bid for the state bidding process. Colorado Video, Inc., Boulder, Colorado, was the successful bidder for the freeze-frame equipment and was awarded the contract in Spring 1982. Other contracts for monitors, carts,

cameras, slide projectors, etc., were also awarded then.

Our goal was to have the freeze-frame equipment ready for programming by September 1982. Ideally, we would have had a semester to experiment with the hardware and train users. Instead, we had less than two months.

In Wisconsin, programs are scheduled on the teleconference networks nine to eighteen months in advance. The SEEN programming year began September 1st, with 25 courses scheduled first semester. Rather than having both the electrowriter and freeze-frame equipment up simultaneously during first semester, we planned to remove and permanently retire the old equipment in summer 1982 as we installed the new. Given the sink-or-swim situation, we made it work.

To successfully meet our goal, we compiled a giant "To Do" list. That list defined 21 major tasks that staff agreed were necessary to successfully implement the freeze-frame network by our deadline.

Here's a sampling of the project teams:

--freeze-frame transmit table design, production, installation

The freeze-frame instructor's transmit table features three
closed-circuit television cameras, monitors, video switcher
panel and control box. One wing of the custom-designed
table also houses a slide projector and the transceiver.

--control room and studio changes
  in Radio Hall headquarters

--telephone system changes,
  including new site installation
  and jack moves and additions

--training of staff and instructors
  on operation of system, how to
  design effective visuals, program
  design techniques unique to
  freeze-frame

--trouble-shooting procedures and
  location technical support

--community survey of existing SEEN
  sites and changes, sites to be
  added, sites to be removed

--equipment and site security

--publicity and promotion

FREEZE-FRAME EQUIPMENT

Twenty-six Wisconsin locations
are outfitted with freeze-frame
equipment. Each location has two 23-
inch black and white television
monitors on audio-visual carts, an
audio unit and microphones, and a
Colorado Video Transceiver 250 which
is locked in one cart. When
participants arrive at a SEEN site,
they simply turn on one on/off
equipment switch (if an aide has not
already turned on the equipment.)

Seven transportable instructor
tables were custom-designed to meet
our requirements. Two of these tables
are dedicated to studios in our Radio
Hall headquarters. The others are
moved each semester depending on
programming needs. Since September,
for example, programs have originated
from UW System campuses in Milwaukee,
West Bend, Whitewater, Superior and
Eau Claire.

Each transmit table features
several closed-circuit television
cameras--one to focus on a head and
shoulders shot of the instructor,
another to focus on graphics or
objects, another to focus on the
audience at that particular location.
The instructor and audience cameras

are pre-set, while instructors adjust the zoom lens on the graphics camera for their specific visuals. Another visual resource available on the table is a 35 mm slide projector.

An instructor selects the desired visual resource and then has two basic send options--either one requiring 35 seconds for the image to develop from left to right onto the TV monitor at each site, or the memory system which allows one to talk about and view visual while sending another into memory at each site. The later means that an instructor can instantaneously switch between two stored pictures-- even though it still requires 35 seconds to transmit each visual.

In addition to a video switcher panel and a remote control box, the table also includes three 9-inch monitors. One previews only graphics, one previews graphics and any other outgoing visuals, and the third is a network monitor. Classroom monitors on carts, a transceiver, and audio speakers and microphones complete the equipment at a transmit location. Most instructors wear a lavalier microphone clipped to their shirt or blouse.

When the image on the preview monitor is the one desired, an instructor hits the control panel's "freeze" button. This snaps the picture. If it's satisfactory, it can be transmitted to all locations by pressing the "transmit" button. In this mode of operation the visual will develop from left to right on the monitors at all sites, taking 35 seconds for the full frame to be transmitted.

To eliminate this scanning process, an instructor can use the memory feature. This involves using several more buttons on the control panel in order to store the upcoming visual in the "opposite" memory.

Another mode of operation is an automatic transmission mode, taking and sending a picture every 35 seconds.

The table allows seated instructors to easily work with documents and other graphics. All instructor's tables and the receive-only equipment at each site are

identical. This standardization not only helps in the troubleshooting, repair and replacement process, but it also makes it easier to train instructors who originate from sites all over the state.

There are only two equipment exceptions to this standardization. One is a Colorado Video Digital Disc Image Storage System 930 which is currently only available at the Madison headquarters. The system allows us to store up to 242 different visuals onto a hard disc. The disc itself is controlled by an Apple computer.

In advance of their program, instructors can come in and store a variety of visuals from book illustrations and slides to typewritten text and hand-drawn graphics. Use of the disc storage system means that instructors cut down on the amount of camera switching, focusing and manipulation of visual materials they need to do during the actual program. Visuals are stored in the order they will be used. Thus, the disc becomes an electronic slide projector. An instructor simply uses forward and reverse controls to bring the visual up on a preview monitor, and then transmits it to all sites.

The other equipment exception is a large screen video projection system which we use in our large studio in Madison to accommodate large audiences of 12-50 people. Both the audio and video signals from all programs are recorded on standard audio cassette tape for replay to students or for evaluation by the instructor. Programs can also include pretaped audio and video segments, followed by discussion.

It should be noted that we do not have facsimile equipment at any of our sites--nor are we using any kind of video printer at each site to reproduce hard copy of the freeze-frame visuals. Print materials that are necessary to accompany a course are sent directly to each registrant.

Nearing completion are a transportable freeze-frame origination unit, housed in a large case, which will allow remote program originations from any non-network site that has two telephone lines. Also available soon will be a suitcase unit which an instructor can pack in a car for origination from any current network site.

SEEN FREEZE-FRAME PROGRAMS

Nearly 50 different courses will be offered on freeze-frame during the 1982-83 programming year to more than 2,000 participants. Most have been non-credit continuing education courses in engineering, small business, history, communication and environmental resources. Two undergraduate credit courses have been offered in engineering; one undergrad/graduate course has been offered in music.

One of the most successful SEEN courses this semester was an eight-week session on "Aviation Weather". Directed to pilots of small airplanes, the course linked more than 170 pilots gathered at 23 SEEN classrooms throughout the state.

Forty percent of all fatal accidents in aviation are generally related to weather, and the course was aimed at giving pilots a better chance for survival. The video medium of freeze-frame helped meet that goal.

The half-dozen guest instructors made full use of the freeze-frame system, using pictures of cloud formations, air masses, satellite transmissions, charts and text outlines to supplement their presentations and audience discussion.

The graphics camera also gave the instructors the opportunity and flexibility to answer questions with illustrations of maps or photographs from a textbook.

A good example of a credit offering on freeze-frame was a course on Engineering Mechanics. The three-credit undergraduate course reached 70 freshmen and sophomores each semester at two-year UW Centers around the state. The course originated three times a week from the UW Center at West Bend, located approximately 60 miles east of Madison. Freeze-frame facilitated presentation of equations and diagrams and also problem-solving activities. A further dimension of the course was optional tutorials held via the system twice a week.

Although most of this year's freeze-frame programs have been one-way video, two-way audio, some instructional sessions and meetings have transmitted visuals from two or more sites during a program session. At this time five sites, in addition to Madison, could transmit visuals during a single program session.

USER TRAINING AND REACTIONS

At ICS, we have always placed a high value on the importance of user training. This is true no matter what the hardware or transmission system.

Freeze-frame was a large-scale project, if only in the area of instructor training! The challenges were many.

We felt we needed to begin acquainting users with the new technology in June 1982. This was to ensure we could work with all the first semester instructors—some 40 of them—by September or October. Unfortunately our prototype instructor's transmit table was still on the drawing board and would not arrive for another month.

Some of those who would be teaching in the fall had used the electrowriter. Others had only audio teleconferencing experience. Still others had no teleconferencing experience. It is important to note also that these instructors had varying backgrounds and commitments to the system. Some were Extension or campus faculty while many others were hired as ad hoc instructors from other state agencies or the private sector. Many were based in Madison, but some were as far away as 350 miles north in Superior. Some were responsible for one two-hour "guest" slot. Others were the sole instructor for a three-credit course.

Our training approach has consisted of various aspects: print materials, one-on-one orientation to the equipment and designing visuals, hands-on practice sessions, and visual workshops.

In advance of having freeze-frame equipment to work with, we sent all instructors a two-page introduction to Extension's new freeze-frame network and its features.

Following this print introduction, and upon arrival of our first instructor transmit table, we scheduled each instructor for a 1-2 hour one-on-one orientation to the system and designing visuals. Most of these sessions took place at our Madison headquarters. The reactions of instructors to these sessions varied. Some left with glazed eyes. Others immediately got very excited about the visual opportunities that the new freeze-frame system offered. Following these sessions, all instructors were encouraged to return for a practice session before their first program. We felt this practice would give them the chance to actually work with visuals they would be using and perhaps prerecord a segment for their own evaluation.

At their orientation session, instructors were given a series of detailed print materials on how to design good visuals. This information focused on the visual criteria of shape, amount of information, legibility and contrast. Other visual guidelines, as well as step-by-step instructions on how to operate the instructor's transmit table, were also included. These materials have been refined several times during the year as we have gained more experience with the system and how to introduce others to it.

Instructor reaction to the flexibility of freeze-frame has been positive. Instructors responsible for only a one or two hour SEEN segment have provided us the most challenge in regard to the design of our transmit table and operation.

Originally, we intended that the instructor transmit table be self-operated and that our staff not become camera operators in a television studio environment. A technician is scheduled, however, for each SEEN program to monitor program technical quality from the control room and assist any sites when needed.

We have modified this approach as the year has progressed. We have learned that we cannot expect some instructors to invest time in preparing their content, preparing freeze-frame visuals, and learning how to operate the hardware. Asking them to do all three has resulted in some

frustrated professionals early in the year. The last thing we want is for technology to get in the way of the message. In these kinds of instructor situations we now encourage the program moderator to operate the transmit table for the instructor, or we assign a technician to provide camera operation.

In some instructional sessions, such as a music course on "Group Strategies for String Teachers" a camera operator was necessary to provide pictures of the musician/instructor demonstrating instrument techniques.

Another training challenge has been to encourage instructors to think "visually" and make full use of freeze-frame system. Feedback from some participants indicate the need for better designed visuals and more visuals. The number of visuals used during a freeze-frame session varies considerably with the personal teaching style and visual resources of the instructor, as well as the content detail of each visual. One 1 1/2 hour session, for example, used 60 visuals. Other sessions of the same length have used as few as 15 visuals. Our general recommendation for most continuing education sessions is one visual every one to two minutes.

To stretch instructor's thinking about visuals and to encourage preparation of better visuals, we have developed and held three visual workshops. These face-to-face small group sessions include a sample tape of successful freeze-frame visuals, an interview with an instructor and discussion about the four visual criteria.

LOCAL SITE SUPPORT

Support at each of the teleconferencing locations is as critical to the operation of SEEN as it is to any of the ICS teleconferencing services. In Wisconsin, most of this support comes from UW-Extension faculty and staff or UW campus personnel who perform these administrative duties in addition to other full-time jobs. The LPA job involves keeping track of all the program information announcements that come for teleconference programs, serving as an information base for questions about teleconferencing, making sure equipment is working and doors are open, and promotion.

The promotion aspect involves such tasks as distributing a newspaper tabloid to county public places and individuals. The tabloid, with a printing of approximately 68,000 copies each semester, includes listings of all courses available to the general public during a semester, features about upcoming programing and also includes a registration form.

LPA's, as well as campus media specialists, also provide valuable assistance in troubleshooting audio and freeze-frame equipment. Defective audio equipment is returned to ICS and a replacement unit or mics are sent out via UPS the same day trouble is reported. An ICS technician from Madison swaps out any video equipment that needs repair.

SUMMARY

Freeze-frame has added an important new dimension for UW-Extension outreach. Network programs, such as the "Aviation Weather" course described earlier, are meeting critical needs. This highly visual program, for example, could not be offered to such a georgraphically scattered audience in any other way but freeze-frame.

Our goals during the next year will be to increase awareness and usage of the system among UW-Extension faculty and System campuses. Further goals are to refine some controls on the instructor's transmit table, continue to refine user training approaches and materials, and analyze course evaluations.

At this point in time it is uncertain how the pending private line restructure case in Wisconsin—and the upcoming divestiture—will affect the SEEN system. Whatever the outcome, we are certain that we have only begun to tap the potential of freeze-frame.

The Teleconferencing Resource Book: A Guide to Applications and Planning
Lorne A. Parker and Christine H. Olgren (eds.)
Elsevier Science Publishers B.V. (North-Holland)
© Center for Interactive Programs, University of Wisconsin-Extension, 1984

CMITS: COMMUNICATION AND CRAFT

Anne Niemiec
Assistant Project Director
Central Maine Interactive Telecommunications System
Medical Care Development, Inc.
Augusta, Maine

Video teleconferencing is like folkcraft. Both are used to meet everyday needs. Both depend on good design and can incorporate technology without losing the craftsman's touch, as a potter using a motorized potter's wheel or electric kiln. Like many folkcrafts such as quilting, video teleconferencing involves those who use the product in its cooperative production. The Central Maine Interactive Telecommunications System is demonstrating that one does not have to wait for standardized centrally distributed education products to come to them. They are producing their own with craftsmanship that builds the community in the process of responding to local needs.

DESCRIPTION

The Central Maine Interactive Telecommunications System (CMITS) is a user-operated, two-way telecommunications microwave system which links five hospitals, a family practice residency program and a university to meet a variety of educational and patient care needs. These include meetings, continuing medical education, courses (credit as well as noncredit), nursing programs, and other educational events including programs via satellite. The System became operational in August 1977 and was originally funded under Exchange of Medical Information Grants from the Veterans Administration to Medical Care Development, Inc. (MCD), a nonprofit health care delivery organization. It is now paid for by the participants and programs are currently running on the average of 35 hours per week--almost all live and interactive.

The CMITS interconnects Augusta General Hospital and the Family Medicine Institute (the practice unit of the Central Maine Family Practice Residency) in Augusta; Central Maine Medical Center and St. Mary's General Hospital in Lewiston; Thayer and Seton

Units of Mid-Maine Medical Center in Waterville; the VA Medical and Regional Office Center at Togus; and the University of Maine at Augusta. The signals from each site converge on the Augusta relay point at Sand Hill where a program may be routed to only one or any combination of the facilities.

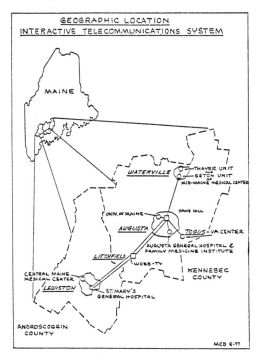

The System is oriented toward actual user operation with a minimum of technical intervention. Each participating facility has been provided with identical equipment so that programs may originate from any location with equal ease. Controls and layout of the equipment were made as simple as possible to minimize some of the tension which accompanies the initial use of new technology. The System also provides for exceptional confidentiality since only stations specifically selected by the originating site can view or hear programs in progress. A site may send a picture to one, all, or any combination of the sites. It is possible for three separate two-way television conversations

to take place simultaneously. We have
recently started a limited amount of
double programming, one involving two
sites broadcasting to each other and
the other interconnecting the remaining
sites.

Design

As with any craft, the design is
of prime importance. With the CMITS,
the design process included both human
and technological elements. Prior to
implementation, Robert Cowan (until
recently, Director of CMITS), with a
background in communications and satellite experience, did extensive research
on both ongoing and past unsuccessful
telecommunications systems and designed
the CMITS to avoid some of the people
and technology problems that plagued
the other systems.[1] He involved key
hospital people in the planning and was
careful not to introduce the System as
something that would solve all problems.
He let it evolve slowly. There were
face-to-face meetings so people got to
know each other before the technology
was in place.

The overall objective of the System is the development of cooperative
educational programs in which no institution either receives or transmits all
network programming, but shares in the
development and implementation of the
network content. The CMITS attempts to
create an educational balance of payments in which each institution that is
a member of the System both exports and
imports educational programming equally.
Each institution has its educational
strengths and weaknesses, and it is the
function of the CMITS to share those
strengths. There is no involvement of
a major medical center or teaching hospital. The participants are of relatively equal size and capability. While
most of the programs use resources from
within the participating institutions,
speakers from the wider community are
shared as well. With its accessibility
and range of resources, the CMITS helps
to establish an information environment
that encourages the integration of continuing professional development into
the daily life style.

USES

The prime ingredient of this coop-

eration is the user group which is composed of one or more representatives of
a specialty area from each hospital,
such as directors of social work, medical records personnel, and plant operations' directors. There are 28 different user groups employing the System to
facilitate education, communication, and
consultation. The user group meets on
the System to discuss areas for programming based on needs which they have identified. The group is then involved in
the design, implementation, and evaluation of the programming. The process,
as well as the product, is important.
It has become truly inventive planning
as described by Ziegler in Planning as
Action. "Inventive planning is grounded
in a belief in the intrinsic worth of
human beings who, by their nature and
irrespective of their official position
in society, possess the potential--and
often the actual--to act competently
with a view to their own future."[2]
Ziegler emphasizes that the social setting for this process should be local--
homes, churches, workplaces, neighbor-
hoods, communities. In this setting
they can be encouraged to "bring their
practical wisdom and critical imagination to play."[3]

Considerable planning, or impetus
for planning, comes from the inservice
education directors, who formed the
first user group. This group meets
monthly to initiate or develop educational programs applicable to professions that are not themselves user
groups and other programs applicable to
the broad spectrum of hospital staff.
The range of programs is increasing as
hospitals become more involved with community health education and preventive
medicine for workers in the hospital and
in other areas. Tai Chi, a combination
of dance and exercise, is one example of
a program developed for more than one
audience. It can also be made available
to patients by connecting the CMITS to
patient television systems.

Nursing furnishes an example of one
profession using the CMITS in a variety
of ways. The CMITS delivers continuing
education programs planned by nurses in
the five hospitals and the University.
Some of these programs are planned in
the individual sites, some by nurses in
more than one site using the CMITS as a
vehicle for cooperative planning.

Nurses also attend continuing medical education programs on CMITS and take
courses, such as social gerontology,

which can be used as electives towards a university degree. The System is used for monthly meetings by professional nursing organizations such as the Association of Operating Room Nurses of Maine and the Critical Care Nurses Association. Nursing students on clinical rotation in the hospitals are able to attend programs giving them information and an example of the professional development of nurses in action.

The use of the CMITS specifically for and by nurses began with a four-part series which sought to provide a forum for the exchange of ideas and to increase the access to continuing professional development through the sharing of resources. This series included clinical conferences, management in nursing sessions, nursing updates, and specialty perspectives. As an outgrowth, user groups in nursing specialty areas formed and have planned their own programs. The latest use is by the night shift nurses who have recently started using the System at 3:00 A.M.

are underway. These efforts are part of cooperation which results in an educational balance of payments.

Some user groups make particularly good use of the visual component of the System. One of the most successful programs in this regard (as well as others) is the monthly meeting of infectious disease specialists. Cases in various states of resolution are presented using charts, x-rays, etc. Opinions are exchanged to arrive at the diagnosis of a current case, or in completed cases, how treatment might have been improved. Physical therapists have demonstrated various methods concerned with gait training, using the equipment available to videotape examples for live discussion. Care does have to be taken in order to make the visuals effective.

Courses needed by employees are offered on a credit and non-credit basis. Resources within the hospital are used for both the planning and the implementation. A modified instructional

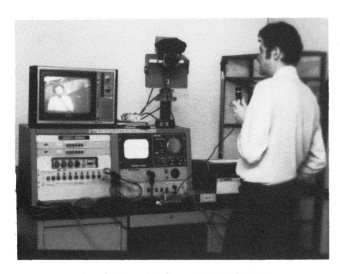

An interactive User Group meeting on CMITS.

Two-way conversation between media specialists.

Sharing on a larger scale is also taking place. Certain courses need to be repeated every year for a variety of specialties, such as coronary care nursing, neurological nursing, clinical pharmacology, etc. Unitl recently, each hospital has provided all these courses for their own staff or sent people to other locations. A plan is being developed by the inservice education directors whereby each hospital will provide only one course, sharing it over the System, and receive four others in return. Two of these courses

development approach was used in a university anatomy and physiology course in which each of the five hospitals contributed audiovisuals and/or instructional resources; among others, a pathologist provided slides and health care specialists taught about the body system in their fields of expertise. This multiple resource approach in which the instructor becomes a manager of the educational environment will be used in other courses.

Courses which were originally in-

tended to be taught at one location are
now taught on the System, thereby cutting travel. For example, certification courses for respiratory therapy
technicians were developed by the directors of respiratory therapy who had
used the System for their own professional development. When the price of
gasoline started to escalate, they suggested putting the courses on CMITS,
saving two trips to Waterville and 220
miles of driving per week for Lewiston
participants. The need for alternatives to travel is increasingly on peoples' minds.

GROUP COMMUNICATION OVER DISTANCE

    We are increasing our awareness of
the differences between group communication over distance and face-to-face
and learning from our own and others'
experiences to improve communications.

    --Over distance there is gen-
      erally more attention to the
      task, though if the individuals
      in a group know each other well
      and the meeting is very infor-
      mal, this is not necessarily the
      case.

    --Consensus can appear to be
      reached when in fact it is not
      there. Conflicting opinions
      appear later in separate con-
      versations and the issue needs
      to be dealt with at the next
      meeting.

    --There is more potential for con-
      trol of group interaction by
      monopolizing the screen. If the
      group feels at ease with the
      equipment and with itself, in-
      teraction comes easier.

    --There are not as many cues from
      non-verbal communication. Some
      people are better at giving
      these cues; they are able to en-
      courage questions or receive
      clarification by their gestures
      and expressions. Each site is
      able to control another site's
      camera, with permission, so that
      they can zoom in and out and
      thus get a closer look at who-
      ever in the group at another
      site is speaking. Having con-
      trol of the camera helps to in-
      crease the amount of non-verbal
      information.

    --In the user groups the communi-
      cation pattern is most often ra-
      dial, sometimes leader-centered,
      but rarely "Y" or hierarchical,
      resulting in more lateral than
      vertical communication.

    --The seating arrangement can make
      a difference. Sitting in a row
      in front of the camera is not
      effective. The best results have
      been reported when the camera is
      placed as if it were a member of
      the group.

    The technology obtrudes less when
motivation is high, be it for information or overcoming a feeling of isolation. Some of these considerations can
be seen as drawbacks, but overriding all
is the fact that there is more access to
remote resources--to people, to viewpoints, and to information.

    Although much of the communication
takes place within professions in the
user groups, some cross-disciplinary
groups exist, such as those concerned
with implementing interdisciplinary team
work in hospitals and a networking group
which is just getting started. The latest group, composed of a number of different professionals from institutions
within and outside of the System, is
meeting to discuss the construction of
alternative futures for at least 25
years ahead.

    We have found that communication
is also enhanced within institutions because of the System. Department heads
at one hospital, after completing an
assertiveness training course on the
System, continued to meet formally
and informally to give each other the
feedback and reinforcement much needed
in the process of trying to change behavior. The System serves a catalytic
function, introducing people and ideas.
One person, new to the area, received
a job as a result of making a presentation on CMITS. Another side effect is
the increase in self-esteem for some
people which occurs from overcoming the
fear, not only of the technology but of
making presentations and successfully
communicating at meetings. This carries
over to other aspects of their jobs.

TRAINING

    The media specialist at each site

is responsible for seeing that the users are trained to operate the System on their own. There are various training modes and varying degrees of skill reached. We list five competency levels ranging from basic operation of the equipment to moving the console from one location to another. In this area, as well as others, we are trying to pay attention to individual learning styles so that a person can learn from the media specialist, or alone using the competency lists and a manual, or from peers, individually or in groups. We are encouraging more peer training. Carey has pointed out, and our experience corroborates, the potential advantages of peer training: "...; common language...less concern about 'looking foolish'...ready access to peers for repeat demonstrations."[4] An introduction to the System is part of the orientation program for new staff at each hospital. To help with this and as an adjunct to other training, we are working on a videotape which will include how to operate the System and how to enhance presentations through the use of audiovisuals.

On the average it takes three interactions before a level of comfort is reached. We have seen some dramatic changes, for instance, someone who at first was not comfortable even being in the room when the System was on is now giving presentations and encouraging others to do so.

In addition to training, the media specialists act as liaison between the hospitals and the System. Their diverse backgrounds in media, psychology, education, and radio enable them to provide the individual hospitals and the CMITS with technical, media, and program planning assistance. The media specialists are employed by the participating institutions rather than by the System. Although this decreases their control by the System, it increases their ability to be effective within the institutions.

MARKETING

Although much of the programming is user group oriented, hence the users are involved in the planning process and aware of what is coming, we are becoming increasingly aware of the importance of marketing the programs to those who may have the need but who are not part of the user group. This has particular im-

plications for career development or careers in transition. For example, a secretary interested in becoming a pharmacy technician can take a pharmacology course. The accessibility of continuing medical education programs allows different health care professionals to attend and provides the opportunity for discussion with physicians and colleagues about mutual patients.

Weekly announcement sheets are sent to the hospitals for distribution to departments and individuals. Courses and special programs appear in special announcements. We are using attendance records to develop a computerized index of individuals interested in certain areas and sending out special mailings on future programs in their interest area. Individual contact, although time consuming, has been especially effective. We definitely keep in mind the "bridge role" that our marketing efforts have in relating staff needs and resources.

EVALUATION

Evaluation has been an ongoing process since the beginning of CMITS. An attendance form and an encounter form are used to gather utilization and satisfaction information. The encounter forms are short and ask questions on relevance, accessibility, visuals, overall rating, and have a space for future program suggestions and comments. The forms work as a source of information for ongoing improvement of programs and a way of noting needs. The suggestions for future programs are passed on to the appropriate user group or brought before the inservice education user group for consideration. The forms are used for troubleshooting; feedback, both positive and negative, is passed on to the instructors. We encourage immediate feedback via the System by students to instructors, but the forms provide a mechanism for those who are reluctant to do that. More in-depth evaluations are conducted following each course and follow-up evaluations are done six weeks later. Evaluation meetings with instructors are held during and after the courses.

There are personal as well as institutional costs that are affected. The System can be seen as a way of conserving human resources as well as material energy resources. The following equation helps to illustrate this point:

$$\text{Margin} = \frac{1 - \text{Load} + \text{Cost}}{\text{Power} + \text{Benefits}}$$

Gibson describes its use in predicting whether an individual will engage in a learning experience.[5] The two factors that primarily affect the learner because of the CMITS are accessibility (location is at the worksite which eliminates cost of travel) and availability in terms of content and resources. In addition, there are the social reinforcement benefits of learning in a group. The System works to effect a cost savings system of education and communication leading to an increase in competence. "Competent people are those who can create valuable results without excessively costly behavior."[6]

SLOW SCAN DEMONSTRATION

Medical Care Development, Inc. has recently been awarded a grant from the Office of the Assistant Secretary for Planning and Evaluation, Department of Health, Education, and Welfare, for a telecommunications demonstration project in Aroostook County. This will allow an interface between the broadband two-way system and slow scan technology which converts a moving image to a still picture and then transmits the visual material to other locations via telephone lines. The participants in this project are four hospitals and a mental health center in Fort Kent, Caribou, Presque Isle, Houlton, and Fort Fairfield. This audio and still-image video conferencing system will provide: (1) talk back and interaction with program originators in the central portion of the State, (2) the ability to utilize slow scan equipment for intracounty conferencing, and (3) interconnection with the CMITS so that Aroostook health professionals may lecture and participate more fully in the educational activities outside the County. Aroostook County, the northernmost county in Maine, is poor, rural, and has severe manpower shortages in some health fields. Involvement of these persons in educational programs that meet their professional development needs, as well as provide peer contact, will greatly reduce the isolation factor that contributes to the manpower shortage problem.

The slow scan expansion to Aroostook

involves the same philosophy of user self-sufficiency as CMITS. In these beginning months before the equipment arrives, we are working to develop and strengthen the communications bonds between the participating sites and to identify risk-tolerant individuals as first users. Enthusiasm and interest is high. The hospitals have for some time been receiving the CMITS weekly program announcements. One of the inservice educators referred to it as a "Christmas wish book." Efforts are being made to develop the interest that already exists in the central Maine area. They, too, are looking forward to meeting their counterparts.

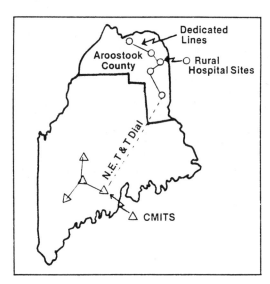

The demonstration will provide people in the County with the opportunity to experiment with new forms of telecommunications technology and develop a system which most closely fits their needs, demonstrating that "the greater the involvement of the user and the greater the focus on participatory education, the more transparent the communications system becomes."[7]

CONCLUSION

We are working to increase the amount of choice and control over the kinds of information and education that is needed. The next step will be to choose from an array of communication technologies. One combination that is possible is described by Glassmeyer as a "wedding of the computer and video technologies" that "will open up enormous opportunities for the development of interactive training systems and other two-way communications applications."[8] We are looking at other combinations, always keeping in mind the cooperative craft in what we are doing. "...choices of media for a given communication need will not be simple choices: they will require a careful blending of communication resources and needs. Communication--electronic or otherwise--is a craft. As with other crafts it cannot be taught mechanistically. It must be developed through personal experiences."[9]

REFERENCES

1. Cowan, Robert. The diffusion of innovation and self-management of telecommunications. In M. C. J. Elton (Chair), Social applications: Fitting the technology to its users. Symposium presented at Electro '79, the international meeting of the Institute for Electrical and Electronics Engineers, New York, April 1979.

2. Ziegler, Warren. Planning as Action, 1972.

3. IBID.

4. Carey, John. Implementing interactive telecommunication projects: A baker's dozen of issues and problems. In M. C. J. Elton (Chair), Social applications: Fitting the technology to its users. Symposium presented at Electro '79, the international meeting of the Institute for Electrical and Electronic Engineers, New York, April 1979.

5. Gibson, Terry L. Designing instructional systems for the adult learner, (n.d.).

6. Gilbert, T. F. Human competence. New York: McGraw Hill, 1978.

7. Medical Care Development, Inc. Aroostook County telecommunications demonstration grant application, August 1978.

8. Glassmeyer, Gerald E. How to design programs that encourage response. Educational and Instructional Television, August 1979, pp. 52-53.

9. Johansen, R., Vallee, J., & Spangler, K. Electronic meetings: Technical alternatives and social choices. Reading, MA: Addison-Wesley, 1979.

The Teleconferencing Resource Book: A Guide to Applications and Plannin
Lorne A. Parker and Christine H. Olgren (eds.
Elsevier Science Publishers B.V. (North-Holland
© Center for Interactive Programs, University of Wisconsin-Extension, 198

INTERACT
A MODEL INTERACTIVE TELEVISION NETWORK

Marshall L. Krumpe
Manager
INTERACT TV Consortium
Hanover, New Hampshire

The overall objective of this publication is to help facilitate people's interest and awareness of two-way television. Specifically, however, it is the interest of the author to:
a) help in communication system analysis;
b) selection of a proper media to meet pre-determined needs; and, c) if two-way television is evaluated as the ultimate medium, to help users implement such a system based on the INTERACT TV Model.

My thanks and acknowledgement to Dr. Dean J. Seibert ("A Decade of Experience Using Two-Way Closed Circuit TV for Medical Care and Education")for his contribution to this report.

Why a Network -- What to Look For...

Why a Network -- whether it be one-way, two-way, cable, slow-scan, microwave or satellite? Availability of funding is unfortunately too often the answer to the above question; and an answer that more often than not leads to an inefficient, underutilized "skeleton in a closet".

The correct answer to the question of "Why a Network?" is - recognized and documented need. The following are a few of the thoughts which have crossed my mind if I were evaluating the rationale for investment into a communication system.

a) Needs assessment prior to choice of technology
b) What technology fits the need
c) Program - Type and Audience
d) Affiliations
e) Budget - How to survive
f) Barriers

Interactive Television is a vehicle for programming which in order to succeed must be flexible and respondent to the changing needs of existing "users" and sensitive to new and innovative uses (and users).

The following outline is an expansion of the points listed above.

A. Needs Assessment Prior to Choice of Technology
   1. Is there a need - identify and define
   2. How relevant - can it sustain capital depreciation and operating costs
   3. Compatibility with existing systems
   4. Relative advantages - one system to another
   5. Is there a functional relationship already existing among prospective users
   6. Physical location - be as mutual and convenient as possible in order to be attractive for alternate users

B. What Technology Fits the Need
   1. Two-way audio and video (cable, duplex microwave, satellite)
   2. Two-way audio, one-way video (cable, ITFS microwave, satellite, slow-scan)
   3. One-way audio and video (broadcast TV)
   4. Two-way audio (teleconferencing)

C. Sample Programming
   1. Teaching - continuing education, continuing medical education, student education
   2. Consultations - all professions
   3. Patient education
   4. General services - business/staff meetings
   5. Digital information
   6. TV Guide as means of disseminating information

D. Affiliations Helpful in Solidifying Health Based Communication Network
   1. Local hospital associations, AMA, State Medical Society, Continuing Medical Education, Blue Cross/Blue Shield, etc. These organizations may also prove to be

useful as programming and funding sources.

E. Budget - How to Survive
   1. Broad programming base - need to be flexible - do not limit to one speciality
   2. Prior dollar commitment - sub-scriber, ad hoc
   3. Funding sources - private foundation, federal government
   4. Full utilization of equipment and manpower by expanding on videotape production, utilizing existing equipment and man power and, by taping for distribution programs aired over the Network
   5. Integration of system(s) for maximum dispersion of information - microwave to cable for local redistribution - microwave to tape for national redistribution

INTERACT - A Model Network

INTERACT - The Time

INTERACT, a network which serves rural Vermont and New Hampshire was conceived ten years ago. The late 1960's and early 1970's were a time of turmoil for this nation. Concerns for greater access to better health services were translated by Congress and the Executive Branch into a multitude of programs which emphasized the needs of the rural and urban disadvantaged. Federal sharing of the cost of medical services for the elderly and the poor was initiated through Medicare and Medicaid. Inequities in the distribution of care were addressed in community programs of the Office of Economic Opportunity and through the Partnership in Health Act (C.H.P.) and regional cooperative efforts, focusing on heart disease, cancer and stroke, were stimulated through Regional Medical Programs. Emergency Medical Services became a major thrust. Federal interest in the innovative application of technology was expressed through the establishment of the Lister Hill National Center for Biomedical Communications and the National Center for Biomedical Communications and the National Center for Health Research and Development. Some of these initiatives have proven to be ill fated, yet almost all have left a lasting impact at the grass roots level.

During this period, the nation's new priorities had a substantial effect on our medical schools. As emphasis on health care shifted from biomedical research to the delivery of care, medical schools increased their enrollment, training of new categories of health

workers was begun and many schools became deeply involved in service programs.

Although the impact of all of these develments in New Hampshire and Vermont would be difficult to quantitate, at least one study suggests that the effects have been profound. The medical care system of this region a decade ago, when INTERACT was conceived, was characterized by an almost total commitment to laisse-faire health services, but today cooperative efforts for the common good no longer trigger a reactionary response. This is not to say that a sense of fierce independence has been totally lost or that interference in the region's autonomy is no longer an issue, but the rationale for cooperative endeavors is far more readily acknowledged. INTERACT emerged during that era as a means of distributing and sharing in a region where supply is limited.

The problems and successes of INTERACT over the past eight years must be considered in the total context of the period. INTERACT has facilitated change and has in turn been influenced by its milieu.

INTERACT - The Place

INTERACT extends from the northwestern border of New Hampshire to the northwestern border of Vermont, running diagonally across the northern part of the latter. The country north of INTERACT is rural, with populations of towns and cities averaging slightly more than 1000.

Health services, like the communities, tend to be isolated and fragmented. Hospitals in the area are relatively small and fiscally poor. Need for communication is substantial but the ability to develop and support new systems without subsidization is questionable.

The region is characterized by mountainous terrain which restricts travel, particularly in an east-west direction. Paranthetically, these mountains also provide a barrier to the use of microwave television, which requires line of sight transmission although as relay points are established by the State's educational television network, this constraint is gradually diminishing.

While primary care is provided in the Northeast Kingdom of Vermont and the North Country of New Hampshire by local physicians and community hospitals, the need for more sophisticated services is filled by the Medical Center of Vermont at Burlington and the Dartmouth-Hitchcock Medical Center in Hanover. In addition to the joint North Country catchment area, the Medical Center of Vermont serves as

the referral point for most of western Vermont
and bordering New York towns, while the Dartmouth-Hitchcock Medical Center relates to the
Connecticut River Valley, encompassing most of
southwestern New Hampshire. Consequently, there
have been functional links between all facilities in the network. Nevertheless, because of
geographical isolation, the historical necessity
for self-reliance, personal characteristics of
reserve, often bordering on suspicion, and very
few dollars to catalyze change, the pace of
developing cooperative regional relationships
has been predictably slow.

Early Planning and Policy Formulations

The late 1960's encompassed a dramatic
period of growth for the Dartmouth Medical
School as it began its reemergence to an M.D.
degree granting institution. The relationship
of the School to the other components of the
Medical Center (Mary Hitchcock Memorial Hospital,
Hitchcock Clinic, Whiter River Junction Veterans
Hospital), the relationship of the School and
Center to the region, and the relationship of
the established basic science departments of the
institution to the new clinical segments generated a dynamic and not infrequently chaotic
atmosphere which stimulated a search for new
solutions to problems both old and new.

A principal concern of the medical school
was, and is, to find effective ways for a small
institution to relate meaningfully to a large
rural area. A number of studies were conducted
under various auspices in the region in the
1960's which helped to define the needs of
those rural areas of New Hampshire and Vermont
served by the School. These studies encompassed
Sullivan County, New Hampshire as well as the
North East Kingdom and North Country of Vermont
and New Hampshire. The latter study recommended
"massive involvement of the medical schools
(Dartmouth and University of Vermont College of
Medicine) in the region, "but it provided few
new insights on how this involvement could be
realistically implemented.

Studies were conducted by several different
groups in the 1960's to determine the feasibility of meeting some of the region's needs via
closed circuit television. The "New England
Land-Grant Network," published by John Bardwell
in 1968, considered the feasibility of establishing information links between six land grant
universities in New England. Dartmouth Medical
School funded an Engineering Study in 1967 which
was conducted by Raytheon to determine the cost
of extending a network into portions of the
North Country, an area then and now considered
to be of high priority. A two way closed circuit television system, leased from the telephone company, joined the Dartmouth Medical

School to the Claremont General Hospital in
1968.

The capital cost of building a microwave
network was of major concern but it was recognized that cost of mountaintop construction,
maintenance and repair of equipment could be
minimized by sharing existing educational television equipment at mountaintop relay points.
It was also recognized that the location of the
then existing E.T.V. towers would define the
configuration of the network to a significant
degree. A compromise was required between
reaching those facilities where extreme isolation created maximum need and maximum cost and
those facilities which could be reached with a
more "reasonable" capital outlay.

Discussions concerning the individual needs
and resources which might be shared via closed
circuit television, were initiated among representatives of the educational television
stations of Maine, New Hampshire and Vermont,
health educators from Dartmouth Medical School,
the University of Vermont College of Medicine
and the University of New Hampshire and with
those administrators and staff of community
hospitals which lay in the path of the E.T.V.
microwave backbone.

At the suggestion of the Director of Continuing Medical Education of the American
Medical Association, the interests of these
individuals, who were a-priori early INTERACT
planners, were brought to the attention of Dr.
Ruth Davis, then Director of the Lister Hill
National Center for Biomedical Communications.
A series of discussions and the development of
a specific proposal to the Lister Hill National
Center for Biomedical Communications led to the
funding of a technical feasibility study conducted by the Atlantic Research Corporation and
to the exploration of additional uses of the
existing phone company microwave link between
Dartmouth and the Claremont General Hospital.

The technical feasibility study included
twenty hospitals and educational institutions
that were located in areas where there was
direct line of sight with or without a modest
tower to one of the E.T.V. mountain top relay
points. Upon completion of these studies and
a further assessment of interest among potential users, a number of critical decisions with
far-reaching implications were made both at the
federal and local level.

It was necessary for the Lister Hill Center
Center to decide on the magnitude of their
investment in INTERACT, a decision which would
be influenced by competing priorities and their
assessment of the program's intrinsic value.
Should the entire network be built in one phase
or should it be developed in a step-wise
fashion?

Construction of the entire network, extending from the Maine Medical Center in Portland to Burlington, Vermont had the virtue of involving a great diversity of users, long-term supporters, both in rural and metropolitan areas. The investment would have been high, however, for a demonstration of unproven value. On the other hand, too small a demonstration might not establish the critical mass of usage and experience which would be required to justify the network's future. It has been stated that the telephone was not a particularly helpful device until it linked many individuals and that the automobile was considered an expensive novelty until enough were owned to justify the design and construction of roads which could accommodate them. Too small a demonstration might similarly compromise the acceptance of interactive television. A network of four to six institutions seemed to be appropriate compromise.

Since it was not deemed feasible to build the entire network in one step, it was necessary to select a limited number of institutions from the many that had expressed a desire to be involved. Factors which influenced this selection included:

1. The perceived level of institutional interest.
2. The distinguishing features of each hospital or educational institution which would foster a difersification of use.
3. The nature of the basic funding for the hospitals and schools; public, private, state agency, etc., and how their diverse fiscal base might contribute to the potential for long-term support.

A decision was made to join the Medical Center of Vermont which includes the University of Vermont College of Medicine and School of Allied Health Science, both state supported institutions, Dartmouth Medical School, a privately endowed educational institution and the Claremont and Central Vermont Hospitals which had a close working relationship with the two medical schools.

Plans were also finalized to link three quite different facilities to the network on a part time basis using a mobile unit, including a Vocational Technical College which emphasizes allied health training, the Windsor State Prison, Vermonts only maximum security facility at that time and the Rockingham Memorial Hospital, a small community hospital.

The compromise meant that the network would not reach the more populated and prosperous areas of New Hampshire—Concord, New Hampshire's capitol and political hub and Manchester, the state's modest industrial center.

Extension into the North Country of New Hampshire and the Northeast Kingdom of Vermont, the areas of greatest need were also excluded by the funding constraints because of the mountainous terrain, the lack of E.T.V. relay stations and the unlikelihood that the very small hospitals of the region would ultimately be able to sustain the operating costs of the network.

Dollar limitations forced another decision which had a significant impact on network development. Sufficient money was available to 1) complete construction of the two ends of the network, (Burlington--Central Vermont and Hanover--Claremont) and employ the principal investigators at U.V.M. and Dartmouth or 2) to complete the network from Burlington to Claremont with the expectation that the two principal investigators would derive their support from other sources. It was the decision of the two principal investigators to pursue the latter course.

It was impossible at the time and difficult in retrospect to estimate the impact of that decision on INTERACT's development over the succeeding five years. Dartmouth's principal investigator, the Assistant Dean for Regional Medical Affairs, and U.V.M.'s principal investigator, the Director of Continuing Medical Education, were responsible for developing the relationships of their respective institutions to the community in a variety of ways. This diversification of responsibility proved to be both an asset and a liability. It provided avenues through which institutional policy relative to the region could be influenced was well as working relationships with many individual institutions who were persuaded to consider INTERACT as a mechanism for implementing regional programs. The diversity of responsibility of top management on occasion, however, compromised their ability to provide sufficient leadership to INTERACT especially when other sources of their support became unstable. At such times forward momentum was slowed and sometimes stiffled.

Planning for INTERACT has been characteristically flexible and "opportunistic intervention" as the demonstration evolved from a single link to a complex network. This flexibility was its early strength; without it, successful implementation would have been impossible. While the ultimate goals remained unchanged when opportunities to explore new avenues occurred, short-term plans were often modified in order to capitalize on interests and resources which unexpectedly became available. While the principal of "opportunistic intervention" became an acknowledged and useful modes operendi, often overriding more deliberate planning, the efforts to capitalize on special circumstances on occasion led to an inefficient scattering of resources and fragmentation of effort. There is no sure way to avert conflicts between long-range planning and the implementation of short term objectives in a domonstration which is experimental in nature and derives strength from flexibility.

For INTERACT, the conflict between rapidly changing short term plans and the steady pursuit of long range goals, including that of self-sufficiency, was greatly complicated by a sub-stantially shorter period of stable funding than had been anticipated. The time frame originally envisioned by the LHNCBC is unknown but a decision was made to terminate support within eighteen months of full network operation regardless of INTERACT's stage of development.

Arguments presented to support requests for a continuation of funding which would have provided stability while new users had an opportunity to explore its potential were not persuasive.

After a relatively brief period of operation, the overriding goal became one of gaining self-sufficiency. Development of new network programs dropped markedly and by 1976 had come to a virtual standstill. Efforts at evaluation, except in the most superficial sense, were abandoned. New areas of activity, particularly videotaping for "in house use," were developed for immediate cash return. Profound uncertainty resulted in the loss of key personnel and the momentum which had been building for several years was perceptibly slowed at the potentially most rewarding phase of the demonstration.

A number of vital steps were taken to respond to the emphasis on organization and management rather than program development. Each of the participating facilities was required to make a substantial fiscal committment to the network. In the case of the community hospitals in particular, this committment far exceeded any previous support that had been allocated to continuing medical education activities. Fortunately, federal economic stabilization controls which had placed a five percent ceiling on annual hospital expenditures in 1971, terminated on April 1st, 1974. If that constraint had not been removed, almost certainly the demonstration would have been required to terminate.

In order to give each of the participating institutions equal representation in formulating operating policies, Station Councils and a Network Advisory Board were established. Currently, a more autonomous organizational structure which reflects the fact that INTERACT has become a consortium is now under consideration.

Despite the necessity for focusing on immediate issues of self sufficiency, planning for the future has not been neglected. Only

a small fraction of the network's potential has been developed and efforts to identify resources which will allow new users to explore the utility of the network remains a high priority.

Consideration of future network expansion goes forward. It is anticipated that two new stations will be added in 1980-81 and extension of the network into both the underspread and more populated areas of Vermont and New Hampshire remains a high priority. A second dimension of INTERACT, the distribution of tapes recorded from live programming to hospitals which are not connected to the microwave network has evolved from a 1976 feasability study. This new program, Media Outreach, while not providing the ease of access to medical information that is fostered through live programming, fills a gap which will remain until the cost of closed circuit networks can be borne by even the most isolated facilities.

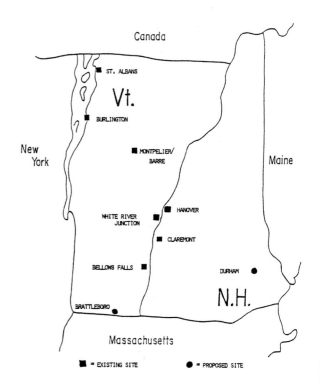

Administration

One of the distinguishing characteristics of INTERACT is its organizational status. The end result of federal research and demonstration monies and local support is an organizational chimera. It is a mix and match of private and public; university and hospital; the big and not-so-big; the volunteer and the salaried employee; the old and the new; the committed and the indifferent. It is the outcome of planning in an experimental framework in which the component parts are independent and whose ultimate common interest in INTERACT could only be surmised. As the network became financially dependent on the user institutions, it became imperative to formalize relationships between them which would provide broad representation and accountability. Figure 1 indicates the current organization structure. It should be noted that the Consortium Board of Directors is the true governing and policy-making body. Administrative links parellel to that level simply reflect the fact that as holder of the F.C.C. licenses, the Trustees of Dartmouth College are held legally accountable for the system.

It should be noted at this time that INTER-ACT management, and Dartmouth College all agree that the future of INTERACT's development somewhat hinges on its relationship with Dartmouth College. There are many reasons for maintaining an affiliation with Dartmouth College, however, it is believed by the majority that INTER-ACT could be more flexible in its growth if it were to go beyond the establishment as a regional consortium, to a free standing legal entity.

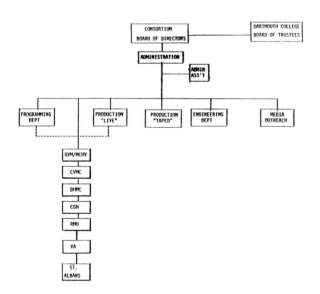

Figure 1

Consortium Board of Directors

The Consortium Board of Directors was created to set policy and advise Network management and Local Station Coordinating Councils in the total operation of the network. Board membership consists of primary delegates from each local council.

The Board is viewed as an organizational

necessity for influencing decisions by network management and local stations. Its primary task is to provide a total network perspective which encourages individual stations to work together for the stature and growth of the Network.

Local Station Coordinating Councils

Local Station Coordinating Councils, as the name implies, support the local station manager in coordinating programming between the Network and the local station. As such, the councils review programming and the impact of such programming. The membership of each council represents the administrative, financial, medical, nursing and continuing education interests of the participating station.

Local Stations

Local stations are the facilities which receive and/or originate programming throughout the Network. These stations are united by a common carrier: INTERACT. The organizational tie is made via the Local Station Coordinating Councils and the Board of Directors. The financial bond is found in Memorandums of Understanding (open contracts) between INTERACT and the institutions representing the Local Station.

Network Staff

The final line in the organizational chart is Network staff. It is appropriate that everything finally comes down to this category where it all happens. Network staff provides the foundation upon which the organization is built. Currently the staff consists of twelve (12) full-time professionals, four (4) part-time high school work-study students and four (4) part-time production and management resources available as needed to meet demands of programming.

Network staff are currently employees of Dartmouth College, and each position is graded and salaried according to policies established by Dartmouth Personnel. It becomes necessary to detail the general descriptions provided by Personnel in order to reflect the team concept of production-engineering-management services. All staff provide services according to these areas with an obvious emphasis on individual professional training and background.

The question often arises - Should INTERACT employ station personnel at subscribing hospitals; or should the job or station manager be incorporated into the daily tasks of an existing hospital employee(s)?

INTERACT has had firsthand experience in both cases, by virtue of the fact that our Rockingham Memorial Hospital station is manned by hospital staff. And, even though our Claremont General Hospital station manager has responsibility for Rockingham Hospital (and spends two days a week at that location), the effect is not the same as having a full time INTERACT employee. If, however, a well-identified need/relationship exists between institution(s), thus eliminating the program development responsibility, a hospital employee may be able to function adequately.

Money Management

One of the most difficult financial tasks has been the securing of operational funding. The first and possibly most major obstacle has been overcoming the initiation "try it -- you'll like it" concept; whereby, the users of the Network had free access. Some preparation had been made with user institutions to secure their financial support while still on soft money -- but not enough.

We have over a period of years transitionalized to the point where our institutional subscribers actively budget our costs into their fiscal year plan. In addition to subscriber income, which varies from 7,000 - 20,000 per annum per institution; we have encouraged use of our facilities by non-health, non-institutional users. In addition to revenue generated, via the use of the microwave network; we also secure income for video productions and engineering services.

The following represents INTERACT's rate structure: (Noting that our airtime programming is a Network priority because it is the means for providing interaction among all stations, and represents the primary source of income.)

1. INTERACT subscription fees, set by the Board of Directors, represents a base fee plus anticipated use per station.

2. All ad hoc airtime programming will be charged for on a $50.00 per hour, per station basis.

3. There will be NO reductions or changes in rates for:
   a) originator
   b) long term vs. short term user
   c) inside vs. outside user

4. The above hourly rate includes "normal" set-up and take-down. "Normal" equalling approximately 1/2 hour per program of one person's time.

5. Additional charges per any airtime programming will be reflected on:
   a) Job Order Ticket and be charged as follows:
      1. Additional "set-up/take-down" time @ $15.00/person/hour
      2. Additional production equipment @ rate schedule
      3. Additional production personnel @ $15.00/hour
      4. 2 and 3 above rates are computed from set-up through take down.

6. An evaluation of perspective inside users' source/quality of funding will be made in order to determine if that user could support our rate with "outside" money.

7. An outside user, if evaluated favorably by all relevant station managers and the Network Manager, may be able to charge a program(s) against prepaid "inside" money. This case would only arise if an outside user, without funding support, had a program which would realistically benefit any or all prepaying institutions.

Non-Airtime Rate Structure

1. Non-airtime programming production rates are established by computing:

   a) $15.00 per hour, per person for all INTERACT personnel, plus
   b) pertinent rate(s) from the production/equipment rate card
   c) Tape stock not included
   d) Travel time, mileage and miscellaneous expenses not included.

Rental

Due to the fact that 90% of INTERACT's equipment is government purchased, it cannot be rented. However, a maintenance charge of 2% of replacement value is charged for the use of loaned equipment. No equipment is loaned without an INTERACT operator unless approved by Network Manager.

Personnel Cost Rates are determined as follows:

$5.00 - reflects to Administration
$5.00 - reflects to Engineering
$5.00 - reflects to Production

$15.00   Total Charge

This procedure is followed in order to provide revenue to sections of INTERACT which, under normal business environment, would be generated by Equipment Rental at 5-10% of Market Value.

The following reflects INTERACT's FY 79 fiscal summary. For further clarification, the expense portion of this summary does not reflect equipment depreciation (per the rules of government-owned equipment).

INTERACT FISCAL SUMMARY - FY79

Revenues	Total
Subscriptions	$ 101,000
Ad Hoc Airtime	13,654
Tape Production	51,180
Other	65,467
TOTAL	$ 231,301

Expenses	
Personnel	$ 168,250
General Op	30,420
Other	20,215
TOTAL	$ 218,885
Surplus %	12,416

The last avenue for securing revenue is through grants and contracts. Although I am not a supporter of this type of revenue support for operating expenses, I fail to see any other alternative than "soft money" support for capital expansion and equipment replacement.

Another point worth noting is the recent willingness in our region of third party insurers to openly condone reimbursement to health institutions for their involvement with INTERACT.

Expenses are divided into "inside" - institutional subscribers and "outside" - all others. Both are recorded on a monthly financial report to each station manager, with the latter also being processed through a formal billing prodcedure.

In order to insure all tasks are properly assigned a rate and a customer, each person performing "outside" work completes a "work tag" which is processed through a system of accounts and billed accordingly. An accounts receivable schedule has been established to insure payment.

How Does the System Work - Programmatically

The value of INTERACT Network lies in programming and its relative interaction and impact on the user. INTERACT's current programming falls into the following categories:

1. Consultations
   Examples: a) dermatology
              b) client - lawyer

2. Specialty Conferences
   Examples: a) ENT (UVM - DHMC)
              b) Human Services (SRS)
              c) Vermont Public
                  Service Board

3. Education
   Examples: a) Continuing Medical
   Education for Health professionals
   (to include AMA, Category I Credit)
   in conjunction with the Office of
   Continuing Medical Education
              b) Continuing General
   Education (inc. hospital administration and general public)
              c) Sharing of faculty
   resources among network educational
   institutions (Anthropology)

4. Direct Service (Staff Meetings)

5. Patient Education

There are some who theorize that a buy/sell philosophy can apply to programming in relationship to the size and type of participating institution. The assumption that all medical centers (in our case - Medical Center Hospital of Vermont and Dartmouth-Hitchcock Medical Center) sell programming and that all small rural institutions buy programming is false within the INTERACT system. One must take into consideration the personality of any given exchange, i.e. is it professionally threatening to ask for help, or I don't give a damn what those country 'doc's' know. We have evidenced many situations where the large resource centers are very willing to disseminate information. Likewise there are examples to reflect the willingness of a small country doctor to ask for educational or professional assistance from the medical center(s). There is also evidence of a strong relationship among the smaller hospitals (RMH, CGH, and CVH), exclusive of the medical centers and vice versa.

I don't believe such buy/sell generalization should be made. Each program and each institution is often independent of all other considerations - the aesthetic part of program development.

In order to avoid copyright problems with program presenters, we send to the sponsoring department or individual participants a release form, stipulating that we are allowed to distribute said programming in an educational, professional sitting.

Responsible Personnel   The Network adopts a 'team' approach in providing the service requested by any user. Expertise in production, engineering and management is provided by staff according to the requirements of the production. Position descriptions for each staff member is defined according to responsibilities with production, engineering and management. Obviously, those responsibilities vary in degree by each position and the qualifications of individual staff. Emphasis is placed upon the group effort utilizing the most and best each person has to offer which can result in the best possible product/program. This is better understood with a summary of the process of staff response to user request.

Process   The following chart is a summary of such a process which is applicable to all programming.

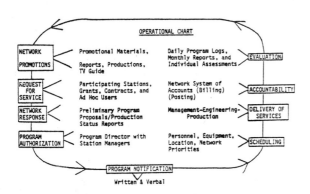

It is obvious from the above chart that an essential component of the Network's program activities is labeled coordination. However, the efforts subsumed under this label are much greater and more diverse than one might ordinarily assume. Those individuals bearing the title "Station Manager" at each institution in the system's link fulfill a three-fold role: (a) germination and nurturing of viable working relationships between community, institutional and medical center personnel who are dealing with similar problems; (b) perform the function of a television producer, cameraman, and technician; and (c) coordination itself, which is composed of needs assessment, resource identification, and bringing together of the two at a specific time and place for a mutually agreed upon purpose. In order to successfully accomplish the above, the Station Manager must effect a successful liaison between all other Station Managers and Network Administration.

Culmination of the above occurs on a weekly basis, at which time all station managers and

the network manager meet to discuss programmatic concerns of the Network. This meeting provides timely response to program needs, evaluation of new ideas; and, in general, makes each INTERACT 'link' aware of the Network's total program requirements. In addition, yearly program planning sessions are held each June to organize programming which will be aired on a regular (weekly/monthly) basis throughout the following (program) year.

Formal organization and distribution of the programming efforts, noted above, culminate monthly in the form of a TV Guide. This Guide is distributed to 1,000 individuals and institutions in our area of operation. Its purpose is to attract new users, provide awareness of existing programs and to relate items of interest.

INTERACT's stations during the course of ten years have been the originating points for hundreds of programs and thousands of interactions. Interactive programs generally involve small groups of individuals at each of two stations. Additional stations may be included but they usually relate passively. Although rapid switching between locales is possible, it tends to add an element of confusion to the session and detracts rather than contributes to the degree of interaction unless a pre-arranged switching "sop" is formulated with the program sponsor.

Over the past three years the INTERACT Network has averaged 2,200 hours of institution programming to over 21,000 participants. For FY 80 we anticipate 3,000 hours of Network programming to 25,000 individuals.

Programming - Finding the Need

INTERACT was created with the idea of facilitating medically related information to rural New England. Prior to INTERACT's inception, there was not an "in depth" programmatic needs assessment or utilization study performed. Consequently, several years later (and without Federal funding as a subsidy) INTERACT is having to step back and re-evaluate its usefulness in terms of program needs.

Since the Federal "subsidy" has terminated and "fee for service" begun, programming quantity has decreased and content changed. A cursory evaluation of current program needs reflects that the following are important for the future success of INTERACT TV.

a) Further diversification of programming.
b) More emphasis should be placed on Continuing Medical Education (with Category I Credit) and Continuing

General Education.
c) Initiation of programs utilizing state and local governments and private business
d) Interface with cable TV network
e) Interface with New Hampshire and Vermont ETV Networks
f) The sharing of faculty with network educational subscribers. (i.e. Dartmouth, UVM, and UNH).

Production

Production as related to 'live' telecasting is a multifaceted responsibility. Besides the standard activities of pre-production coordination, airtime implementation and post-production (i.e. administrative) follow-up, the following are additional characteristics to be considered in 'live' two-way television.

a) Studio set-up - in order to promote audience interaction, cameras and monitors must be arranged to accommodate such and not deter the local environment. The environment is influenced by the number of participants, room size, and the type of presentation.

b) Class Size vs. Interaction:

Situation	% of Audience Interaction
1. one-to-one	100
2. sm. group to sm. group	80
3. one-to-group	75(depending on instructor)
4. sm. group to larger group	50
5. large group to large group	20

The Program Director, if consulted, will advocate to a perspective user the situation above which best reflects his/her needs and still promotes interaction.

d) Lighting - In most live production cases, low light (silicone-diode), black & white cameras are utilized. Most users advocate an unobtrusive profile by production personnel. Using the above-referenced cameras, INTERACT production is able to televise in normal room light with little more than a camera, microphones and cameraperson.

From the inception of INTERACT, production or a combination of production and engineering have played an integral role in the implementation of 'live telecasts'. It is only within the last two (2) years that a greater emphasis has been placed on the role of (non-airtime) production.

This later emphasis was created out of the need for more full utilization of INTERACT personnel and existing equipment in order to secure additional revenue. These non-airtime services include:

a) equipment rental (with and without operator)
b) simple videotaping and/or playback
c) complex scripted videotape productions
d) b & c above can be accomplished "in-house and remote"

These services are available, per a production rate schedule, to subscribing and ad hoc users. It is interesting to note that in the first full year of offering this service, it accounted for 27% of INTERACT's operating budget.

Engineering

How does the system work - technically

The Network is a dedicated, closed-circuit, duplex microwave system. The Network consists of a trunk or backbone system, operating in the 6 GHz business band, extending from Mt. Mansfield, Vermont to Mt. Ascutney, Vermont and seven spur links on the 12 GZz band, each connecting a point in the trunk system to a user location. The trunk system shares facilities with Vermont Educational Television, and all spurs are single hop.

Switching between stations is accomplished remotely by use of a touch tone panel located in a Master Control Center at Hanover, New Hampshire. Each participating station can access the network in order to establish audio and video contact with any other staion. Any station may transmit to all other stations simultaneously in a network mode.

Channels provided are:

a. NTSC video
b. Program aural
c. Voice grade intercom
d. Remote camera control subcarriers
e. Physiological telemetry

Audio and video are switched independently of other channels. The system complies with EIA RS-250-A, "Electrical Performance Standards for Television Relay Facilities," other appropriate standards and regulations relating to antenna construction and marking, and F.C.C. rules and regulations.

INTERACT has evolved over the past five (5) years from early planning and the engineers drafting table, to a smoothly operating user oriented system.

It became painfully obvious, during initial operation, that the remote switching, controlled by the TT signals was not reliable. Any non-linearity in the receivers or an operator whose voice was of the required pitch would cause the switching matrixes to be <u>randomly</u> changed. This was always frustrating and sometimes humorous.

On one such occasion, two conferences were being conducted simultaneously using opposite ends of the network. Someone sneezed while wearing headsets, and the effect was chaos. All four groups instantaneously found themselves looking at total strangers. The solution to the problem proved to be a 20KHZ FM subcarrier which was added to the 6.17 channel. The 20KHZ subcarrier was modulated with the touch tone signals and demodulated only at the switching locations. The modulator was designed using an NE 566 IC, the demand is a NE 560 phase locked loop.

Another problem which was successfully solved concerned conferences which involve more than two stations. The system is designed so that while two stations have complete bi-directional video and audio, additional stations have received video and audio capability only.

In order to promote interaction however it was considered desirable to be able to switch rapidly to a passive station so that they would have full bi-directional transmission and would become one of the active parties. It was found however that during the process of switching parts of sentences (often questions) were lost, destroying the communication. In order to obviate this, a new 7.5 KHZ "passive" audio channel was added to all stations.

At each station the main audio program channel (6.8 MHZ) signals, both transmitted and received, are combined and retransmitted on the passive 7.5 MHZ. When a "passive" station switches the input of their audio amplifier from 6.8 MHZ (active) to 7.5 MHZ (passive) they are able to receive audio from both of the "active" participants and thus follow the conversation with ease.

If the "passive" station wishes to interrupt the "active" participants in order to comment, the individual operating the camera at the passive station notifies the central control room through the "head set" intercom.

Systems Design and Equipment Reliability

INTERACT's microwave radio system was designed to match the existing Vermont ETV microwave TV network. Vermont ETV sites

provide INTERACT with developed mountaintop facilities in a basic North-South relationship. This routing provides access to the participating INTERACT stations with very minimal costs for site development.

Both the radio equipment and the origination equipment have proven to be quite reliable. The duplex radio equipment in use was manufactured by Microwave Associates of Burlington, Massachusetts and includes:

		(Quantity)
a)	Model 7BX	(8)
b)	Model 12BX	(8)
c)	Model 12G	(2)

After some initial problems in the BX series transmitters, the system has stabilized to approximately 64 hours total down time per year. This figure includes access time to reach unmanned sites to make repairs. In the winter, some mountaintop relays can be reached only by skimobile and snowshoe and in the summer by hiking.

Problems Encountered in Maintenance

The greatest source of maintenance problems encountered at INTERACT have been caused by the "ravages of nature." Damage to antenna structures caused by icing conditions, high winds, electrical storms, falling ice, loss of commercial power for extended periods of time, (i.e. greater than twelve (12) hours) have all been encountered. We seem to be defenseless against this type of disaster.

The origination equipment installed consists of Telemation 2100 series cameras. These have proven to be very reliable. INTERACT engineers modified these cameras to accept Silicon Diode tubes. The RCA 4532 series perform very well under the low-light conditions typically encountered in conference rooms and auditoriums.

As the Network became operational the need for additional equipment to meet user requirements became apparent. In an effort to maximize the available dollars to meet these immediate needs some used origination equipment was purchased. The life expectancy of much of the equipment when purchased new is estimated to be between ten to fifteen years. The useful life of second hand equipment naturally depends on age and original quality.

The problem of allocating dollars for the eventual replacement of this government owned equipment simultaneously with the rapid phasing of the demonstration from federal support to self sufficiency is critical and as yet unresolved.

Because the equipment is government owned or purchased one cannot establish a rental fee which includes capital costs or amortization to other government contract/grant recipients of which INTERACT has many. In addition because the equipment is not owned by the 'custodian' one is not allowed to depreciate or establish a contingency fund for replacement.

Given these circumstances, an element of "self destruction" is built into a demonstration that has a very large technical component.

Currently INTERACT, the ward of approximately $500,000 worth of equipment when originally purchased is unable to generate income from it, cannot depreciate it even if revenue was available to effect such an expense and according to recently enacted F.C.C. rules must replace seventy five percent of it by 1985.

Engineering Manpower Requirements

The number of engineering staff has varied from a high of three to a low of one. An average of two seems to be adequate for typical operations. The logistics encountered in such a system make it impossible to consistently maintain the reliability of the system with just one man. The north and south terminus of the Network are over three hours apart by automobile.

Summary - INTERACT Overview

INTERACT (the Vermont-New Hamshire Medical Interactive Network) is a two-way closed-circuit television system which links together hospitals and educational institutions in northern New Hampshire and Vermont. The facilities currently joined include the Dartmouth-Hitchcock Medical Center (Hospital and Medical School) in Hanover, New Hampshire; the Claremont, New Hampshire, General Hospital, a community hospital of eighty-four beds; the Rockingham Memorial Hospital in Bellows Falls, Vermont, a community hospital of fifty-three beds; the Central Vermont Medical Center in Berlin, Vermont, a community hospital of three hundred and twelve beds and the Medical Center of Vermont (Hospital, Medical School and School of Allied Health Sciences) in Burlington, Vermont; the Vermont Diagnostic and Corrections Facility-St. Albans, Vermont; and the Veterans Administration Center, White River Junction, Vermont. Until its closure in June 1975, the Windsor, Vermont, State Prison, the State's only maximum security facility was also joined to the Network.

The first portion of the Network was funded by the NIMH and in 1968 linked Dartmouth Medical School in Hanover, New Hampshire, to the Claremont, New Hampshire, General Hospital, thirty miles away. The initial demonstration tested the feasibility of providing psychiatric consultations via television.

In 1970, the Lister Hill National Center for Biomedical Communications supported exploration of a variety of health education and service programs between the two stations and simultaneously undertook an engineering study to determine the feasibility of constructing a network which would reach across northern New England. This study included twenty hospitals and schools extending from Portland, Maine to Burlington, Vermont.

In 1971, construction began on a four station duplex network, the backbone of which is carried via three mountain relay points. This system became operational in 1972. In 1973, three additional facilities served on a scheduled basis by a mobile van were joined to the Network and operation of the seven station system was initiated in November of 1973. Capital costs of the Network totaled $783,502.

Three principal factors in addition to a superficial needs assessment were considered in selecting potential facilities to be served which included: the geographic location of the facility, most particularly whether or not line of sight for microwave transmission was available to a mountain top relay point; the unique character of each facility which provided opportunities to explore many different uses; the likelihood that the institutions involved would, in the long run, be able to afford the operation and maintenance costs of the Network. As INTERACT has evolved from a field trial to a fee-for-service operational Network, organizational ties between member stations have been developed in order to assure equal representation among the paying users.

Current FY 80 operating costs are approximately $250,000. Income is derived from member institutions and ad hoc users. A serious problem concerns a lack of funds for equipment replacement. Federal regulations prohibit amortization of federally purchased capital equipment or a rental fee. Thus it was not possible to establish a contingency fund during the experimental years.

Staff responsibilities have been clearly delineated between management, engineering and production. Each of the initial four stations have had a full-time coordinator who has responsibility for defining local community health needs which might be served via INTERACT, identifying resources at other INTERACT stations to meet these needs and catalyzing the necessary programming. We believe that this is one of the most critical aspects of interfacing the technology with the disparate needs of the community.

Internal evaluation has been focused principally on an analysis of user attitudes and acceptance coupled with feedback to management, engineering and production when barriers have been encountered. This trouble shooting and problem solving evaluation has been found extraordinarily useful in gaining acceptance of the Network.

Short-term funding following initial operation of the full Network required a drastic curtailment of new porgram development and evaluation and necessitated an almost exclusive focus on the issue of self-sufficiency. While valuable momentum and opportunities for exploration were thus compromised, substantial strength and unity of purpose evolved from the necessity that users bear the total burden of cost. A non air time dimension of INTERACT has emerged - Media Outreach, which is a valuable byproduct of live programming. Media Outreach videotapes air time INTERACT programs, edits them and distributes them to many hospitals in Vermont and New Hampshire that are not currently linked to the microwave Network.

The future of INTERACT appears promising. Soon another station will be developed in Brattleboro, Vermont, which will incorporate the local hospital, a psychiatric retreat, a child development center and the State Office of Human Services. Also, active planning continues for extension of the Network into southeastern New Hampshire and into the more rural North Country. Linkages with other terrestrial networks via satellite is becoming a more defined goal for the 1980's. And, in the future we may claim the first success at interconnecting two existing networks - INTERACT and the Central Maine Interactive Television System.

Much has been learned, much remains to be discovered. Interactive television is a communication media which is unique. It has its own strengths and constraints. It is not yet fully integrated into our health care and education system. Continued exploration of its full potential, both as a substitute and supplemental way of sharing resources and extending services, provides exciting opportunities for the future. And, as the cost of energy increases, moving information instead of people via technology will be an objective of many organizations and institutions.

The Teleconferencing Resource Book: A Guide to Applications and Planning
Lorne A. Parker and Christine H. Olgren (eds.)
Elsevier Science Publishers B.V. (North-Holland)
©Center for Interactive Programs, University of Wisconsin-Extension, 1984

LIVE INTERACTIVE TELEVISION: THE MEDICAL GRAND ROUNDS

Julia C. S. Keefe
Producer-Director
Health Communications Network
Division of Continuing Education
Medical University of South Carolina
Charleston, South Carolina

*Televised grand rounds are a wave
of the future. With the increasing
expense of attending conferences
and the current emphasis placed on
continuing medical education, these
broadcasts provide the most cost-
effective and simplest way to keep
up-to-date with medical progress.* --
Lawrence L. Hester, M.D., Chairman,
OB/GYN Department, Medical University
of South Carolina

Health professionals in South Carolina
have a unique opportunity: each month they
may attend grand rounds in Psychiatry, Family
Medicine and OB/GYN presented by hospitals in
four distant locations within the State. Yet
the farthest they will have to travel is to
their own hospital auditoriums. They will not
have to interrupt a full day's schedule of office and hospital duties. Instead, these busy
medical professionals can interact with their
colleagues for an hour of medical grand rounds
through the State's live interactive broadcast
system. These live videoconferences represent
an innovative and economically feasible approach to continuing medical education.

The Health Communications Network (HCN)
is a full-color broadcast facility dedicated
to the development of effective continuing education programs for South Carolina's health
care professionals and hospital support personnel. The fundamental mission is to anticipate and respond to the needs of health professionals in a manner that will enable them
to stay abreast of established information and
to offer information and instruction in new
methods and techniques that have proved to be
meaningful. The ultimate purpose is to contribute to the achievement of optimum quality
health care for the people of South Carolina.

HCN was established in May 1969 with a
link to four hospitals and the Medical
University of South Carolina (MUSC) via telephone lines and the use of videotaped programs
at each participating hospital. The initial
purpose was to provide physicians and other
health care personnel with educational opportunities concerning pediatric cancer. HCN now

links 30 hospitals, the University of South
Carolina's School of Medicine and MUSC via
closed circuit television. Currently, over
80 hours of continuing education programming
are provided on the closed circuit Network
each month. Also, health information programs
designed for the public are broadcast weekly
throughout the State on open circuit educational television and FM radio. For several
years a leader in national continuing medical
education broadcasts, the Network has recently
emerged as a pioneer in the field of live
multi-site teleconferencing.

The Health Communications Network is
continually expanding its services in response
to the increasing responsibilities, demands
and challenges faced by today's health professionals. Increased involvement in live
interactive programming, utilizing the Network's two-way interactive video capability
and Network-wide telephone talkback system, is
setting a state-wide trend and provides an important link between teaching institutions and
community or regional hospitals throughout
South Carolina.

HEALTH COMMUNICATIONS NETWORK

Map shows location of participating S.C. hospitals.
*Indicates 2-way interactive video and audio sites.

For staffs in smaller hospitals located at a distance from any major metropolitan center, opportunities for top-quality continuing education are often severely limited by time and financial restrictions. Live interactive televised seminars offer a feasible alternative to health professionals in institutions such as these, enabling them to receive updated continuing education in their own hospitals with less time away from work and at a minimal cost. To cite an example, the Health Communications Network broadcast a seminar on Quality Assurance live to its 31 member institutions for three hours a day, October 13, 14 and 15, 1981. Two hundred and eighty-six health professionals in South Carolina registered for the seminar which was conducted by Dr. Richard E. Thompson, a nationally recognized authority on this timely subject. The average cost per person was $30.10. To attend an equivalent seminar in Columbia, South Carolina, the average cost per person would be $310.68, and for the same group to attend a seminar in Atlanta, the average cost per person would be $655.98.

COMPARISON OF COSTS
FOR EQUIVALENT QUALITY ASSURANCE SEMINARS

Atlanta

Air Fare Atlanta	Registration Fees $195 each *	Lodging, Food, Taxi, Parking
$47,042	$55,770	$84,799

Total cost - $187,611
Average cost per person - $655.98

Columbia

Travel Cost To Columbia	Registration Fees $195 each *	Lodging, Food, Taxi, Parking
$2,704.80	$55,770	$30,381

Total cost - $88,855.80
Average cost per person - $310.68

Broadcast Over TV by HCN-DCE

Total cost to hospitals - $8,610
Average cost per person - $30.10

* $195 registration fee is the average for seminars scheduled by the "Education Division of the Joint Commission for Hospital Accreditation."

The Network-wide telephone talkback system and two-way interactive video capability in five of the Network sites enable audiences and faculty members to actively participate with one another through questions and discussion. These five sites form the production core of the TV grand rounds series and include Spartanburg General Hospital in Spartanburg, Greenville General Hospital in Greenville, Richland Memorial Hospital and the University of South Carolina's School of Medicine in Columbia, and the Medical University of South Carolina in Charleston, where the Health Communications Network is based. The interactive system was completed in 1979 and its headquarters are centrally located in Columbia, the home of the South Carolina Educational Television Network (SCETV). In close cooperation with SCETV, HCN produces its TV grand rounds series in Psychiatry, Family Medicine and OB/GYN on a monthly basis.

For the production of live medical grand rounds, incoming signals are transmitted from the five originating sites that are responsible for the visual portions of the programs. The signals are selected, as program content requires, by a director in Columbia. These signals are then transmitted to the 31 Network member institutions. Questions and comments from the interactive sites in Charleston, Columbia, Greenville and Spartanburg can be asked from the floor in any random order, with the questioner on camera and visible to the entire Network. Audiences at the 26 hospitals that cannot originate video can view the live broadcast and participate through an audio-only talkback system which is then re-transmitted to the Network.

A second talkback circuit is used by the director at SCETV to coordinate such technical matters as camera shots and audio-video levels with each site director. The director at SCETV has a bank of picture monitors which are used to view all incoming signals at any given time. This facilitates precise switching among the originating sites, an effect that is essential to the program's continuity.

From the community hospital perspective, medical grand rounds are an integral part of the continuing education program. Spartanburg General is one of the five Network member institutions in South Carolina that has two-way video and audio capability. Physicians in the Spartanburg area have played an important role in TV grand rounds. Their OB/GYN department, headed by Harold R. Rubel, M.D., initiated the first case presentation in September 1979. Through the close cooperation of Lawrence L. Hester, M.D., chairman of the OB/GYN department at MUSC in Charleston, this pioneering effort was the first use of SCETV's interactive television facility for professionals in the medical field.

The TV grand rounds series is incorporated into a typical academic year--September through May. Each summer a planning session is held in Columbia. The physicians from departments of Psychiatry, Family Medicine and OB/GYN who act as program coordinators at each of the five institutions meet to assign case topics and dates for the coming year. Evaluations of the previous year's series are discussed and suggestions for any change in schedule or content are made. Also attending these meetings are the video coordinators from the presenting sites who work closely with the participating physicians in program production, as well as a technical advisor from SCETV. Before the new fall series begins, the group meets once again over the live interactive broadcast system to finalize details.

Two weeks prior to each presentation, a case description and announcement are sent to all interested physicians and health practitioners in the State. Thus, everyone has an opportunity to study the medical data before viewing the case. Audiovisual materials are reviewed before air time with the video coordinator to assure that slides, videotape inserts or x-rays can be successfully viewed by the television audience. Program content is often prepared and presented by a resident and may include such general topics as Home Births, Postpartum Psychiatric Syndrome and Rape Examinations, or a specific diagnosis and management, such as Ruptured Liver During Pregnancy. The content specialist then has the opportunity to reveal the actual diagnosis of the case in question, and an open discussion follows, calling on all participants throughout the State for their comments, opinions and questions. Spontaneity, therefore, is not limited to a personal encounter; the live interactive grand rounds create a stimulating and often challenging learning environment.

In its two-year history, the live interactive grand rounds series has produced over 70 case presentations. To date, these programs are viewed by an average of 300 physicians per month, and with a production cost of $900 per broadcast, physicians can earn an hour of CME credit for approximately $6.

OB/GYN TV GRAND ROUNDS
TOTAL ATTENDANCE REPORTED
September - December 1981

September	295
October	408
November	293
December	276
TOTAL	1,272

Reporting Sites

Anderson Memorial Hospital, Greenville General Hospital, MUSC, Richland Memorial Hospital, Spartanburg General Hospital

Through the combined technical efforts of the Health Communications Network and the South Carolina Educational Television Network, 31 health care institutions in the State are provided with face-to-face medical conferences which physicians may attend every month without the loss of time and money normally associated with such meetings.

Dr. Harold Rubel, now a veteran of TV live broadcasting, clearly describes the growth potential of this innovative approach to medical grand rounds:

When we first presented these (OB/GYN TV) grand rounds, I thought everyone was too stiff and the interaction was definitely limited. Now, as the participants are becoming more comfortable with the cameras, sets and lighting, a much nicer spontaneity has developed. This sharing of ideas on sometimes controversial methods of management can be extremely useful to the television audience. For example, the most recent grand rounds presentation presented (among the discussants alone) over 200 years of cumulative clinical experience. For the young house staff members present, this provides a unique learning experience. For the practicing physicians around the State, this certainly represents a source of expertise immediately available to them.

Contributors:

Joan K. Hunt   Richard Wrigley
Audiovisual Coordinator   Engineer
Spartanburg General Hospital   SCETV

The Teleconferencing Resource Book: A Guide to Applications and Planning
Lorne A. Parker and Christine H. Olgren (eds.)
Elsevier Science Publishers B.V. (North-Holland)
© Center for Interactive Programs, University of Wisconsin-Extension, 1984

132

TELECONFERENCING: TEN YEARS OF EXPERIENCE IN SOUTH CAROLINA

Timothy A. Prynne, MS, MPH, EdD
Director of Media Services
South Carolina Department of Health and Environmental Control
Columbia, South Carolina

In South Carolina the current teleconferencing system began when the closed-circuit educational television system linked up the main campus of the University of South Carolina in Columbia with its then five regional campuses, and included talk back capabilities. The College of Business Administration was the first academic area to offer credit courses over CCTV. Today both undergraduate and graduate credit can be earned via telecommunications in virtually every discipline. In 1971 this same type of telecommunications link was established between ten regional hospitals and the Medical University of South Carolina in Charleston. For statewide distribution, the video was transmitted on long-line from Charleston to ETV headquarters in Columbia and then throughout the rest of the ETV system. Thirty-two hospitals are now linked to the system, using one of four video channels with an exclusive talkback system for health-related subject matter. Fifteen public health district offices also participate in this health-related system, as does the new School of Medicine at the University of South Carolina Columbia campus.

Original seed money for the health communications system was provided by the South Carolina Regional Medical Program. The RMP's were established nationally in 1967 and poured many millions of dollars into individual state systems principally to develop and expand medical services in community hospitals, at the same time encouraging outreach continuing medical education from medical teaching centers to a scattered population of health professionals. The latter concept was continued and expanded by the Carnegie Area Health Education Centers in the 1970's. The neighboring state of Georgia used its grant monies to produce health-related programing at studios in Atlanta at Emory University School of Medicine and in Augusta at the Medical College of Georgia. These programs were videotaped and distributed to community hospitals for continuing education purposes. In both Georgia and South Carolina, although the emphasis had been on medicine and continuing education for physicians, increased recognition was being given to the fact that other professions are involved in health care, namely nurses, pharmacists, physical therapists, etc., and that

these groups have a need for professional continuing education as well as physicians. Significantly, in 1981 only 4.6% of programing on the South Carolina health communications system was for an audience of physicians alone.

The two case studies of teleconferencing presented here illustrate professional continuing education efforts for nurses and other public health personnel. These studies demonstrate cost-effective programing and the value of numbers when comparing investment to return in multi-site telecommunications. Both participant groups represent relatively large numbers of people with a similar learning goal -- that of professional credit for completion of the course work involved. Another component of major importance is the shared use of human and technical resources from a number of state agencies and academic institutions. State budget cutting has brought about a dramatic increase in similar cooperative efforts.

Our first case study is of a series of six one and a half hour programs for the professional development of clerical workers in health-related areas. Although the series was primarily aimed at clerical staff of the South Carolina Department of Health and Environmental Control (DHEC), over seventy staff members from a number of South Carolina hospitals participated in the series, bringing the total audience to 453 individuals. In constructing a theoretical model for comparison, only the DHEC personnel were used, so that the series of six programs given over a six-month period was assumed to be the equivalent of providing a similar program for 383 registrants within a two-day period at one location under normal conference conditions. It would be impractical to give one two-day workshop for such a large number of DHEC's clerical staff because effective administration of the entire state public health service would be difficult with most of the clerical staff away from their offices for two days. Table 1 shows a comparison of costs between a two-day workshp and a teleconference. Accommodation is estimated as $10,724; mileage (claimed at $0.18/mile, 1978 level of reimbursement) is estimated at $2,998, based on an average distance of 174 miles with four persons sharing each vehicle. Another major cost of such a workshop would be productive time lost, here estimated as $9,578. Admittedly these costs are hypothetical but, whatever criteria are used, there is a wide disparity between the costs of the face-to-face meeting and the teleconference program where the production costs have been calculated as follows: broadcast time is $83 per hour (based on DHEC's average use of twenty-one hours of broadcast time per month; total annual TV/talkback circuitry costs $20,988). A TV production crew (one person at about $10/hour and two at about $7.50/hour, 1980 salary level) totaled $675 for the 27 hours of program development and transmission/record time for the series. In addition, graphics, slide reproduction and other A/V costs were calculated to be approximately $450.

versus presenting the same curriculum in three cities for two weeks each. Since the hypothetical tour would be held in only three locations, travel and accommodations would have to be provided for faculty, staff and students. The costs for teleconferencing the 12-part series include the $83/hour for program time as calculated for the first case, and also include collaboration with the College of Nursing at Clemson University. Clemson provided faculty, study guides, reference lists, and the pre- and post-testing procedures. The programs were produced and recorded at the Clemson University studios, with a DHEC Office of Nursing representative and myself as faculty advisor and production consultant, respectively.

Table 1: Clerical Development Series

Two-day seminar in a central location for 383 DHEC clerical staff:

Cost of productive time lost (including average travel time of 4.1 hours/participant) based on median salary of about $6/hour	$ 9,758
Average mileage reimbursement (four persons/vehcle) based on 174 miles per trip at $0.18/mile	2,998
Accommodation and subsistence at $28 each	10,724
TOTAL	$23,301

Teleconference - six 1½ hour programs over a six month period:

Broadcast time, 9 hours at $83/hour	$ 747
Staff time, 27 hours program development/production/recording	675
A/V costs	450
TOTAL	$ 1,872

Table 2: Nursing - Family Health Series

Three-city loci for 150 participants for 2 weeks in each city:

1978 costs for faculty, travel,etc.	$18,000

Teleconference - 12-part series:

Clemson faculty, media resources	$ 6,000
Program time, 18 hours at $83/hour	1,494
DHEC staff travel, media resources	1,000
TOTAL	$ 8,494

The second case study consists of a 12-part course in Family Health for public health nurses given over a full year. To serve the professional education needs of public health nurses not holding a bachelor's degree in nursing (which is preferred for the state service rank of Community Health Nurse), a curriculum was designed for teleconference presentation. At the end of the course, each of the 150 nurses in the class would receive professional certification and 2.2 continuing education credits. Table 2 gives a very simple comparison of teleconferencing costs

It must be emphasized that CCTV production costs are based on the use of the system for an average of twenty-one hours per month ($20,988 / (21 x 12) = $83). The long lines of transmission within the state are leased from Southern Bell, with most of the costs being borne by the public school CCTV system (900 schools connected at this time). Health-related programing is only a small part of SC-ETV's total closed-circuit transmission time.

Some additional details of the SC-ETV system are relevant at this point and are especially pertinent to the future of teleconferencing in South Carolina. Currently the state public school system is presenting a total of 141 courses over CCTV and radio, with a total enrollment of over one million children. An additional 75 courses are presented in support of higher education. Broadcasting to the general public over open-circuit only accounts for approximately 10% of SC-ETV's total programing schedule. As a result of a major study of projected needs for the system over the next twenty years, SC-ETV is now proposing to purchase its own transmission system, thus eliminating the need for leased telephone circuitry. The intention is to construct an Instructional Television Fixed Service (ITFS) system consisting of a number of omni-directional transmitters to provide television and radio coverage of the entire state, in addition to a number of tape-delay and repeater centers. A capital investment of approximately $25 million for this state-owned system is calculated to save over $40 million in line charges over the next ten years (the telephone companies -- and there are over 50 independent telephone companies in South Carolina -- have raised their line charge rates an average of 76% annually since 1979). At the current rates, by 1990 the annual costs for SC-ETV transmissions would be in the region of $18 million. In addition, the proposed ITFS system would expand the capability of SC-ETV in covering 100% of the school systems, in contrast to the 56% coverage today. No new reception sites, including hospitals, health departments or other state agencies, can be added to the present system because of the crowded circuits and prohibitive costs -- line charges and telephone company termination liability agreements. The new SC-ETV transmission system would be free of charge for state agencies which, in the case of DHEC, would eliminate the current annual cost of $20,988. An additional four channels of video, for sole use by state agencies and institutions of higher education, would be added to the four existing channels which are now shared by the entire system.

In South Carolina we have new entered an era of shared resources, especially among public service agencies and educational institutions. This situation is not unique to South Carolina; it has been precipitated and accelerated by the current economic crisis. The figures given in the case studies presented here are not intended to provide a detailed cost-benefit analysis of teleconferencing but are adequate for giving serious consideration to this form of communication as compared to the traditional face-to-face meeting, conference or workshop.

There are questions to be asked as to the effectiveness of teleconferencing versus the traditional approaches when evaluating fulfilment of a program's goals and objectives. There is a vast amount in the literature on this subject; some should be required reading for organizers of teleconferences.

Interactive television and teleconferencing is still in its infancy as far as general acceptance is concerned. Most people, if given a choice, would still rather go through the tedious and relatively expensive process of traveling many miles for face-to-face meetings instead of interacting with each other over a multi-site communications system.

If you are considering cost-cutting by installing cheap equipment, you will be well-advised to budget a maintenance agreement into your system. Those of use who are not engineers have to assume that the hardware is going to work -- we have enough "people problems" as it is without worrying about technical breakdowns. But equipment still breaks down!

In 1968 when a new basic sciences building was going up at the Medical College of Georgia, the TV studios were to be included in the same building. Having neglected to properly work out the location of the large transformers for the building's power distribution, the engineers spotted an area on the plans with significant ceiling height and promptly hung the transformers in the television production area -- where they may be to this day!

When the television studios were being constructed at Grady Memorial Hospital in Atlanta (affiliated with Emory University), the administrators of the hospital insisted that the studio match the general decor of the building -- the studio walls were duly lined with especially mined imported green marble! Communications problems can even occur in the planning and design of communications systems.

Although program origination locations may be next door to receiving locations, the techniques for production and presentation remain the same for systems that may cover thousands of miles. The ground rules for effective presentation techniques remain the same. Teleconferencing has not created a new discipline in communications, only variations on very well-tried principles and themes. This does not mean that expertise in audiovisual production will automatically give one the credentials for teleconferencing, although it might help. A higher-than-average ability to withstand considerable stress might be the ideal characteristic for the teleconference organizer. On the operational side of teleconferencing, we are dealing with a great variety of disciplines, expertise and job strata but each and every link is as important as another.

In South Carolina, with more than 50 independent telephone companies responsible for the circuitry, and hundreds of microwave hops between transmitter and receiver sites, and no two receptions sites the same in design, and accommodation varying from site to site so potential audience participation is a very important consideration, it is vital that the organizer(s) of a teleconference be prepared to run the risk of generating more antagonism than gratitude. The different duties, schedules and, in many cases, priorities for participation in teleconferencing on the part of site managers must be taken into account when

planning any multi-site meeting. The recurring
nightmares, of course, are breakdowns in transmission at any point in the system -- whether
or not it affects the whole system. The people
having problems at the end of a particular link
will let the whole network know very quickly
and considerable frustration is caused by the
inability to do anything at the time. The problem becomes more complicated when either the
presenter or the audience is taking part in
teleconferencing for the first time. It may
take a great deal of energy to persuade these
disgruntled people to participate in teleconferencing again. The problem of dead air is
most unnerving for first-time presenters;
"seeded" questions not only bolster the confidence of the presenter but encourage the
"hidden" audience to participate as well.
Somebody has to maintain interactive control
throughout a teleconference.

As in so many other activities, the key
words of successful teleconferencing are PLAN-
NING, PREPARATION, EDUCATION/FAMILIARIZATION,
and COORDINATION. In most cases the longer the
preparation time the better. The commercial
networks plan overall programing schedules at
least a year in advance. Teleconferencing
needs the same lead time in planning and preparation. The more receiving stations there
are, the longer the preparation time to be
allowed.

In conclusion, I recommend that all
schools of communications, media, television,
and especially journalism, include the principles and practices of teleconferencing in
their curricula. At least one of the pre-
requsites should be fundamentals of curriculum
and instruction. In addition to the main
course work in media production, students
should be able to present different media
strategies to their prospective clients and
should be able to support one recommendation
over another with case examples and, above
all, pertinent evaluation instruments. In
most cases, the actual teleconference itself
would only be one element in the communications
process -- not a single concept form of communications strategy. Clear writing of goals
and objectives would be essential. Also,
teleconferencing should enter into the language of those who write up job descriptions,
sufficient to ensure adequate remuneration to
match the responsibilities and the work of
those or organize teleconferencing activities.
Lastly, teleconference organizers should not
be confused or intimidated by the use of the
word "telecommunications" by computer and data
processing people, to whom the word means the
process by which machines talk to machines.
I believe that there is more professional,
and perhaps personal, satisfaction to be
gained from encouraging and helping people to
talk to people.

The Teleconferencing Resource Book: A Guide to Applications and Planning
Lorne A. Parker and Christine H. Olgren (eds.)
Elsevier Science Publishers B.V. (North-Holland)
© Center for Interactive Programs, University of Wisconsin-Extension, 1984

THE IRVINE INTERACTIVE CABLE TELEVISION PROJECT
1974-1981: A CASE STUDY

Virginia Boyle, Ph.D., and Kimberly Burge
University of California, Irvine
Irvine, California

Background

In 1974, the Irvine Unified School District, in Irvine, California, used two cable television channels to link two schools. The result was that district fifth-grade students "talked back" to their TV sets for the first time. This two-way Interactive Video system was designed by UCLA Professor Mitsuru Kataoka in collaboration with school district staff under the leadership of Superintendent A. Stanley Corey, and Community Cablevision Company technicians.

The system was designed and installed with the intention of providing the following capabilities for IUSD students, teachers, and administrators:

1. Cost-effective person-to-person interaction among learners and teachers.

2. Accessibility to learning resources beyond the physical limits of the school site, and eventually beyond the limits of the school district, the community, and potentially the state.

3. Personal control of and responsibility for the learning process through hands-on access to the system equipment, and in developing student-created programming.

4. Experience in using television as an active, rather than passive, medium of expression, the "electronic classroom."

In testimony before the House Subcommittee on Science and Technology:

"Mr. Corey then proceeded to describe the Irvine project which had a philosophical base which implied that 'in the application of technology to the problems of education there has been too much big-brain-little brain thinking.'

"It is our assumption that all learners have things to contribute to their own learning and to the learning of others, and that if we look in the larger context at America's social problems, our concern is not one of failure of education in the informational sense. If there is a failure, the failure lies in the ability to produce self-actualizing, thinking young people who believe enough in themselves to look into an area of interest, gather what data they can, reach some decisions about what data they can, reach some decisions about what should happen and then have the courage to act."

Each year, as the community and school district grew, additional interactive video sites were added to the system. The simplicity of the system was such that programming development, scheduling, system operation, and site-level equipment maintenance was coordinated in the IUSD central offices by administrators and classroom teachers who had little or no technical training. Access to the system was decentralized, and control of the interactive channels was maintained by the users at each location. Classroom teachers and their students were encouraged by this to produce their own programming.

Soon, the system was programmed an average of four to five hours daily through the academic year. In 1978, the UCI Office of Teacher Education joined the system which, by this time, had expanded to an educational institutional network of more than twenty sites. Funding for the Office of Teacher Education project was provided for two years by the UC Irvine Committee for Instructional Development. After two years, operation of the UCI-ITV project was funded out of the Office of Teacher Education budget. Programming is now coordinated by Dr. Virginia Boyle, Assistant Director, Office of Teacher Education, and Kimberly Burge, Campus Television Coordinator.

Technical Design

The Irvine Interactive Cable Television System is a multi-institutional network providing teleconferencing between thirty locations in public schools, city government offices, the public library, a nonprofit science foundation, and University of California, Irvine. The network will soon include a local art museum and community college. The Community Cablevision Company, serving approximately 20,000 subscribers in Irvine, parts of Newport Beach and Tustin, has designated two channels for two-way educational access cable teleconferencing between participating institutions. Full-motion video and audio signals are carried via two cable TV channels upstream to the company head end from any two of 30 interactive TV cablecasting sites. Here the signals are processed, amplified, and sent downstream to each ITV location and to home subscribers. Each cable-drop location in a network site is a potential cablecasting access point. Each of the 28 locations is equipped with cablecasting and receiving equipment. TV site equipment is usually contained in a cart and consists of two TV receivers with channel selectors, a black and white and/or color consumer-grade camera, a microphone mixer with assorted mikes, and RCA color modulator with interchaneable drawers and miscellaneous cables. The video equipment is office-operated by elementary level students and is durable, low in cost, easy to maintain, and is functional with existing light conditions. Cost for cart and equipment varies with desired features. It can be as low as $3,500. The carts are self-contained and an operator student, teacher, secretary, administrator needs only to roll the cart to a location, classroom, office, laboratory, etc., equipped with power and cable tap, to access the interactive video system. Control over the use of the two interactive channels is decentralized. There is no central switching capacity. To avoid conflict and programming interruptions, a master schedule for use of the system is maintained by the Irvine Unified School District. In addition, it is possible to cablecast, from sites with two-way television capabilities, to the larger community via a third public access channel. Interactive video equipment for each site is purchased and maintained by participating institutions. The interactive video system is installed and maintained by the Community Cablevision Company. Further information about the system design and specifications may be obtained by writing to Mr. Tom LaFourcade, President of the Community Cablevision Company, 1061 Camelback Street, Newport Beach, CA 92663.

Irvine School District Programming 1974-Present

The Irvine Cable Teleconferencing Network is, by definition, a shared operation. Two-way television programming origination is shared between two or more locations within a participating institution; as an example, between two or more schools or administrative offices in the Irvine Unified School District (IUSD), or between two or more locations in two or more participating institutions. As an example, between the UC Irvine campus and the Irvine City Hall. The master programming is maintained, published and distributed each week by the Irvine Unified School District Video Department. The system is programmed an average of six to eight hours each day during the academic year and community information programming is cablecast to the community on a regular basis.

Interactive video programming generated by students, teachers and administrators in the IUSD is varied and extensive. Students from the second grade and up have learned to operate the site-level equipment and have created much of their own programming.

Some examples of student generated programming have included: "Animal Lovers," a club for 4th grade pet owners; "Pen Pals," regular meetings between second graders who write to each other; "The Book Lovers Club," sixth graders share their favorite books on a regular basis; and "The Working of Video," cross-age tutoring by a seventh grade video technician who teaches elementary level students how to operate portable videotape recording and editing equipment. Two programs, "Anatomy: The Dissection of a Cat," and "Microscope Study" have been taught for the past two years by high school students for elementary level students.

IUSD teachers have used the system extensively for teaching courses in geometry, the use of calculators, fine arts, and languages. For two years, two IUSD Resource Teachers have provided more than 150 students, located at up to fifteen schools with English as a Second Language instruction for up to five hours each day. Occasionally, a student participating in a course taught by a teacher will assume responsibility for teaching that course to other students. This happened in a course taught by the IUSD Video Department Director, Craig Ritter. Craig taught several lessons in the art of Paper Geometry to sixth graders at three schools. One of the students became so proficient in creating the shapes that he began volunteering on a regular basis to give directions to the other students. After a couple of months the student was teaching the course.

Administrative uses of the system have included programming by and for district secretaries, and regular meetings for principals with the superintendent. The benefits of using the system as a substitute for face-to-face meetings usually outweigh the disadvantages. Secretaries and principals alike have been able to meet conveniently, without being required to leave their school sites to travel to a central district location.

When the university joined the system, Irvine students and teachers had access to a variety of resources. Programming with the university has included dialogues with visiting lecturers, courses and lectures provided by UCI faculty, career counseling, and staff development courses.

Visitors to the school district and the university have used the system to talk with student teachers and administrators at multiple locations without the inconvenience of having to physically move from one location to another. These visitors have included: Governor Jerry Brown, Ralph Nader, the Honorable Shirley Hufstedler, and California Superintendent Wilson Riles.

The system has been used to test and evaluate the effects of commercially produced television programming. Sally Baker, KHJ-TV Producer, showed portions of programming from "The Froozles" to IUSD students, then questioned them about their impressions and responses to the videotaped segments.

Each interactive site also provides the means to cablecast to the community via the local public access channel. Principals have used the channel to report test results to parents in the evening. The twice monthly IUSD Board of Education meetings are cablecast by student technicians to Irvine residents, and the school district has sponsored a variety of community information programs utilizing the personnel resources of the district to discuss issues in family guidance, health, nutrition, and the future of education.

The Office of Teacher Education, UCI, has used the system extensively in its teacher training program for methods observation of IUSD classroom teachers, dialogue sessions with IUSD students, teachers, and administrators, and for practice teaching.

UCI interactive television programming includes: "Mastery Lesson" series, instruction by UCI faculty for Irvine School District staff development programs, two-way TV dialogues for community cablecasting by UCI guest lecturers, national teleconferencing participation, and UCI faculty participation in IUSD-sponsored community information evening programming. Each of these programming areas are briefly described below.

The Human Factors

The first step in implementing a program is the warm-up session, the purpose of which is to assist students in overcoming their inhibitions about appearing on television.

Warm-up Sessions

Warm-up sessions are usually held as an initiator to the use of the interactive video system. The interactive TV coordinator arranges to introduce teacher education students to the technology by meeting with them through the interactive system. Asking each student his or her name and allowing them to ask questions about the system "breaks the ice." Students see themselves on monitors and become familiar with their television image and with the electronic classroom atmosphere. Here, we are focusing on a television screen receiving an image and a monitor displaying their own images. Learning to relate to others through television media is a factor in this process. Some students may feel uncomfortable and shy at first. Usually, they are pleased with their image and soon become accustomed and, in many cases, enjoy their new role. Switching back and forth between the regular classroom atmosphere and the electronic atmosphere does take some preparation for and consideration of students.

Letting participants know the interactive video schedule in advance, planning a warm-up session, and preparing for each individual session by a discussion of possible topics and questions, are all preparatory steps to using the cable teleconferencing system.

"Mastery Lessons"

Office of Teacher Education faculty have future teachers in their courses observe, via the cable teleconferncing system from the UCI campus classroom, demonstration lessons conducted in Irvine Unified School District classrooms. These observations are preceded and followed by "two-way cable television dialogues with the school district "master teacher." In this manner, fifteen to twenty-five UCI education students and their instructor can observe the same lesson without leaving campus. The portable and relatively inconspicuous ITV equipment, usually operated by students in the observed class, fourth and higher grade levels, does not significantly disrupt the normal classroom atmosphere. The equipment does not require the use of additional lighting, and can be quickly removed from the classroom at the end of the demonstration lesson and dialogue session. Instructional methods, observation lessons in math, spelling, reading, study skills, and English as a second language, are regularly scheduled during the academic year.

Television and Teaching

An Office of Teacher Education course, "Applied Technology in Education: Television and Teaching" will be offered in the spring of 1982. Future teachers will be introduced to the fundamentals of videotape equipment operation and functions, and will be trained to use the interactive television system for instruction. Each student will develop and present a series of lessons in the subject area of their choice via the interactive television system for students in the IUSD.

Curriculum Dialogues

Dr. Virginia Boyle's Secondary Curriculum students meet approximately four times each quarter with IUSD students, teachers and administrators for a candid discussion of course content, instructional techniques, classroom management, and issues in secondary education. These interactive television exchanges serve to integrate theory and practice.

Irvine Unified School District Staff Development

A UCI Extension credit course, "Applied Technology in Education: Computers" was offered fall of 1981 through the Office of

Teacher Education. The course was taught by UC Irvine's Dr. Alfred Bork, Professor of Physics and Director of the Educational Technology Center. Some of the sessions were conducted via two-way television to reduce the number of times teachers needed to travel to the UCI campus. It provided them with the capability of working during class time with computer hardware and software available in their respective schools. Additional inservice and instructional staff development programs utilizing UCI academic resources are in the development stage, with priorities given to upper elementary and middle school science technology and fine arts areas.

Irvine Unified School District Student Instruction

UCI faculty from a number of campus departments have conducted informal lessons and dialogue sessions with IUSD students from second grade through high school levels. As an example, Martha Lou Thomas, UCI Librarian, has conducted several thirty-minute lessons for IUSD second graders entitled, "Story Time." Each week, Mrs. Thomas read a short story to groups of students located at two elementary schools. At the close of the lesson, each student created a picture illustrating an aspect of the story. The pictures were gathered from the schools and Mrs. Thomas used the student drawings to illustrate the same story as it is read via two-way TV for the second time. Sometimes, the students write and illustrate poems. The program, designed to stimulate interest in reading, also stimulated the writing and illustration of poetry.

Other programs have included "American Indian Arts and Culture," for elementary studies, by Billie Masters, UCI faculty member; courses in creative writing and library research skills for high school students, and a series of discussions about career opportunities with staff from the UCI Career Planning and Placement Center.

UCI Visiting Lecturers

The two-way CATV system provides UCI with a convenient method for the sharing of UCI guest lecturers with the IUSD and with the local communities of Irvine and Newport Beach. Last year, the Honorable Shirley M. Hufstedler, then U.S. Secretary of Education, visited the Irvine campus. Following a campus luncheon she walked to a classroom on the UCI campus equipped with two-way Cable TV equipment, and dialogued with IUSD students and staff located at fifteen district schools. The program was simultaneously cablecast to the community via the local public access channel. Other programs of this nature are being planned for the 1981-82 academic year.

Teleconferencing

In June, 1981, faculty in the Office of Teacher Education participated in a national teleconference broadcast from the studios of WGBH, Boston. UCI participants were linked via the two-way CATV System to community participants located across town in an IUSD office conference room. Both groups watched the WGBH program via the public access channel, then used the two-way channels to formulate responses for audio return from the district location. By using the Irvine system, a large number of participants could be accommodated at numberous locations for a community teleconference within a national teleconference.

Community Information Programming

Numerous University of California faculty members have been guest panelists on a series of IUSD-sponsored community information programs. These live phone-in programs are cablecast from IUSD media centers or conference rooms using the two-way CATV equipment to access the local public channel. IUSD middle school students operate the video and audio equipment and answer the phone. The one-hour weekly programs (forty-two to date) cover a variety of topics of local interest. Some examples are: Health and Nutrition, Family Counseling and Guidance Services, and topics about the future of local development, transportation, education, technology, and government.

The Future

Two more cablecasting locations, one in the local community college, and one in a local private contemporary art museum, will increase system programming capabilities. When functional, the community college link will provide IUSD high school students with opportunities to attend college courses during the day, without having to waste valuable time traveling to and from the lower campus. Irvine residents will be able to watch evening lectures telecast from college classrooms, and it will add new dimensions to the UCI teacher training program.

The art museum link will provide IUSD students with increased opportunities to meet well-known contemporary artists, to visit museum exhibitions; and community cablecasts of museum openings will encourage residents to visit these exhibits. UCI student artists will be able to use the system to experiment with a new medium, and to share their work with artists in-residence at the art museum.

The university is exploring the possibility of extending the interactive television system to other campus departments for direct access to the IUSD public schools and community locations. The County of Orange has initiated a study of the feasibility of interconnecting cable systems for county-wide sharing of community information and educational programming. It is not unreasonable to predict that one day all county education and government institutions may be linked together by 2-way cable television for teleconferencing between remote locations.

Meanwhile, interest among educators in the use of teleconferencing systems to meet instructional and communications needs, and to cut costs at the same time, has steadily increased interest in the Irvine project. It is viewed by many as a model use of an available technology to meet identified instructional needs in a cost-effective manner.

The Teleconferencing Resource Book: A Guide to Applications and Planning
Lorne A. Parker and Christine H. Olgren (eds.)
Elsevier Science Publishers B.V. (North-Holland)
© Center for Interactive Programs, University of Wisconsin-Extension, 1984

HOTEL-BASED VIDEO TELECONFERENCING VIA SATELLITE

James H. Black, Jr.
Executive Vice-President
VideoStar Connections, Inc.
Atlanta, Georgia

Although the concept of closed-circuit televised meetings is not new, it has, in the last two years gained much greater prominence as a communications technique for business, government and other institutions. This widespread usage may be attributed to two recent developments: The availability of inexpensive transponder (channel) time on commercial communications satellites, and the development of high-quality large-screen video displays.

As distinct from smaller-scale, highly interactive meetings, this paper will focus on larger meetings held primarily in hotel facilities. These closed-circuit meetings, or tele-conferences, may be characterized as:

° Full motion, color, one-way video, two-way audio telecasts.
° Large-scale meetings, usually involving thousands of attendees gathered in regional meeting centers.
° Primarily for the purpose of disseminating information, rather than problem-solving sessions.

Teleconferencing's Role in the Meetings Industry

Meeting planners are increasingly discovering that large-scale, multi-city teleconferences are a practical tool available today to supplement more traditional meeting formats.

Of the roughly 1.2 million "off-premises" meetings held annually by corporations and associations, over 50% are of a type that can benefit from teleconferencing. These include national sales meetings, new product introductions, training seminars, stockholder meetings, and others.

Most of these meetings are presently formatted as a one-way information flow from a few people to many hundreds of people. This pattern is carried over to "point-to-multipoint" teleconferences, in which the necessity of gathering everyone in a central location is giving way to a preference for regional multi-city meetings in premium hotels with teleconference capabilities that link the regional locations to any origination point.

Formerly, corporate employees and national association members who could not afford the time or money to fly to the central location for a major meeting were excluded from many activities. Today, through teleconferencing, these people are involved.

In addition to the out-reaching point-to-multipoint format, the capability to "inward teleconference" prominent speakers into large meetings or conventions is becoming increasingly popular.

In such a situation, a recognized industry or government figure has been invited to address a gathering, but time constraints or geographic remoteness make it impossible for that speaker to attend in person. The speaker can now be brought into that meeting electronically, without having to leave his or her home base. Furthermore, additional speakers can be brought in from other locations, to create a panel discussion with the opportunity of audio interaction among the various participants. The featured speaker may debate an issue of concern to the convention, for example, then take questions from attendees.

It should be emphasized that the majority of teleconference applications will occur as part of a pre-existing meeting or convention. The unique features of this communications tool make it a valuable supplement to more traditional meeting formats and techniques, particulartly where there is a time urgency to disseminating important information to many people. Rather than becoming the entire meeting, teleconferences will be an integral part of many larger meetings, as a way of involving members and speakers who previously were excluded.

Advantages of Large-Scale Teleconferencing

The most commonly cited advantages of teleconferencing include:

° Greatly enhanced "reach", to include many more participants in important meetings.
° Immediacy and impact of the video medium.
° Uniformity of message.
° Time savings for key professionals.
° Cost effectiveness.

The subject of extended reach has already been introduced. The reason for its importance is that American institutions - corporations, associations, government - have become both more complex and geographically dispersed. In order to maintain in their members a sense of belonging and commitment, these organizations have sought new ways of communicating. For many requirements, video teleconferencing allows executives and other leaders to put themselves in contact with people they would otherwise not reach.

The impact of a live video presentation, often on larger-than-life screens, should not be underestimated. The video teleconference format is not intended as a substitute for face-to-face communications; rather, it makes possible communications that would otherwise have been handled through much less meet direct means. For example, if a corporate executive wished to announce a major organizational change, introduce a new business policy, or unveil a new product, previous approaches might have included a simple memo or possibly a video tape. These techniques, although inexpensive, don't have much personal impact and certainly don't allow for live question and answer interaction. A live teleconference, on the other hand, is much more vivid and involving.

In addition, uniformity of the message being delivered is an important factor. Rather than having important information conveyed second hand and risking distortion or changed emphasis

in the re-telling, an organizational leader can simultaneously reach many levels of his organization (and many locations), not just a small group immediately around him.

White collar productivity has become a major concern in recent years. This is a reflection of two concurrent developments in the business environment: the increasing complexity of the decision-making process, which requires involvement by more people in a shorter time-frame; and the trend toward more decentralized, widely scattered organizations. Communications techniques such as teleconferencing enable key managers and other professionals to spend more of their time actually giving or receiving information and acting on it, rather than travelling long distances in order to communicate. This advantage certainly applies to such events as policy announcements, product introductions and training seminars.

Companies with nationwide sales forces or distributor/dealer networks are increasingly using regionalized teleconferences in 20-40 cities in place of one centralized gathering or a "travelling road show" approach. This format can save valuable time for both information presenters and the field personnel. At the same time, participants still have the advantage of the meeting with their local colleagues in an attractive off-premises setting.

A large-scale video teleconference can be extremely cost-effective, particularly when viewed on a cost-per-participant basis. Because of the large number of participants in a nationwide meeting, often many thousands of people, the cost of distribution and presenting a teleconference can frequently be as low as $10-$20 per person.[2] Another factor contributing to the cost-effectiveness of this technique, of course, is that it is used on an ad-hoc basis. Therefore, users need not invest in an expensive fixed network which is only used occasionally.

These various advantages are consistent with the underlying forces driving the growth of electronic communications in general:

° The need to communicate more information accurately to more people on a timely basis.

° A growing reluctance to travel great
  distances for relatively brief,
  non-problem solving meetings.
° Very high transportation costs,
  particularly for long-distance air
  travel.
° Decreasing costs of communications
  services, particularly those that are
  satellite-based.

More specific to hotel-based teleconferencing is the fact that most of
the large scale events currently taking
place are already normally held in hotel
facilities. National sales meetings,
product introductions and press briefings typically require a "neutral" site,
food and beverage services, and other
amenities not available in the normal
office environment. If only a very
small percentage of these existing
off-premise meetings include a teleconference segment, the volume of such
events could eventually be in the
thousands per year.

This projection is supported by the
fact that many teleconference organizers
have been repeated users; indeed, some
have greatly increased their frequency
of usage. Ford Motor Company, Chrysler,
Merrill Lynch and Lanier Business
Products are prominent examples of
companies that have repeated, and that
can be expected to further accelerate
their usage.

Keys to Successful Implementation

The keys to successful use of the
large-scale teleconferencing medium
include:

° Applying this technique to the
  right kinds of meetings.

° Having a well-conceived and
  professionally produced program.

° Using a reliable networking
  supplier to ensure fault-free
  distribution of the video signal
  to all receiving locations.

The need to select "the right kinds
of meetings" has already been addressed.
The key qualifications are that the
meeting content be primarily a one-way
communication (but with interactive
audio), that the message have a time
criticality, and that a relatively large
number of people be involved.

In order for a teleconference to
succeed, its basic content and structure
-its substance - must be well planned and
executed. Most organizations that would
consider using this technique are already
working with an outside video production
firm for sales meetings, trade shows and
other "high impact" functions, many of
which are distributed on a limited
closed-circuit televised basis. These
same production firms are now designing,
scripting and producing satellite-
delivered telecasts. Typically, such a
firm's responsibility ends with bringing
a high quality video signal out of the
originating facility (TV studio,
corporate office, hotel ballroom etc).

Once this signal has been originated, it is essential that the networking and distribution functions be handled
by a reliable, experienced supplier.
Such a firm should have have experience
in both satellite and terrestrial
communications, and a nationwide field
engineering force to man the receiving
equipment at regional meeting sites.
This comprehensive service assures the
meeting planner and producer that all of
the creative efforts in planning and
creating a first-rate program will not be
jeopardized by technical problems
anywhere in the network.

The diagram below illustrates the
major steps in distributing a satellite
teleconference and highlights the role of
the network coordinator.

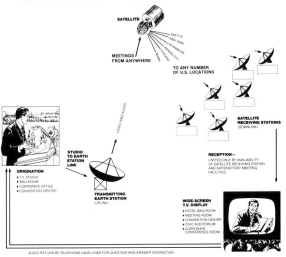

SATELLITE TELECONFERENCING

VIDEOSTAR TELE-MEETING™ NETWORK

Beginning in the lower left corner, the network coordinator's first responsibility is to transmit the live signal from the orgination point to the transmitting earth station (uplink). In most locations, this is accomplished by way of a temporary video landline from the studio to the uplink. Transportable uplinks have also been used for this purpose, but usually end up being more expensive. In the future, it is likely that some permanently installed earth stations at large convention hotels may be equipped with a transmit capability, thereby eliminating one link in the chain.

Next, the network coordinator will have arranged to transmit the signal to the satellite and booked the transponder time for the duration of the meeting (and some pre-meeting test time). The satellite should be chosen on the basis of scheduling flexibility, technical performance and signal privacy; an experienced coordinator will have a well-developed sense of the trade-offs involved here.

The video signal is then transmitted by the satellite to the various receive sites. (In fact, the signal covers the entire U.S. and adjacent areas; it can be received by any satellite antenna pointed at the right satellite and receiver tuned to the right frequency.)

The coordinating firm provides quality earth station equipment, either permanently installed or transportable, at the designated meeting sites. Reliability is the key here. A successful teleconference requires that all the meeting sites receive a high quality telecast.

Transportable earth stations have served most teleconferences to date, since relatively few premium hotels are equipped with permanent receiving equipment. However, it should be pointed out that transportable stations are not as reliable as permanent installations. For example, severe weather (wind, snow, heavy rain) can knock a transportable out of service, and there is also risk of damage or delay in transporting the equipment to the receive site. Knowing this, teleconference planners should give preference to those hotels with permanent receiving equipment. In the long run, the use of transportables will likely decrease due to their higher cost and lower reliability.

The final step in the network is bringing the video signal into the meeting room and displaying it on a monitor or wide screen. The choice of display equipment should be based on the size of group viewing the program and on the nature of the meeting. In general, the larger the group and more elaborate the production, the larger should be the displays. The equipment is normally provided by either the network coordinator or the video production firm.

Case Examples

Perhaps the best way to illustrate the successful application of hotel-based teleconferencing is to describe a few actual examples.

Ford Motor Company, in February, 1981, made marketing history with its first use of satellite teleconferencing. It introduced its new EXP and LN7 sport coupes to over 20,000 dealer and sales personnel in 38 cities-the largest teleconference held until that time. The teleconference was structured as part of Ford's Mid Winter Dealers' Meeting. The televised portion included two live 90-minute telecast segments during the day, plus an hour-long nationwide press conference in the evening. Wide-screen projectors and telephone hot-lines for questions-and-answer interaction between the receiving cities and origination site in Detroit were used for the meeting and press conference.

"Through teleconferencing, we reached more dealers than we ever have in the past, said Jim Olsen, Public Relations Manager of the Ford Motor Division. When asked if Ford would conduct another teleconference, Olsen replied: "Most definitely. There was a tremendous saving of our executives' time. Remember that Ford has 34 sales districts, so our alternative would have been to run a 'dog and pony show" in each district. When you consider the extra dealers we contacted and the impact of the event, we consider these factors a fair trade-off for the lack of face-to-face contact."

In fact, Ford has since held three other satellite teleconferences, the most recent being the 1982 Mid Winter Dealers' Meeting in February. Again, as in 1981, the company used the occasion to unveil a new product, this time the Ranger pickup truck. The Company also repeated the nationwide press conference, which included Toronto this year.

Merrill Lynch first used satellite teleconferencing in September, 1981. In a program transmitted to 30 cities around the country, the investment firm presented a seminar on implications of the new tax law recently enacted by Congress. This telecast, involving both Merrill Lynch brokers and their clients, really had a dual purpose: to educate both groups on the specifics of tax law changes; and, more subtly, to encourage investors to use Merrill Lynch's services in capitalizing on opportunities presented by the new law.

One interesting aspect of the program was that it coincided with a national economic address by President Reagan. Because of the obvious relevance of the President's talk, it was presented in its entirety mid way through the teleconference, with the seminar program continuing afterwards.

After the teleconference, Merrill Lynch indicated it would use the medium again, and did so within the next couple of months to conduct an informational seminar for its brokers.

Avon Products' initial experience with teleconferencing represents a somewhat unique application. On December 11, 1981, the company telecast a portion of its annual "President's Celebration" from Hawaii back to seven cities on the mainland. The top sales representatives and district managers who had earned trips to Hawaii participated in a program designed to inspire the nearly 1000 runners-up who were viewing the proceedings in their respective regional locations.

This event, rather than being primarily an informational session, was designed for motivational purposes. It was done in conjunction with a special selling and promotional program the company holds annually in late summer, traditionally one of the industry's toughest selling periods. The success of the teleconference is summarized by Al Gershlak, Director of Representative Motivation: "From an impact point of view, the boradcast was as successful as anything we've ever done ... The medium offers broader opportunities for the way we communicate to our field people."

A good example of "inward tele-conferencing" of recognized speakers into a large convention is the November 1981 event by the Southeastern Telecommunications Association (SETA). A panel of communications experts, including Bernard Wunder, Assistant Secretary of Commerce, discussed issues relating to the telecommunications industry regulatory environment.

The discussion was telecast to about 500 SETA delegates meeting in Atlanta, who had an opportunity to ask questions of the panel. This format is typical of many programs that will be designed to add a further dimension to existing conventions or conferences. In this particular case, the medium and the message were closely related.

Questionnaires distributed to the SETA delegates indicated that the teleconferenced session was the most popular session of the three-day conference. Based on this demand, SETA will again include an inward-teleconference on their program for 1982.

The Hotel Industry's Role

Hotels themselves, naturally, are playing an increasing role in the development and growth of large-scale teleconferencing. With the exception of Holiday Inns, most hotel companies have, until recently, taken a cautious approach. With the significant increase in activity over the past year, however, many are beginning to take a much more active role.

There is a growing realization that, in order to be viewed as a full-service meeting center, larger hotels will need to offer a teleconference capability. Many conventions and other large meetings will, in the future, include a teleconference segment. Without the facilities to serve this need, a hotel may have difficulty retaining some of its meetings business. A permanetly installed satellite receiving station will be viewed as an essential resource, an adjunct to more traditional capabilities such as audio-visual equipment.

In addition to having a commercial quality earth station installed, it is important for a hotel to be part of a professionally managed nationwide

satellite network. As discussed
earlier, the satellite communications
company which operates the network will
coordinate all the facilities for
teleconferencing and will man the
receiving stations with qualified field
engineers during a telecast. Because
teleconference planners and producers
want to reach numerous cities
simultaneously, they will select a
network which provides nationwide access
to first-class meeting facilities, and
which is easy to use. From this
perspective, the partnership between the
hotel company and a qualified,
experienced satellite communications
company is key to successful use of this
medium.

Footnotes

1. Derived from data reported in
    Meetings and Conventions; "Meetings
    Market Report"; January, 1982; page
    45.

2. Because the costs of preparing and
    producing a teleconference can vary
    so widely, these are excluded from
    this cost-per-participant figure.
    Also, these costs would often be
    incurred for a large-scale meeting
    of this type anyway, and are not
    always incremental costs. Included
    are the costs of sending the video
    signal from the studio to a satel-
    lite transmitter, uplinking the
    signal to the satellite, the satel-
    lite time itself, and arranging the
    receiving stations and wide-screen
    displays at regional meeting sites.

The Teleconferencing Resource Book: A Guide to Applications and Planning
Lorne A. Parker and Christine H. Olgren (eds.)
Elsevier Science Publishers B.V. (North-Holland)
©Center for Interactive Programs, University of Wisconsin-Extension, 1984

USE AND PROMOTION OF TELECONFERENCING BY THE ONTARIO GOVERNMENT:
A PERSPECTIVE

Neeru Biswas, P. Eng.
Supervisor, Systems Development
Telecommunications Services Branch
Ministry of Government Services
Toronto, Ontario

Wendy Cukier
Project Leader
TEMP Teleconferencing Task Force
Ministry of Transportation and Communications
Downsview, Ontario

Introduction

The Ontario Government uses teleconferencing to deliver essential services to remote locations within the province and to control travel costs, improve productivity and save energy. In addition, the Government has played a unique role in promoting teleconferencing as a substitute for travel, acting as a catalyst to both supply and demand in Ontario.

Background

The province of Ontario covers 413,000 square miles and has a population of over 8.6 million. A majority lives within one hundred miles of the U.S. border: over 3 million in the Toronto area alone. The rest are scattered in small communities through the north. The Government of Ontario is responsible for the delivery of a variety of services, such as health care, education, transportation, communications, and social services to all residents of Ontario.

Furthermore, the quality of these services must be consistent throughout the Province. To accomplish this, several Government of Ontario departments (called "ministries") are highly dispersed geographically. Regional offices are established in most of the larger population centres in the north, such as Thunder Bay at the head of Lake Superior, and district and area offices cover smaller townships such as Trout Creek and White Fish. Headquarters in Toronto are more than a thousand miles away for some of these offices.

The Ontario Government, in its search for innovative and cost-effective ways to deliver timely services to the North, began investigating teleconferencing in the early seventies. At that time, specialists and consultants routinely travelled around the province, providing support and expert advice to local staff in the area of health care and various social services. The Government was often faced with unfilled job vacancies in many professional areas because skilled specialists were not willing to relocate to the "isolated" north, even with the attraction of significant salary differentials. The Government was also challenged by the feeling of isolation experienced by many of the northern tax-payers.

Owing to the geographical, and, in some instances, administrative decentralization of Ontario ministries, management in various offices across the province needed to meet regularly. This traditionally necessitated travel with all its associated costs - time, money, and energy.

System Development

Defining teleconferencing requirements for the Ontario Government was a complex task. Most of the technology was untested and potential users were unfamiliar with its applications. In addition, the need for teleconferencing emerged at two levels - some services could be shared by all ministries, but others had to be tailored to specific needs. Consequently two approaches to system development were taken. Teleconferencing was explored on an experimental basis through pilot projects and various studies were conducted to assess user needs.

Satellite Pilot Projects for Common Service Applications

Communications Technology Satellite (Hermes). In 1972 the Government of Canada and the U.S. were developing plans for applications of the 12/14 Gigahertz satellite communication technology that was "launched" as the experimental Communication Technology Satellite (CTS), later renamed Hermes. The benefits of the higher frequency and higher power of the satellite were that earth stations could be smaller and more portable.

The Government of Ontario decided to participate with the Government of Canada on their Hermes experiment to test the viability of the "teleconference for service delivery" concept. Satellite communications links were judged to be the perfect vehicle for this sort of pilot trial. Not only could the links be moved around readily between geographical communities of interest, but satellite communications could also allow link-up with townships that were not served by any other reliable means of telecommunications. For instance, in 1972, many northern communities had only high frequency radio communications available. Thus, those very communities, such as Sioux Lookout in the far north, that felt isolated from the south could easily participate in the pilot program with Toronto. Another benefit to testing the viability of the concept through satellite communications was that capital costs in terms of expensive land line construction were minimized.

The Ministry of Government Services, within its mandate to provide services to all ministries in the Ontario Government, selected sites for the pilot link-up via CTS based on the following criteria:
  . geographical distance from Toronto (head office site) was large enough to justify, in the minds of potential users, a substitute for travel
  . location of the sites allowed access to as many Ontario Government ministries who had responsibilities for program delivery in the north as possible
  . communities that were isolated because of unreliable existing communication links
  . enthusiastic user community in a township.

Satellite earth stations (some for audio only, some with full audio/video capability) were sited at Thunder Bay, Dryden, Red Lake, Pickle Lake, Sioux Lookout, Winisk, and several base camps in northwestern Ontario.

Applications in the areas of health care, environment control, fire fighting, administrative meetings, police information, agriculture and educational communications were examined for suitability via teleconference.

Audio only, audio supported by facsimile, slow scan video and full motion video - all modes were tried in various configurations for the above applications and auxiliary activities.

The CTS experiment by MGS established both the viability and desirability of using teleconferencing as a tool for the delivery of government services to the north.

CTS TRANSMIT/RECEIVE EARTH STATION ON THE ROOF OF MGS BUILDING IN TORONTO

It also established the need for further and more refined trials of both freeze frame and full motion video teleconferencing.

As a result of the CTS experiment, basic operational teleconference centres offering both audio and video teleconferencing were established in both Toronto and Thunder Bay. Permanent freeze frame video links between Sioux Lookout and Toronto, and a full motion video link between Woodstock and London, Ontario, were established for telehealth applications.

Experiments Via The Anik B Satellite. The Anik B satellite was then used to try colour video teleconferencing (only black and white technology had been used in the CTS experiments) both in freeze frame and full motion modes. The major findings from this experiment

that was conducted in 1979/1980 were at two
levels:

At the system level it was found that full
motion video links, though more expensive than
colour freeze frame by about a factor of three
(1979 estimates), were most desirable for the
overall government requirements. This was
because:
- full motion video provided the best eye
  contact for applications such as psychiatric consultations.
- once full motion video service was in
  place at a site, all graphic support requirements other than hard copy could be
  taken care of via that service, without
  going to the additional expense of installing freeze frame video equipment.
- freeze frame video technology in 1979/1980
  was such that different suppliers' equipment was not compatible. There were also
  problems with regard to availability of
  service depots for the equipment in the
  remote north of Ontario.
- full motion video equipment, being similar
  to broadcast equipment, was easier to install, maintain, and mix and match from
  various suppliers for best cost performance.

The most important findings at the human
factors level were:
- users were most comfortable with full motion video because it was, in their words,
  "closest to a face to face meeting".
- users responded well to "colour television".
- minimal training was required for full
  motion video teleconferencing.
- users did not like to leave their premises for a teleconference; so teleconference facilities should be located centrally for best utilization.

Operational System. Following the CTS and
Anik B Satellite Pilot Projects, a full motion
colour video system with associated audio was
established between Toronto and Thunder Bay 900
miles away for the use of all ministries.

The video teleconference facilities are
located in government buildings where most personnel have offices for easy access. In
Toronto the equipment is transportable. Because of space limitations, a dedicated room
was not set up. Instead, several meeting
rooms were wired for teleconferencing. The
equipment is wheeled into whatever room is
available.

A natural meeting room environment has
been maintained. The addition of high intensity flourescent lights and carpeting (for
acoustics) were the only changes.

ONTARIO GOVERNMENT PORTABLE
AUDIO/VIDEO CONFERENCING
FACILITIES

User friendliness was a major concern in
designing the system. To keep the technology
as transparent as possible, a technician is
available to set up and operate the cameras
remotely. When privacy is a concern, the cameras can be pre-set or participant controlled.
Two cameras are used. Regular meeting aids -
flipcharts, blackboards, slides, films and
video tapes - can all be used. The cameras
have pan and tilt, zoom, focus and close up
controls. The cameras can be used flexibly
depending on the nature of the meeting. Both
cameras can be used for "people shots", or one
camera can be dedicated to graphic support for
the whole or part of the meeting. In addition
to the incoming video monitor in the teleconference room, there is a small video monitor
provided, allowing participants to view the
outgoing picture.

Voice can be sent with the video signal or
on a separate audio link for added privacy.
Occasionally, participants elect to use video
teleconferencing for the first part of the conference and use audio teleconferencing for the
rest. Audio is always available as a backup or
for exclusive use. Two way full motion teleconferences can be arranged between two points,
or one way video can be "broadcast" to several
points with fully interactive audio.

Transmission in the demonstration phase
was via Anik B satellite, but is now over terrestrial microwave. Transmission facilities
are rented on a pay as you use basis from Bell
Canada.

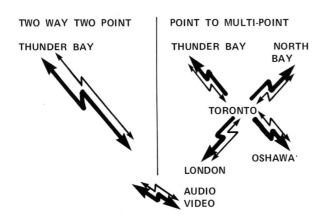

VIDEO TELECONFERENCE CONFIGURATIONS

Costs. The variable cost is based on the price of a return flight between the points in question. Toronto to Thunder Bay, for example, is $280. The system cost was minimal - $25,000 per location for hardware (1983 dollars), $3,000 for room conditioning, and $1,000 per month fixed rental for lines.

Audio Teleconferencing for an Individual Ministry

User Needs Assessment. In addition to expanding services, teleconferencing was regarded as a potential means of reducing travel to conventional meetings to reduce costs, improve productivity, and save energy. Individual ministries began to explore its feasibility as a travel substitute in the late 1970's. The Ministry of Transportation and Communications (MTC) carefully assessed needs in terms of key travel linkages, meeting requirements, and attitudes to teleconferencing and travel.

Travel patterns within the government varied from ministry to ministry. At MTC, for example, with 3000 people in head office, 3350 in 5 regional offices, and 3650 in district offices, a survey of travel records indicated that in 1978 over one third of travel expenditures were on trips between head and regional offices, 6% on trips within Toronto, and the remaining 60% was expenditure on trips throughout the province.

	TRAVEL COSTS	
HEAD OFFICE -REGIONAL OFFICE	$135 000	34%
WITHIN METRO TORONTO	25 000	6%
OTHER	240 000	60%

TRAVEL LINKS (MTC 1978)

As regular meetings were thought to be most easily teleconferenced, a survey of committees was undertaken to identify meetings which could be teleconferenced. Again, head to regional office communications appeared to be primary. At least 200 committees met regularly between those locations. Finally, as teleconferencing was a new concept, an effort was made to assess attitudes towards travel and teleconferencing within the ministry. The results were not encouraging. There was little awareness of teleconferencing and little interest in reducing travel. Of travellers surveyed, only 8% were interested in reducing travel, and 21% wanted more. However it was clear that heavy travellers were more amenable to teleconferencing than occasional travellers, and that the greatest potential lay with them.

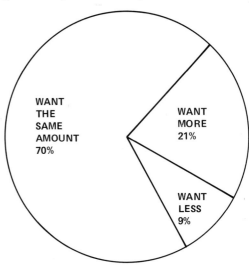

ATTITUDES TO TRAVEL (MTC 1978)

The conclusions of MTC's assessment of user needs were :
. that audio teleconferencing was sufficient for most meetings with supporting documents mailed or transmitted by facsimile in advance.
. that a high quality network was required to link head and regional offices, which could allow interaction between 5 or 6 locations.
. that the telephone network, being flexible and universally available, could be used for other meetings.
. that a significant training and promotional effort would be required to gain acceptance for teleconferencing.

Operational Systems. The user needs assessment convinced MTC of the potential for teleconferencing. Two complementary audio systems were established - a CNCP Broadband Dedicated Network linking head to regional offices, and a telephone-based system employing Conference 2000s*, Speakerphones*, and the Bell Conference Operator.

It was felt that the volume of traffic between head and regional offices and the size of meetings justified installation of a higher quality, full-duplex network. It was thought that while voice switching, present in Conference 2000s and Speakerphones, was tolerable for 2 or 3 location meetings, it would be unmanageable in 6 or 7 location meetings. Moreover, while the Bell Conference Service had a low fixed cost, the variable costs were high on multi-point calls. CNCP Telecommunications (analagous to Western Union in the USA) provides a switched, four-wire data communications network called Broadband. A voice teleconferencing option is offered with this service. Subscribers may obtain a bridging service which can be preprogrammed to allow them to directly access certain combinations of locations by automatic dialing. Alternatively, they can directly dial up pre-configured groups of locations; or they can use the CNCP conference operator to set up the conference.

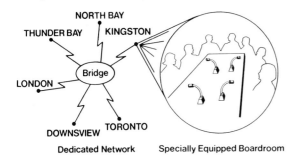

Dedicated Network   **Specially Equipped Boardroom**

While establishment of this system involved a relatively high fixed cost, the variable costs were low. Given the anticipated volume of use, it was more cost effective than using the Bell Conference Operator

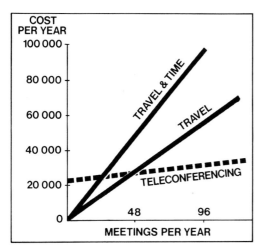

BREAKEVEN ANALYSIS OF MTC DEDICATED
NETWORK (48 CALLS PER YEAR)

Meeting rooms were equipped with push-to-talk speaker-microphone units produced by Tech 5.

SPECIALLY EQUIPPED BOARDROOM

For meetings involving the regions and districts and other locations, the telephone network was used. Speakerphones were installed in district offices and a Conference 2000 was placed at head office. Add-on and the Bell Conference Operator were used for multi-point calls.

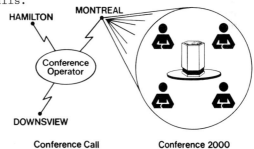

Conference Call   **Conference 2000**

CONFERENCE 2000 IN USE

User Training. Based on the theoretical literature and on the evidence of attitude surveys within the Ontario government, it was clear that user training was a key factor in determining the success of teleconferencing. Sessions were conducted to:
. persuade potential users of the benefits of teleconferencing and its applications for them.
. to show them how to arrange teleconferences and to use the equipment.

* Trademark: Northern Telecom, Montreal

. to teach them how to effectively conduct a
teleconference.
The importance of advance preparation, of protocol and adapting meeting format and style to
teleconferencing were emphasized. Sessions
ranged from 1-1½ hours in length and included
an audio visual presentation on the benefits,
a how-to demonstration, a live teleconference,
and the distribution of printed guides including charts and checklists.

Promotion. Promotion was closely allied
with training, and was just as important. Senior management endorsement - letters from and
examples set by senior management were important. Most recently, interest among managers
seeking alternatives was generated because of
budgeting constraints. Internal marketing
through newsletters, brochures, posters, and
slidewhows also was used. Finally, personal
selling, word of mouth, and live demonstrations
generated interest in teleconferencing.

Evaluation Teleconferencing over 18
months was closely monitored. Participants
completed evaluation forms at the end of each
session and in depth interviews were conducted
at the end of the period. It is estimated that
during the trial period $50,000 in travel costs
and 2078 hours of management time were saved
through use of the CNCP Dedicated Network.
Most conferences lasted 2-2½ hours, involved
5-6 locations and 15 participants. Use of Bell
Conference Calling and Conference 2000s was
more difficult to monitor, but meetings using
them tended to be smaller and shorter. Over
88% of teleconferences using CNCP facilities
were thought to have replaced travel, and the
rest were meetings that would not otherwise
have been held. Satisfaction among users was
high. Many opponents of the system were converted. Most frequently cited advantages
were time savings, cost savings, increased
efficiency (meetings were shorter), increased
participation of regional staff who would not
normally travel. Loss of face-to-face contact
and non verbal communication was the most
frequently cited disadvantage. In addition,
users found that teleconferencing was not suitable for all meetings, and that the additional
preparation required was inconvenient. Still,
96% of users indicated that they were completely satisfied or very satisfied with the
medium.

TEMP Teleconferencing Task Force Program

Given its own success with teleconferencing and the potential benefits of travel substitution, the Ontario government has adopted
a proactive role in promoting teleconferencing
in Ontario. In early 1980 the TEMP Teleconferencing Task Force was created as part of the
Government's Transportation Energy Management
Program (TEMP), including representatives from
the Ministries of Government Services, Energy,
and Transportation and Communications.

Government Promotion

Since then the TEMP Task Force has been
working actively to promote the use of teleconferencing in the Ontario Government - with
considerable success. The Task Force pub-
licizes the benefits of teleconferencing with
brochures, articles, memos from senior management, displays, presentations, and demonstrations. In addition, assistance is provided to individual ministries in designing
and implementing systems and training users.

Expansion of Facilities. The Ministry of
Environment and the Ministry of Northern Affairs have implemented CNCP dedicated networks
linking their head and regional offices. Several ministries have installed Conference
2000s, Speakerphones, and other devices.

Shared facilities have also increased.
MGS has established audio/video teleconferencing centres in Oshawa and Sudbury, and further expansion is planned to North Bay and
Saulte St Marie.

For multi-point audio teleconferences, a
teleconferencing bridge was installed in January 1983 which allows the Government of Ontario Central Switchboard operators to link up
to 10 points. This service complements the
services available from Bell allowing government users to teleconference using tie lines
or foreign exchange lines. The bridge was
custom designed by Rockwell/Wescom. It offers
superior line quality, having automatic gain
control on each port. It allows operator intervention in case of problems encountered by
users and to ensure security. The operator
verifies each participant's name against a
list provided by the chairman. Both meet-me
and dial out modes can be used, and the service can be expanded.

Increased Usage. In some ministries, formal evaluation procedures were put in place:
in others usage was not thoroughly monitored.
Nevertheless, it is clear that within the
Ontario Government:
  . most ministries use some form of telecon-
    ferencing.
  . some ministries use teleconferencing
    extensively.
  . most ministries could use teleconferencing
    more.

For example, at each of the Ministries of Transportation and Communications, Northern Affairs and Environment, about 8 teleconferences are held each month using CNCP Broadband. Use of telephone based systems is harder to measure but seems to be even more extensive. The full motion video service is presently used 4 hours each month for meetings and 4 hours each month for telehealth applications; but with the recent expansion of the facilities, projected usage will more than double.

Applications. The applications of teleconferencing in Ontario range from regular committee meetings to the delivery of social services. Several ministries use it for monthly or weekly meetings between head and regional offices to plan, coordinate, and review programs. Others use it between regular meetings to deal with urgent issues. Teleconferencing is also used for meetings with the Federal and other provincial governments to discuss joint programs, and for meetings and progress reviews with outside consultants. In addition, new types of meetings have developed to take advantage of the media, including career development and training programs. It is also used for "events", for announcements by Ministers and press conferences. Recently the Minister of Industry and Trade used video teleconferencing to open several high technology centres in Ontario. Job interviews have also been successfully conducted via teleconference.

Finally, teleconferencing is used to deliver much needed services to northern Ontario. Several hospitals have systems providing education and consultation to small centres, and the Ministry of Government Services facility is used regularly for consultations between psychiatrists in the south and patients in the north, and for training seminars.

Benefits. Teleconferencing has both reduced travel and improved communications and services. The Ministry of Environment estimates a 30% travel reduction has been achieved through teleconferencing. The time savings are just as important as the transportation cost saving. The Ministry of Energy recently conducted a nation wide teleconference that would have cost $6,000 and 114 hours to hold face-to-face.

While dollar and hour savings are most easily quantified, users cite other benefits as well. About half of the teleconferences held in the Ontario Government replace face-to-face meetings: the other half are meetings that would not have otherwise taken place. The improvement in delivery of services which has resulted from the use of teleconferencing is no less important than the reduction in travel.

Problems. Most of the problems with teleconferencing in Ontario have been technical. Because of the newness of some of the technology used, the limitations of the network and so forth, technical problems have not been uncommon. Most, however, have been resolved. The principal problem remains: changing human behaviour. The results to date have indicated that considerable potential is still untapped and that continued promotion is needed to maximize use.

Private Sector Promotion

Originally established to promote the use of teleconferencing within the government, the mandate of the Task Force was expanded to include the private sector. In Ontario, use of teleconferencing was limited and suppliers were unconvinced of its potential, as well as its industrial benefits in terms of improved productivity.

Approach. The TEMP effort was aimed at stimulating both the supply and demand sides of the market. TEMP provided information, demonstrations, training and implementation assistance to users. Demonstration projects were established with Polysar, a petrochemical company, and the University of Ottawa. Bell Canada was also encouraged to establish an internal teleconferencing project to assess the benefits of teleconferencing, and to more actively market its services. In addition, the Task Force assisted other suppliers in developing and testing new products, including the Rockwell/Wescom bridge and portable video terminals. Finally, the Task Force researched use, applications and techniques for teleconferencing, and recently conducted a survey of teleconferencing use in Ontario.

Achievements. Information on teleconferencing has been distributed to 1500 other organizations in the province and numerous seminars have been held. Bell Canada's internal demonstration proved very successful, and a wide range of services are now being actively marketed. The TEMP survey revealed that while 72% of large Ontario companies use some form of teleconferencing, most feel use could be increased. While relatively few (23.4%) report that a substantial travel reduction has been achieved through teleconferencing, most expect that both use and substitution will grow. Given the current activity in the Ontario Teleconferencing market, it is expected that the TEMP Teleconferencing Task Force will conclude its activities in 1984.

CONCLUSION

The use of teleconferencing is expected to
grow substantially over the next few years,
both in the Ontario government and private sector. The Ministry of Government Services will
continue to expand and promote services for the
use of other ministries in the Ontario Government. Voice and text messaging were recently
introduced, and are used on a limited basis.
"audio plus" teleconferencing, using the Canadian videotext technology, Telidon, is expected
in the next few months.

The common carriers and equipment suppliers are beginning to develop and actively market new teleconferencing products and services.
A meet-me conference service, customer bridges
and audio-plus-Telidon conferencing are recently announced offerings. Technological improvements, deregulation and increased interest in
improved productivity will undoubtably facilitate the growth of teleconferencing in Ontario.

BIBLIOGRAPHY

Bell Canada, TEMP Teleconferencing Program
    Report. February, 1983.

Biswas, Neeru, Voice Message Service Pilot
    Trial at Ministry of Government Services:
    An Evaluation Report. Toronto: Telecommu-
    nication Services Branch/Ministry of Gov-
    ernment Services, August, 1982.

Cukier, W.L., Teleconferencing Within the On-
    tario Government: presented to the Insti-
    tute for Graphic Communications Conference.
    Carmel: April, 1982.

Cukier, W.L., Teleconferencing and Travel Sub-
    stitution : prepared for IEEE Conference
    on Vehicular Technology. Toronto: May,
    1983.

Gorys, Paul K.J. and Brian C. DesLauriers,
    Audio Teleconferencing: Internal Needs.
    Survey for the Ministry of Transportation
    and Communications. Downsview: Ministry
    of Transportation and Communications,
    1978.

Gorys, Paul K.J., An Evaluation of the Audio
    Teleconferencing Demonstration Project at
    the Ministry of Transportation and Commu-
    nications. Downsview: Ministry of Trans-
    portation and Communications, 1980.

Telecommunication Services Branch, Summary Re-
    port: Ontario Government Multi-Ministry
    Hermes (CTS) Satellite Administrative/
    Operational Experiments. Toronto: Minis-
    try of Government Services, Revised

January 9, 1979.

Telecommunication Services Branch, Ontario Gov-
    ernment Multi-Purpose Anik B Satellite
    Communications Network Project: Report
    on Phase I. Toronto: Ministry of Govern-
    ment Services, August, 1981.

Telecommunication Services Branch, Ontario An-
    ik B Telehealth Pilot Project Report.
    Toronto: Ministry of Government Services,
    March, 1983.

Transportation Energy Management Program
    (TEMP), Teleconferencing: The Problem
    Solver. Downsview: Ministry of Trans-
    portation and Communications, 1982.

Transportation Energy Management Program
    (TEMP), Survey of Teleconferencing Use in
    Ontario. Downsview: Ministry of Trans-
    portation and Communication, 1983.

Transportation Energy Management Program
    (TEMP), Teleconferencing Systems Guide.
    Downsview: Ministry of Transportation and
    Communications, 1983

Transportation Energy Management Program
    (TEMP), Teleconferencing Program Guide.
    Downsview: Ministry of Transportation and
    Communications, 1983.

Telecommunication Services Branch, A Prelim-
    inary Cost-Benefit Analysis of Audio
    Teleconferencing Using the Public Switched
    Telephone Network, Toronto, Ministry of
    Government Services, January 1979.

The Teleconferencing Resource Book: A Guide to Applications and Planning
Lorne A. Parker and Christine H. Olgren (eds.)
Elsevier Science Publishers B.V. (North-Holland)
©Center for Interactive Programs, University of Wisconsin-Extension, 1984

ROOM DESIGN AND ENGINEERING FOR TWO-WAY TELECONFERENCING

Robert E. McFarlane
Associate
and
Robert J. Nissen
Vice-President

Hubert Wilke Inc.
Communications Facilities Consultants
New York, New York

Typical Approaches

Executives anxious to jump on the "tele-conferencing bandwagon", often hand down mandates without knowing the limits of the state-of-the-art. When one realizes where an executive gets most of his information, this is understandable. Business publications regularly make it appear that competitors have every form of "whiz-bang" technology ever envisioned for "Buck Rogers" or "Star Wars". Even the best magazines print article upon article lauding the benefits wrought by "high-tech" corporate facilities which aren't even done yet, or have never actually worked. Video conferencing, of course, is a favorite topic these days.

So the well-meaning executive, with the idea that these marvels can be bought at the local "electronics supermarket", assigns the project on a rush basis, to the department which first comes to mind. In the case of video conferencing (full-motion, of course) the directive may go one of several places; Television, Telecommunications, Data Processing, or Advanced Office Systems (Office Automation). All of these departments are skilled in their own disciplines. Few, if any, have experience in all the communications, electronics, and human engineering cross-disciplines let alone in the fields of architecture, acoustics, electrical engineering, and air conditioning required to design the actual room.

In an attempt to fulfill Management requirements, one of two approaches is often taken:

1) Put out a Request For Proposal (RFP), addressed to every vendor of equipment, systems, transmission links, and consulting services that can be located, asking for breakdowns of each alternative, option or service which might be proposed.

2) Assemble an in-house team of engineers. Have them learn all they can about teleconferencing, design a system and, perhaps, even build it themselves.

Both of these approaches have serious pitfalls. The RFP must usually be put out several times, because each set of responses provides more insight into what should have been asked in the first place. The in-house design team, typically having little comprehensive experience, usually starts looking at equipment and piecing together a system design. In so doing, they are likely to repeat the mistakes which others have made before them. Without "hands-on" experience in this new medium, it is difficult to understand why "logical" designs so often don't work. In either case, the results are usually long in coming, tend to become more "experimental" than useful, and are very expensive if staff time is properly accounted in the system cost. None of these factors should keep the system from eventually being workable. However, an improper room design is not easily corrected. Failure to deal adequately with this vital element may well undermine the efficacy of the entire project.

The Missing Element - The Room

In either approach, emphasis is usually placed on equipment. The room is likely to be given little attention, either because its importance is not realized, or because it is the least understood and, therefore, the most easily avoided. Whatever the reason, if a teleconferencing room is not carefully designed, the finest equipment in the world cannot save the day.

The Room Design Sequence

Proper design of any special facility must follow proven architectural design sequences:

- Programming (Needs Analysis)
- Concept Design (Sufficient for Cost Estimation)
- Detailed Design (Sufficient for Bid & Construction Documents)
- Implementation (Supervision and Checkout)

Programming

The "Needs Analysis" is fundamental to the entire process. It must be done thoroughly, and its value cannot be overemphasized. It should cover the functions of the rooms, who will use them, how, why, how often, and for what purposes. Needs analysis programming is a systematic probing of the requirements and financial implications of the medium for a given company. It should provide the answer to two main questions; does it solve our communications problems, and is it cost effective? It must also provide sufficient information to begin the Concept Design.

It is not the purpose of this paper to delve into this part of the work, other than to note that remaining designs cannot be properly accomplished without a valid study as a foundation.

Concept Design

At this stage, the results of the Needs Analysis should be used to determine the numbers of active participants required, the types of graphics which must be accommodated, the types of meetings the room will serve, and whether Full-motion Video, Freeze-frame Video, Augmented Audio, or Audio Only will be required at each site. Needs may differ, and sites may be equipped compatibly, but differently. e.g., a site equipped for Freeze-frame Video can still conduct an Audio Only or Augmented Audio teleconference with a site not needing video support. The determinations made at this stage will allow selection of a general table shape, establishment of room sizes, and creation of a generic listing of the various pieces of primary equipment needed in each room (e.g., color cameras, lens types, monitors, codecs, switchers, hard-copy printer, etc.). Unfortunately, this is where many design efforts end, and equipment aquisition starts. In reality, it is only the beginning.

Detailed Design

This is where the Concept Design is proved out. At this stage every aspect of the room, the equipment, and the human interface should be examined, and committed to drawings. The Systems Design drawings should include Equipment Layouts, Table Details, Rack Elevations, Functional Block Diagrams showing Audio, Video, and Control Interconnections, Schematic Diagrams of special circuitry, Control Panel Layouts, and every piece of equipment required, including accessories, identified by Manufacturer and Model Number. The Room Design should consist of full architectural drawings, showing Floor Plans, Wall Elevations, Reflected Ceiling Plans, Lighting Plots, Wall Construction Details, Acoustical Treatments, Electrical Provisions (power, lighting, empty conduit), HVAC (Heating, Ventilating, and Air Conditioning) designs, and the Design Standards in terms of Power and Heat Loads, Allowable Temperature and Humidity Ranges, Lighting Levels, and Acoustical Criteria.

It is during the Detailed Design effort that every aspect of the design is worked out on paper; the limitations on lensing, the precise locations of all equipment, the ventilation system solutions, the architectural and structural designs, special interfacing for dissimilar equipment, the particulars of acoustical treatments, the room finishes, and all ancilliary equipment required for a complete operational system.

There is an inordinate amount of effort, knowledge, and information involved in accomplishing a thorough Detailed Design. If it is done in-house, the hours should not be underestimated. The time involved to gather information on unfamiliar equipment, and to learn the myriad building materials, acoustical treatments, lighting fixtures, and the like, will exceed the time necessary to translate that information into finished drawings. Remember also, that drawings detailing the room must utilize symbols and formats readily understood by the architect, electrical engineer, mechanical engineer, and/or construction trades that will work with them. If such drawings are not your daily fare, there will either be a lot of additional learning time involved, or a lot of time explaining and interpreting the plans to construction professionals.

With a knowledge of what a Detailed Design encompasses, it is easier to understand the dilemma of a Systems Contractor who is asked, in a Request for Proposal, to bid on a Concept Design. A good Contractor will realize the magnitude of work which is missing. But he will also recognize that he is bidding against vendors who would gladly avoid most of that work to "low-ball" the job, and assemble in the field on a "cut-and-try" basis. A Contractor of the latter type will probably also ignore most of the room construction details, throwing that design responsibility right back on the Owner. In short, if you are writing an RFP which is meant to include Design Services, make certain you spell out those services in detail (and insist that the Bidder do so as well), and be prepared to pay for the work. Preferably, you should require the design fees to be stated separately from the equipment and installation costs. This not only keeps everything above-board, but allows a more valid comparison with the charges of independent engineering firms which you might also be considering for the design phases.

Implementation

This is the phase of work everyone wants to get to on day-one. If the detailed design has been done well, and a competent Systems Contractor has been chosen, most of the equipment will have been pre-fabricated in the Contractor's shop while the room construction was in progress. This keeps the equipment and technicians out of the way of the carpenters and electricians, avoids equipment damage and dirt contamination, and allows thorough pre-testing of most of the system in a shop environment where the "bugs" can most readily be worked out. If you are using a design consultant, he should thoroughly check the systems in-shop at this time, before shipment to the site.

If you are building the system in-house, you would be well advised to follow the same procedures. Avoid the temptation to build on-site. Testing widely separated systems is difficult enough, even when you know they once worked together. Without this step the problems can become insurmountable, and the time schedule will stretch enormously.

Once the systems are in, be prepared to commit the manpower at each site to thoroughly test the systems. Also be prepared for the time and cost for room and system revisions to correct details which don't quite mesh. Keep in mind that much of the application of technology to teleconferencing is still more in the realm of "art" than "science". There are not, as yet, thoroughly proven design rules for the layman to follow. Because of the newness of the technique, it may be many years before good standards evolve.

The Detailed Design

The remainder of this paper will deal with the Detailed Design, since it is at this stage that most jobs get into trouble. And, because Room Design seems to be less understood than Systems Design, most of the information will pertain to that element of the work. If you are designing in-house, these are many of the problems you will have to consider and solve. If you are writing an RFP, these are the things your chosen Vendor or Design Consultant must take care of.

This discussion will assume a design for two-way Video Conferencing. For Augmented Audio, or Audio Only, only specific considerations will apply. Do not overlook, however, the importance of designing for upgrade to video, if that is a future possibility.

The "Driving Elements"

A big problem with any design is where to begin. In two-way video conferencing, there are two elements which "drive" the design:

For the systems design it is, of course, meeting the functional requirements of the system as determined by the Needs Analysis. But beyond this obvious requirement, the driving element is the design of the control panel. It must allow easy control of the conference equipment by the participants without the intervention of technical personnel. The control functions it contains dictate many of the system details and circuitry. Control panel design for video conferencing is a post-graduate exercise in ergonomics (the relationship of humans to machines).

For the room design, it is the table configuration. Everything else in the room, including the walls, must relate to it. Because we are concentrating on the Room Design, we will begin by examining table considerations.

The Table and The Room

Table Shape

The debates which rage over table shape are reminiscent of the Paris Peace Talks. Different approaches have their own devotees, and their own advantages and disadvantages. The final decision must be based on a combination of user preference, suitability for the purpose (as determined by the Needs Analysis), and physical considerations. It is not the purpose of this paper to advocate any one design over another. Rather, it is to provide an understanding of the advantages, the limitations, and the tradeoffs of alternative designs. We will illustrate the considerations with two quite different designs; the "semi-circular" table typified by AT&T PMS Rooms, and the "wedge-shaped" table common to SBS facilities.

The "Semi-circular" Table

The arced shape of this table satisfies two important optical parameters; a) every participant is approximately equi-distant from the cameras, and b) every participant is approximately equi-distant from the video monitors. (See Figure 1)

This arrangement allows participants to be an appropriate distance from the monitors, and avoids subjective image distortion introduced when lenses must encompass both very near and relatively distant objects. It also tends to give virtually equal importance to each participant and approximates the feeling of a contin-

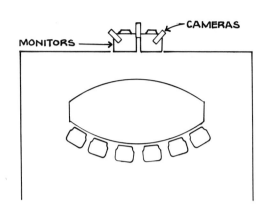

Figure 1.

within the room than the semi-circular design, allows better overview pictures, and avoids the debate over "switching algorithms". Since it requires fewer cameras to cover, it also tends to be less expensive.

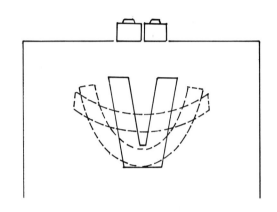

Figure 2.

uous conference table, divided only by a transparant "hole in space". Unfortunately, this extended conference table illusion can be achieved only with large screen television projection, which has other disadvantages as will be discussed later. Another disadvantage of this table design is the difficulty of simultaneously viewing all participants using a single front-mounted camera. This problem will also be discussed later. The voice-selected cameras provide excellent "people pictures", which support the needs of executive conferences very well, but the switching algorithm which determines which camera is selected, when, and why, will probably never satisfy every user in every situation.

Another objection sometimes voiced about the "semi-circular" design is that it tends to create a formal setting, and is not conducive to open discussion among participants within the room. The positive aspect is that this table shape orients participants toward the monitors, which in turn places prime emphasis on the conference relationship with those participants at the external remote location. For this reason, we generically identify this table type as an "External Oriented Configuration". The design is particularly suited to full-motion video systems where face-to-face teleconferencing is of prime importance.

The "Wedge-shaped Table"

One of the development principles behind the "wedge-shaped" table was the intent to break down the "hole-in-space" barrier while using standard monitor sizes. The table can be looked at as the end result of "bending" the "semi-circular" table inward until it forms a wedge or "V" shape. (Figure 2).

Proponents of this design say it provides a meeting setting more conducive to discussion

Critics point to the variety of optical and mechanical problems which must be addressed. The person at the table apex is usually the Chairperson. However, the lens tends to emphasize participants nearest the camera. The Chairperson is also farthest from the monitors. In other words, the main participant gets the worst seat in the house. Another comment sometimes voiced is that, because the table promotes interaction among those seated around it, it tends to make those at the remote location "observers" of in internal conference, rather than participants in a teleconference. It is for this reason that we have labeled this table shape an "Internal Oriented Configuration". This configuration requires participants to perform an almost 180-degree head swing in order to look from an adjacent participant to the incoming video monitor. (See Figure 3).

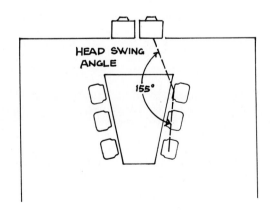

HEAD SWING
ANGLE

155°

Figure 3.

Whatever table configuration is chosen, be it one of these two extremes, or an intermediate design, the room design must meet a number of physical requirements which the table design dictates.

The First Decision - Distance To Monitors

A videoconferencing facility is intended primarily to enable transfer of aural and visual information. Visual information tends to be heavily graphics-oriented in most systems, rather than "people pictures". Even if people closeups are important, it is the need to read graphic information which governs much of the room design.

The limiting factors in reading graphic displays are three-fold:

1) The resolving capabilities of the human eye. Generally speaking, for good readability, a character height should subtend no less than 17 minutes of arc at the eye. Put in another way, character height should be no less than 1/200 of the viewing distance.

2) The density of graphic information. Ideally, there should be no more than 15 lines per page with each line not exceeding 40 characters across. This assumes that a standard 525 scanning line system is being used.

3) The size of available video monitors (generally 25", 27", or 30" diagonal measurement).

Figure 4 illustrates the factor defined in item 1 above. Assuming 15 lines of text per page, and using the 1/200 criteria, it can be calculated that a viewer should be no farther from the monitor screen than five times the screen width. We refer to this as the "5W" rule. Given these conditions, the normal eye can easily and comfortably resolve the text or equivalent graphics. As a rule of thumb, the <u>width</u> of a television image is approximately 0.8 times the <u>diagonal</u> measurement.

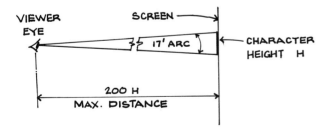

Figure 4.

If, for example, we design for a 27" monitor (which has an image width of 21.6"), and use the 5W rule, we find that a viewer should be no farther from the monitor than 9.0 feet.

Figure 5 illustrates this distance for the semi-circular table.

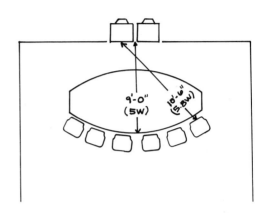

Figure 5.

It is also undesirable to have the nearest viewer less than 2W from the screen. Figure 6 illustrates the 2W and 5W rules for the wedge-shaped table, and the resulting problems.

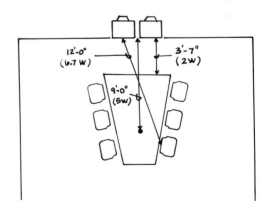

Figure 6.

This, of course, assumes that the graphics are sized as suggested. Should one try to show half of a typewritten page, there will be about 25 lines of print with about 70 characters to the line. While it is possible to resolve this amount of detail it is certainly not ideal. Under these conditions, the effective viewing distance would be about 8W. Most people will be unable to <u>easily</u> read the material at this relative distance, <u>particularly</u> with the image quality limitations of the 525-line television picture, and the bandwidth limits of most transmission media, both of which reduce displayed resolution.

The Lens Problem - Covering The Table

If the table position is set by the viewing distance, then camera lensing must be chosen to obtain pictures of the participants <u>at that distance</u>. If we plot the required lens angles for full coverage, we quickly find that both the "wedge-shaped" table and the "semi-circular" table require very wide angle lenses to simultaneously cover all the participants. These lenses produce subjective image distortions. We suggest that the widest <u>practical</u> lens that should be considered under these circumstances is a 12.5mm on a 1-inch camera tube, or a 9mm lens on a 2/3-inch camera tube. Both provide a horizontal coverage angle of approximately 52-degrees.

The plot of these angles for the semi-circular table is shown in Figure 7 and those for the wedge-shaped table in Figure 8.

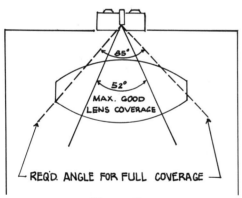

Figure 7.

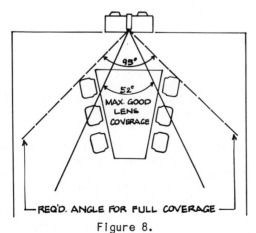

Figure 8.

To cover all participants at the table with one camera (and assuming distances to the camera/monitor wall previously determined) requires a lens with a very wide angle. This sort of lens creates subjective image distortions that are usually unacceptable for video conferencing. Designers of wedge-shaped table facilities will normally compromise the design by moving the table farther from the monitors, thus decreasing the subjective sizes of the displayed images, or cover the table using a camera on a pan-tilt mount. Those who design semi-circular table facilities have frequently put an "overview" camera to one side as shown in Figure 9. The angle of the over-view camera modifies the visual balance of importance among the participants, as well as the angle of apparent viewing, which some people find objectionable. With a semi-circular table three cameras are automatically voice-switched to cover groups of participants.

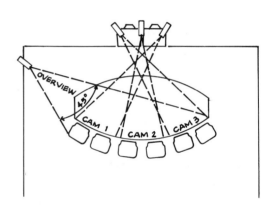

Figure 9.

Camera Position

If "people pictures" are at all important, it is vital that the camera be kept as close as possible to the monitors. It should also be slightly above eye level. The reasons have to do with establishing a sense of eye contact. If the camera is to one side, a participant looking at the incoming monitor (the other person in the conversation), will appear at the other end to be looking off to the side. This is most disconcerting for most viewers and is not conducive to good communication. It is for just this reason that large projected images are such a problem. There is no way to place the camera close to the image center.

The "above eye level" placement is a well known tenet of professional television. The viewer always feels more comfortable, and less threatened, if he is looking <u>slightly</u> down on the person talking to him. Only in a video conference can all parties feel they are in a superior relationship to their counterparts!

Satisfying these requirements requires that the cameras be as closely spaced as possible, between, and slightly above the monitors. Small deviations will not be noticable to many, but may take the "edge" off communication in a way everyone feels, but few can enunciate. A good configuration is illustrated in Figure 10.

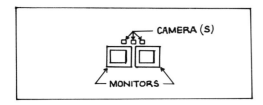

Figure 10.

Lighting

Television lighting is both art and science. It embodies the principles of photographic lighting, which allow a three-dimensional object to appear realistic in a two-dimensional medium.

The ability of lighting to alter realism can be easily demonstrated with the classical "halloween flashlight" gimmick. A flashlight, held directly below the face, produces a grotesque appearance. The reason is the unfamiliar shadowing caused by lighting from below.

We are accustomed to seeing people in lighting which originates from above (the sun, ceiling lights, etc.). To create a natural appearance on television, lighting must produce a similar effect, but must emphasize particular shadows, and de-emphasize others, to produce the illusion of depth without sunken eyes, shiny heads, or triple chins.

It is appropriate at this point to review classic television (or film) lighting principles. These are the standards - frequently modified for particular circumstances - that provide for both the technical requirements of the medium and the desired aesthetic results. Figure 11 illustrates, in plan view, a simple light plot for lighting a fixed single person.

The key light provides the main apparent source of illumination for the picture. In a studio environment it is typically a fresnel lens instrument which emits a spectral type light that causes sharp shadows. It is this shadowing that details the form and shape of the lighted object. The fill light is normally a soft-light instrument, such as a scoop, that partially fills in the dark shadows caused by the key light. The back light provides the all

essential rim lighting on heads and shoulders to give a three-dimensional illusion to the reproduced two-dimensional picture.

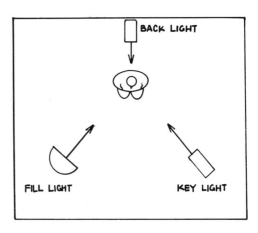

Figure 11.

Figure 12 illustrates, in an elevation view, the desired vertical lighting angles. The key and fill lights should preferably be angled at no less than 45 degrees below the horizontal. This criteria is based on two factors. First, such lighting produces a picture that is aesthetically pleasing and normal. Second, if the front lights are angled substantially less than 45 degrees, the light entering the eye of the talent becomes very annoying and causes squinting. We refer to this angle as the "glint" or "glare" angle of the eye. Conversely, the angle should not be substantially greater than 45 degrees or the face shadowing will be too deep.

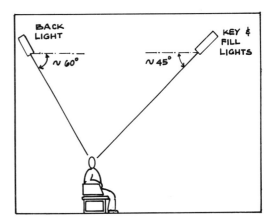

Figure 12.

A back light placed so as to angle approximately 60 degrees below the horizontal will provide the desired visual separation between subject and background. The angle should not be more than this angle or the visual effect of a top light will be created which is particularly a problem with balding people.

Video Conference Room Lighting

So much for the simple classical lighting plot. Now, what do we do for video conference room lighting? The answer to that question is not so simple and in fact an optimum solution has not, in our opinion, been achieved. The difficulty lies in the inherently contradictory requirements of such a room. In addition, the field is too new, and sufficient experimentation has not taken place under full operating conditions.

We must first establish the criteria for the lighting of such functions. We see the main criteria as follows:

1. The intensity of the lighting should be such as to meet the technical requirements of the color cameras with the selected lenses set at the desired f/stop. For typical designs, 100 footcandles is usually adequate.

2. The illumination should preferably be specular in nature, rather than diffuse, to delineate the form and substance of the picture.

3. The illumination should be as inobtrusive as possible. Lighting angles should be outside the typical glint angle of the eye. Furthermore, the participants should not get the feeling that they are in a studio environment.

4. Back lighting must be provided to accent the third dimension of the picture.

5. Ambient light falling on the monitor screens must be reduced as much as possible to maintain a good picture contrast ratio.

6. If lighting is provided from mixed types of luminaires, they should be corrected so that each has the same color temperature.

Using these criteria as design guidelines we can now consider several possible approaches to the lighting of video conferencing rooms.

The first approach (a very simplistic one) is to provide standard flush-mounted ceiling fluorescent fixtures in sufficient quantities to provide the intensity of illumination required. This approach can provide adequate illumination but little else. The lighting will be very diffuse, ambient spill light is difficult to control, and most of the illumination is top light. It is not recommended.

A modification of this approach is used in many typical AT&T PMS rooms. Front lighting is provided by a bank of ceiling-mounted fluorescent fixtures which are aimed toward the participants either by tilting or by internal reflectors. This directionality helps to control the ambient light spill on the monitors. Back lighting is provided by incandescent down-light instruments. The mixture of flourescent and incandescent luminaires can be handled by placing color correcting filter material within the fluorescent fixtures. This approach can adequately satisfy most of the above criteria except for the one relating to specular lighting. The lighting from the front fluorescent fixtures is by nature diffuse and provides a flat looking picture. On the whole, however, it can be considered a feasibly alternative.

A third approach uses the classic lighting techniques of television or film production. All instruments illuminating the participants are incandescent with specular characteristics. Additional ceiling mounted flourescent fixtures (with directional louvers and proper color correction) can be used to provide ambient room illumination outside of the participant area. Properly designed, this approach can meet all of the criteria cited above. The main problem is how to design the system without giving participants the feeling of a television studio.

From an aesthetic standpoint, it is preferable to use flush-mounted architectural down-light fixtures. Unfortunately, these fixtures seldom provide the degree of directional control attainable with studio type instruments.

An alternative approach is to use architectural light tracks with moveable fixtures. A wide range of fixtures and accessories are available for light tracks which can provide relatively good control of light distribution. If the tracks are surface mounted on the ceiling, these protruding instruments tend to convey an undesirable studio type atmosphere. One solution, is to recess the tracks in ceiling coves so that the instruments are not so obvious.

Figure 13 illustrates one possible light plot for lighting the semi-circular table configuration using incandescent instruments on light tracks. The placement and adjustment of each fixture must be carefully controlled in order to approximate the key and fill lighting functions.

Figure 14 illustrates a possible light plot for lighting the wedge-shaped table configuration with incandescent instruments on light tracks. Clearly, lighting the wedge-shaped table is more complex because people are facing different directions.

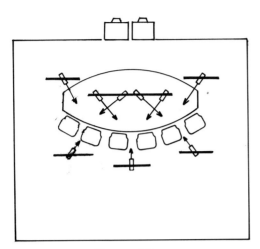

Figure 13.

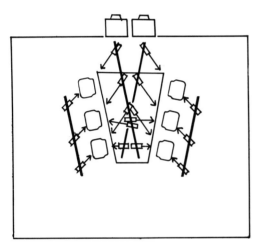

Figure 14.

Choosing fixtures of any kind requires careful study of the photometric data. But, if a problem develops, track-mounted fixtures can easily be changed or adjusted in the field.

We do not recommend that dimmers be used for video conference lighting systems. Unless the dimmers have expensive radio frequency filtering they can cause interference with the video and audio systems. Furthermore, since the function of the room does not require changes in the light plot, once properly adjusted, there is little use for dimming systems. Illumination intensity can appropriately be handled by selecting the proper lamp wattages.

When considering light track, don't overlook the effective reduction in ceiling height. If you are working with an 8-foot ceiling, the bottom of the lights will be at about 7-feet. This is too low for both optical and aesthetic reasons. Track lighting should be restricted

to rooms with high ceilings or where the track can be recessed up into ceiling coves.

Light Spill and Contrast Ratio

One of the most important considerations for readability of a display on a television screen is the contrast ratio between the characters and the screen background. It can be demonstrated that a contrast ratio of at least 10-to-1 offers a highly readable display. Since a properly adjusted television monitor can provide a contrast ratio of up to 40-to-1, any lack of contrast at the display will not be occasioned by inherent characteristics of the television monitor. Rather, it will be caused by ambient light falling on the screen.

It takes only a surprisingly small amount of ambient light spilling onto the face of a monitor screen to diminish the contrast ratio to an unacceptable level. For example, if the lighted portions of the screen (the characters or graphics) emit 40 footlamberts of light (a typical value for a properly adjusted monitor) and ambient light spilling onto the screen causes the background to reflect 10 footlamberts, the contrast ratio has been decreased to 4-to-1. It is important to emphasize that this is _not_ an inherent problem of the television monitor but, rather, is caused by ambient light falling on the screen.

One of the unique design requirements of a video conferencing room relates to this problem. Adequate illumination must be provided on the participants for camera pickup but must be kept off of the television monitors.

Graphics Illumination

Since the majority of visual information utilized in most video conferences is "graphic" in nature, rather than "people pictures", it is important that the materials to be used are properly illuminated.

Unfortunately, the graphic materials participants wish to transmit may range from as formal as a detailed chart prepared by graphic artists, to as basic as a hastily drawn pencil sketch. Graphics may be on professionally prepared 35mm slides, or on overhead transparencies poorly reproduced from a typewritten draft on a photocopy machine. The requirements for illumination vary widely, but in a video conference, users should not be expected to adjust lighting or lenses for each situation.

Two preliminary steps can simplify the problem:

1) Identify in the Needs Analysis, those graphics most often used in meetings.

Obtain samples, and design for them. Do not expect to handle every possible graphic with equal quality.

2) Set "norms" and "standards" for graphics, publicize the standards, and design to optimize complying materials. For example, include a mask for typewritten pages, delineating how much text can be shown at one time. Design for white paper, not colored. Provide flip chart paper in a fixed room location. Provide a "slide chain" if 35mm slides are to be frequently used.

When designing the lighting for graphics, there are three major considerations:

1) Illumination level should be essentially the same as that for the participants, i.e. about 100 footcandles.

2) The illumination for graphics should be essentially flat over the entire field of the largest size graphic to be used.

3) It should be possible to use graphics material in a normal manner without casting large shadows from the users head, body, or hands.

When using rear-illuminated displays, such as are sometimes designed into a conference table for the transmission of transparancies, one must remember that the overhead lighting which illuminates the table surface also illuminates the top of the transparancy, reducing the contrast ratio. If the particular table design, or needs analysis, requires such a display, it may be necessary to incorporate room light dimming for selected fixtures to reduce the top light during the time that the transparancy is displayed.

Room Color Treatments and Finishes

While one might wish to decorate a video conference room with "splashy" colors, such decor can cause serious problems with the televised image.

It is important to use relatively neutral colors on walls, carpet, and upholstery so as to avoid problems with colorimetry. Highly saturated colors can produce a colored "pall" on the image, particularly with lesser quality cameras. Inappropriate background colors can also clash with the participant's face or clothing color. Such backgrounds can also call attention to themselves and detract from the primary focus which should be on the participants or the graphics.

It is also imperative that surfaces in the

room produce no spectral reflections. This means no polished tabletops or chrome chairs, among other things.

Audio Systems and Room Acoustics

Most teleconferencing users will agree that the audio link is the most important part of the system. If the video equipment breaks down, it is usually possible to hold a worthwhile meeting by audio only. The reverse is not true.

For teleconferencing, one wants to engage in two-way conversations which are as close to natural as possible. Ideally, this would mean that the audio would be "high fidelity" with all parties able to talk and listen simultaneously, just as if everyone were in the same room.

Transmission Bandwidth

Wide-band audio (as much as 15kHz bandwidth) removes the "tinniness" of the normal telephone call. But wide-band circuits also cost money, and since speech intellegence is contained within the normal telephone bandwidth of 3kHz, standard "voice grade" telephone lines are generally utilized. If user-owned wideband facilities are available, they should certainly be used. Normally, however, one will have to assess the value of the more costly circuits in light of the expected end-results, particularly where widely separated locations are involved.

One must also consider the acoustical implications of using wide-band audio, since such a system will require more critical room design. Overlooking this fact could easily negate any advantages of the better circuits, and could even make them less effective than standard voice-grade transmission.

The "Open Audio" Concept

The ability to talk and listen simultaneously implies circuitry known as "full duplex" (FDX). FDX systems allow signals to move in both directions at the same time, as opposed to "half duplex" (HDX) systems in which the line must be "turned around" each time information goes the other way. "Full duplex" and "half duplex" circuits are functionally illustrated in Figure 15.

In teleconferencing, the "full-duplex" concept has become known as "open audio", because it leaves the lines "open" in both directions. This allows natural, two-way conversation, with performance similar to that of a standard telephone handset, except that microphones and

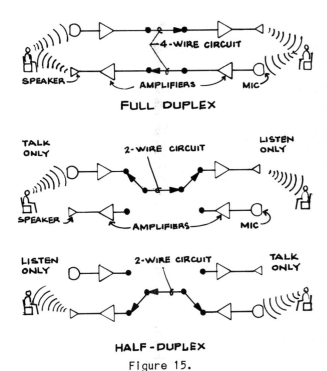

Figure 15.

loudspeakers are used to make the system "hands-free".

Open audio contrasts sharply with the familier "Speakerphone" which requires each user to wait until the other is completely finished before beginning to speak. Failure to wait before talking generally cuts off the first and last parts of what was said. Normal, open conversation cannot take place over Speakerphone systems. Even a simple acknowledgement can interrupt the conversation, instead of reinforcing it. This forces the conversation into a "half duplex" format, which most people find objectionable.

Open Audio Problems

If open audio is so desirable, then why is something like the Speakerphone even used? The answer is that most offices are not acoustically capable of supporting an open audio system. The reason is illustrated in Figure 16.

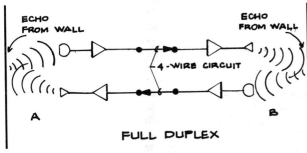

Figure 16.

In this situation, sound originating at the "A" location comes out of the loudspeaker at location "B", is reflected by the wall, re-enters the microphone at "B", and returns to location "A", where the poor acoustics cause it to go around again. Since it has been delayed by its round-trip (as much as a full second if via satellite), it is very bothersome, even to the point of inducing temporary stuttering in the person talking. In the extreme, it will set up the familier public address system howl known as acoustic feedback. The term comes from the fact that the "loop" is completed acoustically, rather than electrically, feeding the signal back to its origin.

To break the feedback loop, a Speakerphone employs a device known as a "gain shifter" which, in essence, cuts off the microphone or loudspeaker at each end, depending on who is talking. The gain-shift circuitry (a rapid, automatic volume control) reduces the amount of audio returned from the other end, preventing the feedback from occurring.

Room Acoustics - The Two Concerns

If severe gain shifting is not utilized, the only way to stop an echo from causing feedback is an acoustical environment which reduces echo to where it is no longer bothersome. To do this requires a very good <u>interior</u> room acoustic design. We will examine the interior acoustic requirements later, after first considering the problem of <u>intrusive noises</u> which originate from outside of the room.

Keeping Noise Out Of The Room - 4 Steps

Sound originating outside a room can enter the room in three ways, as illustrated in Figure 17; through the structure (Structure Borne Noise), through an air leak in a wall or door (Air Borne Noise), or through the wall, ceiling, or floor construction (Barrier Transmission).

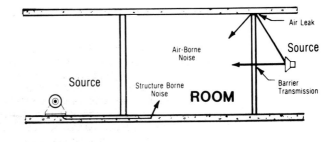

Figure 17.

Step #1 - A Good Location

Regardless of the type of noise, the best way to get rid of it is to avoid it. Locating a room away from outside walls, bathrooms, elevators, shops, loading docks, mechanical rooms, water pipes, waste drains, and the like, can avoid the need for very expensive sound insulation. An effective approach is to surround the teleconferencing room with relatively quiet spaces, such as storage rooms, other conference rooms, or limited access corridors, which provide a contiguous sound buffer between the high noise areas and the teleconferencing room. Don't overlook the space <u>above</u> the room. It can be the source of very objectionable noise, particularly from footfalls and rolling carts. Padded carpets on the floor above will usually isolate footfall noises.

Step #2 - Isolate The Source

Noise may enter a room by a vibrating building structure. This structure-borne noise is typically caused by heavy motors and machinery. The solution here is to isolate the noise <u>source</u> from the structure via springs or pads. For a teleconferencing facility it would be too expensive to consider the only other alternative - a floating floor.

Step #3 - Determine The Room Requirements

Expensive acoustic wall designs are often put on paper before there is any knowledge of the level of noise outside the room, or the maximum noise level which will be permissible within the finished room. Without this information, a wall or ceiling design will be either over-done, or less effective than required.

In determining the maximum permissible noise level for teleconferencing rooms, one must keep two things in mind: a) low frequency noise may be annoying to room users, but will not be transmitted to the far end if voice-grade circuits of limited bandwidth are used for the audio. b) Mid-range sounds will be annoying to people at both ends. Because microphones "hear" differently than people, and prevent us from "tuning out" bothersome noises when we are listening to loudspeakers, it is probable that certain mid-range noises will be more bothersome to far-end listeners than to those in the offending room.

The relative importance of different noise frequencies is one of the factors incorporated into a family of curves called Noise Criterion (NC) curves (See Figure 18.) Each NC curve specifies the maximum permissible intensity of ambient noise at various frequencies. It is thus possible, with a single NC figure, to specify the maximum noise intensity allowed for a group of frequencies.

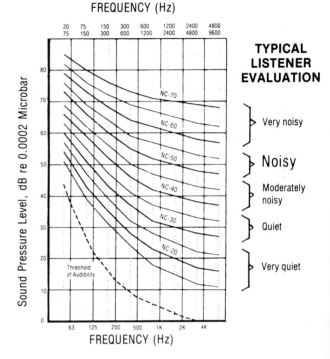

NOISE CRITERION (NC) CURVES

Figure 18.

For a teleconferencing facility, the ambient noise inside the room should not exceed the NC-30 curve. NC-25 would be preferable, but probably too expense. Note that this is for <u>total</u> noise, including that from Heating, Ventilating, and Air Conditioning (HVAC) systems and room equipment <u>as well as</u> that from intrusive noise.

Step 4 - Noise Transmission Control

If the maximum allowable noise level at each frequency is set by the selected NC curve, and the noise level outside the room is also known, then determining the amount of sound insulation required in each band is a matter of subtraction. Unfortunately, designing sound barriers (walls, floor, ceiling, doors, and windows) to achieve these results is not so simple. That is the job of a competent acoustical engineer.

For illustration, some sample sound barrier constructions are presented below, along with definitions of terms necessary to understand them.

Sound Transmission Loss

When airborne sound strikes a sound barrier, it causes the barrier to vibrate. The amount it vibrates, (a function of its mass), and the degree to which the two surfaces of the barrier are coupled together, determine how much of the original sound is re-radiated into the next room. The amount that airborne sound is reduced as it passes through a sound barrier (measured in decibels (dB) at each frequency), is called the Sound Transmission Loss (STL). STL is, therefore, a measure of the sound insulating efficiency of a sound barrier. (See Figure 19.)

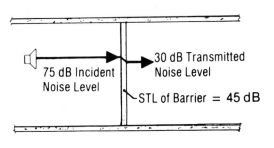

Figure 19.

Because sound transmission through barriers varies with frequency, a number of STL figures are required to rate a single barrier. This is an accurate method, but not a convenient one. In 1961, the American Society for Testing Materials (ASTM) devised a "single figure" method of rating the Sound Transmission Class (STC) of barriers. The STC is determined by measuring the STL at 16 test frequencies, and comparing the resulting curve with a standard reference contour. (See Figure 20.) While this method of rating barriers is convenient and common, it should be viewed with some caution. First, it attaches different importance (or weighting) to different frequencies. Second, it permits considerable deviation from the reference contour. Therefore, two barriers with the same STC ratings can perform rather differently. In designing teleconferencing rooms, one should be careful to specify the desired end results (NC curve), and let an acoustician decide what barriers are needed.

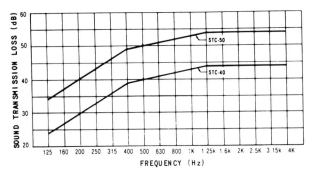

Figure 20.

Examples of some wall constructions are given in Figure 21, along with their STC ratings. Note that there are two basic types of construction; homogeneous, in which the construction is essentially the same throughout, and non-homogeneous, in which two or more elements are combined and/or acoustically isolated from each other.

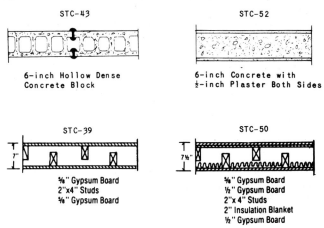

Figure 21.

Homogeneous walls depend entirely on mass to attenuate sound. However, there quickly comes a point beyond which constructing thicker, heavier structures is impractical. Homogeneous barriers are perfectly adequate for most situations. When they are not, non-homogeneous construction must be utilized to achieve the desired sound isolation.

Non-homogeneous walls use multiple materials which vibrate differently at each frequency. When combined, the multiple layers tend to equalize the performance of each material, creating a more uniformly attenuating wall.

When the sections are fully separated by a dead air space, the barrier is known as a "cavity wall". This decoupling makes it much harder for the vibration of one surface to be carried to the other, reducing re-radiation, and improving the STC of the barrier.

As can be seen in the examples, non-homogeneous walls provide excellent sound isolation (high STC figures) using relatively thin, low-mass construction. However, one should note that the construction is rather complex. This is an important point. If a non-homogeneous barrier is not built exactly as designed, its performance may be severely degraded. A small air leak, or a nail in the wrong place which couples the two sides together, can negate the purpose. A qualified acoustician should make the design determinations and inspect the construction at critical stages.

It's Also Important To Keep Sound In

The principles set forth above are also employed to maintain conference room privacy. This "isolation in reverse" is often overlooked in room design, but is very important for many types of meetings. Budgets, policy, new engineering designs, and the like can all be compromised if secrecy is violated, and a room which does not provide good control in this regard cannot be expected to be used for meetings requiring security.

Noise Within The Room

It does little good to spend money stopping intrusive noise from entering a room, only to generate noise inside the room. Common sources of internal noise are the HVAC system, buzzing fluorescent lights, equipment motors and blowers, clicking control buttons, shuffling feet, tapping pencils, etc. All of these can be controlled; most can be controlled inexpensively.

Fluorescent lights should have sound-rated ballasts, or the ballasts should be separately located. Equipment with motors can be shock-mounted, or put inside drawers or cabinets. Control panels can be mounted so as to avoid a "sounding board" effect when controls are used. Carpet will stop foot-fall noise, and a soft table surface will reduce noises originating there. Only the HVAC system requires special, and possibly expensive, treatment.

Assuming that the teleconferencing room has been isolated from fan room or compressor noises, the problems with HVAC systems will probably be rushing air, creaking ventilation ducts, or the transmission of noise from - or to - another room.

The sound of rushing air is due to high velocity air turbulence within the ducts, and to air movement past grills. Duct velocities should normally not exceed 300 feet per minute in a good ventilation system. Standard grills should _not_ be used. They have a tendency to create turbulence and, hence, noise.

Sound transmission from or to another room can be stopped with "dog-legged" ducts and good internal acoustic lining. This problem should _never_ occur in a properly designed system. A good mechanical engineer, together with your acoustician, can help you avoid these important problems at the outset.

The Final Item - Reverberation

We are finally back to where we started this subject. "Reverberation" is the multiple reflection of sounds by surfaces within the room. In nominal amounts, it provides the "naturalness" we like to hear. Excessive reverberation will cause severe problems, especially in teleconferencing situations where speech intelligibility is paramount.

In teleconferencing rooms, where open audio is desirable, too much sound reflection, at the wrong frequencies, will cause sound from the loudspeaker to re-enter the microphones, setting up undesirable echos or acoustic feedback. We want the room to be relatively "dead" or "dry" acoustically, rather than "live" (See Figure 22.)

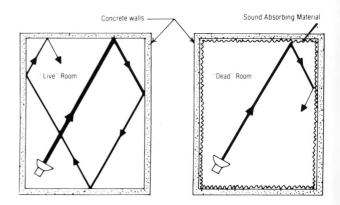

Figure 22.

How Is Reverberation Controlled?

Room reverberation is controlled by _what_ materials are applied to the interior surfaces, _how much_ material is used, and _how_ it is mounted. As with so many things, a little knowledge is a dangerous thing. It is quite common to see rooms, or room designs, calling for carpeting on the floor, and thin fiberglass panels, or fabric on ceiling and walls. A room thus constructed will seldom provide good internal acoustics.

The reason has to do with the sound absorbtion characteristics of different materials at different frequencies. Sound absorbers perform better at high frequencies, just as do sound barriers.

Materials such as carpet, acoustic tile, and thin fiberglass panels are relatively inefficient sound absorbers at lower frequencies. When applied directly to a surface, they absorb high frequencies only. The result is a "boomy" room. This is not only bothersome to users; it does not solve the acoustic feedback problem either. It is also a common mistake to use sound <u>absorbing</u> materials for sound <u>insulation</u>. It won't work, since absorbing materials - by their very nature - leak like a sieve.

The ability of a material to absorb sound is called the sound absorption coefficient. The coefficient can have a value between 0 and 1. When multiplied by 100, the coefficient indicates the percentage of sound absorbed. For example, if a material has a coefficient of 0.7, it absorbs 70% of the sound energy which strikes it at the frequency for which it is rated. (See Figure 23.)

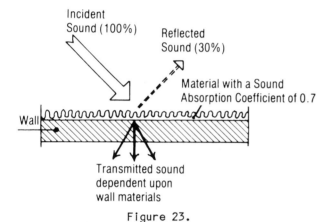

Figure 23.

A number of values must be examined to determine the effectiveness of any material in each situation. Table 1 provides the sound absorption coefficients for several representative materials and mounting methods at the six standard frequencies used for rating.

Sound absorbers are frequently rated using a single number known as the "Noise Reduction Coefficient" (NRC). NRC is the <u>average</u> of the sound absorption coefficients for only four frequencies; 250, 500, 1000, and 2000 Hz. For rooms with critical acoustical considerations, NRC is an inadequate indicator because it does <u>not</u> include the very low frequency absorption. The numbers can also be misleading since two materials with the same NRC values can have considerably different sound absorption characteristics. One must also be careful not to confuse NRC with true reduction of intrusive noises from outside the room. NRC has nothing whatsoever to do with insulation of the room from outside noise.

Again, the safest approach for someone trying to design or specify a teleconferencing facility is to specify the desired end result, <u>not</u> the method of achieving it. Let a competent acoustician determine the design.

SOUND ABSORPTION COEFFICIENTS FOR TYPICAL MATERIALS

Materials \ Frequency	125 Hz	250 Hz	500 Hz	1000 Hz	2000 Hz	4000 Hz
1. Gypsum board, ½", on 2 x 4 studs	0.29	0.10	0.05	0.04	0.07	0.09
2. Plate glass	0.18	0.06	0.04	0.03	0.02	0.02
3. Carpet, heavy, on concrete	0.02	0.06	0.14	0.37	0.60	0.65
4. Velour drape, medium weight	0.07	0.31	0.49	0.75	0.70	0.60
5. Typical "ceiling" tile, mounted to hard surface	0.10	0.30	0.56	0.70	0.68	0.50
6. Typical "ceiling" tile, hung on suspension system with 16" air space	0.45	0.48	0.65	0.75	0.72	0.55
7. Mineral wood blanket 2" thick, mounted with 1" air space	0.30	0.70	0.85	0.86	0.87	0.87
8. Fiberglass, 4" thick, mounted to hard surface	0.40	0.92	0.98	0.97	0.93	0.88

(Multiply coefficients by 100 to obtain absorption percentage)

Table 1.

How Much Absorption Do I Need?

The answer to this question (for a teleconferencing room) is alarmingly simple: as much as can be achieved using practical materials. <u>However</u> it is important that the absorption be "flat" over the <u>entire</u> speech frequency range of 125 Hz to 4000 Hz. This is the key to solving the myriad problems which have been discussed previously. Typical conference rooms, using thin acoustic tile, carpets, and drapes, absorb mainly high frequencies. This produces a "boomy" or "hollow" sounding room, because the low frequencies are still bouncing around the room (reverberating) long after the high frequency sounds have disappeared.

Heating, Ventilating, and Air Conditioning

"HVAC", (the common abbreviation in the mechanical engineering field), is responsible for the comfort level which can be maintained in the teleconferencing room. The system must handle not only the heat produced by conference participants (including any "secondary participants" or "observers"), but also the heat introduced by the lighting and the electronic equipment. Even if the room is designed for audio only, consider designing the HVAC to support the needs of video conferencing. A retrofit later will be inordinately expensive.

Another thing often overlooked is the time period over which the HVAC system must operate. In many buildings the air handling systems are shut down after normal hours. Since teleconferencing, by its nature, invites conferences with other time zones, it is possible that the HVAC system will have to be separate from the main building system to keep a proper room environment at all times.

The acoustical requirements for the air handling systems have already been discussed. It is worth noting, however, that the HVAC system must handle more heat in a teleconferencing room than in a normal conference room, because of the heat from the specialized lighting and from equipment. This means more air flow than normal, which can mean more noise if the system is not carefully designed. The acoustical requirements for a teleconferencing room will probably be more rigid than for any other conference room you have.

Lastly, one should make certain that the HVAC system is designed to properly handle smoke removal. Anything less than six air changes per hour is inadequate. Smoke not only bothers many people; it is also not good for equipment. Since it would be unrealistic to prohibit smoking in a teleconferencing room, it is important that the air system be able to cope with it.

Room Environmental Specifications

We suggest that the HVAC system be designed to maintain the following temperature and humidity conditions, with all equipment in operation, and the maximum number of participants in the room:

Temperature Range: 68 to 72 Degrees F, Dry Bulb

Humidity Range: 40% to 55% Relative

The specification for humidity control is necessary because of videotape equipment, computers, and other devices which can be adversely affected by either high moisture or static discharge.

Conclusions

The design of a teleconferencing room is still more of an art than a science. There are many more considerations than most people recognize when the task is first begun, and the room is often overlooked entirely until the equipment is ready to be installed. In view of the fact that teleconferencing is still in its infancy, with most potential users taking a "show me" attitude, it is necessary that installations be done as well as possible. The room design is critical to the success or failure of a system, and its design should not be taken lightly.

The Teleconferencing Resource Book: A Guide to Applications and Planning
Lorne A. Parker and Christine H. Olgren (eds.)
Elsevier Science Publishers B.V. (North-Holland)
©Center for Interactive Programs, University of Wisconsin-Extension, 1984

VIDEOCONFERENCING AND BRITISH TELECOM:
FROM CONFRAVISION TO THE VISUAL SERVICES TERMINAL

Gillian M. Reid, B Sc.
Executive Engineer
Centre for Visual Telecommunications
British Telecom
Martlesham Heath, Ipswich, England

Introduction

At the British Telecom Research Laboratories the Centre for Visual Telecommunications has been set up to explore the use of television systems in the business community. It is responsible for the planning and implementation of a range of visual services.

In Britain in the 10 years since the Confravision® video-teleconferencing service began, there have been many technical advances and social and economic changes. The cost of fuel has risen dramatically and the world has become gripped in a period of recession. The video market has expanded with increased usage of closed circuit television, video tape recorders and video games. All these factors have created an environment conducive to visual telecommunications, with businesses now more receptive to the idea of video-teleconferencing.

The Centre for Visual Telecommunications is responsible for coordinating technical aspects of Confravision®. It is also involved in field trials of systems that provide visual aids to enhance audio-teleconferences. Also being developed is a visual services terminal which is shortly to go on trial with 20 firms in the UK.

The simplest form of audio-teleconference, with participants linked in sound only, is the loudspeaking telephone. It involves no more than a simple phone call and multilocation conferences are possible using conference bridges. Until recently, no visual information could be disseminated at such a conference, but slow-scan TV and telewriting devices are now on trial alongside audio teleconference equipment currently available. Facsimile systems have of course been available for documents but until recently transmission times were rather long e.g. 3 minutes per document.

Cyclops

British Telecom have sponsored a human factors trial of audiographic conference equipment with the Open University[1,2]. This equipment is known as Cyclops and enables interactive drawing on the face of a TV screen by means of a light pen. The equipment is being used for telephone tutorials held with a tutor at one location and students at up to 8 other locations, connected via a conference bridge. The tutors can prepare their tutorial visual aids on a standard audio cassette and send the prepared information down a data line to each remote location. The tutors have use of a scribblepad as well as a light pen and it is hoped that the trial will yield data on the relative merits of each of these telewriting devices.

Orator

A new audio-teleconferencing unit is being brought into service in British Telecom under the name of Orator[3] (marketed internationally as CONFER-TEL). This system requires two telephone pairs either using the public switched telephone network (PSTN) or private wires (PW). This system is microprocessor controlled with automatic adjustment of frequency response, gain and line equalisation. It uses high quality microphones and loudspeakers, and incorporates a shallow voice-switch that is available as a user volume control. A multipoint facility will be available in late 1982.

Slow-scan TV can be utilised to add interactive visual facilities. It can be provided on separate PSTN or PW connections or possibly on the established connections already in use for audio. Of course the latter case results in the loss of speech in one direction while visual information is being sent. It does however save the cost of an extra telephone line for slow-scan equipment.

Slow-scan TV

Slow-scan Television also has a great potential for surveillance and security[4]. Other areas of potential use at present under trial, include observation of radar screens, medical applications (from X-rays of broken bones to neurological information), editorial conferences on advertisement layouts in newspapers and the monitoring of a lorry park for illegal dumping.

A higher resolution version of the surveillance trial[4] slow scan equipment is being developed for working with Orator and will have 625 line definition with the possibility of colour and will be used to transmit images of objects or drawings.

Conventional 625 line monochrome cameras and monitors will be used. The user will be able to choose the resolution appropriate to the visual information being sent. The choices are 256x256, 512x512, or 768x512 picture elements (pels). The higher the resolution the longer the update time for a particular transmission. Table 1 gives the approximate maximum transmission time expected for each type of circuit. Variable length coding can cut transmission time by about a third, depending on the picture content.

Data Rate Kbit/s	Pels. per Picture	Maximum Transmit Time s	Type of Circuit
4.8	256x256	50	Dial-up or PW
	512x512	200	
	768x512	300	
9.6	256x256	25	Dial-up or PW
	512x512	100	
	768x512	150	
48	256x256	5	Local/junction PW
	512x512	20	
	768x512	30	
64	256x256	4	System X Speech path
	512x512	16	
	768x512	24	

TABLE 1
Picture transmission times

Figure 1 summarises the various methods of adding visual information to audio-teleconferences. The number of telephone lines is indicated at the side of each system. The audio teleconferencing system shown has 4 wire working. Ordinary loudspeaking telephones would need only one telephone pair but in this case the other equipment would need extra lines to be provided. A minimum number of two lines are required for successful audiographic conferences. Slow-scan can only provide still pictures and therefore should not be used where full motion videoteleconferencing is thought to be required.

Narrowband TV

A further method of adding visual information to audio-teleconferences is also under consideration where real time movement is required. This involves low definition real time television to 313 lines, 50Hz standard. So far it has only been used for surveillance. It is thought it could be used where facsimile is available for transmission of graphics so that the lack of definition would not be a problem. Standard local network cables are used in this narrowband system but the main disadvantage is that only short distances (up to 5 Km) can be provided, due to crosstalk in the cables.

Confravision®

British Telecom provides a microwave network to carry TV signals for the broadcasters from studio to transmitter etc. These circuits have a stand-by channel (1 for every 7 operational channels) to cover circuit failures. In the late 1960's tests began of a 2 way video-teleconferencing service, using these stand-by channels, between the old PO Research Station on the outskirts of London and the Telecommunications Headquarters in central London. These trials led to the introduction, in 1971, of Confravision®, a public broadband video-teleconferencing service.

Originally there were five public Confravision® studios, situated in London, Birmingham, Bristol, Manchester and Glasgow to which have been added an extra London studio plus Leeds and Martlesham. There are also mobile studios and private studios. Up to three locations can be linked with all participants seeing and hearing each other. More than three studios can be linked but not all could take part in discussions.

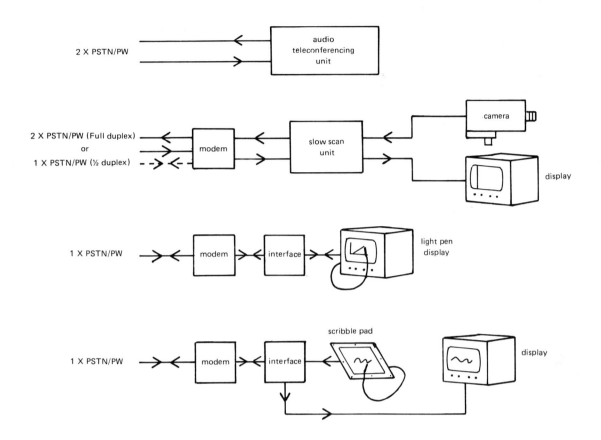

Figure 1
Visual aids for audio-teleconferencing

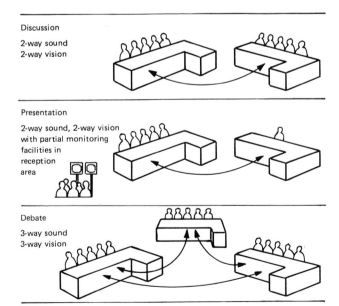

Figure 2
Confravision® configurations

The studios are designed to be as close
to a normal conference room environment as
possible. The lighting level is kept down to
prevent the feeling that conferees are "on
television". No British Telecom staff are in
attendance so the conferences can be held in
confidence and without interference. There
are facilities for displaying graphical
information, slides, documents, solid objects
and for using audio and video tape recorders.
The conference can be controlled from either
the chairman's position in the centre of the
conference desk or from the secretary's
position located near the display area. An
auxiliary facility allows other cameras etc.
to be employed. Overflow facilities, so that
other groups of people may observe the
proceedings, are also available (see Figure
2). Calls can be booked at a minimum of two
hours notice.

Design

The TV equipment works on 625 line,
5.5MHz broadcast standards. At present the
service is in monochrome, but the studios are
in the process of being converted to colour.
Colour TV is becoming 'the norm' in everyday
life and business people have come to expect
it, but it is not thought that colour will add
much to the meetings although it should help

the feeling of 'visual presence' and be an aid
when using the graphics facility. Until now
the addition of colour meant an increase in
overheads with extra lighting, more air
conditioning etc. However semi-professional
colour cameras are now available which,
although not up to the performance standard of
broadcast cameras, are able to operate under
the same conditions as the old monochrome
cameras and give an adequate performance.

Split screen. A further improvement that
is being implemented is the use of a split
screen technique. In an ordinary meeting the
participants are usually seated around a table
with all the visual information obtained from
a field of view bounded by the table surface and
the top of the participants heads. In a
Confravision studio the participants are
seated at the correct viewing distance to
obtain optimum performance and benefit of the
TV display. The camera needs to be as close
as possible to the display in order to
preserve eye contact with the remote
participants. In the conventional studios it
is possible to view either 3 conferences or 5
conferees. There is reduced definition and
detail when viewing five conferees, as well as
a lot of wasted screen space with a large area
above the conferees heads that contains no
useful information. In order to retain the
definition of the 3 person display for a
larger number of people, it is necessary to
use 2 cameras. However the problem then
becomes one of switching. No operators are
present to preserve privacy and voice
switching is considered unsuitable from a
human factors point of view.

A split screen mode of presentation is
being introduced to overcome the above
disadvantages and to make more use of the
available screen area for up to six people,
without incuring the penalty of extra
transmission requirements. This is now the
subject of a CCITT recommendation[6]. The
useful middle portions of the pictures from
the two cameras are combined but displaced in
time so that camera 1 output forms the top
half of the combined picture and camera 2
output the lower half. These combined pictures
are transmitted and can be separated and
displayed on 2 side by side monitors.
Figure 3 shows the format of the two kinds of
presentation. The images on the screen have
the equivalent of 2900 picture elements in the
area of interest, (a participants head),
compared with only 1000 elements per head when
5 people are seen in the conventional format.
The split screen thus gives approximately a
three times improvement in definition. The
Martlesham Studio, shown in Figure 4, is one
such studio.

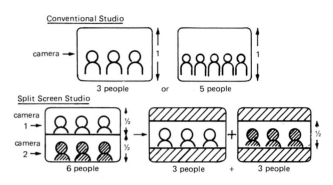

Figure 3
Confravision® display format

Mobile studios

Mobile studios currently in use are
similar to the permanent studios inside, but
only provide three shot facilities. These
mobiles are restricted in size to that which
can be transported along British roads without
a police escort, approximately 9m x 3m x 3m.
They can be transported, as and when needed,
to the customers premises, with access to the
broadband network provided by short hop
microwave links. One studio is being moved to
various sites in turn, to act as a temporary
studio while the existing permanent studios
are converted to colour split-screen working.
Another mobile is being moved to Belfast to
ascertain the market potential for a
Confravision® service. More mobiles are
planned for the future for both use as private
facilities, customer based, and for evaluating
market potential in several areas.

Private studios

Seven private studios are used by British
Telecom staff and contractors involved in work
on new System X telephone exchanges. These
studios are very basic with the equipment
installed in a cupboard. The cupboards can be
placed in a suitable room. The cupboard doors
open to reveal the camera and two monitors.
The microphone and control box are on long
leads and can be put on the conference table.
There is only one camera but documents can be
positioned on a board on a table in front of
the cupboard and the camera zoomed in to
enable participants at the remote end to read
them. The sound is not of broadcast quality
but is comparable to telephone quality.

Figure 4
Confravision® Studio Martlesham

International Calls

Experimental connections have been made
to similar systems in Australia, Holland and
Sweden and demonstration calls have been made
to Canada and Italy. However the high cost of
wideband satellite transmission has precluded
so far the establishment of an international
service.

Usage

Data on the use of the Confravision®
service is restricted to numbers of meetings,
meeting size, and type of use. Up until
mid 1979, there was steady growth of the
service with the usual seasonal fluctuations,
showing maximum usage in the early part of the
year with a drop during the school summer
holiday period. Statistics show a number of
commercial users have made contractuaral
arrangements for regular bookings and
considerable use is made by British Telecom
staff for service calls. The studio at The
British Telecom Research Laboratories is in use
frequently for meetings between local staff
and those who work in London. This saves a two
hour journey and enables more efficient use of
staff time. The studio at the Research
Laboratories is also open to the public.

Disadvantages

The system requires broadband circuits which are very expensive with a TV channel approximately equivalent to 1000 telephone channels. Confravision® also suffers from a limited network and requires participants to travel to a studio. The social aspects of travelling to meetings should not be overlooked. For many, a day away from the office can be looked upon as a treat. The cost of a Confravision® call is high but is dictated by the cost of hire of protection channels. Table 2 gives Confravision® charges that applied at November 1981.

2-way calls varying with distance
                between studios

Calls up to 200km   £ 80
Calls over 200 km   £120

3-way calls

From £170 to £250 depending on the studios involved.

Table 2
Confravision® charges per 30 minutes call

The human factors of conducting meetings over TV have been the subject of much research[7]. Protocol can be a problem but provided there is a strong chairman and the video-teleconference is treated in the same way as a face to face meeting, there should be no difficulties.

Visual Services

History

At the same time as the Confravision® service began, in house trials were being carried out to investigate the visual telephone or 'viewphone' as it was called[8]. Viewphone exchanges were sited at the Research Station at Dollis Hill on the outskirts of London and the former Post Office Telecommunications Headquarters in the City of London. The standard was satisfactory for a single person user but there was no graphics capability.

Current developments

Although not totally successful, a great deal of useful information was obtained from the trial which in 1974 led to the submission of a paper to the CCITT by the Post Office on an evolutionary visual telephone service[9]. With the advent of large capacity digital systems using media such as optical fibres and the development of special digital coding techniques for TV, the field of visual telecommunications opens up. Digital techniques mean that systems are no longer limited in capabilities. The concept of a visual service terminal, illustrated in Figure 5, has been generally accepted. The unit has a modular design and consists of a service unit, that provides for interface with the network, a control unit, a display with camera incorporated (composite unit) and has facilities for the provision and use of a separate camera and/or display etc.

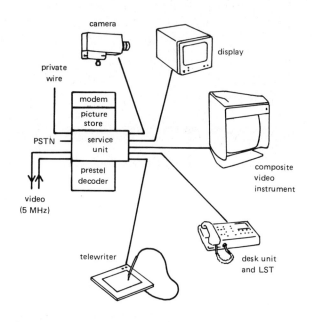

Figure 5
Visual Services Terminal Concept

The terminal uses 625 line, 50Hz, 5.5MHz bandwidth TV technology. It is designed to be located in an ordinary conference room/office environment with no special lighting or acoustic treatment. It should be accessible to all levels of staff with potential need for visual telecommunications as a travel substitute.

Visual Services Trial

In 1980 a market research agency was commissioned to carry out a market survey of a random selection of 246 UK firms listed in The Times Top 1000 companies. Interested respondants were allowed to use a working simulation of the visual services terminal and were shown video tapes of the picture quality they could expect. The demonstrations were followed by questionnaires and telephone calls and in October 1981, the most promising companies were invited to a live demonstration of the terminal with 2Mbit/s transmission with a view to selecting trialists. These potential trialists showed a need for regular communications between dispersed sites. In the trial, commencing in late 1982/early 1983, about 50 terminals will be employed at company locations with access to the digital network[10].

Figure 6
The visual services terminal to be used for the trial in 1983

Visual Services Terminal

The composite instrument shown in Figure 6 is compact and unobtrusive. The display is small (12 inches) to encourage a short viewing distance and easy operation in the graphics mode, as well as to improve picture definition. Two camera lenses are hidden in the dark band above the display, one for each mode of operation. The problem of eye contact is overcome by placing the camera as close to the display as possible. Ideally the camera should be behind the image on the display. In the conference mode the camera provides one or two person views. In the graphics mode a second lens, with mirror arrangement, has a field of view immediately in front of the display unit, for viewing documents and objects. In this mode an electronic zoom facility is provided. A check view facility is provided in both modes.

The control unit has a shallow switched loudspeaking telephone facility but an alternative loudspeaker is located in the display so that sound and vision are co-located.

The service unit, as well as providing network interface, also contains sound in vision transmission equipment and a Prestel® decoder.

The composite instrument is suitable for meetings of up to three people per location, encouraged by research which has shown that most meetings have between four and six participants in total[7]. Larger groups can make use of the alternative camera and display.

Trial Network

The network for the trial is shown in Figure 7. Local transmission to the terminal will be analogue (5.5MHz bandwidth) utilizing either cable, optical fibres or 29GHz microwave transmission. Where necessary analogue concentrators will be used to reduce the number of long local ends. The codecs that enable this 2 Mbit/s transmission are complex and expensive items of equipment. To achieve maximum utilization they are shared by several customers. Booking reservations and switching of the network will be under computer control.

International connections at 2Mbit/s will also be available to Europe via satellite (initially OTS and later ECS and Telecom 1). Small earth stations will be located in London and Martlesham. Connections to USA, Canada, Far East and Australia are possible using 2Mbit/s paths on Intelsat IV or V from large earth stations at Goonhilly or Madley.

2Mbit/s codec

The codec (shown in Figure 8) is the result of collaborative European research[11]. Several hours of OTS satellite time have been spent in international videoconferences with Martlesham linked via the prototype codecs, with laboratories in Italy, France and West Germany.

The 2Mbit/s codec operates in two modes, face-to-face and graphics, as shown in Figure 9. The face-to-face mode (mode A) is essentially a moving picture mode using a technique known as conditional replenishment[12]. Only the moving parts of the picture are transmitted, with the stationary part held in a frame store. For the graphics mode (mode B) a higher sampling

frequency and a transmission time of 1.6 seconds
per frame, means that the received pictures
have better definition than in the face-to-face
mode but movement is blurred.

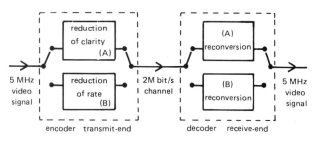

Figure 9
Picture conversion for 2Mbit/s transmission

Multipoint conferences will be possible
using the Visual Services Terminal[10]. It is
considered that it is preferable to see all
the participants continuously rather than have
just the talkers and voice switching
arrangements. However at present, this is not
possible for more than two locations linked so
initially the chairman will be responsible for
selecting who should be seen.

EVE

Teleconference trials in the European
Video Experiment (EVE) are due to begin in
April 1982 between the PTTs of Britain,
Netherlands, France, Italy and West Germany.
They will culminate in 1983 in commercial use
with the business in the UK Visual Services
Trial being able to experience international
teleconferences to parts of their own
organisations in some cities in the other
countries.

Conclusion

Video-teleconferencing in Britain is no
longer restricted to the public Confravision®
service. The approach as described in this
paper allows for fully integrated
teleconferencing services. Thus public
Confravision® studios and any of the private
studios can be accessed by the visual services
network. International connections are also
provided from any of the teleconferencing
systems operated by British Telecom. In our
view by providing on site visual telecommunications using ordinary conference rooms, with no
need for environmental alterations, strong
growth patterns will be established for
teleconferencing within the UK. Increasingly
expensive travel, travelling time as well as
the inconvenience and stresses caused to the
individual who has to travel regularly, all
help to promote teleconferencing as a
substitute to travel.

Figure 7
The Visual Services Trial Network proposed for
1983

Figure 8
2Mbit/s codec to be used in the Visual
Services Trial

Acknowledgement

The author wishes to acknowledge W Blyth
and Dr J Thompson, for their help in the
preparation of this paper, and T Duffy of
McMichael Ltd., for the photograph of the
2Mbit/s codec. Acknowledgement is also made
to the Director of British Telecom Research
Laboratories for permission to publish this
paper.

References

1. CLARK W J "Cyclops - A field evaluation",
1982 International Zurich Seminar on Digital
Communications, Zurich, 9-11 March 1982

2. SHARPLES M "Cyclops Audiographics System",
Teleconferencing and Interactive Media '82,
Madison, Wisconsin, 19-21 May 1982

3. GROVES I S "'Orator' - The Post Office
Audio Teleconferencing System", IEE Conference
Publication 184, IEE Communications 80
Conference, Birmingham, England, 1980

4. KENYON N D "Slow-scan TV goes on trial",
British Telecom Journal, Vol 1, No 4, Winter
1980/81

5. HAWORTH J E "Confravision", Post Office
Electrical Engineers Journal, Vol 64, No 4,
January 1972.

6. CCITT "Visual Telephone Systems", Final
Report to the VIIth Plenary Assembly, Document
75, Recommendation H.61, Study Group XV,
Geneva 1980

7. TYLER M, CARTWRIGHT B COLLINS H A,
"Interactions between Telecommunications and
Face-to-Face Contact: Prospects for
teleconference services", Post Office Long
Range Intelligence Bulletin 9, May 1977

8. HILLEN C F J "The face to face telephone",
Post Office Communications Journal, Vol 24, No
1, Spring 1972

9. CCITT "An Evolutionary Visual Telephone
Service", Document Com XV, No 119, January 1975
Contribution to study group XV, Question 4

10. THOMPSON J E "Visual Services Trial - The
British Telecom System for Teleconferencing
and New Visual Services" Post Office Electrical
Engineers Journal, April 1982

11. LIMB J O, PEASE R F W and WALSH K A,
"Combining Intraframe and Frame to Frame
Coding for Television", BSTJ Vol 53, No 6,
pp1137-1173, 1974.

12. THOMPSON J E "European collaboration on
picture coding research for 2Mbit/s
transmission", IEEE Transaction COM-29 No 12
pp 2003-2004, December 1981

The Teleconferencing Resource Book: A Guide to Applications and Planning
Lorne A. Parker and Christine H. Olgren (eds.)
Elsevier Science Publishers B.V. (North-Holland)
© Center for Interactive Programs, University of Wisconsin-Extension, 1984

"PICTUREPHONE® MEETING SERVICE: THE SYSTEM"

David B. Menist - Supervisor
Visual Communications Service Planning
and
Bernard A. Wright - Supervisor
Visual Communications Systems
Bell Laboratories
Holmdel, NJ

INTRODUCTION

Video teleconferencing can provide more effective communication among personnel at dispersed locations and reduce the need for business travel. Its use will thereby contribute to two goals of public concern: the improvement of managerial and professional productivity, and the reduction of energy consumption. The results of market studies, experience from the See-While-You-Talk Service trial, and the availability of digital technology have all contributed to AT&T's decision to offer a video teleconferencing service (subject to F.C.C. regulatory approval). The service will be marketed as PICTUREPHONE Meeting Service (PMS).

PMS will provide two-way interactive audio and video, on a reservation basis, between any two compatible conference rooms that are connected to the transmission network. Conference rooms may be located either on customer premises or in public locations. The service will utilize digital encoding and bit-rate reduction of the video signal to achieve economy of transmission.

This paper gives a general technical description of the overall system and components that are planned to provide this service. A brief overview of the PMS system is presented first. Subsequent sections cover in greater detail the conference rooms, picture processor, network facilities, and operations support system. The network layout for initial service is described last.

SYSTEM OVERVIEW

An overview of the system is shown in Figure 1.

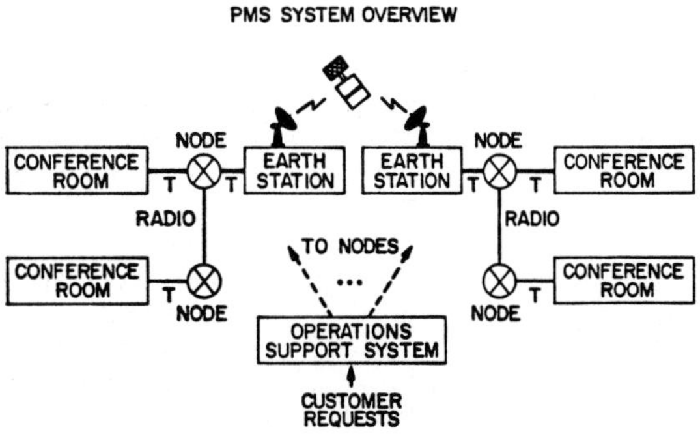

FIGURE 1

The conference room contains video and audio equipment that generates a 4.2 MHz analog NTSC color video signal and a 7 kHz analog audio signal. A picture processor located at the conference room converts the analog video and audio signals to digital, encrypts them, and combines them with processor-to-processor control signals into two DS1 signals (1.544 Mb/s each). These signals are then transmitted, via local and exchange facilities, to a metropolitan node. At the node, the signals are routed either to another local room for local conferencing or to long-haul facilities for transmission to a distant node. The internodal network is comprised of both terrestrial and satellite digital transmission facilities.

A central computer-based network control and operations support system is provided to process call requests. The system receives customer requests from an attendant who is accessed by customers over 800-Service lines. Maintenance personnel also have access to the system to enter facility outages and to reserve facilities for testing.

CONFERENCE ROOM AND PICTURE PROCESSOR

Conference Room

Room Design and Configuration. A PMS conference room provides a comfortable distraction-free environment for video teleconferencing. In the conference room shown in Figure 2, a typical public PMS room, six conferees may be seated at the table.

FIGURE 2

Up to five additional conferees may be accommodated with a second row of chairs behind the table conferees. To satisfy other customer needs, the room may be easily configured for as few as 2 or as many as 10 table-seated conferees.

The Bell System provides a complete end-to-end videoteleconferencing system including extensive guidelines to aid customers in the planning and design of their conference room. These guidelines cover such important topics

as site selection, room floor plans, lighting, acoustics, HVAC requirements, cabling requirements and table and cabinetry construction.

The design of the video coverage for a PMS conference room simulates a natural face-to-face meeting where the conferees in the distant room appear to be seated across the conference table. For this reason, an elliptically shaped table is used and the display of the remote conferees appears on monitors located in the long wall across the table. This arrangement achieves good eye contact with conferees in the other room. It also allows the conferees to view the distant room without rotating their chairs or turning their heads. Other popular video teleconferencing system designs seat the local conferees around a table facing each other. This requires the conferees to rotate their chairs or turn their heads to view the other room which is located on the short (end) wall. The PMS arrangement is more convenient and natural for the user and allows better camera coverage for the conferees.

Room Equipment Overview

An overview of the PMS room equipment is shown in Figure 3.

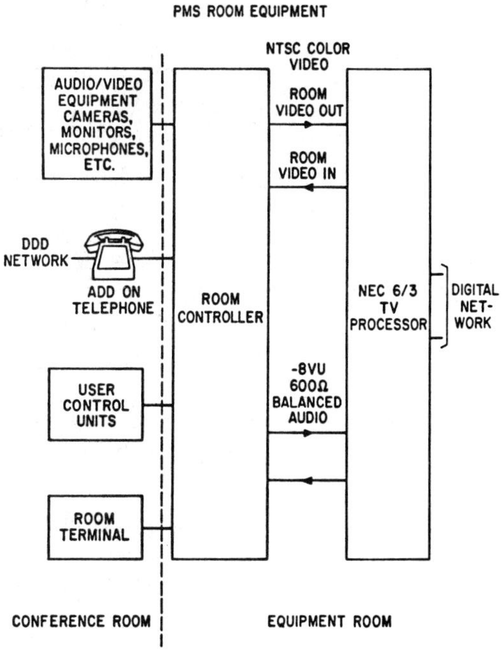

PMS ROOM EQUIPMENT

FIGURE 3

The audio and video equipment (cameras, monitors, microphones, speakers, etc.) connect to a microprocessor-based room controller located in the equipment room. This room controller provides the interface between the room equipment and the picture processor and allows customer control of the room equipment.

The customer controls the room equipment with two easy-to-use control panels. A room terminal is provided for use by craft personnel or the room attendant for specialized functions such as maintenance, equipment alignment and encryption key entry.

The room controller also provides the audio and video communication system for the room. This includes such functions as the switching and distribution of video signals and audio system echo suppression, microphone voting, and conference (add-on) bridging.

Video System

A typical PMS conference room contains three close-up cameras, an overview camera and three graphics cameras. Each close-up camera produces a head and shoulders image of a pair of conferees seated at the conference table and is automatically selected as the conferees speak. The overview camera provides a wide-angle view of the conference room and conferees. The three graphics cameras provide the capability to show graphics, slides and other objects. A ceiling-mounted graphics camera equipped with a zoom lens is used to display either transparent or opaque table-top graphics. A multipurpose camera with a zoom, pan, and tilt capability is typically used to cover stand-up presentations at an easel, but may also be used to display other room objects and/or people. Three monitors are used for display. One monitor displays the picture from the distant room, a second monitor displays the picture being sent from the local room and a third monitor is used to preview graphics material prior to transmission.

A hard copy unit is available to make monochrome copies of pictures received from the distant room. A

video cassette recorder is available to record the conference or to locally play back previously recorded cassettes. All video room equipment is commercially available. The cameras, monitors and video casette recorder are NTSC color equipment. A customer may provide his own room equipment.

The equipment complement in a given room can be matched to individual customer needs. A room equipped with a minimum equipment complement would have one table-top graphics camera, one camera for viewing conferees and one monitor for viewing the distant room. The typical room described above, when equipped with 5 close-up cameras, has the maximum equipment complement. To keep user operation simple, the control panel is arranged to only provide for features associated with equipment which is provided in a given room.

Audio System

An essential and often neglected part of a videoteleconferencing system is the audio subsystem. For PMS, careful attention was given to the design of the audio system in order to achieve natural sounding and easily interactive communication.

Up to 12 conference participants may each be equipped with a chest-equalized lavaliere microphone. These close-talking microphones are used instead of table-mounted microphones to maintain fidelity and to minimize acoustic feedback and room noise. Four ceiling-mounted speakers provide audio coverage for the room. The audio system bandwidth is 7 kHz and full duplex transmission facilities are used between rooms. Additional conferees may be added to the audio portion of the PMS conference by dialing them through a specially controlled room add-on phone.

In a teleconferencing environment with open microphones and speakers, acoustic feedback (speaker to microphone) causes echos. These echos require a higher degree of suppression or cancellation when using satellite facilities because of the delay of the echos. When conventional half duplex voice switched circuits, such as speakerphones, are used for echo cancellation, the communication between rooms is poor. Because of the satellite delay, the conferees lose their clues as to when the other room has stopped talking. This causes them to interrupt at inappropriate times which cuts out the other party. Highly interactive communication is thus very difficult over these half duplex systems.

To eliminate these difficulties, the PMS audio system uses a newly designed voice gated form of echo suppression. Extensive human factors tests have shown that this voice gated system is preferred over normal voice switched speakerphones and that easily interactive communication is achieved.

Picture Processor

PMS uses the NETEC-6/3 TV processor, modified to Bell System specifications, produced by the Nippon Electric Corporation. Customers may provide their own picture processor, as long as it is$_1$compatible with the network interface. The interface between the conference room and the picture processor is composed of a two-way audio and a two-way video channel, as shown in Figure 3. The video channel format conforms to NTSC color television standards. The audio is equivalent to a 7 kHz channel.

The output of the picture processor combines digitized audio and digitized, bit-rate-reduced video together with a low-speed auxiliary data channel. The auxiliary data channel is used to transmit control signals between picture processors. The combined signals are output as two standard 1.544 Mb/s signals including framing and synchronization.

The picture processor contains a subsystem which encrypts the video and audio information using a cipher stream derived from the Data Encryption Standard (DES) algorithm published by the National Bureau of Standards. Prior to a call, the encryption key is entered locally at each end.

In the receive direction, a decoder accepts the two DS1 signals as inputs, corrects errors, and recovers audio, video, and control information by performing the inverse of the encoding operations.

NETWORK FACILITIES

T1 Local Access Channels

Two dedicated T1 lines provide the full-time local access channels between the room and a node. A backup T1 line and automatic protection switching are provided to control outage levels between the conference room and the node (see Figure 4).

PMS ACCESS ARRANGEMENT

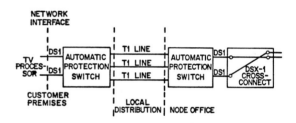

FIGURE 4

The picture processor connects to the local access channels via standard DS1 interfaces to the protection switch.

Cross-Connection and Switching

The local access channels in each serving area terminate at a nodal office on a DSX-1 cross-connect along with internodal channels and/or end links to a satellite earth station. The node thereby serves as: (1) the entry and concentration point from dedicated local access facilities to the shared internodal network, (2) a cross-connect point for trunks on contiguous segments of the internodal network, and (3) the cross-connect point for calls between conference rooms within the local serving area. Cross-connections are established, by reservation, on a per-call basis.

Manual patching at the DSX-1 cross-connect is utilized for initial service. Connections are set up and taken down by craft personnel who receive instructions generated by the operations support system. As traffic grows, it is planned to replace manual patching with automated cross-connect systems operating under remote control.

Earth Station and Satellite

T1 lines or other digital facilities provide end links between a node and the earth station. The earth station transmits digital information to the satellite and receives satellite transmissions which have been sent from other earth stations. Using Frequency Division Multiple Access (FDMA) techniques, a single transponder can be loaded with 24 one-way DS1 signals, i.e., 12 one-way 3 Mb/s video teleconferencing signals, simultaneously. A video teleconferencing signal is transmitted as 2 individual DS1 signals by modulation in 2 FDMA modems. The outputs from the modems are combined and connected to the earth station high frequency equipment for satellite transmission.

Initially, each station will transmit to only one transponder. This will minimize the total number of up-links required by the network and will, therefore, minimize the cost. To ensure that complete interconnectivity will be possible throughout the network, each station will be capable of receiving signals transmitted over any transponder used for video teleconferencing. In later years, some high traffic stations may require uplink access to more than one transponder.

Fixed trunking FDMA will be used initially. Permanent DS1 trunks in sufficient numbers are established between all pairs of nodes. There is a permanent connection via a DSX-1 cross-connect of an end link to a transmit modem which is fixed tuned to a particular transponder channel. At

the receiving earth station, the receive modem is fixed tuned to the particular transponder channel and permanently connected to the end link to the node. Reconfiguration is by engineering and circuit orders.

Terrestrial Radio

Long-haul microwave radio will be used for video teleconferencing transmission between many cities. The system used for this initially will transmit 20 Mb/s digital signals over the existing TD radio network via FSK modulation.

Multiplexers convert up to 12 DS1 signals into one 20 Mb/s digital stream in the form of a four-level 10 Mbaud signal. This signal is restricted to a 7.5 MHz bandwidth and is in a form suitable for input to the standard TD FM transmitter and modulator. The digitally modulated signal is then transmitted on the existing TD analog plant.

Automatic hardware protection for the 20 Mb/s terminal is provided on a one-for-one redundancy basis. In the event of fading on a transmission channel, the existing analog TD protection switching (1 for n or 2 for n) is used to switch the fading channel to the protection channel. The 20 Mb/s terminal employed is the TRW VIDAR DM-12A.

T1 or other short-haul digital facilities are required between a network node and the long-haul radio system if the node is not colocated with a radio system. Interconnection is through a standard DSX-1.

OPERATIONS SUPPORT

Customer reservations, maintenance, network control, and administration operations are provided through the PMS operations support system. The interfaces to this minicomputer system are shown in Figure 5.

PMS OPERATIONS SUPPORT SYSTEM

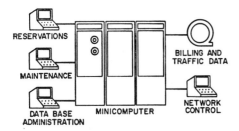

FIGURE 5

Reservations

Reservation requests are received by attendants at a centrally-located bureau via 800-Service lines. The reservation attendant queries the system to determine the availability of network facilities. The system makes the necessary update in information if the reservation is accepted and, at the appropriate time, generates the call setup information for the nodes. Records of call details are generated by the operation system for processing by a billing center.

Reservations are accepted from six months to one hour before a call. Calls are reserved in half-hour increments and may begin or end on any quarter hour. A call may be extended, if facilities are available, if requested at least one half-hour before the scheduled end of the call.

Maintenance

Maintenance personnel have access to the operations system on a dedicated link through a terminal. The maintenance position can request loop-backs at various points to localize reported failures. Also, the maintenance position can take transmission facilities out of service and reserve specifically identified facilities for testing.

Network Control

Call setup and takedown information for all calls for the current and next day is generated daily by the operations system for each node. Directions for the cross-connects which must be performed are transmitted from the system to terminals in the nodes or by personnel in a video teleconferencing network Control Center. Directions for call reservations that are received on the day of the call and for call extensions are transmitted to nodes as soon as the operations system processes these requests.

Administration

Administrative changes in the operations system minicomputer data base necessitated by network additions, deletions, or rearrangements, are accomplished from a data terminal.

NETWORK LAYOUT

The initial eleven city network for PMS is shown in Figure 6.

EARLY PMS NETWORK

FIGURE 6

Five cities are served directly by earth stations (New York, San Fran-

cisco, Chicago, Atlanta and Dallas).
Terrestrial intercity trunks complete
the connectivity of the eleven city
network.

A total of forty-two cities are
currently planned for PMS by AT&T by
the end of 1983. Public conference
rooms will be established in each of
the forty-two cities.

These plans are based on current
forecasts of traffic demand and are,
of course, subject to change.

FOOTNOTES

1. See AT&T Technical Reference PUB
   61511, "Video Teleconferencing
   Service Network Interface
   Specifications," July 1981.

The Teleconferencing Resource Book: A Guide to Applications and Planning
Lorne A. Parker and Christine H. Olgren (eds.)
Elsevier Science Publishers B.V. (North-Holland)
© Center for Interactive Programs, University of Wisconsin-Extension, 1984

CANADIAN PLANS FOR

AN INTERNATIONAL HYBRID TELECONFERENCING SERVICE

By: James W. Johnson
    Manager, Advanced Business Services
    Teleglobe Canada
    680 Sherbrooke Street West
    Montreal, Quebec H3A 2S4

Introduction

Teleglobe Canada is Canada's international telecommunications carrier responsible for planning and implementing services between Canada and all external points. This paper describes Teleglobe's efforts to develop a cost-effective international teleconferencing service between Canada and selected European points in response to a number of customer inquiries and as part of the Corporation's overall plans to develop new services for business communications. The terms "hybrid teleconferencing", as used in this paper, mean "two-way interactive video, audio and graphic communications between three or more persons in two or more locations.

A number of studies have served to show that up to 50 percent of all business meetings could be effectively handled by teleconferencing of one method or another; either audio (telephone conferencing), audio/graphic (e.g. telewriting), or video. Current plans within Teleglobe call for the development of an experimental "hybrid" teleconferencing service which will combine the best features of the various conferencing methods (audio, video and graphic) to provide a "full-service" conference capability to the international business community. Teleglobe's plans evolved from an early recognition of the role teleconferencing could play within the realm of corporate communications in the near future. A number of market surveys and informal discussions with major corporations that have headquarters/ subsidiaries in Canada and Europe have indicated that significant opportunities exist in the development and provision of a point-to-point intracorporate hybrid teleconferencing service.

The Relative Economics and Service Fit

Recently, considerable interest in teleconferencing has been aroused by the continued rise in the costs of travel and accommodation. Major multinational corporations are beginning to request an economical teleconference service between their international operating entities. Here the term "economical" clearly implies that the service must provide the opportunity for savings in comparison to the alternative cost of travel. Potential customers view this as the single most important criterium to the service's success, followed by the benefits of convenience, efficient use of time and improved intra-corporate communications.

However, the emphasis on the economy of teleconferencing relative to travel should not be construed to mean that teleconferencing will necessarily replace international travel or enable all users to reduce current travel expenditures. It must be noted that, in general, teleconferencing is felt to be suitable for up to 50% of all business meetings. The question is, what percentage of **international** business meetings fall within that 50% which is suitable for teleconferencing? -- and the answer to this is not known.

Although the strongest interest in teleconferencing is currently being generated by companies who wish to use it as a substitute for travel, a significant number of potential users see applications which augment rather than replace travel. For example, many will use teleconferencing for advance preparation to ensure more productive trips; others will use it to assess the need for travel, and many desire to use it to conduct meetings they do not currently conduct because travel is simply not the answer, or the costs and/or inconvenience are prohibitive.

In simple terms, the need for teleconferencing to be economical relative to travel stems from the fact that all customers, regardless of application, will tend to use travel costs as a "benchmark" to assess the viability of teleconferencing.

The Need and Economic Constraints of Video

While potential users are requesting an economical service, they are also insisting that the service provide features which fully simulate a face-to-face meeting. Hence the market requirement is for a cost-effective, two-way interactive video, audio and graphic service that places the corporate conference room in an international teleconference context.

Of the current methods of teleconferencing, video requires the greatest channel capacity (frequency bandwidth) and hence is the most expensive. The single greatest impediment to the use of video-based conferencing systems is the cost of transmission. Nonetheless, both telecommunication carriers and potential teleconference users are deeply interested in video due to the considerable advantages it provides over other teleconferencing methods as an alternative to "face-to-face" meetings. With video, users have clear identification of participants and the ability to present graphics all in one medium. One perceptive user has indicated that video is necessary to ensure that conferees are awake and attentive to his presentations. The first challenge is to reduce the cost of video transmission so that it may be placed properly in a teleconference role and adequately assessed as a new business communications service.

Experiments with videoconferencing are not new, although to date they have been confined to special-event teleconferences and national experiments/trials conducted in isolation from one another. Today, a number of trial services exist within Canada, the United States and Europe. Most of these trials, such as Bell Canada's Videoconference service and Britain's 10-year-old Confravision service, are based on standard broadcast TV technology and use public studios to which the customer comes to conduct a conference with counterparts at another similarly equipped studio. Currently the charges for these national trial services provide, at best, only a marginal cost advantage over the alternative cost of business travel within the country since the distances and hence travel costs are not great. As a consequence, these trials have met with limited success.

International Service Constraints

Intuitively, the economic viability of videoconferencing should increase with the distance that would otherwise have to be travelled. Data collected on existent trial services supports this in that the conference link that is the longest is the link in greatest use. This implies that videoconferencing will find greater application in the international arena as an alternative to business travel, particularly between continents. Teleglobe, as the Canadian signatory to Intelsat, currently provides satellite facilities under the "occasional use" television tariff in support of international video conferencing. However, this has not been found to be a cost-effective offering for teleconferencing. The primary impediment is the end-to-end price of the two-way video (TV broadcast quality) service under existing space-segment and terrestrial tariffs. In addition to high transmission costs, the major constraints confronting the development of an international videoconference service stem from:

- Dependence on broadcast video (i.e. television) standards and facilities.

- Different video standards in various countries of the world (PAL/SECAM/NTSC) and the need for conversion.

- Lack of available international wideband video links between customer premises.

- Preemptibility of video conference links because they are implemented on the normally unused "protection" channel of systems carrying telephony and/or standard broadcast video.

- Time zone differences that essentially limit intra-corporate conferences between North America and Europe to a three-to-four hour common work period (Toronto/London) each day.

- Current standard broadcast video transmission facilities and analog technologies which limit the number of simultaneous conferences that can be conducted during this time period.

. Conservative and often nationalistic policies governing the implementation of new technology and the pricing of new services.

Developments in New Technology

Today, substantial technical advances are being made in teleconferencing that will enable companies with overseas operations to implement a private teleconferencing network and thereby benefit from the "economic crossover" between decreasing communication costs and increasing travel costs. The primary challenge here is to implement new technologies and facilities which are radically different from those in current use. The major advances being made today can serve to counter most of the foregoing constraints and significantly reduce the cost of videoconferencing. Examples of these developments are:

. New digital signal processing techniques, now becoming available in commercial products (CODECS) which enable "compression" of the video signal by reducing the redundancy of its components. As a consequence, a limited-motion, full-colour video signal suitable for teleconferencing can be sent through a channel having 1/40th the capacity of that required for standard broadcast TV.

. Higher power satellites and power-concentrating "spot beam" antennas on board satellites operating in the less congested frequency bands above 10 GHz permit the use of smaller, lower-cost earth stations within urban centres in close proximity to users. This in turn reduces the dependency on high-cost, long-haul wideband terrestrial links that are often preemptible for alternative or emergency use.

. Advances in the automated office, digital communications, the Integrated Services Digital Network (ISDN), and satellite technologies are coming to enable the integration of various communication services (voice, data, video) over a single communications link to the business user.

Whereas today, an interactive (two-way) videoconference between Toronto and Paris under existing facilities and tariffs would cost approximately $10,000 per hour, these new technology developments offer the potential of holding the same two-way conference for $2,400 per hour or less. This level of charging end-to-end would indeed be economical for most international intra-corporate teleconferencing applications as an alternative to face-to-face meetings.

Teleglobe's Two-Phased Approach

Intuitive reasoning, on the substitutability of teleconferencing for international business travel, and conventional market research techniques are unreliable sources for service definition given the considerable risks and high costs inherent in the technologies involved. Research by others has shown that demonstrations of videoconferencing are of little value in assessing service features and market viability, particularly where such demonstrations are provided free of charge. Experience gained in practical applications is needed before the service can be identified and competing alternatives evaluated. Hence, in order to examine the characteristics of a teleconferencing service and the associated new technologies in the international market, and to ultimately define a long-term commercial service offering, Teleglobe has devised plans for a two-phased market trial/experimental service to be offered in cooperation with selected European administrations and Intelsat.

The principal justification for the market trial/experimental service approach herein is to enable potential customers to gain "practical-applications" experience through use of the service over a period of time. To launch this two-phase program, Teleglobe has already sought and gained the cooperation of Intelsat and a number of European Administrations.

Phase I Market Trial

Phase I of Teleglobe's international teleconferencing program is a short-term expedient approach fashioned after the current national offerings within Canada and elsewhere. It is based on existing terrestrial and satellite TV transmission facilities and technologies, and makes use of the public conference studios which are currently used in trial services within Canada and Europe. The specific objective of the Phase I trial is to provide an early demonstration of the characteristics of an international video-based conference service, and to gain experience in the market demand and acceptance of this type of service as a possible alternative to international travel for face-to-face business meetings.

Phase 1 is highly restricted due to the limited capacity and preemptibility of facilities throughout the communications link. In addition, this phase is oriented solely to video and does not include many enhanced features requested by customers of a full-fledged teleconferencing service. Nonetheless, with little or no capital investment required, Phase I represents a least-cost approach for overseas administrations to conservatively assess the current needs and potential of international videoconferencing. Commencing this year, a group of Canadian customers will use studios in Toronto, Ottawa, Quebec City and Montreal to conduct videoconferences with counterparts in London, England and Paris, France.

In view of the diseconomy of the existing standard broadcast TV tariffs for application to videoconferencing, a special "cost recovery" charge is being levied on users of the Phase I service. As an example of the charges under Phase I, a user would pay approximately $3,700 (Canadian) per hour for a two-way videoconference between Toronto and London, England.

Phase II Experimental Service

Phase II of the Teleglobe program, targetted for Fall 1983, has been devised to overcome the shortcomings of Phase I by incorporating additional enhanced service features and advanced technologies. This phase will be implemented under an applied research and development effort that will use the results of basic research in the areas of digital video compression, tele-writing, freeze-frame TV, and 14/11 GHz satellite transmission for the purpose of creating a new cost-effective, full-service teleconferencing package.

The primary objective of Phase II is to provide a teleconference service that is cost-effective by virtue of lower end-to-

end tariffs based on lower cost facilities (i.e. earth station and terrestrial links) and the use of new digital video compression techniques to achieve lower space segment costs consistent with more efficient bandwidth utilization. This Phase will also enable an assessment of the complementary/competitive aspects of the alternative new conferencing methods of video and audio/graphic (including telewriting).

The Conference Room of the Future

For Phase II, Teleglobe is constructing a multi-purpose (hybrid) conference room co-resident with its sales office in The First Canadian Place in downtown Toronto. Teleglobe's Phase II studio in effect represents the corporate conference room of the future, incorporating both local conferencing and international teleconferencing capabilities. Service features have been selected to include all amenities of the present corporate conference room within a teleconferencing context--telewriting by means of an "electronic" blackboard, document exchange by means of facsimile, and presentation aids using freeze-frame TV and high resolution graphic display. Customers of the Phase II service will be able to select the best combination from among these features.

Although on the surface Teleglobe's Phase II studio appears similar to that of the more avant-garde Corporations in the United States and Canada, the Teleglobe technical design is quite unique in incorporating a number of additional service features. Also, special care has been taken to design a compression subsystem that incorporates a CODEC which allows all these features to be placed along with the compressed video into a single digital stream. A "fall-back" mode of operation has been designed to enable users of the studio to conduct a conference in the audio/graphic mode in the event of a failure in either the video compression subsystem, the satellite earth station, or any point in the satellite link. In effect, the customer has all conferencing capabilities except full-motion video in the fall-back mode of operation.

Digital video compression and transmission techniques will be utilized to reduce the bandwidth and hence the recurring costs (i.e. transmission) to the end user. A small, low-cost 3.7 metre earth station, will be mounted on the roof of The First Canadian Place and once interconnected to the studio, will provide direct access to the Intelsat "spot beam" antenna at 14/11 GHz to provide a further reduction in transmission costs and elimination of the dependence on high-cost, preemptible wideband terrestrial links.

To operate a satellite link in an urban centre such as Toronto or Montreal, the upper frequency bands of 14/11 GHz (14 GHz transmit and 11 GHz receive) are preferred over the highly congested bands of 6/4 GHz which are common to both satellite and terrestrial microwave systems. In addition, the 14/11 GHz band enables utilization of a small earth station suitable for rooftop mounting. To take advantage of the digital compression of video in achieving efficient use of the satellite, special digital modems will be used.

Other Considerations

Use is made of public studios in both phases of Teleglobe's plans. Although it is felt that the longer term market for teleconferencing will see wide-spread use of private systems and terminals within corporate conference rooms and the automated office, the public studio will play a major role for sometime to come. The use of public conference studios is predicated on the idea that most potential users of videoconferencing are unknowledgeable of the technology/applications or are either reluctant or unable to incur the up-front investment in private studio facilities. These public studios are further required due to the scarcity of local video wideband networks to the customer premises. For those users who are advanced enough to implement private facilities, the Phase I and II plans have been devised to allow them access over leased lines. Depending on the success of the Teleglobe studio in Toronto, it is possible that the Corporation could establish similar studios in other major centres such as Montreal and Vancouver, as justified by market size.

In view of the fact that the success of a teleconference can depend upon such factors as (1) prior knowledge of participants and roles, and (2) establishment of a common purpose or objective for each session, the limitation of the initial service to intra-corporate (within the same

company) applications is considered essential to success. One factor common to all current national videoconference trials is that the greatest use of these trial services is being made by the telephone company or communications carrier itself, i.e. an intra-corporate application by the service provider.

192

The Teleconferencing Resource Book: A Guide to Applications and Planning
Lorne A. Parker and Christine H. Olgren (eds.)
Elsevier Science Publishers B.V. (North-Holland)
© Center for Interactive Programs, University of Wisconsin-Extension, 1984

CONFERENCE 500[TM] - "TODAY'S WAY TO TELECONFERENCE"

Louise Roberge
Section Manager - Product Management
Bell Canada
Toronto, Ontario

INTRODUCTION

Visual aids play an important role in our face-to-face meetings. Viewgraphs, 35mm slides, flipcharts are as pervasive as the ever-present cup of coffee. A frequent scenario consists of several presentations made by different participants, each having prepared his material prior to the meeting. The visual support plays several important roles. It can be used to summarize or emphasize points being covered by the speaker, to inject structure to a meeting with an outline or agenda, or to illustrate data with charts and graphs.

"Meetings by telephone" are becoming increasingly popular. The requirement for visual support still remains. Full motion video may not always be available or affordable; nor may it be necessary. Audio-plus or audio-graphic teleconferencing is recognized as a cost effective way to teleconference. It provides the means to disseminate a visual message to locations participating in a teleconference. The terms "audio-plus" and "audio-graphic" include a wide range of possibilities: facsimile, electronic blackboards, freeze-frame, to name a few.

Bell Canada's Conference 500 uses videotex technology to distribute "electronic slides" to remote locations. The slides, which can include text and/or graphic information, are prepared prior to a meeting by the various presenters, stored in a computer and during the meeting displayed at each of the meeting locations.

WHAT IS CONFERENCE 500?

Briefly, Conference 500 is a computer-based service that provides for the creation, manipulation, storage, retrieval and broadcasting of non-interactive colour slides. Users are equipped with Telidon terminals connected to the switched network.

Conference 500 offers a user two ways to prepare his material for a presentation. If the material consists of text, he can create his slides from a Telidon terminal equipped with a keyboard. This is done on-line with commands similar to those used in typical word processing systems, although additional commands are available to specify colours and character sizes. These slides can be subsequently revised on-line, up to minutes prior to a meeting - a capability which is greatly appreciated by those presenting highly volatile information.

Conference 500 will also accept more elaborate pages of graphic information generated by Telidon page creation systems. These pages are created off-line, stored on a floppy disk, and transmitted or mailed to the Conference 500 computer.

Slides, whether prepared on-line or off-line, are stored in the computer and reside in a user's account. The user in preparation for a meeting can arrange and re-arrange his slides into any number of sequences (referred to as slide trays) on-line using a Telidon terminal and keypad.

At the time of a meeting, each participating location establishes an audio link and connects its Telidon terminal (via a second telephone line) to the Conference 500 computer. Conference 500 acts as a "data bridge" which allows conferees to present slides to the other locations also equipped with a Telidon terminal and television monitor. The retrieval and broadcasting of slides is controlled by a presenter via keypad commands (see Figure 1). Any location can "take the floor" and present its slides to the other locations. A presenter controls the display of slides with keypad commands such as "next", "back" or the actual position of a slide (Example: slide no. 4). He can preview a slide without broadcasting it to the other locations. Having privately verified a slide, he can then broadcast it or return to the slide still being viewed by the other locations.

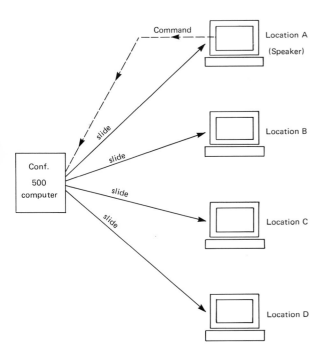

Figure 1 - The presenter is controlling the display of slides to his location and locations B, C, and D. Later, in the meeting, location B, C, or D could take control and present their slides.

At any point during a meeting, a participant has the status of presenter or conferee. As conferee, he can in no way control the display of slides, but he can request a "help" page which reviews Conference 500 commands. It is also possible for a conferee to have the current slide retransmitted in the event of a transmission error.

Conference 500 provides several levels of security to prevent unauthorized access to presentation material. First, each slide in the system must have an owner. In order to retrieve, edit or display a slide, the user must specify his account number and password. The second level of security is the conference key. A conference key is assigned by the Conference 500 administrator and given to the chairperson or organizer. The key is activated for the period of time for which the Conference 500 system is reserved. The chairperson is responsible for informing all participating locations of the conference key. (This is most conveniently done on the audio link as meeting begins.) Each participating location enters the conference key with a keypad and is thus admitted to a meeting's visual component. During a meeting it is possible for a speaker to verify the number of terminals accessing a slide, via a keypad command. To exclude intruders, the speaker can implement an "electronic lock" during a meeting, by choosing a random number, entering it with his keypad, and communicating the new password to authorized participants. All participating locations will be prompted by the system to respond with the new lock number. Failure to enter the lock number will result in disconnection from the system. After this point, any one joining the conference will have to specify the conference key and the lock number in order to view the slide presentations. The lock number can be de-activated at any time by the speaker with a simple keypad command.

BELL CANADA'S INTERNAL EXPERIENCE

Like many other organizations, various forms of teleconferencing have been available to Bell Canada personnel for many years. Add-ons, PBX conferencing, and handsfree sets are common throughout the organization. The operator-handled conference call has been available to employees and customers since the 1930's. Until September 1981, very little was done to promote the internal use of these basic teleconferencing capabilities or to monitor their usage. Video studios have been available in the major cities in Bell Canada's territory since 1972. Their use by Bell Canada personnel has not been widespread.

In 1980, the Ontario Government established a Teleconferencing task force within its Transportation Energy Management Program (TEMP) whose mandate was to promote the use of teleconferencing as an alternative to travel. (Note: Bell Canada services most of the territory in the provinces of Quebec and Ontario). Bell Canada has actively participated in TEMP by promoting teleconferencing to its personnel and to the market place. It has expanded its portfolio of teleconferencing products and services it offers its customers. Internally, a major promotional campaign in early 1982 increased the awareness of teleconferencing, from 61% to 85%. Conference rooms were equipped with group audio terminals; automated audio bridges were installed in 3 major cities; new products and services, still under development are being experimented with. Typically, Bell Canada managers have had some experience with audio conferencing, a little experience with video conferencing and virtually no experience with computer conferencing. In 1982, Bell employees demonstrated a "pro-teleconferencing" attitudinal change as more of them seek new ways to deal with budget cutbacks and increased emphasis on higher productivity. It is in this environment that Conference 500 has been introduced for internal trial use.

In July 1982, Conference 500, in a "pubescent" stage of development, was demonstrated to a few committees. Conference 500 received enthusiastic support and three conference rooms used by Bell Canada executives were equipped with terminals. By late July, Conference 500 was ready for its first "live" conference. The inaugural meeting was one held by a committee which meets quarterly and includes members from Canadian telcos dispersed from British Columbia to Newfoundland. A two-day meeting entails significant travel expenses and losses of productivity. On July 28th and 29th, committee members met in 10 conference rooms across the country equipped with Telidon terminals. They covered their items of business in 2 three-hour sessions. They commended the experience and have used Conference 500 for their subsequent meetings.

Committees with members from Bell Canada and other Canadian telephone companies have joined their ranks. By November, Bell had equipped another three conference rooms and had obtained the support of its Administrative Services group to prepare "electronic slides" for Conference 500 as they prepare viewgraphs and 35mm slides for other "more pedestrian" meetings.

Questionnaires have been sent to meeting participants and although the user group is still too small to provide solid statistical conclusions, it is possible to derive some insight into the suitability of Conference 500 for certain types of meetings and the benefits it generates.

Most usage has been among upper level management and in meetings where project status or budget requirements are presented. Presentations are often prepared by lower echelons of management. Meeting participants were typically located in Montreal, Ottawa and Toronto. About 70% of users have indicated that they would have travelled if a teleconference had not been held. Generally, a Conference 500 meeting is judged as being not totally as effective as a face-to-face meeting, however more effective than an audio-only teleconference where hard copy of a presentation is distributed prior to the meeting. Presenters valued the control they had over the pace and sequence of slides and discussion. Eighty-four percent said that the quality of the slides exceeded their expectations. Not surprisingly many of the responses pertained to non-visual components of the teleconference. If speakers don't identify themselves, if one location has a poor quality audio terminal, if a meeting room has poor acoustic treatment, or if seating arrangements are inadequate, general satisfaction with the audio-plus teleconference deteriorates. One distinct advantage over the face-to-face meeting and the audio conference is the fact that visual support is conducive to better focus of attention. Comprehension is quicker, discussion remains pertinent. Conference 500 meetings are shorter than face-to-face meetings and shorter than a comparative audio-only conference.

Two of the Conference 500 rooms are being re-designed as "audio-plus" rooms with both Conference 500 and freeze-frame. A third room is being equipped with freeze-frame only. Careful attention is being paid to improved audio terminals, acoustics, lighting, air conditioning, and seating plan. It is expected that freeze-frame will find greater popularity among working level meetings where visual interactivity is required as well as audio. An interesting point to observe will be the portion of meetings in which the two audio-plus alternatives are used concurrently.

APPLICATIONS

The successful use of a medium (whether it be in the teleconferencing arena or not) is highly dependent on the correct identification of applications. One would not expect to find a blackboard and white chalk in a boardroom, nor would you expect a professor of Calculus I to teach with the exclusive use of a slide projector.

Once the need for audio-plus has been
recognized, its implementation must begin
with the identification of users, their communication requirements, meeting habits,
locations, etc. A meeting profile can be
matched with the growing number of audio-plus
options.

A videotex-based audio-graphic service
such as Conference 500 has the following
important characteristics which would be
considered in an application study:

- high quality display of colour text
  and graphics
- material can be revised or rearranged by a user until minutes
  before a meeting
- interactivity during a meeting is
  restricted to audio-only
- universal availability through
  regular voice network

Conference 500 can be used to replace face-
to-face meetings or to enhance teleconferences having the following objectives:

To train or educate geographically
dispersed employees or students. Material
can be stored indefinitely and used whenever
required. An instructor can easily prepare
a package and subsequently customize it for
a specific audience by revising and re-
arranging slides. Conference 500 has been
used to launch two new services. Without
Conference 500, training of representatives
from 9 telephone companies would have meant
that an instructor would have travelled
across the country and spent a half a day at
each of 9 locations. With Conference 500,
two or four training sessions (to accomodate
the various time zones), each lasting from
two to three hours did the trick.

Educational institutions were quick to
marry videotex and teleconferencing to serve
their remote education programs. The Open
University in the UK is involved in a two-
year trial of the Cyclops videotex terminal
as a telewriting addition to standard audio
teleconferencing. Dr. Barry Ellis of the
University of Calgary has implemented a highly successful program of educational teleconferencing which began in the fall of 1980.
Visual materials for the earlier courses
were sent by mail or courier to each of the
local teleconference centers prior to a
class. The logistics were cumbersome and the
costs were significant. Experimental classes
have been held, using Telidon pages stored
in a University of Calgary computer. Evaluations have shown that students and instructors "felt that the quality of instruction

was superior to the audio-only method, noting
that Telidon improved student motivation, increased flexibility, added the capability to
highlight material and aided in concept development".

To disseminate information. In meetings
such as budget meetings, it is important that
all participants have access to accurate up-
to-date data. With Conference 500 slides can
be revised up to minutes before a meeting to
reflect the very latest available figures.

To discuss conceptual areas. An electronic slide greatly improves the understanding
and impact of communication. A picture is
truly worth a thousand words when dealing with
conceptual topics such as policy development
or long term planning.

To project a professional image. The
application of state of the art technology
and the use of high quality graphics greatly
enhances a presentation's credibility and impact. Conference 500 is a powerful medium
ideally suited for boardrooms and customer
seminars.

THE VEHICLE: VIDEOTEX

Of key interest in the Conference 500
trial and other similar endeavours is the
question, "How suitable is videotex as an
audio-plus medium?"

As mentioned earlier, Conference 500
slides are in fact Telidon pages of text and/
or graphic information and are viewed on
Telidon terminals. Telidon is the videotex
technology which was developed by Canada's
Department of Communications and is currently
being used in commercial and trial applications across the country. Telidon has been
praised around the world for its effective
application of colour and graphics and it can
add a high quality visual dimension to any
meeting. However, inherent in Telidon and
other videotex technologies are certain constraints. A sound understanding of the
medium - both its strengths and weaknesses -
is critical to its effective use.

Telidon uses Picture Description Instructions (PDI's) to define a picture in terms of
basic geometric elements: point, line, circle,
rectangle, arc and text characters. The results are "alpha-geometric" pictures which
are composed of pixel-dots as opposed to
pixel-blocks which are made up of 2 x 3 pixels.
Pixel-blocks produce "alpha-mosaic" pictures.
A Telidon terminal displays 256 x 200 pixels.
Each pixel-dot is associated with a memory
location in the Telidon decoder.

This dot structure can accomodate higher resolution graphics. Although superior to alpha-mosaic videotex technologies, Telidon cannot be used to create extremely precise graphics such as a detailed floor plan or electrical drawing.

Typically, in the Conference 500 environment, a speaker prepares a presentation in longhand and rough sketches. From this, he may input text slides himself or have a clerk create the slides. Any graphic content will be sent to a specialized Telidon page creation system operator who may be an in-house resource or with a supplier of page creation services. Because these pages are to be viewed in the context of a meeting, both the presenter and the page creator should be cognizant of certain factors.

Slide lay-out. The end product will be viewed in standard TV format (3:4) and is therefore horizontal.

Size of screen. A Telidon terminal can include an integrated screen or it may feed an RGB-modified set or a TV monitor with composite video input. The audience could be viewing slides from screens varying from 13 inches (diagonal) to 45 inches. If participants at your meeting will view from the smaller screens, avoid fine detail.

Limited text capacity. A Telidon text grid is 40 x 20 characters (smallest size) but 38 x 18 is recommended to avoid a cramped effect.

Colour selection. The 699E and 709R standards offer 8 colours, 6 shades of grey, black and white. The 709PLPS standard supports 16 colours and 16 greys from a palette of 32,000 shades. (This standard is compatible with the North American Presentation Level Protocol Standard.) Colours should be chosen carefully. Poor selection can distract the viewer and reduce a slide's effectiveness. A black background is recommended for pages with a large quantity of text. Text should be displayed in colours which have a high luminance, such as yellow, cyan, or light greys. Red, purple and dark blue should be avoided for text. White should not be used except for highlighting since its extremely high luminance can cause eye fatigue. In most cases, more than three colours per page produce eye fatigue and a kaleidoscope effect.

Speed of page development. Slides are transmitted at 1200 bps. The complexity of a page will affect the time required for it to develop completely. An overly lengthy page may be distracting to viewers, and may even take longer to develop that your audio commentary.

Visual consistency. Consistent use of colour, titles, page lay-out, etc., within a presentation are recommended for a cohesive effect.

Highlighting. Several alternative methods of highlighting a piece of information are possible: a different colour, character size, positioning on the page, underlining, even (in extreme cases) flashing.

Pagination. One page should deliver a complete message. Never break a point or thought between two pages.

Upper case characters. Especially, when a small character size is used, upper case characters are recommended. They cover more surface and are therefore easier to read.

Overlays. Overlays can be used to produce interesting effects by building upon an existing page. For example, a presenter may want to show resource requirements for Phase I and subsequently show requirements for Phase II without clearing or erasing figures for Phase I. (Overlays should be used very carefully, if a lot of movement between slides is expected during a presentation. The second layer could be displayed upon an unappropriate background.)

CONCLUSION

Bell Canada is eagerly awaiting results from its continuing use of Conference 500 internally. It is also planning a 12-month market trial which will involve the participation of 8 to 10 major customer organizations. The service is expected to have high appeal to organizations already using other forms of teleconferencing. The trial will provide customers an opportunity to experiment with an exciting new audio-plus alternative. Bell Canada, in turn, will derive insight into the service's market potential and its desirable configuration and set of features.

BIBLIOGRAPHY

Bachsich, P., "Audio-videotex teleconferencing", Viewdata 82, London, Oct. 1982 (Online Conferences Ltd.: Middlesex, U.K.)

Ellis, G. Barry, "Telidon holds promise as audio enhancer", Telecoms, Aug./Sept. 1982

BIBLIOGRAPHY (cont'd)

Hurly, Paul, Hlynka, Denis and Hurly, Janet, "Videotex - An Interactive Tool for Education and Training," Teleconferencing and Electronic Communications: Applications, Technologies and Human Factors, University of Wisconsin - Extension, 1982.

Johansen, Robert, "Social Evaluations of Teleconferencing", Telecommunications Policy, 1977, pp. 395-419

Kelly, Tim, "How to Integrate New Meeting Technology", Successful Meetings, March 1982, pp. 51-59

Sonneville, Walt, "Teleconferencing Enters its Growth Stage", Telecommunications, June 1980, pp. 29-34

Stockbridge, Christopher, "Multilocation audiographic conferencing", Telecommunications Policy, June 1980, pp. 96-107

The Teleconferencing Resource Book: A Guide to Applications and Planning
Lorne A. Parker and Christine H. Olgren (eds.)
Elsevier Science Publishers B.V. (North-Holland)
© Center for Interactive Programs, University of Wisconsin-Extension, 1984

TELECONFERENCING HUMAN FACTORS TESTING

BY

D. J. EIGEN

ABSTRACT

The Bell System is developing a new network teleconferencing service. This service enables the customer to set up a conference with or without an operator, and it incorporates state-of-the-art signal processing technology to enhance the audio quality. Human factors testing for this new service concentrated on characterizing the customer, optimizing the perceived audio quality, and making the conference set up procedures easy to use and error free. A coordinated series of interviews, surveys, laboratory studies, and field studies was implemented to increase customer satisfaction, performance, and usage. This sequence of testing is culminating in a Controlled Product Test, which is a dress rehearsal of a new service, using Bell System employees as the first customers.

1. INTRODUCTION

The Bell System is developing a new network teleconferencing service[1] to provide multipoint conferencing for audio, analog graphics, and Group 4 facsimile protocol graphics equipment. This service will be offered via a Network Services Complex (NSC),[2] which provides the bridging capability,[3] Group 4 protocol compatibility,[4] a human-machine dialogue for conference setup, and the operator connection for assistance. See References 5 and 6 in these proceedings for additional information and discussion on the NSC teleconferencing market, service, and equipment.

Teleconferencing is an intrinsically complex service, and human factors testing is an essential ingredient in making the service usable. Human factors testing for teleconferencing focuses on three areas:

1. characterization of customer need, use, and perceived audio quality,

2. maximization of audio quality, and

3. optimization of call setup and control.

By matching customer needs with service capabilities, improving audio quality, and making the service easy to use, service usage customer satisfaction, and performance are maximized.

Several methodologies are being employed to evaluate Teleconferencing. Customers of the current Bell System operator assisted (cordboard) Teleconferencing service were interviewed to identify their service usage characteristics, their needs for new services, their problems with the current service, their willingness to set up conferences themselves, their utilization of graphics devices, and so forth.

Laboratory studies were conducted to design the best user interface, including announcement wordings, dialed sequences, and error handling strategies.

Other laboratory studies were conducted to choose the best audio bridge parameters, and values such as, automatic gain control, echo cancellation, and speech detection.

Further, a Controlled Product Test (CPT) is being conducted as a dress rehearsal of the new service using Bell System employees as test customers. During this test, customer interface parameters will be adjusted, operations plans will be tested, and marketing materials will be evaluated. Finally, the new Teleconferencing service will be monitored as it is introduced to commercial users.

2. CUSTOMER CHARACTERIZATION

To help understand what customers prefer in a new network teleconferencing product, two different interviews of current "cordboard" teleconferencing users were conducted. With both studies combined, a total of 1600 customers in Houston, Dallas, Washington DC, New York, Chicago, St. Louis, Los Angeles, and San Francisco were interviewed.

Some general conclusions of these studies were:

1. Cordboard customers feel that audio quality is the most important attribute of teleconferencing.

2. Many would prefer to set up a conference without an operator. This preference appears to be sensitive to price, but a few customers would prefer operator setup in any case.

3. Almost all current conference users have access to TOUCH-TONE® phones.

4. Many customers use speaker phones during conferences, but other equipment or aids were used infrequently (see Figure 1).

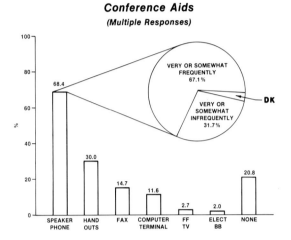

Conference Aids
(Multiple Responses)

Figure 1

5. Most of the conferences involve only several locations. The number of conferences of each size is shown in Figure 2.

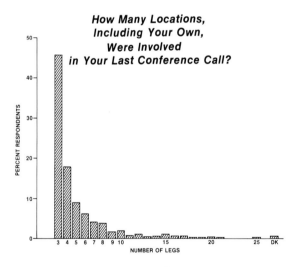

How Many Locations, Including Your Own, Were Involved in Your Last Conference Call?

Figure 2

6. Although customers are generally satisfied with cordboard service, they reported a number of problems as shown in Figure 3.

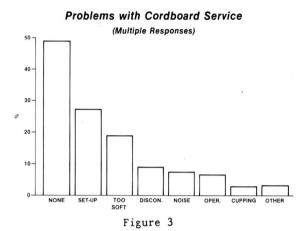

Problems with Cordboard Service
(Multiple Responses)

Figure 3

The NSC bridge uses state-of-the-art technology to address the problems cited by cordboard customers. This technology includes echo cancellation, speech detection, signal processing, and automatic gain control designed to equalize sound levels, reduce noise, minimize clipping, and allow multiple simultaneous speakers. Problems with the operator setup of cordboard conferences are eliminated by providing a self-dial capability.

In summary, the NSC has features or capabilities which address each of the problems that current teleconferencing users reported.

The NSC service also appears to meet a significant need of a large segment of the current market, as defined by the data collected. The bulk of teleconference calls involves only several locations, and many of those customers prefer to set up a conference without an operator.

The customer characterization studies point to the need for improved audio quality and a self-dial setup capability, both of which are specifically addressed in the NSC design. Therefore, NSC conferencing service is expected to be well received in the market place.

3. DIALING PLANS

To allow customers to set up conferences themselves without assistance from an operator the NSC has a human-machine dialogue capability. This dialogue consists of a sequence of announcements directing the caller to dial dual tone multifrequency digits to set up a conference. The customer may also dial one of several conferencing options.

Human factors engineers designed the initial human-machine dialogue and provided requirements for developers.

The design objective is to make the service simple and self-explanatory, so that a customer needs no prior knowledge or written instructions. The dialogue is designed for both the inexperienced and experienced dialer. The inexperienced dialer may wait and listen to complete prompts and explanatory announcements, while the experienced dialer may speed conference setup by dialing before or during announcements.

3.1 Conference Setup and Options

A customer originates a conference by dialing a particular NSC. This allows the customer to choose the bridge which is most advantageously situated for a particular conference, as well as to select the NSC suitable for the conference type, i.e., audio, nonprotocol voiceband graphics, or CCITT Group 4 graphics.

A welcome announcement identifies the NSC site. After the welcome announcement, the customer must take the following steps to set up a conference:

1. enter the number of locations,

2. dial a location,

3. add the location,

4. repeat steps 2 and 3 for each additional location until all conferees have joined the bridge, and

5. add self to start the conference.

From the customer's perspective, NSC teleconferencing has three different modes:

1. control - the conference controller interacts with announcements to set up and control the call,

2. privacy - the conference controller talks to a location privately and separately, and

3. conference - the conferees and controller talk with one another on the bridge.

A number of conference options are available with NSC teleconferencing. These options provide the more sophisticated caller the flexibility to handle a variety of conferencing needs. The options include:

1. dialing through or before announcements to curtail or preempt them,

2. controlling the conference without participating (for a teleconferencing clerk or secretary),

3. speed dialing disconnected legs,

4. having an operator set up the call,

5. self-correcting errors and eliminating timings,

6. dialing international calls to foreign locations,

7. using nonprotocol voiceband graphics devices like the electronic black board or slow scan or freeze frame TV, and

8. using CCITT Group 4 facsimile protocol graphics equipment at 4.8 kb/s or 56 kb/s speeds.

The capability to add up to 59 conferees, the multiple controller modes and the numerous options provide a rich set of capabilities for the teleconferencing user. However, this feature abundance places tremendous demands upon the human-machine dialogue to make the service usable.

3.2 Human-Machine Dialogue Design

The design process for human-machine interfaces used here is best described as: Design, Test, Redesign... . A basic set of design principles and a good initial design were used to begin this iterative process.

The design principles used are based on experience with a number of different voice prompted services. Still, the overriding principle is to design the dialogue in such a way as to maximize flexibility for later test - redesign iterations.

This flexible design is embodied in the concept of a transaction.[7] A transaction is the set of dialogue components necessary to collect one digit string. In teleconferencing, the digit string could be the number of ports requested, an address of a location, a control option, or the command to add a party to the bridge or start the conference. The transactions are sequenced to construct the entire dialogue needed to set up a conference and select the appropriate options.

The dialogue for a transaction consists of:

1. prompt,

2. error handling for misdialing,

3. error handling for no dialing,

4. call disposition for repeated errors, and

5. call (dialogue) disposition for successful dialing.

The prompt should be interruptible to allow the more experienced dialer to dial through the announcement. The prompt should incorporate some silence to serve as a dialing window for customers.[8]

Error handling for misdialing usually consists of announcements and subsequent chances to dial the correct digits. Experience indicates that two to three chances is enough to enable customers to succeed. It is often useful at this point to make all or part of the error announcement uninterruptible so that customers will hear the announcement and do not mistakenly interrupt it.

If a customer does not dial when prompted to do so, after a suitable timing period, the customer should be reprompted and given subsequent chances to dial. After two to three unsuccessful attempts, call disposition must be determined. For teleconferencing, an operator is connected to assist the caller after several errors or time-outs.

When a customer successfully enters a valid digit string, then the next digit string is prompted (or the conference is started).

To help users to set up a conference a design capability to use the # and * was included for test purposes. The # symbol is to be tested as a signal to proceed (curtail timing, end dialing, request more information, or indicate yes), and the * symbol is to be tested as a signal to back up (cancel entry, drop leg, or indicate no). Steps are being taken to consider this use of * and # for other services.

3.3 *Dialing Plan, Announcements and Instructions: Laboratory Experiments*

Making teleconferencing simple to use is a complex endeavor. For example, users are especially sensitive to changes in announcement wording, and there are over 80 different announcements employed in teleconferencing setup procedures. Experience with earlier and much simpler services like Calling Card Service[9] has pointed to the difficulty in constructing dialing sequences, announcements, error handling strategies, and timing structures which give acceptable user performance. Laboratory studies were used to help derive an initial set of instructions, announcements, and dialing procedures.

Human Factors Laboratory

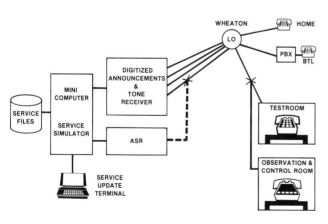

Figure 4

Figure 4 illustrates the laboratory arrangements. A commercial announcement machine and four digit receivers were used in a configuration that allowed up to four callers to interact simultaneously with different dialogues from human factors laboratories located in Holmdel, New Jersey, and Naperville, Illinois, or, for some studies, from any phone at home or in the office. While subjects interacted with many conference control announcements and procedures, actual conferences were not set up.

Table 1 shows success rates for the initial dialing plan, a later dialing plan, and the last plan tested in the laboratory. Written instructions were present for each of the experiments. Compared with experiments using no written instructions, the instructions appear to improve performance chiefly on the first attempt. It also appears that overall success rates for simple setups should be acceptable with the dialing plan and announcements alone, that is, without instructions. Instructions are required for initiating special options.

T/C Lab Studies
Success Rates

	1ST TRY	2ND TRY
Initial Announcements & Instructions	48%	70%
New Announcements & Instructions	75%	92%
"Simplified" Plan	83%	92%

Table 1

The announcements and instructions were refined throughout the series of laboratory studies. Still, the final product of these studies only serves as the initial set of announcements and instructions for the Controlled Product Test (CPT). In the CPT the dialing plan and instructions will continue to be refined.

A CPT allows the dialing plan to be tested under realistic conditions with real users. In the laboratory studies, subjects never actually experienced a conference, instead they just interacted with announcements in the control mode. Thus, many dialing conditions and situations could not be adequately tested in the laboratory. A CPT provides the opportunity to evaluate and adjust the complete dialing plan as well as many other aspects of the service.

Audio quality is another important facet of the user interface. Since a conference was never actually set up in the dialing plan

laboratory experiments, separate studies had
to be initiated to evaluate and optimize the
NSC audio quality. Like the dialing plan, the
number of possible factors potentially
affecting customer perception was sufficiently
large to warrant a sequence of laboratory
studies before a Controlled Product Test was
initiated.

4. *AUDIO QUALITY*

The NSC audio bridge incorporates
state-of-the-art digital signal processing
technology in order to improve audio quality.
The audio quality objective of NSC
teleconferencing is to be equivalent to that
of comparable 2-point DDD connections.

Audio quality is, in part, subjective. Human
factors testing and NSC design were developed
to measure and enable the optimization of the
perceived audio quality. The NSC signal
processing algorithms entail many parameters
which can be adjusted in software. Human
factors testing used a coordinated sequence of
laboratory and field tests to choose the best
set of these audio bridge parameters to
maximize perceived audio quality.

The audio quality testing sequence included:

1. listen-only tests to pick test bridge
   configurations,

2. initial conversational tests with a
   simulated network,

3. simulated conference tests,

4. field bridge comparisons, and

5. the Controlled Product Test.

4.1 *Listen-Only and Conversational Tests*

Six configurations of bridge parameters were
chosen from the listen-only tests. The
maximum number of simultaneous talkers and
minimum number of ports enabled, the audio
signal delay to reduce clipping, the use of
echo cancelers for echo control, holdover
periods to reduce clipping, and automatic gain
control adjustment rates were among the
parameters considered. The six bridge
configurations chosen were evaluated with
conversational tests.

In the conversational tests 40 subjects read
parts from scripts and then rated the audio
quality of the bridge in question, noting
communication problems related to speech
level, noise, echo, and voice switching. See
Table 2 for the audio quality rating question
and problem check list.

Response Sheet
(Simulated Conference Tests)

Study Date: _____
Location: _____
Participant: _____
Session: _____

The box below contains a list of problems that occasionally occur during a teleconference.
Please check off any problems that you experienced during the last teleconference connection.

Problem	Occurred During Conference
Speech level too low	
Speech level too high	
Speech levels too different	
Speech levels too similar	
Speech hollow or reverberant	
Speech muffled or distorted	
Missing syllables or words	
Too hard to interrupt remote talker	
Noisy interference on audio line	
Other (specify):	

Please rate the overall quality of the last teleconference connection by circling the number that best reflects your opinion.

Very Poor	Poor	Fair	Good	Excellent
1	2	3	4	5

Comments:

Table 2

Different network conditions were also
simulated and tested. Figure 5 shows the
laboratory simulation arrangement for these
tests. From the conversational tests, three
bridge configurations were chosen for further
study in the simulated conferences.

*Per-Port Functional Block Diagram of Audio Bridge
and Network Impairment Simulation*

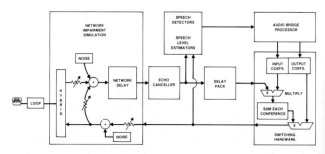

Figure 5

4.2 *Simulated Conferences*

Sixty people participated in sequences of
simulated six port conferences using the
Desert Survival Task. The participants were
split between Naperville, Illinois, and
Holmdel, New Jersey, and were connected to the

laboratory NSC and network simulation used in the conversational tests. Besides the three bridge configurations tested, two network conditions and the use of a handset versus hands-free devices were studied. The two network conditions simulated legs of equal length (symmetric) and the more realistic case of legs of different lengths (nonsymmetric). Hands-free and handset devices were studied to determine the effect of terminal voice switching on perceived audio quality. The three bridge configurations were 1) an all feature bridge (A) with optimized parameters, 2) a cost-reduced bridge (B), e.g., without echo cancelers, and 3) a reference bridge (C).

In the more realistic nonsymmetric network, bridge configuration A – the full featured bridge – performed significantly better than the other bridge configurations. This result was evident even in the presence of hands free devices (see Figure 6). Given the proliferation of hands-free voice switched devices used in current cordboard conferences, (see Figure 1) it was important to test whether significant improvements in audio quality were masked by terminal device voice switching.

Mean Opinion Score for Two Networks, Three Bridge Configurations, and Two Terminals (Simulated Conference Test)

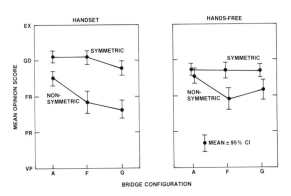

Figure 6

4.3 *DDD and Bridge Comparisons*

The next step was to compare the audio quality of the NSC teleconferencing service with a two-point DDD call and current cordboard conferencing service. To make the test more realistic, and therefore more generalizable, an in-field production model of the NSC and real network connections were used. The NSC and cordboard were located in Los Angeles. Eight conferences with four legs were held for each bridge – two short legs in Los Angeles and two long legs to Naperville, Illinois. Subjects were unaware of which bridge they were using and, again, the Desert Survival Task was used to provide grist for the conference. Subjects were asked to rate the quality of each conference and check any problems that occurred.

To provide a reference for comparing the two bridges, each conferee in Los Angeles was connected on a DDD call to his or her counterpart in Naperville and asked to rate the audio quality and check the problems with the DDD connection. Subjects used only handsets. All connections were measured for noise. Connections which had very high noise or very low noise were not used.

Figure 7 illustrates that the NSC meets its objective of DDD equivalence on both measures. The NSC is also significantly better than the current cordboard bridge.

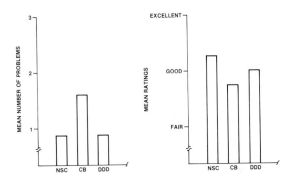

Figure 7

Optimization and evaluation of audio quality will continue during the Controlled Product Test. With the CPT, people's perception of audio quality with real meeting situations can be evaluated.

5. *CONTROLLED PRODUCT TEST (CPT)*

A Controlled Product Test (CPT) is a dress rehearsal of a new network service. Laboratory experiments provide an empirical means to eliminate alternatives and bound the design. But the laboratory experiments could not provide a real service under real conditions. Laboratory experiments also suffered from other limitations. For example, error handling was especially difficult to evaluate in the laboratory because equipment limitations prevented having all of the error handling procedures and announcements present on any one test call.

A CPT provides the fidelity (realism) necessary to make generalizations about the real service. It is therefore a logical and necessary next step in evaluating a new service and in optimizing the human-machine interface. Since the CPT uses the real product, the limitations present in the laboratory are avoided. In addition, the CPT provides the means to address operational and marketing issues.

Bell System employees are used as the first customers during the CPT. The CPT for NSC audio teleconferencing began in February of this year with designated Pacific Telephone and Telegraph locations in San Diego,

Los Angeles, and San Francisco. Other Bell System companies and locations will be added to the test. The CPT for CCITT Group 4 teleconferencing is also planned. The service will continue to be monitored for some period after the service is introduced.

During the CPT, marketing, operational, and human factors issues are being studied and the service is being fine-tuned to maximize service usage, customer performance, and customer satisfaction.

5.1 *CPT Service*

CPT teleconferencing is due to be offered initially from NSCs in Los Angeles, California. Later in the test, NSCs in Chicago, Illinois, and White Plains, New York, will be added to provide additional CPT service. See Figure 8.

T/C CPT NSC Sites

Figure 8

Callers will be able to originate CPT teleconferencing calls only from designated locations in participating Bell System companies. Phones with Dual Tone Multifrequency (DTMF) signaling with * and # are required to originate a call. Conference originators can add up to 59 locations anywhere in North America.

CPT teleconferencing assistance operators will also be provided. They can be reached in one of four ways:

1. before or after the conference by dialing an 800 number,

2. during the conference setup (control mode) by dialing 0,

3. during the conference setup by making several errors or by not dialing, and

4. during the conference (conference mode) by dialing #0.

Figure 9 illustrates the network configuration of teleconferencing and the network paths to the teleconferencing operator.

Operators are keeping a detailed log of all calls and responses. Call-back contacts are made by CPT experts when operators cannot resolve the difficulty.

T/C Controlled Product Test

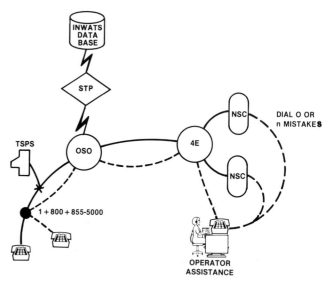

Figure 9

5.2 *Publicity and Instructions*

To make potential users aware of the CPT service and encourage them to use it, fliers will be distributed desk-to-desk and posted on bulletin boards. See Figure 10.

Issue 1 (1/83)

Beginning February 1983

WHAT	CPT Teleconferencing—A new network service which lets you connect 3 or more locations by telephone without an operator. This service is being field tested with a Controlled Product Test in the Bell System.
WHY	Teleconferencing—Meet Instantly Speed Decisions Save Time Save Travel Be Productive
HOW	Dial it yourself—Anytime—Anyone

Information - Problems: Instructions:
Call 1+800+855-5000   See Back

CONTROLLED PRODUCT TEST

Figure 10

Also, caller awareness interviews will be made to potential users of the service to determine if they have heard of the service and to identify potential users or problems.

A design objective is for the service to be fully prompted and simple enough to use so that a user would not require written instructions to set up a basic conference call. Nevertheless, laboratory studies have shown that instructions can enhance first use performance. The pocket card provided to initial users is shown in Figure 11. A more detailed instruction booklet has been produced and also will be tested and revised.

Figure 11

5.3 *CPT Data Collection, Analysis, and Adjustment*

The most critical and sensitive aspects of the human-machine dialogue, such as announcements, timings, error handling strategies, and digits to be dialed, have been parameterized to allow rapid changes to the service. To provide objective data on human performance, e.g., accuracy and speed, the NSC also time-stamps and stores customer and NSC actions. A sequential time log of every digit dialed, every announcement provided, and the state of the conference is the result.

Besides this objective record of caller system performance, the interviews and questionnaires are used to measure customer reactions to and satisfaction with all aspects of the service. Each day NSC data and interview data are sent to the CPT analysis system. Figure 12 illustrates the network configuration of the NSC and the CPT analysis system.

T/C Controlled Product Test

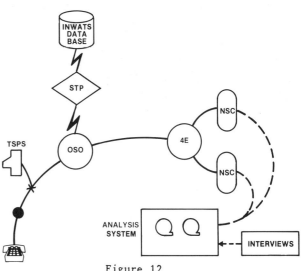

Figure 12

The analysis system runs under the UNIXTM operating system and consists of several software packages. These packages include NSC data communications, Polaris relational data base manager, S statistical language, data processing and analysis routines, graph and report generating functions, issue tracking, and newsletter capability. See Figure 13.

Analysis System

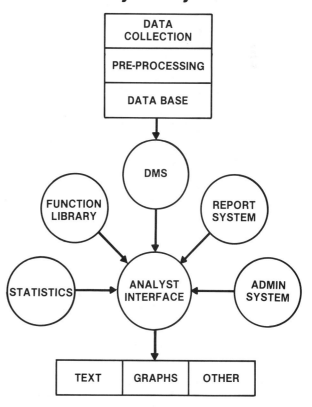

Figure 13

The analysis system produces a number of on-line reports each day, provides the basis for statistical analyses, and generates detailed human readable event histories. See Figure 14. These data and analyses provide the bases for determining the nature of a problem and the effect of service adjustments to correct it.

Sample Human-Machine Interaction Protocol

elapsed time	action	digits or announcement
DATE: 07/28/82	TIME OF CALL 08:14:56	
0.0	caller connected	
0.0	digit collector allocated	
4.8	announcement machine allocated	
4.9	announcement played	Welcome
8.4	digit entered	*
8.6	announcement interrupted	
9.2	digits entered	267
9.8	announcement played	Voice Stations
11.7	digit entered	5
11.9	announcement interrupted	
14.6	announcement play	Voiceback
14.6	announcement played	Five
16.8	announcement completed	
23.5	announcement played	Dial Station
46.8	announcement completed	
52.7	digits entered	13125551212
62.7	announcement played	Add Station
64.4	digit entered	2
64.6	announcement interrupted	
64.8	announcement played	Station Added
67.5	announcement completed	
67.5	announcement played	Next or Join
101.1	announcement completed	
101.1	digit entered	2
101.1	announcement played	Conference Control
118.3	announcement completed	
118.3	announcement played	Thank You
132.2	announcement completed	
132.2	announcement machine deallocated	
132.3	digit collector deallocated	

Figure 14

In fact, the CPT process can itself be characterized as an iterative feedback system: design, test, redesign... . A simple model is depicted in Figure 15.

Design – Test – Redesign...

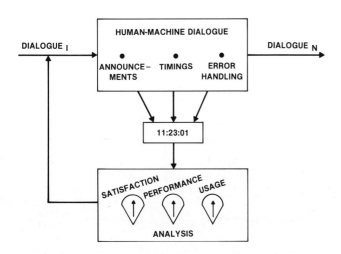

Figure 15

The knobs (dots) in the figure represent the NSC parameters; the meters depict the NSC, and interview measurements. The object of varying the service by turning the knobs is to register a beneficial change in service on the meters. If the meters are considered to show customer satisfaction, usage, and performance, then the object is to find the combination of knob positions which maximizes the readings on the meters. The analysis system provides the quick analysis of results necessitated by this design-test-redesign... method.

To allow us to interpret the effect of the changes during field tests, several strategies are employed. New customers are being phased in to provide new users unbiased by prior experience with previous versions of the service. Logs are being kept of all coincidental events (holidays, bad weather, etc.) that could conceivably alter dialing and usage behavior. Repetitions of adjustments will corroborate that changes indeed increase or decrease performance.

Near the end of CPT, the best set of announcements, timings, and error handling strategies will be tested without adjustments for a month to obtain engineering estimates and check final results.

6. *SUMMARY*

Human factors testing for teleconferencing focuses on three areas: 1) customer characterization; 2) audio quality, and 3) user control. An iterative approach to design and testing is being applied to maximize customer satisfaction, performance, and usage. A coordinated series of caller interviews, laboratory studies, and field studies is being used which culminate in the Controlled Product Test.

Customers rank audio quality as the most important aspect of teleconferencing. Audio quality testing has indicated that NSC teleconferencing is better than the current cordboard teleconferencing service and equivalent to a two-point Direct Distance Dialed call.

Customers are willing to set up conferences without an operator. Changes in instructions, announcements, and procedures as a result of laboratory and field testing, has improved the success rate of customer conference setup.

At the time of the writing of this paper, the Controlled Product Test had just begun. Continued usage and measurement of the teleconferencing product is expected to provide ample data to fine-tune the service.

7. *REFERENCES*

[1] J. O. Boese, R. J. Jaeger, Jr., R. C. Thanawala, and L. A. Tomko, "Audiographics Teleconferencing," IEEE International Conference on Communications, Philadelphia – June 13-17, 1982, pp. 4E.1.1-4E.1.4.

[2] J. H. Huttenhoff, W. R. Shleicher, L. E. Suk, and P. R. Wiley, "Network Services Complex Architecture," IEEE International Conference on Communications, Philadelphia June 13-17, 1982, pp. 4E.2.1-4E2.5.

[3] J. H. Bobsin, Mohamed Marouf, and P. J. Rutowski, "Network Services Audio Bridges," IEEE International Conference on Communication, Philadelphia June 13-17, 1982, 4E.3.1-4E.3.5.

[4] D. E. Herr, R. Metz, A. J. Risica, and L. A. Russell, "Network Services Teleconferencing Data Bridge," IEEE International Conference on Communications, Philadelphia June 13-17, 1982, pp. 4E.4.1-4E.4.4.

[5] J. Hayes, "Overview of Market for Network Bridging," Conference on Teleconferencing and Interactive Media. May 18-20, 1982 - This issue.

[6] J. O. Boese, J. H. Huttenhoff, A. B. Mearns and R. Metz, "Networking Services Audiographics Teleconferencing," Conference on Teleconferencing and Interactive Media, May 18-20, 1983, - This issue.

[7] D. J. Eigen, "Human Machine Telephone Service Protocols - Design and Analysis," International Conference on Digital Communications, March, 1982.

[8] E. C. T. Walker, "Human-Machine Dialogues: Design, Test, Redesign..." Second Annual Phoenix Conference on Computerized Communications, Phoenix, Arizona, March 14-16, 1983.

[9] D. J. Eigen and E. A. Youngs, "Calling Card Service - Human Factors Studies," BSTJ, September, 1982, Vol 61, No. 7, Part 3, pp. 1715-1736.

208

The Teleconferencing Resource Book: A Guide to Applications and Planning
Lorne A. Parker and Christine H. Olgren (eds.)
Elsevier Science Publishers B.V. (North-Holland)
© Center for Interactive Programs, University of Wisconsin-Extension, 1984

QUORUM™ TELECONFERENCING:

A MEETING OF MINDS

Christopher Stockbridge   and
Customer Systems
 Research Laboratory
American Bell Inc.
Holmdel, NJ 07733

David R. Fischell
Teleconferencing Performance
 Objectives Studies
Bell Laboratories
Holmdel, NJ 07733

ABSTRACT

American Bell markets the QUORUM™ Teleconference system, a growing range of audio, audiographic and full-motion video products, made especially for fully interactive group-to-group telecommunicated meetings, or teleconferences.

This paper describes QUORUM teleconference equipment available now for worldwide use.

INTRODUCTION

Corporate travel expenses are climbing an average of 20 percent each year. This trend not only reflects rising transportation and lodging costs, but also suggests a continually growing need for meetings among business colleagues. In fact 75 percent of all business travel is to and from meetings.

Large corporations have long recognized the expense and inconvenience involved in bringing employees together from places hundreds or thousands of miles apart. And when employees travel,

businesses assume not only the cost of transportation, meals, lodging and living expenses, but also pay for a great deal of unproductive time — time which traveling employees spend waiting or in transit instead of at work.

In addition to these corporate expenses, employees themselves also lose when traveling. Some business trips are exhilarating, but after awhile they become exhausting. Business trips take many otherwise-free hours away from employees —

Getting there. Today, three out of every four business trips are for group meetings. Though such meetings can be invaluable, the time and expense involved in traveling to them can undercut business productivity significantly. Electronic information technology now allows business colleagues to meet and communicate without traveling.

hours that could be spent at home with family and friends.

Aware of both the attractions and the disadvantages of business meetings, scientists and engineers formerly at Bell Laboratories, now at American Bell Incorporated, have used their acknowledged expertise in voice communication technology to develop QUORUM Teleconferencing, a more efficient, cost-effective — yet still natural — way for business colleagues to communicate. The QUORUM Teleconference System contains a growing line of sophisticated customer premises products that can sometimes raise productivity, certainly decrease wasted time, reduce travel costs, and perhaps improve the quality of employees' lives.

Studies on both sides of the Atlantic agree that at least half of all business meetings can be replaced by some form of a teleconference. 'Teleconference' conjures up different images to each of us. To some it means three or more people in different places talking together on telephones. To others it means a person or group facing a television screen, looking at, talking to, and interacting with people hundreds or thousands of miles away, rather like an interview on the evening television news.

Both are accurate sketches of a teleconference, but each is at the opposite end of a range of electronic capabilities that overcome distance as a barrier to holding a meeting. Between these extremes of three-way calling and full-motion video conferencing lie many audiographic options in this new spectrum of fully interactive group-to-group communications.

The QUORUM Teleconference System is a range of products, from audio and audiographic equipment to top-of-the-line PICTUREPHONE® Meeting Service, which meet the needs of business people to teleconference. We shall now describe each in turn.

AUDIO TELECONFERENCING

The most basic form of teleconferencing[1] is interlocation voice communication: three or more people talking together while at separated places. The QUORUM Teleconference System includes several products that make this both natural and economical.[2]

The Bell 4A Speakerphone System[3] is perhaps the most familiar. Developed at Bell Labs, the speakerphone is the first step on the road to natural audio teleconferencing. It allows two, three or four people in the same room to all hear and talk to remote people, and still have their hands free for typical meeting activities, such as writing, shuffling papers, or checking calendars.

Easily attached to a standard desk-top telephone, the speakerphone consists of two components: a loudspeaker unit which also contains the speakerphone electronics; and a pedestal control unit which contains the electret pressure zone microphone, on/off switch, and a volume control for the loudspeaker.

With over 800,000 speakerphones currently in service, this popular audio teleconferencing tool allows small groups of people to get together easily over the telephone. When used properly — that is, when participants talk within 18 inches of the pedestal microphone — it provides high-quality audio.

Another QUORUM Teleconference product appropriate for larger meetings is the QUORUM Portable Conference Telephone.[4] Also developed at Bell Labs, this telephone comes with a built-in loudspeaker and directional cardioid microphone, as well as two extension omni-directional microphones. It can easily accommodate up to eight people seated around a conference table

Being there. Electronic meetings — or teleconferences — are an increasingly popular substitute for in-person meetings, especially among businesses that have many locations throughout the country. Here, ABI employees in New Jersey teleconference with colleagues at a distant location, using the QUORUM™ Teleconferencing Microphone, Group Audio Teleconferencing Terminal, GEMINI® 100 Electronic Blackboard, **as well as special still-video television**

since it has three microphones and since its loudspeaker provides up to 7 dB more volume than the speakerphone.

This unit can be brought in and out of a conference room as needed. It is simply placed in the middle of the table, and plugged into a power outlet and standard telephone jack.

ACOUSTICS IN TELECONFERENCING

Room acoustics is an element of teleconferencing too often neglected. Ideally, the room should not be symmetrical and the walls should be non-parallel, covered with both sound scattering and noise-absorbing panels. At the very least, a room for teleconferencing must be quiet with a carpet and window drapes. Noisy equipment with fans, such as facsimile machines or the memory unit for the GEMINI 100 System, should be installed in a soundproofed cabinet or outside the room in a closet. And slide or transparency projectors with noisy cooling fans must be replaced. These precautions help control the two most common acoustic problems in teleconferencing: equipment noise from within and outside the room, and room reverberation (the hollow, echo sound often associated with microphones).

In addition to the Speakerphone and Portable Conference Telephone, the QUORUM system includes a very high-quality sophisticated audio terminal for dedicated teleconference rooms: the QUORUM Group Audio Teleconference Terminal (GATT). Originally designed for NASA by AT&T Long Lines, and subsequently redesigned by Bell Labs for standard use, the GATT terminal is unique; it is the only terminal which can be fine-tuned to take full advantage of the good acoustical properties of a well designed room. While teleconference rooms should be designed to recommended acoustic specifications (see the panel on acoustics), the GATT can even compensate for some of the effects caused by poor acoustics. By taking full advantage of good room acoustics, it allows people to converse naturally, to interrupt easily, to question and comment as they wish. This is achieved by reducing switched loss when aligning the GATT during installation to achieve no more than the required 25 dB Echo Return Loss. This alignment process requires adjusting amplifier gains and loss switching circuits, using a speech-weighted noise generator (USASI) and an internal VU meter, both of which are built into the GATT. In short, the GATT is much more than a speakerphone; and **in the proper acoustic environment,** a GATT will give nearly transparent operation, in which natural conversation is typical on all but high loss and/or noisy connections.

The GATT is controlled from a modified six-button telephone with the bulk of the terminal electronics normally placed out of sight in an adjoining equipment closet. The GATT accepts up to 16 microphones, it has a powerful amplifier capable of driving several loudspeakers; it is the terminal of choice when connecting to public address systems for teleconferencing in auditoria, lecture halls, or executive conference rooms.

NEW ARRAY MICROPHONE

All of these teleconference terminals — the speakerphone, the Portable Conference Telephone, and the Group Audio Teleconferencing Terminal — make it very easy for small or large groups of people to meet with others using telephone circuits. But the capabilities of all three can be extended further with a recent addition to the QUORUM product line: the QUORUM Teleconference Microphone and Loudspeaker.

These unique devices were specifically invented and developed at Bell Labs for teleconferencing. Designed like an antenna, the QUORUM Teleconference Microphone[5] frees teleconferencing participants from the constraints associated with table and lavalier microphones — like talking close up or wearing a lanyard around one's neck. And since only one QUORUM Microphone is needed for 20-30 people, the table-top clutter of conventional multi-microphone systems needed for large groups is eliminated. And at the same time, this brilliant invention reduces the hollow "echo" sound often associated with speakerphones, especially those misused in acoustically inappropriate places.

A 30-inch vertical column, the QUORUM Teleconference Microphone is actually an array of 28 tiny electret transducers. Together they act as a directional acoustic antenna that can pick up speech within a 12-foot radius. Mounted on the QUORUM Teleconference Loudspeaker, the microphone can be connected to a 4A Speakerphone System, Portable Conference Telephone, or a GATT. It is simply placed in the center of the conference table out of the way of conference activities. Conferees can then speak in normal voices, unencumbered by lanyards or cords, and move freely about the conference room without losing the ability to hear or be heard.

When the QUORUM Teleconference Microphone (and Loudspeaker) is used with a Group Audio Teleconference Terminal, superior performance is realized when most of the loudspeaker power comes from a ceiling loudspeaker mounted at least four feet **directly** above the shaft of the microphone. This

Voice terminals. The QUORUM™ Teleconferencing System includes a variety of voice terminals through which the audio portion of a teleconference may be conducted. Shown, from left, are the Speakerphone, the QUORUM Portable Conference Telephone, and the control telephone for the QUORUM Group Audio Teleconferencing Terminal.

installation allows at least 12 dB of switched loss to be removed during alignment (see above); thus it provides more nearly perfect conversational transparency.

SOCIAL DYNAMICS OF ELECTRONIC MEETINGS

A group of people talking together differs from a one-on-one conversation. An intimate chat with a friend is different from giving a speech. These extremes help teleconference equipment designers understand that group-to-group teleconferences are significantly different in extent as well as kind from person-to-person telephone calls.

Indeed, teleconferencing must account for human factors completely different from those involved in standard telephone conversations. For example, in a teleconference, people talk at different levels, and start to speak at unpredictable times, depending more on cues from the context of the conversation than on who spoke last. Who nodded agreement, who grunted dissent may be very important in determining the outcome of a meeting - these human dynamics place new design requirements on telephony. They are made more obvious when the electronic system is inadequate, by each conferee's being immediately able to make interperson and interlocation comparative value judgements of how well the equipment performs. Not well understood yet, constrained vocal dynamics in group teleconferences can lead to unfortunate social consequences. One example must suffice here - the location with low signal level will be perceived as inferior, will be ignored, and may feel itself manipulated to disadvantage. Such circumstances may cause the people at this location to withdraw from the meeting in electronically caused anger.

AUDIOGRAPHIC TELECONFERENCING

Although audio teleconferencing is certainly an effective way to conduct many business meetings, in some cases it is just not adequate.[6] Many times, people must not only hear about each others' ideas, but they must also visualize them. They need some way to exchange written and graphic information.

The QUORUM Teleconference System includes a unique component that allows customers to do that. Conceived and developed at Bell Labs, it is called the GEMINI[R] 100 Electronic Blackboard System. This "blackboard" presents one common writing space to any number of teleconference rooms; it requires only normal voice-grade telephone circuits which can of course be bridged worldwide. Much like a real blackboard, the GEMINI 100 enables participants in all locations to see, write, and erase items in a common electronic space. But unlike normal blackboards, the GEMINI allows its users to be hundreds or thousands of miles apart. As a person writes on the blackboard with ordinary chalk, the writing appears on the television monitors at all participating locations. With a video hard-copy unit, paper copies can be made of what has been written on the GEMINI 100. In the Bell Labs trial [see panel] people would often sign on the Gemini as an attendance roster, then a paper print is made at each location while the conference proceeds.

In addition to this graphic device, teleconference rooms may contain TOUCH-TONE controlled, random-access optical projectors as well as special still-video television and facsimile equipment, which allow people to exchange photographs, people pictures and prepared graphic images, or paper copies of written documents. [Note: We prefer the term 'still-video' to what others have called freeze-frame and/or slow scan television.]

Top-of-the-line full motion video conferencing is offered in PICTUREPHONE Meeting Service (see the panel on PMS).

Construction. The electronic blackboard is about 1¼ inches thick. It consists of two plastic sheets in front of an aluminum-honeycomb/aluminum-sheet "sandwich." Writing causes the front plastic to touch the rear plastic, which is cemented to the sandwich. Contact between the plastic sheets generates voltages representing the position of the chalk. (Drawing not to scale.)

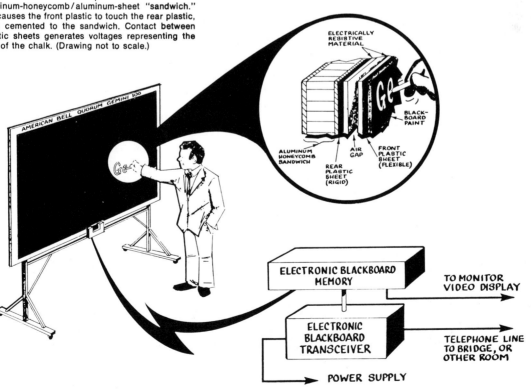

SETTING UP MULTIPOINT TELECONFERENCES

With about 30 percent of all teleconferences involving more than just two locations, an effective way is needed to interconnect several locations simultaneously — to get everyone "on board." The QUORUM System includes two advanced conference bridges that permit customers to schedule and set up teleconferences among many locations: the QUORUM Teleconference Bridge, and the QUORUM Teleclass Bridge.

Derived from the Universal Conference System designed by AT&T Long Lines, and further developed by the Adaptive Engineering Group at Western Electric[8], the QUORUM Teleconference Bridge is especially practical for business teleconferences. It is a microprocessor-controlled analog device with a membrane switch control console designed by human factors specialists[9] now at American Bell.

Setting up. Leon Weinglass, of Western Electric, uses the QUORUM™ Teleconferencing Bridge to connect locations for a teleconference. The newest offering in the QUORUM product line, the bridge allows up to 28 locations to meet electronically.

The bridge is divided into four groups of seven ports, which in turn are connected to telephone lines. By linking the four groups together, as many as 28 locations can join a teleconference. Any of the groups of seven ports may be assigned to connect **either** audio **or** graphic devices in a teleconference. For example, one group may interconnect seven GATTs and speakerphones, another may interconnect seven GEMINI 100 electronic blackboards, and a third seven still-video systems. For larger conferences, the seven-port 'bridges' can be linked to interconnect up to 14, 21, or 28 voice terminals.

WHAT'S SO HARD ABOUT BRIDGING?

A conference bridge[10] is a device for connecting together many telephone lines. The major difficulties in providing good bridging are 1) compensating for network loss, 2) controlling echo from 2 to 4 wire hybrids, and 3) preventing excess noise. The network objective of 14 dB end-to-end loss is no problem in 2-location calls. When bridging three locations, if the bridge is at one of the end locations, the effective loss from one remote phone through the bridge to another remote phone will now be at least twice the previous loss or 28 dB, resulting in a listening level far below that needed for adequate reception, i.e. the remote parties can not easily hear one another. The process of amplifying a signal coming into a bridge so that it leaves the bridge at its original level is called level restoration or loss compensation. Even if the bridge is in the network[12], near the virtual loss center, where the expected loss is halved, there is still the need for level adjustments to compensate for circuit vagaries and soft-spoken talkers.[13]

Echoes create real problems in bridging. In 2-location calls the signal may echo back along a single path but it is usually attenuated enough not to be a problem. In bridges however, the signal from one location can echo back from all other locations and in some cases cause distortion, or even "singing." Echoes in bridging systems are controlled in two ways. Analog bridges use echo-cut circuitry, a type of voice switching, which places loss in the receive paths from all locations except the one that is talking. Digital bridges use echo canceler chips which subtract out the echoes without the need for voice switched loss, thus allowing "full-duplex" multiperson telephony.

The second bridge, the QUORUM Teleclass bridge is tailored to the needs of teaching and training. Specifically any combination of its 20 ports can be placed (transferred) into any of three sub-bridges. This allows a class to be split up into 3 smaller groups and perhaps reassembled later. When required, signals (speech or graphics) can be inhibited either from or between remote locations, or the whole bridge can be used completely open to conversation in any direction.

Both QUORUM bridges can be used to set up a teleconference in either of two ways. A bridge operator can **dial-out** to each remote location, and after speaking to the correct party, add this location to the conference, just by pressing the 'conference' button on the console. Or, more simply, each location can **dial-in** to the bridge and be added to the teleconference automatically. This dial-in or "Meet-Me" feature reduces the time taken to set up a teleconference; it also means that each location is billed separately for its participation in the conference call. With first-time users, particularly overseas, or when vocal screening is required, the dial-out method is recommended.

Picturephone meeting service, using a digital network combining satellite and land-based facilities, provides two-way, full-color video, voice and data transmission between specially-equipped meeting rooms. Inaugurated in 11 cities in 1982, the service is planned for 42 cities.

TELECONFERENCING TRIAL

There is no better way of testing out one's design than to really use it, for example in conducting project reviews or making a presentation to one's own management. Thoroughly believing this maxim, in 1980 we set up a joint trial of audiographic teleconferencing rooms at five Bell Labs engineering locations and two marketing locations at AT&T. The rooms contain exploratory models of the QUORUM Teleconferencing Microphone, GATT conference sets, GEMINI 100 Electronic Blackboards, and commercially available still-video systems and various displays. Nearly 2,000 working conferences have shown: 'the system works,' 'it is needed' (we had 26 conferences in one week from Denver), and 'users keep coming back' (some as many as sixty times). Room layout and equipment designs are nearly correct; however, there is more work to be done in making the equipment easier to use. Larger, crisper, brighter visual displays and fast, quiet graphics which can be bridged are the items most needed to improve teleconferencing overall.

It was the success of the prototype array microphone in this Trial which achieved its rapid transfer from research to QUORUM product development. A tribute to the success of the trial: audiographic teleconference rooms are now being installed at 23 Bell Labs locations for company-wide service.

PICTUREPHONE® Meeting Service

PICTUREPHONE Meeting Service[11] is American Bell's most advanced teleconference offering. Through two-way interactive full-motion color video it provides the dynamic face-to-face interaction not available in audiographic teleconferences.

PICTUREPHONE Meeting Rooms have seven cameras: three focussed on conference participants, one for room overview; one for 35 mm slides; one for graphics; and a seventh with remote pan, tilt, and zoom features. Each room also has three television monitors, which look very much like home television sets.

The three cameras focused on conference participants; are talker-activated. When a participant speaks, his or her image is displayed automatically on monitors in both conference rooms.

PICTUREPHONE Meeting Service can be provided between any two compatible conference rooms connected by digital transmission facilities.

A PATH TO IMPROVED PRODUCTIVITY

The QUORUM Teleconference System which we have only briefly described, is a growing line of advanced audio and audiographic communication equipment that can bring business colleagues together electronically. QUORUM teleconferencing allows groups of people hundreds or thousands of miles away from each other to meet without ever leaving the office. As business time becomes increasingly valuable, and as travel costs escalate, QUORUM teleconferencing provides a certain means of economy — in time, in money, and in human resources. No longer the wave of the future, its time is **now**.

REFERENCES

1. C. Stockbridge "The Basic Principles of Teleconferencing",

COMMUNICATIONS NEWS, December 1978.

2. D. R. Fischell, C. Stockbridge, "The QUORUM™ Teleconferencing System: Going the Distance for Business Customers", Bell Laboratories RECORD, December 1982, p. 278-283.

3. G. W. Reichard, Jr., R. L. Breeden, "The 4A Speakerphone — A Hands-Down Winner", Bell Laboratories RECORD, September 1973, p. 233-237.

4. J. A. Alvarez, W. W. Grote, C. E. Nahabedian, "Conferences and Classes Via PCT: If You Can't Come, Call", Bell Laboratories RECORD, April 1973, p. 99-103.

5. D. R. Fischell, "A New Tool For Teleconferencing", TELECONFERENCING and Electronic Communications, UWEX, Madison, WI, 1982, p. 146-150.

6. C. Stockbridge "Multilocation Audiographic Teleconferencing", TELECOM POLICY, June 1980, p. 96-107.

7. L. E. O'Boyle, P. H. Shah, G. P. Torok, "Have Blackboard, Needn't Travel", Bell Laboratories RECORD, October 1979, p. 255-258.

8. D. R. Fischell, L. Weinglass, R. Kriete, "Bridging the Needs of the 80's: The QUORUM™ Teleclass and Teleconferencing Bridges," TELECONFERENCING and Interactive Media '83, UWEX, Madison, WI, 1983.

9. H. Zenner, "QUORUM™ Teleconferencing System: Human Factors in Bridge Operation," TELECONFERENCING and Interactive Media '83, UWEX, Madison, WI, 1983.

10. R. W. Ruedisueli, "Audio Teleconferencing Transmission Systems and Considerations, "Technical Design for Audio Teleconferencing, UWEX, Madison, WI, 1978, p. 53-67.

11. D. B. Menist, B. A. Wright "PICTUREPHONE® Meeting Service: The System", TELECONFERENCING and Electronic Communications, UWEX, Madison, WI, 1982, p. 173-179.

12. J. H. Bobsin, M. Marouf, P. J. Rutkowski, "Network Services Audio Bridge", Proceedings ICC, Philadelphia, PA, 1982.

13. H. S. Schultz, "System Testing for Audio Bridge Quality," TELECONFERENCING and Interactive Media '83 UWEX, Madison, WI, 1983.

Authors

Stockbridge   Fischell

Christopher Stockbridge (coauthor, *QUORUM™ Teleconferencing: A Meeting of Minds*) is supervisor Exploratory Audiographic Teleconferencing in the Customer Systems Research Laboratory, Engineering Design and Development division of American Bell Incorporated. He joined Bell Laboratories in 1959, and has since worked on surface chemistry, medical data transmission, pattern recognition and trials of visual communications in health care and criminal justice delivery systems. Since 1975 he has focussed on voice-band audiographic teleconferencing.

Dr. Stockbridge received the, B.A., M.A., and Ph.D. degrees in metallurgy all from the University of Cambridge, and the B.A. in liberal arts from the University of Chicago. He won the Christie Prize from Magdalene College, Cambridge, and has published 34 scientific papers.

David R. Fischell (coauthor, *QUORUM™ Teleconferencing: A Meeting of Minds*) is supervisor Teleconferencing Performance Objectives Studies at Bell Laboratories. Prior to his promotion January 1, 1983, he worked with Stockbridge designing exploratory audiographic teleconferencing equipment, and he provided technical support to the audiographic teleconferencing trial at Bell Labs.

Dr. Fischell joined Bell Labs in 1979, and initially worked on acoustical testing of linear array microphones including the QUORUM Teleconferencing Microphone.

Dr. Fischell received the B.S. in engineering physics, and the M.S. and Ph.D. in applied physics, all from Cornell University.

The Teleconferencing Resource Book: A Guide to Applications and Planning
Lorne A. Parker and Christine H. Olgren (eds.)
Elsevier Science Publishers B.V. (North-Holland)
© Center for Interactive Programs, University of Wisconsin-Extension, 1984

ATMC, THE MULTI-MEDIA TELECONFERENCING SYSTEM

Dr. Clifford J. Hoffman
University of Dayton Research Institute
Dayton, Ohio
and
Mr. William A. McKinney
United States Air Force Human Resources Laboratory
Brooks Air Force Base, Texas

ABSTRACT

The ATMC (Advanced Technology - Multimedia Communications), teleconferencing system being developed by the University of Dayton Research Institute (UDRI) under the sponsorship of the U.S. Air Force Human Resources Laboratory, is designed to provide quality real-time teleconferencing over the switched telephone network. Color Business Graphics, Freeze Frame Video, Cursor Pointing, Electronic Handwriting, and Data Communications are integrated into a single highly modularized system to provide superior quality presentation support.

Full color charts and other presentation materials are prepared electronically without use of photographic media. Local conferences are conducted using CRT monitors and/or large screen video projectors to display the material. Addition of data communications enables the charts to be rapidly transmitted to other ATMC sites, and then the same electronic ATMC presentation equipment used for local conferences is used to support real-time teleconferencing.

ATMC has several advantages over other teleconferencing systems: 1) By using the public switched network, ATMC avoids the high costs and operational restrictions characteristic of full motion video systems. 2) By providing each of the most popular forms of audiographic support in a single system, ATMC adapts to a wide variety of changing user habits and needs. 3) Because ATMC provides high quality local presentation graphics support, a system can be justified independent of teleconferencing needs. Migration to ATMC teleconferencing becomes a natural extension of established user habits.

ATMC systems promise to have a similar impact on presentation graphics and teleconferencing as word processors have had on typing and electronic mail.

ATMC TELECONFERENCING CONCEPT OVERVIEW

The ATMC teleconferencing concept combines quality presentation graphics with quality audio to yield a conferencing system useful for both local and geographically distributed conferences. ATMC presentation graphics encompass most of the popular forms of visual and printed aids used in corporate conferences, formal presentations, or classroom instruction:

1. Business graphics

2. Images (freeze frame, optionally combined with graphics)

3. Handwriting (electronic overwrite of above)

4. Pointing (electronic visual pointing)

5. Printed page information.

The first four aids, displayable on a TV monitor or large screen video projector, are illustrated in Figure 1.

A high speed color computer business graphics subsystem capable of producing professional quality high resolution graphics is integrated into ATMC. Next is an imaging subsystem capable of grabbing single frame images as well as supporting electronic handwriting and certain forms of illustrated graphics. Pointing, implemented as a visual cursor, is provided to complete the list of ATMC video aids. Printed page information printed on a system printer is provided as an alternative to freeze frame or FAX transmission for document distribution.

ATMC teleconferencing requires each site to be equipped with an ATMC system. A point-to-point configuration is illustrated in Figure 2.

C.J. Hoffman and W.A. McKinney

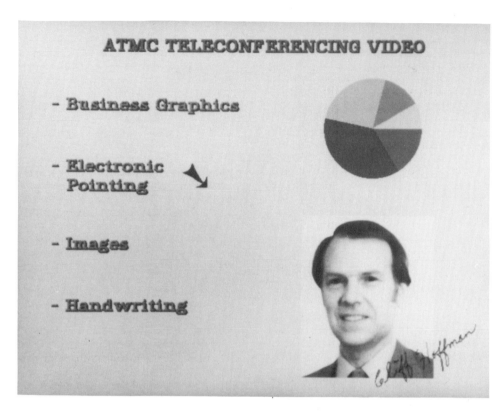

Figure 1. An ATMC Prepared Chart Showing ATMC Video Aids.

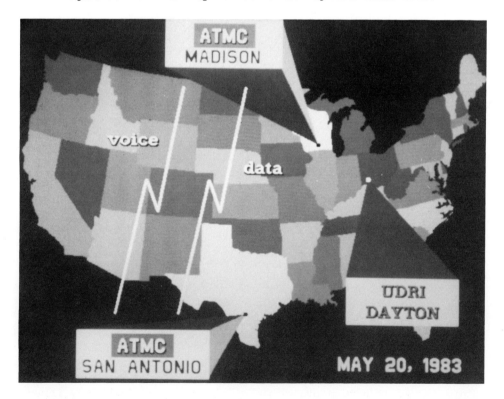

Figure 2. Configuration Used for ATMC Teleconferencing Demonstration at TELECONFERENCING
AND INTERACTIVE MEDIA '83 Conference. UDRI Participated in the Audio Portion
via a Bridge Located in San Antonio.

Because high quality audio teleconferencing poses technical problems not directly related to ATMC presentation graphics, ATMC teleconferencing audio is handled separately from the data network. Both the voice and the data network are implemented using equipment from telephone companies and other suppliers in a manner similar to methods usually used with commercial freeze frame equipment, electronic blackboards, etc. Eventually both data and voice will be integrated into a communications system which will appear to be a single network to users.

HUMAN FACTORS CONSIDERATION

Only two predominant human factor considerations are presented. The first we call "responsiveness," the second we call "user friendliness."

By "responsiveness" we mean the ability of the system to respond to the command of the user. For example, we found that anything longer than about one second to locate and display a presentation chart appears sluggish to a user and that significantly greater acceptance is achieved if the average access and display time is less than 1/3 second. Whenever possible, users must perceive of everything as happening in real-time.

"User friendliness" is a broad term which simply interpreted means that each user, whether graphics designer, programmer, or conferee must be presented with an interface requiring minimal training to perform tasks. The interface must be readily adaptable to accommodate increased user knowledge and user demand for more capabilities.

Services offered by ATMC are new to most users, and, if the average user is presented with too many options at one time, he can easily be overwhelmed. For this reason all services are introduced at an elementary level and only after the need for more services is apparent, are new services introduced.

DESIGN OBJECTIVES

Design objectives frequently emerge from experience or from ingenious hindsight. The ATMC objectives which have been with us from the onset and which continue to be our major guidelines are:

1. <u>Systems Concept</u>. In contrast to many systems which focus on a narrow segment of the corporate communications problem, ATMC was to address a much broader span of communication needs. ATMC was not to be a terminal concept, a media concept, nor was it to be simply a teleconferencing concept. It was to employ a total systems concept encompassing friendly communications needs and user friendly interfaces whether strictly local or geographically dispersed. All ATMC systems are to be communications compatible with each other within limits of user selected options/modules.

2. <u>Switched Network Operation</u>. The primary incentives behind this objective were low operating costs and the need for interactive teleconferencing between ATMC sites. The existing telephone network offers cost effective world-wide communications today, independent of future technologies. If ATMC could be made to operate effectively over the existing network, any improvements to the network will improve ATMC still more. Examples are enhancements to the switched network which offer greater bandwidth or improved reliability.

3. <u>Improved Local Support, then Migration to Teleconferencing</u>. If a system responds first to existing user needs with improved services and products, system acceptance is greatly enhanced. Although teleconferencing is the ultimate user service objective, ATMC is to first provide users with improved local conference services. Then after winning user confidence, the move to the teleconferencing environment can occur with minimal change in user habits.

ATMC SYSTEM DESIGN AND OPERATION

Within system design objectives, ATMC systems can be configured in many ways. Figure 3 illustrates the Multipurpose System as implemented by UDRI to provide all ATMC services. Other compatible configurations can be implemented strictly to support teleconferencing, others to prepare graphic media, etc.

This multipurpose system is explained with emphasis on how each component/subsystem responds to the total conference communications environment.

ATMC Controller

The ATMC controller is housed in a cabinet measuring approximately 19" W x 14" H x 22" D. A general purpose microcomputer controls all system functions, including presentation graphics preparation, teleconferencing, and hard copy media production. The presentation graphics subsystem shares the high speed computer data bus which enables data to pass

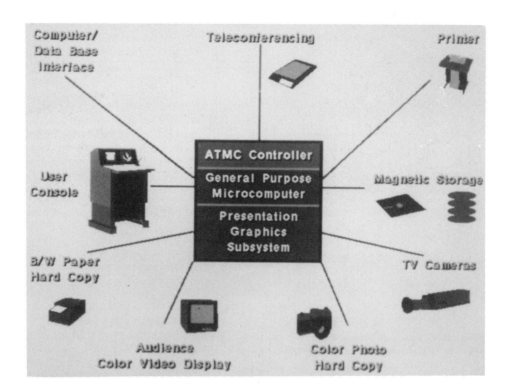

Figure 3. ATMC Multipurpose System. Each System Element Plays an Important Role
in the ATMC Concept.

freely to/from main memory and the magnetic
disk peripherals. The system is self-contained
inasmuch as all software needed for color
graphics slide preparation, freeze frame imaging, handwriting, data communications, etc. is
stored on magnetic disk and is accessed by the
microcomputer as needed for specific tasks.
There is no dependence on a larger central computer.

For a teleconference, slide presentation
material is preferably sent ahead of time to
other sites via the teleconferencing data interface* and placed on magnetic disk. Last
minute changes, including insertions and deletions are transmitted just prior to the teleconference, depending on the extent of the
changes. During the teleconference, presentation graphic slide material is taken from disk
and displayed as needed. Only simple commands,
cursor movement, handwriting, and other extemporaneous information (e.g., impromptu freeze

frame images) is transmitted during a teleconference.**

User Console

The prototype user console illustrated in
Figure 4 was designed and fabricated by UDRI
for multiple uses. In its raised stand-up
position, it is used for formal large group
presentations and conferences. In its lowered
position it is used either for media development or for small group conferences/telecon-
ferences. Because of the versatility of the
console, it will only be described in relation
to how it is used during a teleconference.

---

*Other data networks or public mail of disk-
ettes can also be used.

**Background communications will enable presen-
tation material to be transmitted during conference dead time. In a normally structured
teleconference, natural pauses will enable
presentation material to arrive at remote
sites by the time it is needed.

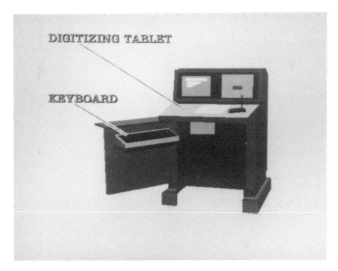

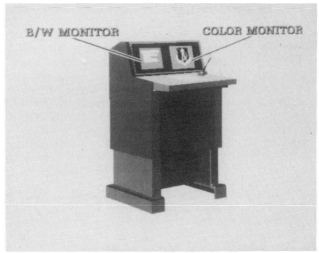

(a)   (b)

Figure 4.   ATMC User Console. (a) Sitting Position. (b) Stand-up Position. The
Keyboard is Normally Stored Out of Sight During a Teleconference.

The B/W monitor and the digitizing tablet are the two standard console components. The B/W monitor is an operator monitor on which all operator prompts, backup notes and backup data, help messages, screen menus, etc. are displayed. If desired, information on this monitor can be shifted to the color monitor and/or large screen video projector for audience viewing.

The total control of the system is accomplished using the digitizing tablet. Overlays such as those shown in Figure 5 are placed on the tablet and alignment points are touched with the stylus. Operating commands are then selected by touching function key areas. Cursor movement is controlled with XY key areas. The combination of digitizing tablet, overlays, and associated control software is called the "ATMC soft keyboard."

Overlays are designed to match operator needs and skills. If only a few features are needed, the overlays are made very simple. As more features are required, the overlays become more complex. For example, the overlay in Figure 5(a) provides only simple slide and cursor control. The HELP key is used to assist the operator in learning functions. If first the HELP key, and then one of the other keys are touched, an explanation of key operation is displayed on the operator monitor.

The more complex overlay in Figure 5(b) is used in an identical manner. It has keys which enable the operator to randomly jump to any slide, access notes and data, make B/W hard copies, etc. The ATMC soft keyboard, in combination with the information displayed on the

B/W and color monitors, offers extensive opportunity for meeting our objectives for user friendly interfaces and also for providing additional functions as they are introduced.

The integrated color monitor adds significantly to the ease of console operation. However, it is only mandatory when there is no other color display viewable from the user position. Information displayed on the color monitor is identical to that seen by conference participants at all sites.

The alphanumeric keyboard is used for system setup, but it is not normally used during a teleconference. In exceptional cases where complex operations require assistance from a programmer, the keyboard is placed along with an additional B/W monitor in a convenient conference room or support room location. Verbal contact with the keyboard operator is then used to handle the requests. An example of such keyboard use may be when it is essential to make an on-the-spot change to an important color graphic slide.

Audience Color Video Display

Good quality high resolution RGB (RS170) monitors or large screen video projectors are recommended. For very small groups, the small color monitor in the user console is adequate. For larger groups of 5 to 8 people, a 19-inch diagonal monitor serves very well.

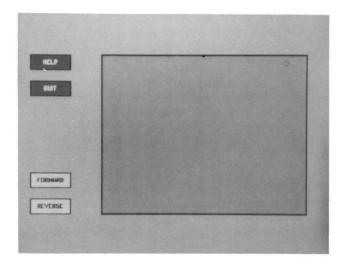

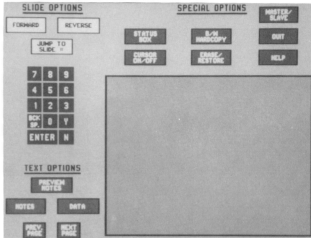

(a)   (b)

Figure 5. Examples of Tablet Overlays. (a) Simple Overlay. (b) Overlay With
           More Functions.

Large screen video projector technology which provides affordable high resolution video images of 5 ft diagonal and larger is progressing at a very rapid pace. As these projectors become more widely available, it is anticipated that they will be preferred over standard video monitors, even for small groups. Conferees appear to prefer large screen video projection in the same way they have traditionally shown preference for large screen projection of 35 mm slides and overhead transparencies.

B/W Paper Hard Copy

The black and white video paper hard copy unit serves two major functions: 1) to provide draft copy of presentation materials, and 2) to provide on-the-spot copies of presentation materials to conferees.

Presentation materials most often originate from author notes and sketches. This input is then converted to presentation charts by an ATMC operator. Upon completion, the original sketches are returned to the author along with B/W video copies for proofreading. Corrections and changes are marked on the B/W copies before being returned for final entry. Color editing is handled at the operator console or through handwritten notes, if needed.

To illustrate the second function, imagine a teleconference where a delivery schedule is being intensely debated. Finally, all parties come to terms and electronic handwriting changes are made on the displayed schedule.

Even signatures can be added to denote commitments. B/W video copies are then made of the "Decision-Time Edited" (DTE) chart and given to conferees. The DTE chart showing the "decision audit trail" has greater meeting value in most cases then an updated color graphic chart.

Color Photo Hard Copy

Users continue to demand photomedia for compatibility with existing projection equipment, publication, and versatility for travel. Many meetings simply do not allow use of anything but photo media. For these reasons ATMC provides system compatible photo hard copy.

ATMC color photo hard copy is separated into two resolution categories: the first is raster video with a resolution of 640 x 480 pixels. The second is professional quality with a pixel resolution greater than 1600 x 1200 pixels. The illustrations published in this paper are all raster video resolution. With certain allowances for Figure 1, each illustration could easily be produced in high resolution format using the same ATMC source code as was used to prepare the published illustrations.

Both resolutions can be produced on a variety of film formats, including 35 mm slides and overhead transparencies. Advantages of the raster video resolution are lower cost and more rapid turnaround. The high resolution media is more suited to graphic arts department production. It offers advantages of greater graphic detail and smoother lines.

TV Cameras

Up to four TV cameras can be connected to the ATMC controller for input of freeze frame images. When needed, each camera is provided with a separate camera control overlay area on the soft keyboard. Only one camera can be used at a time.

Magnetic Storage

The amount of magnetic disk storage provided on an ATMC system depends greatly on anticipated use. Ten to twenty megabytes of Winchester type storage is adequate for smaller systems with limited imaging capability. With increased emphasis on freeze frame image storage, the need for disk storage, including removable cartridges, can exceed 100 megabytes.

The diskette drive is a general purpose interface device used for presentation chart storage and program loading. One diskette can easily hold several hundred color graphic charts, whereas it may only hold four or five freeze frame images, depending on image complexity. The diskette can be used as a transportable media between ATMC sites.

Printer

This general purpose printer can range from an inexpensive character printer to a high speed line printer. It is used for ATMC program and presentation material preparation, general purpose record keeping, and for printing converence materials when necessary.

Teleconferencing Data Interface

This interface is used to transmit presentation graphic materials via the telephone network to other ATMC sites, and then during a teleconference, it is used for control and extemporaneous graphic information such as handwriting, freeze frame, and cursor pointing. Other uses for this interface may be to signal an exuberant talker at a distant location that you would like to interrupt. Printer data (electronic mail?) can also be passed between sites, printed, and then used during a teleconference.

Computer/Data Base Interface (Option)

In many organizations, meetings involve presentation of a great deal of data which is simply updated periodically in standard graphical format. The data frequently resides in a large central computer data base, or it may reside in a small personal computer.

To facilitate use of this data for automatic update of standard graphic charts on an ATMC system, standard data communications interfaces are provided by ATMC. In some cases, the graphic charts can be completely prepared on an external computer and then simply passed to an ATMC system via the interface (or via diskette) for teleconferencing use. Another method is to pass the raw data to the ATMC microcomputer and then reduce it to graphical form using ATMC general purpose graphics software.

In any case, the problem of interfacing to user data bases must remain a user problem. ATMC simply supplies the tools to make the interface as painless as possible.

Other Options

The system design of ATMC using a general purpose microcomputer coupled with the software written by the University of Dayton Research Institute provies the expandability required to add optional peripheral devices such as a flat bed plotter, printer plotter, 35 mm slide projectors, multiscreen capabilities, and other options as desired.

PROGRESS TO DATE

ATMC is an ambitious development undertaken and promoted by the University of Dayton Research Institute in concert with the U.S. Air Force Human Resources Laboratory. Following delivery of the first ATMC system to Brooks Air Force Base in May 1982, and then with the demonstration of ATMC teleconferencing in September 1982, interest in ATMC by the U.S. Air Force and other organizations has increased.

More test sites will be added this year which will provide valuable user feedback from operational teleconferences. Acceptance of the color graphics features so far has been very good and these features alone have generated major interest. Other ATMC features such as freeze frame and handwriting are still in the laboratory developmental phase. Few users have previous experience with communication features such as these, much less in a system where they

are all available simultaneously. Our expectations are to win user confidence with quality
local presentation support and then, using this
learning experience, comfortably convert users
to the teleconferencing environment.

THE FUTURE

Fully interactive ATMC multipoint teleconferencing will be developed in the near future.
Most Air Force commands and laboratories are
geographically located at three or more locations. Whereas periodic briefings traditionally
require travel to a central site, ATMC multi-
point will enable these briefings to be conducted with greater ease and without requiring
travel.

ATMC multiscreen capability is a straightforward outgrowth of ATMC architecture. This
development which will provide control of both
raster video and photo media projectors will
enable strong points of both projection technologies to be utilized.

The ATMC concept, because of its emphasis
on user friendly interfaces, will provide single
action call establishment to remote sites. For
example, once presentation material is prepared,
a single touch of an overlay key will establish
both voice and data channels to remote sites.
Whether these two channels are physically separate or integrated in a voice/data network, the
ATMC controller will insure total link integrity
throughout the teleconference.

Few developments proceed with greater
awareness of the need to establish interface
standards than the ATMC development. The potential problems with ATMC look-alikes which are
unable to exchange information in a teleconference are analogous to the codec incompatibility
problems evolving in full motion video. In some
cases it is a problem of establishing or forcing a standard where none exists. Increased
emphasis will be given to development and dissemination of ATMC standards.

A growing information resource is developing around Videotex standards. Many organizations have already made sizable commitments to
defacto graphic standards such as Tektronix
Plot 10. For ATMC to achieve maximum user
acceptance, friendly interfaces to standards
such as these are almost certainly in ATMC's
future.

If commercialized today, each ATMC system
would sell for $75,000 to $150,000 depending on
features. Because of recent technology improvements and greater understanding of design concepts, a price range of $50,000 to $100,000 may
not be far away. ATMC prices will undoubtably
follow a price reduction path similar to most
other high technology products.

The Teleconferencing Resource Book: A Guide to Applications and Planning
Lorne A. Parker and Christine H. Olgren (eds.)
Elsevier Science Publishers B.V. (North-Holland)
© Center for Interactive Programs, University of Wisconsin-Extension, 1984

THE FUTURE OF VIDEOCONFERENCING:
THERE'S MORE THAN THE MEETING TO CONSIDER

Karen File
Vice President
National Analysts

Robert Forrest
Market Staff Analyst
AT&T Long Lines

Alan Kuritsky
Staff Manager
AT&T Long Lines

Sally Leiderman
Study Director
National Analysts

The thought of the videoconference becoming an important tool for corporate America in the 1980's seems quite logical in these times of expensive energy and lagging productivity.

What better way to enhance white-collar productivity than to take executives off the airplane and into the meeting room for a videoconference with peers a thousand miles away? And, what better way could there be to reduce soaring business travel costs and conserve energy than to let the videoconference replace face-to-face meetings whenever possible?

At first thought, the future of videoconferencing in the United States seems secure. A closer look, however, reveals that there are much deeper considerations -- particularly about the nature of the potential user company -- to consider before any wide-scale implementation of videoconferencing can approach reality. And these issues must be considered even before budgetary considerations can be made.

The following report presents some of the salient conclusions about the potential acceptance of videoconferencing reached in a large-scale nationwide market survey of 2,000 general business managers. The survey was conducted by the National Analysts Division of Booz·Allen & Hamilton, the international management and technology consulting firm, for AT&T Long Lines. National Analysts is a market research organization consisting of research professionals with wide-ranging expertise, including telecommunications specialists.

As part of their training for this assignment, many of the interviewers participated in videoconferencing sessions. And, those who interviewed telecommunications specialists were specifically trained in the language of teleconferencing so they could discuss the subject and all of its technical aspects on a peer basis.

While the survey concluded a number of major findings, it is important to note just some of the highlights that could have a significant impact on the future course of videoconferencing:

. The characteristics of the potential user are more important in predicting a company's likely acceptance of the videoconference than the characteristics of the meetings themselves.

. The usual format or type of meeting that would be subject to a videoconference is not really essential or helpful in predicting the future acceptance of video as a business tool.

. There is an underlying phenomenon that ties together both the importance of the characteristics of the potential user and the usual type of meeting that might be videoconferenced. This leads to a very fundamental question: Is this meeting in this environment with these users videoconferenceable? Until now, research in the field centered almost exclusively on the characteristics of the meetings in question. The new research shows, however, that a greater depth of understanding is essential.

Another major point the survey indicates is that managers believe that videoconferencing is equally applicable for meetings in which all participants are from the same company and for meetings in which at least one participant is from outside the firm. Until now, videoconferencing had been considered almost exclusively as an internal communications tool.

In order to anticipate the market acceptance of videoconferencing, two possible information sources are available:

. The opinions of those exposed to videoconferencing demonstrations (Project Prelude, Bell Canada, Picturephone🔾 Meeting Service)

. The opinions of those not exposed to videoconferencing demonstration projects who nonetheless constitute the primary market for the capability

Most investigators have focused on the first group, and have provided the field with a number of valuable insights. A relative few have chosen to explore the attitudes of the second group.

The Videoconferencing Meeting

The question of whether business meetings are suitable for videoconferencing has been explored extensively. New users of video have typically had positive feelings toward the use of videoconferencing (Duncanson and Williams, 1973; Williams and Holloway, 1974; Champness, 1973; Ellis, McKay, and Robinson, 1976; Reid, 1970). Some studies have described how individuals behave differently when they experience videoconferences than they would have in a face-to-face meeting. Other studies have looked at how this behavior change can be successfully utilized in a business setting.

A number of common business problems, such as exchanging information, asking questions, offering opinions, problem solving and generating ideas, have all been demonstrated to be satisfactorily accomplished using video (Champness, 1973; Williams and Holloway, 1974; Jull and Mendenhall, 1976). The type of group which assembles for a video meeting also does not seem to be a key determinant of whether or not the meeting can be successfully teleconferenced (Noll, 1976 and Strickland, Guild, Barefoot and Patterson, 1975). Furthermore, videoconferencing seems less likely to cause groups to split into leaders and followers than meetings held in person (George, Coll, Strickland, Patterson, Guild, and McEown, 1975).

Individuals do not seem to have a specific problem persuading other people via video (Short, 1972). One problem that can occur in business meetings when using video, however, is the tendency for strangers not to get to know one another as well as they would in a face-to-face meeting (Christie and Holloway, 1975; Jull and Mendenhall, 1976). Also, meeting participants may express themselves less personably (Short, Williams, and Christie, 1976), or may behave more stiltedly than they would act in a business setting (Bretz, 1974).

The issue of meeting efficiency also becomes significant when videoconferencing. Williams and Holloway (1974) found that video meetings seem to be conducted more quickly than face-to-face meetings. In fact, Noll (1976) found video to be perceived as more satisfactory than face-to-face for handling regularly scheduled communications and for giving or receiving information. Moreover, Champness (1973) and others found that video meetings are perceived to be more orderly than face-to-face meetings. Finally, a number of non-published studies conducted by AT&T suggest that the types of meetings where video communications could result in a highly productive meeting included:

. Short meetings where a break was not required

. Meetings that have specific purposes and goals

. Meetings where agendas are preplanned and highly structured

. Small group meetings (2 to 4 participants maximum at each end)

. Meetings in which participants know each other in advance

. Meetings in which the seating of participants is preplanned so that individuals having common interest or problems sit next to each other

The survey analyzes the opinions of a cross section of business people at different levels of business. Participants were asked to present their views of their present meeting behavior and their reaction to using video in these meetings. Before examining their current and anticipated meeting behavior, it is appropriate to describe the study methodology.

Methodology

Data were taken from a sample of about 2,000 businesspeople (middle, upper and top managers as well as technical and professional staff selected from a list of business leaders) who were interviewed by telephone. The managers surveyed represented a thorough cross section of managerial levels, functional areas, and ages. This breadth of rank, function and work experience was deemed essential in order to gather information that would not be biased in advance.

The methodology called for respondents to identify the most recent meeting they had attended with three or more participants, where at least one person had to travel 30 minutes from the office to attend. This type of meeting was assumed to meet the minimum standards for a "videoconferenceable" meeting.

We also asked managers to describe the salient characteristics of this "videoconferenceable" meeting -- the purpose of the meeting, its length and whether documents or other materials were used. Once these characteristics were enumerated, and the meeting event was crystallized in the respondent's mind, we described the videoconferencing capability and asked for an estimate of the probability of using video for that recent meeting. (All interviewers read the same concept statement of videoconferencing to respondents.)

Current Business Meeting Behavior

Before analyzing the amenability of different types of business meetings to videoconferencing and the willingness of managers to choose a videoconferencing mode, it is appropriate first to describe current business meeting behavior.

As one might expect, top, upper, and middle managers attend the largest number of videoconferenceable meetings; first-line supervisors and junior or senior staff professionals attend fewer videoconferenceable meetings in a month. Predictably, individuals in marketing and sales attend more videoconferenceable meetings than do managers in other functional areas. Meeting attendance was also discovered to vary by industry segment and, to a similar extent, by company size. For example, sales and marketing managers attended more recent videoconferenceable meetings than did those in manufacturing

or distribution areas of responsibility. By far, the most frequent reason to hold a videoconferenceable meeting is to exchange information. Other meeting purposes -- negotiating or bargaining, buying or selling, resolving a crisis, and introducing new people to each other -- are comparatively infrequent.

Managers of different levels or areas of functional responsibility appear to attend the same proportion of meetings of each type.[1] The types of videoconferenceable meetings do not vary significantly by respondent or by company characteristics.

For all levels of managers, the most recent videoconferenceable meeting was perceived to be "somewhat" or "extremely" important[2] prior to the time it took place. Slightly more top-level managers regarded the most recent meeting as "extremely important" as opposed to "somewhat important." By contrast, managers below the topmost level chose "somewhat important" over "extremely important" in about the same proportion. The modest difference may well reflect the "buck stops here" attitude at the highest management level.

Similarly, more top-level managers attached a rating of very high or very low risk associated with the most recent videoconferenceable meeting than did their associates lower in the corporate hierarchy. Most responses by first-level to upper-middle managers indicated generally moderate risk.

Not surprisingly, the number of participants at videoconferenceable meetings varies with company size. More people assemble to work together in larger organizations. The length of meetings also varies with the number of attendees; the addition of more people, presumably generating a larger volume of thoughts and ideas and possessing a need for expanded communication and interaction, results in more time needed to complete a meeting.

Meeting length is impacted by the type of videoconferenceable meeting. Typically, buying and selling occupy the shortest time frames, while negotiating and bargaining take longer. Crisis resolution is more evenly distributed across the duration categories.

Slightly more than half of the videoconferenceable meetings involved personnel from other than the respondent's firm. Interestingly, the average meeting length is shorter for those meetings involving managers from different companies than for intra-company meetings. Extremely important meetings are finished quickly.

Graphic aids -- ranging from handouts to movies -- are used in approximately two-thirds of the meetings held. Virtually none of the managers surveyed dismissed graphics as unimportant, and a large number of the respondents termed them essential for accomplishing the goals of the last meeting.

One of the survey questions had to do with how often the respondents who ever attend videoconferenceable meetings must travel overnight in order to participate. Answers show that first-line managers travel overnight seven times per year to attend meetings, while respondents at higher management levels travel twice as frequently.

Curiously, the willingness to travel is independent of the amount of travel one does. A majority of business managers stated that they preferred to travel about as often as they do now, regardless of how much business travel they currently do. A smaller, but by no means inconsequential group, would prefer to travel less often, while only a miniscule proportion of respondents expressed a desire to travel more frequently. The desire to travel or not to travel seems to be explicable by personality variables rather than by the professional level or functional area of responsibility.

Anticipated Videoconferencing Meeting Behavior

Business managers believe videoconferencing could fill a wide range of business needs. While the medium is preferred more for some applications than others, the types of meeting characteristics just described do not appear to substantially affect the likelihood that videoconferencing would or would not be used for a specific meeting. Rather, there seems to be an underlying phenomenon that helps to explain why some kinds of businesspeople would choose videoconferencing for some kinds of meetings. Before elaborating on the specific nature of this phenomenon, it is helpful to look at business managers' opinions about the potential use of videoconferencing.

Business managers are willing to use videoconferencing for business meetings with different purposes, with one exception. While they use the videoconference about as

often for meetings to exchange information, negotiate or bargain, or buy and sell, they would be less likely to use this tool for meetings whose primary purpose is to introduce new people to each other. In this regard, preferences of potential users mirror the opinion of videoconference experts.

In addition to the purpose of the meeting being relatively inconsequential to the acceptance of videoconferencing, the number of participants and the length of the meeting are not important considerations. Business managers believe videoconferencing has equal application for meetings with from 3 to more than 13 participants, although meetings with 7 to 12 participants are slightly more likely to be viewed as appropriate for videoconferencing.

Likewise, business managers feel videoconferencing is equally applicable for meetings where all participants are from the same company, and for meetings where at least one participant is from a different company.

It is also interesting to note the various meeting lengths for which managers believe videoconferencing could be used. Survey respondents thought videoconferencing was best suited for meetings lasting between one and one-half hours and three hours. At the same time, much shorter and much longer meetings were also considered videoconferenceable.

As one would expect, businesspeople think videoconferencing would be used much more frequently for meetings in which audiovisual aids were used than in meetings where they were not. However, the exact type of audiovisual aid makes little difference in managers' perceptions about which meetings would be held on the medium. Thus, while some slight preferences emerge, the specific characteristics of the meeting, such as meeting length, number of participants and the like, are not the determining factors in deciding where the medium could be used.

General managers who are content with the amount of travel they do are less likely to think videoconferencing would be helpful than either those who wish to travel more or those who wish to travel less. Thus, business managers who are discontented with their current business meeting behavior are more likely to see potential for videoconferencing than those who are complacent.

The perception of business managers, then, is that videoconferencing could have been used for meetings of various purposes, with differing numbers of participants, and lasting between a few minutes and several hours. However, certain kinds of businesspeople are more likely to use the medium, regardless of the type of meetings they attend.

Specifically, engineers, professionals and technicians consider themselves likely users, while top managers do not.

Conclusions and Implications

What do these opinions of the mass market mean for the future of videoconferencing? In order to answer that question, it is necessary to understand why businesspeople react to the concept of videoconferencing as they do. Specifically, it is important to speculate on the reasons that characteristics of users are more important than characteristics of meetings.

We believe that the function of business meetings can be described by a continuum between two polar ideas. On one hand, there are meetings with the primary objective of establishing contact between meeting participants, perhaps for the purpose of establishing or continuing a relationship, or reassuring someone that a problem or issue is being "looked into." Or, perhaps the meeting itself serves as a reward. These types of meetings may be attended by persons of differing rank, where status and deference symbols are important. Audiovisual aids may or may not be needed for such meetings. However, these materials would be characterized by lack of task orientation. Meetings at this end of the continuum can be referred to as "contact-oriented" meetings.

At the other end of the spectrum are meetings whose function is to get something specific accomplished -- "task-oriented" meetings. Such meetings will typically involve businesspeople in more lateral positions, who have worked together in the past. Where audiovisual materials are used, they are likely to be important to getting the task accomplished. The meeting will be successful only if it accomplishes its goal, unlike "contact-oriented" meetings that are successful merely by virtue of having been held.

We believe videoconferencing is more likely to be used for "task-oriented" meetings than for "contact-oriented" ones. One reason is that videoconferencing tends to equalize the status of the various attendants. Because all participants in a videoconference are outside their usual office environment, certain status symbols like outsize chairs, awards or other wall ornaments and the like are not available. Further, the standard videoconference room configuration does not allow persons to dominate the meeting in traditional or accustomed ways through physical placement. In a videoconference, participants sit along one side of a table. In non-videoconferenced meetings, leaders gravitate to the "more powerful" end positions. In addition, each person that attends a videoconference must travel at least some distance to participate, even if the room is only down the hall. Thus, the normal deference distinctions between host and guest are obscured. Finally, contact-oriented meetings are often held because someone wants to "press the flesh." These types of meetings have a tactile, as well as an oral and visual, component.

It should be clear from this discussion that videoconferencing is not well-suited meetings that are primarily contact-oriented. What is a barrier for contact-oriented meetings, however, is a benefit for task-oriented ones. First of all, status symbols, tactile interaction and deference patterns are not important for these kinds of meetings. In fact, the absence of these features may act to promote efficiency. Secondly, there is some evidence that videoconference meetings are shorter, more structured, and more to the point, as task-oriented meetings should be. Finally, the very nature of the medium encourages its use for meetings that require the participants to see something, such as blueprints, tables, objects and the like. The need for audiovisual capability is more closely associated with task-oriented meetings than contact-oriented ones.

Business managers' opinions about the likelihood that videoconferencing would be used for their last meeting can be understood better in terms of this continuum. That is, while the continuum is defined in terms of meetings, it does not rely on specific meeting characteristics, such as length, number of participants, or even purpose as previously defined. Rather, a specific meeting falls in its place on the continuum because of an interaction between the goal of the meeting, and the personality or style of the primary meeters, e.g., the person who calls it, structures it and/or runs it. Thus, in answering the question whether "this" meeting (as defined by overt purposes), in "this" environment with "these" users is videoconferenceable, this additional dimension has considerable significance.

Footnotes

[1] Among the exceptions we find: sales and
marketing professionals attend more sales
meetings; top-level managers enter into
more negotiations; and staff professionals
and technicians schedule more meetings for
the purpose of information exchange. Crisis
resolution occurs among the levels of staff
professionals through upper-middle managers;
interestingly, top management has far less
need to mobilize to turn around a crisis
solution.

[2] The relative degree of "importance" of the
most recent meeting compared to other
meetings was rated by respondents as being
"extremely unimportant," "somewhat unimportant," "somewhat important," or "extremely
important."

References

Bretz, Rudy, 1974. Two-Way TV Teleconferencing for Government: The MRC-TV System, The Rand Corporation, Report R-1489-MRC.

Champness, Brian, 1973. The Assessment of Users' Reactions to Confravision, Communications Studies Group, London, England, Paper E/73250/CH.

Christie, Bruce, and S. Holloway, 1975. "Factors Affecting the Use of Telecommunications by Management," Journal of Occupational Psychology, Vol. 48, pp. 3-9.

Duncanson, James P. and Arthur D. Williams, 1973. "Video Conferencing: Reactions of Users," Human Factors, Vol. 15, No. 5, pp. 471-85.

Ellis, Susan, Vince McKay, and Michael Robinson, 1976. A Preliminary Report of the Follow-Up Study of Users of the Melbourne-Sydney Confravision Facility, Swinburne Institute of Technology, Australia.

George, Donald A., D. C. Coll, L. H. Strickland, S. A. Patterson, P. D. Guild, and J. M. McEown, 1975. The Wired City Laboratory and Educational Communication Project, 1974-75, Carleton University, Ottawa, Ontario, Canada.

Jull, G. W., and N. M. Mendenhall, 1976. "Prediction of the Acceptance and Use of New Interpersonal Telecommunication Services," in Lorne A. Parker and Betsy Riccomini, eds., The Status of the Telephone in Education, Madison: University of Wisconsin-Extension Press.

Noll, A. Michael, 1976a. "Teleportation through Communications," IEEE Transactions on Systems, Man, and Cybernetics, pp. 753-56.

Noll, A. Michael, 1976b. "Teleconferencing Communications Activities," Communications Society, Vol. 14, No. 6, pp. 8-14.

Reid, Alex A. L., 1970a. Electronic Person-to-Person Communications, Communications Studies Group, London, England, Paper B/70244/CSG.

Reid, Alex A. L., 1970b. The Costs of Travel and Telecommunication, Communications Studies Group, London, England, Paper P/70220/RD.

Short, John A., 1972a. Conflicts of Opinion and Medium of Communication, Communications Studies Group, London, England, Paper E/72001/SH.

Short, John A., 1972b. Medium of Communication, Opinion Change, and Solution of Problem Priorities, Communications Studies Group, London, England, Paper E/72245/SH.

Short, John A., 1972c. Medium of Communication and Consensus, Communications Studies Group, London, England, Paper E/72210/SH.

Short, John A., 1972d. Telecommunications Systems and Negotiating Behavior, Symposium on Human Factors and Telecommunications, Stockholm, Sweden.

Short, John A., Ederyn Williams, and Bruce Christie, 1976. The Social Psychology of Telecommunications, London, England: John Wiley & Sons, Ltd.

Strickland, L. H., P. D. Guild, J. R. Barefoot, and S. A. Patterson, 1975. Teleconferencing and Leadership Emergence, Carleton University.

Williams, Ederyn and S. Holloway, 1974. The Evaluation of Teleconferencing: Report of a Questionnaire Study of Users' Attitudes to the Bell Canada Conference Television System, Communications Studies Group, London, England, Paper P/74247/WL.

The Teleconferencing Resource Book: A Guide to Applications and Planning
Lorne A. Parker and Christine H. Olgren (eds.)
Elsevier Science Publishers B.V. (North-Holland)
© Center for Interactive Programs, University of Wisconsin-Extension, 1984

MRC-TV - A LARGE SCALE VIDEO TELECONFERENCING SYSTEM
"ANATOMY OF FAILURE"

Albert J. Morris, President
&
Charles Martin-Vegue, Vice-President

Genesys Systems, Inc.
Palo Alto, California

Introduction

The Metropolitan Regional Council (MRC) of New York, New Jersey and Connecticut is one of about 700 Councils of Government in the U.S. The region covers some 5,000 square miles and has a population of about 16 million people. It was formed to provide a forum for discussion of problems which cross State, County and City boundaries: problems such as transportation, the environment, public safety, law enforcement and drug abuse. In the late 1960's, MRC conceived of a two-way interactive video teleconferencing system which would enable members of the participating governments to meet without having to face the time, frustration and even danger of traveling to and from meetings in this most densely populated area of the U.S. This concept led to MRC-TV, which was made operational in June, 1973 and ceased operations at the end of June, 1979. Technically, the television system performed as expected. Failure of MRC-TV was due to lack of a clear understanding of how such a system needed to be used, lack of a plan for full utilization and commitment, lack of expertise and imagination by the system managers, and weak management.

MRC-TV was a technically unique system. The system was even financed in a very unique way. This paper treats what and how things were actually done, along with suggestions of what and how things probably should have been done based on experience elsewhere. The reasons for failure, along with possible cures for this, are pinpointed. Video teleconferencing is in its infancy. Failure of a major system can severely inhibit future progress unless the reasons for this failure are understood and unless possible paths leading to success are clearly delineated.

Background

Teleconferencing is a relatively old concept. Its potential to overcome the inhibiting effect of geographic barriers has been recognized for a long time. And, a number of private teleconferencing systems have been in successful use for years. The application of teleconferencing in general to alleviate urban communications problems and to provide training for government employees has been treated extensively[1,2,3,4]. The metropolitan region surrounding New York City is the most heavily populated in the U.S.; it includes some 550 municipalities in the Tri-State area within a 40-mile radius from the tip of Manhattan. It should have been a perfect test vehicle for a teleconferencing system. This highly congested area formed a Regional Council in 1956. Early efforts to use telephone teleconferencing were considered unsatisfactory. The number of participants in different locations was large and much time was wasted in the speakers constantly having to identify themselves.

The two-way TV network was the brainchild of the first MRC President, Robert P. Slocum[11], who developed the idea several years ago while driving back from a meeting with government officials in Linden, New Jersey. Recalls Slocum: "I got caught in one of those tremendous traffic jams and I thought that there's got to be a better way for us to communicate without leaving our offices or using the telephone. The idea for the Tri-State TV network was born at that instant." Armed with a $60,000 grant from the Department of Housing and Urban Development, MRC was able to conduct site surveys, hire a consulting engineer, and conduct financial and technical feasibility studies.

In those early days (about ten years ago), it was contemplated that TV links would be established with the State capitals in Albany, N.Y., Trenton, N.J. and Hartford, CT as well as Washington, D.C. This was to enable local government officials to have face-to-face meetings with state and federal representatives and receive assistance in obtaining grants. Also, when the TV system was planned, a typical week's program schedule was to look like Figure 1. Note that programming was contemplated from 10:00 a.m. to past 8:00 p.m. all during the week and that the programming was to involve interaction between officials in many different areas of

	Monday	Tuesday	Wednesday	Thursday	Friday
10 AM - 12 NOON	Chief Elected Officials	Social Service Commissioners	Health Commissioners	Narcotics Prevention Programs	Planning Directors
12:30 - 2:30 PM	Sub-Regional Meetings	Sub-Regional Meetings	Sub-Regional Meetings	Sub-Regional Meetings	Sub-Regional Meetings
2:45 - 4:45 PM	District Attys. and County Prosecutors	Federal Aid Programs HEW	Data Processing	State Aid Programs	Federal Aid Programs HUD
5-6 PM	Law Enforcement	Law Enforcement	Law Enforcement	Law Enforcement	Law Enforcement
6-8 PM	Education	Education	Education	Education	Education
8 PM	Emergency Planning and Police	Emergency Planning and Police	Emergency Planning and Police	Emergency Planning and Police	Emergency Planning and Police

FIGURE 1 - TYPICAL PROPOSED PROGRAM SCHEDULE - MRC-TV

government. Education was planned to be less than 20% of the total effort.

MRC-TV System Description

As conceived, MRC-TV was to provide for two-way interactive video teleconferencing. Programs could originate at any remote two-way site or at the Central (command and control) site at the World Trade Center Building in New York City. TV programs originating at any two two-way sites could be seen by all sites at one time by means of split-screen television. A remote two-way site could repeat programs received from the Central site transmitter to enable sharing the programs with one-way receive-only sites in the local areas surrounding each remote site. Alternatively, each remote two-way site could originate its own program which could be shared only with its local community, or shared as well with the Region through MRC-TV or transmitted only to the Region and not locally. The Central site could act as a passive repeater of remote two-way site signals, or it could originate a program and become a primary transmission site, or it could, by split-screening, transmit its own program along with that of any one remote two-way site. The Central site controlled which of the programs were to be shared by all system participants.

Figure 2 shows the geographic distribution of all sites. Figure 3 shows the Central site command and control console. Figure 4 shows the Central site studio. Figure 5 shows a typical remote site studio, and Figure 6 shows a typical operational remote site meeting. A detailed technical description of MRC-TV and its design is given in Microwave System News[5].

MRC-TV Financing

MRC was not a government agency, it was a consortium of governments. As such, it had no tax base and relied on dues paid by the participating municipalities plus income from government grants and contracts to conduct its operations. It had no capital funds with which to buy the TV system ($750,000), but it received commitments initially from eleven governmental units to install two-way interactive sites and pay MRC $14,000 per year per site. These commitments were sufficient to enable MRC to enter into a ten-year lease agreement with Genesys Systems, Inc. of Palo Alto, California to provide and install the equipment.

Due to municipal law, no governmental agency may sign a contract for services for a period of longer than one year. Therefore, all of the contracts between MRC and the local muni-

cipalities were for the period of one year. However, based on the enthusiasm with which local government officials had responded to the new communications system, it appeared highly doubtful that a community would agree to have antennas and towers installed at their building and an entire studio installed within their building and cancel their agreement with the Council after only one year. Genesys was willing to take this risk on the basis that payments received by MRC for use of the TV system would be used to make the lease payments. In practice, none of the municipalities dropped out during the first five-year period of MRC-TV operations.

In addition to the original eleven participants, at least seven more two-way site participants were projected along with at least fifty one-way (receive-only) sites, each of which was expected to pay $2,000/year. Also, MRC committed to use funds from numerous other sources of income to make its lease payments if necessary. In essence then, in Genesys, MRC found an organization willing to provide the front-end financing and capable of designing and installing a sophisticated communication system.

MRC-TV Studies

The MRC-TV concept was studied almost entirely with funds provided by NSF[6,7,8,9,10]. If the several hundred thousand dollars probably expended in "studying" MRC-TV had been instead devoted to making MRC-TV work, the situation might now be different. It is worth reviewing these studies in the context of what actually happened.

The first Study[6] describes a method and plan for evaluating MRC-TV's impact on intergovernmental communication behavior of top public officials and to identify the factors contributing to MRC-TV's performance as an intergovernmental communications mechanism. The intent was to provide a guide for comparable systems in other locales. The basic approach was to compare behavior before and after MRC-TV for both users and non-users. Information was to be derived from questionnaires and interviews. The evaluation objectives were to look at:

Assessing the Impact on Intergovernmental Communication Behavior
- Communication Mechanisms
- Communication Content
- Communication Frequency
- Communication Mode
- Communication Patterns

Impact on Participants' Attitudes and Perceptions
- Communication Incentives
- Communication Obstacles

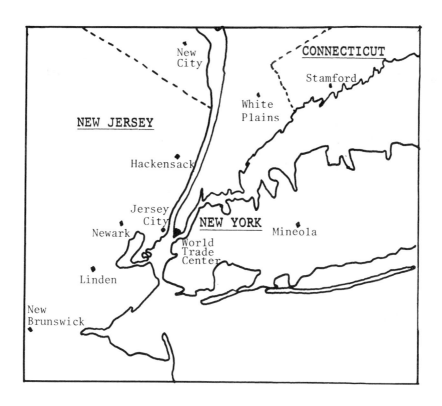

FIGURE 2 - MRC-TV SITE LOCATIONS

FIGURE 3 - MRC-TV CENTRAL CONTROL CONSOLE

FIGURE 4 - CENTRAL SITE STUDIO

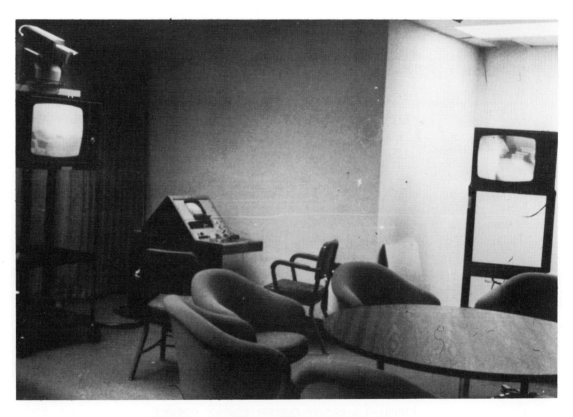

FIGURE 5 - TYPICAL REMOTE SITE STUDIO

FIGURE 6 - MRC-TV REMOTE SITE MEETING

- Feelings about Intergovernmental Contact and
  and Action

Impact on Actions

Unanticipated By-Products

Identifying and Assessing Behavioral and Technological Aspects of System Operation and Effectiveness

The MRC-TV effort from June,1971 - March,
1972 was reported by Rand[7]. It was summarized
in the form of an MRC report[8] for MRC officials
and others. Responses were received from 648
officials, including: elected executives; legislators; clerks and professional administrators
in education, finance, personnel, public safety,
health, welfare, community relations, housing,
planning, public works, transportation, environmental protection, and general management; officials in federal, state, county and municipal
governments, as well as officials from school
districts, special districts, and regional agencies and authorities. Key results were:

- The most important (existing) communication
methods in the region for public officials are
meetings of informal peer groups, professional
organizations, and government associations.
These informal meetings apparently seldom cross
State or County lines. There is a demonstrated
need for tying together and improving the effectiveness of these communications.

- How government officials contacted each other
was of significant importance. It was found
that impersonal modes of communications were
generally in the form of telephone calls and
some face-to-face contact, although not as much
of the latter as officials desired. There are
no systematic means for data transmission and
communications on a region-wide basis. Conference calls are not used extensively. Officials
rely on correspondence, one-to-one telephone
calls and such impersonal media as newsletters.
Newsletters and journals were widely read by
officials. Respondents listed membership in 552
professional and governmental organizations,
each with its own publications.

- It was found that there was a significant
difference between the type of communications
taking place within a government, e.g. inter-
departmental, with that taking place among different governments. There was a further differ-
entiation in communication among different
governments based on the level of the governments communicating. It was found that 70% of
the respondents communicated with other departments or agencies in their own government at
least weekly. This type of communications dealt
primarily with administrative, personnel or
financial matters. Most respondents felt that
such intra-governmental communications were
necessary to improve coordination.

- The substance and reasons for inter-govern-

mental communications are more complex. Contacts
among governments at different levels (e.g. Mu-
nicipality-County or State-Federal) deal almost
exclusively in some form or other with financial
aid. Contacts among different governments at
the same level (e.g. Municipality-Municipality
or County-County) deal with problems held in
common as opposed to problems of regional concern. When asked which of seven possible statements best typified the respondents own experience in inter-governmental communications, 58%
listed "It usually results in cooperation and/or
resolution of differences".

- When asked about training opportunities,most
respondents felt that the training effort of
their government was inadequate. When asked
what they would do with 45 minutes a day for job
related training or education: 48% said they
would prefer talking with knowledgeable persons
in their respective fields; thirty-four percent
would rather read books or articles on their own
and 14% would attend college classes. Most of
the respondents had little or no contact with
college or research staff in terms of discussing community or administrative problems. The
implication here for MRC-TV is apparent. Quality programming aimed at the need for training
curriculum would be well received. The ability
to bring the very best resource persons via
television to the many counties and major cities
on a regular basis would be a major contribution
to upgrading skills and understanding among public officials. Would government officials at
all levels be resistant to this type of electronic communication with its inherent multitude of practical applications? Probably not,
since the majority of questionnaire respondents
felt that the "climate or receptivity for
change" was good in their offices or agencies.

- Obstacles to intergovernmental communications in the region included foremost, the difficulty in knowing who to contact in other
governments, followed by the fact that many
people in other governments are unresponsive.
Ironically (for local public officials),Federal
and State respondents also listed these points
as being difficulties in intergovernmental communication.

- A majority of respondents conducted joint
programs or operations with other governments.

- It was concluded that a real and immediate
need for a regional communication system existed in the metropolitan region and that with the
joint effort of governments MRC-TV could fill
the communication need and offer government substantial benefits in terms of efficiency and
responsiveness.

The entire period 1969 - January, 1974 is
reviewed by Rand[9]. This must be read in the
context of the positive conclusion previously
drawn[7] about the role of MRC-TV. The Report

Summary states:

- The MRC-TV system is not the first use of television to join together two widely separated groups for a single meeting. It is the first, however, to join as many as ten together at once. It is not the first TV system to be called interactive; but it is the first to join so many locations in a two-way (symmetrical) system: audio and video to and from each site. This unique capability, allowing people to both see and hear each other in a natural conference relationship, appears particularly well suited to the communication needs of government officials. The motion video component allows for a degree of interpersonal rapport similar to face-to-face interaction. When used for instructional purposes, the two-way feature encourages an interaction between teacher and learner that is highly prized but not always achieved in face-to-face classroom teaching.

- MRC-TV consists largely of conferences, meetings, and seminars, whose planning and development are probably similar to what goes into conventional face-to-face conferences. Twenty-five meeting groups of functional counterparts, such as a group composed of personnel directors, and another composed of purchasing agents, have been organized and have used MRC-TV on a regular basis to discuss common problems. In addition, regular in-service training courses for county personnel have been instituted.

- Generally, the MRC-TV system has 20 to 30 hours a week of regular use. Its physical operation requires a basic staff of one communication director at the World Trade Center headquarters and one maintenance technician for the entire system, plus part-time county-employed operators at each outlying location. One MRC staff member is responsible for all instructional courses. The equivalent of four staff members plan meetings, each staff member preparing two to four specials or intergovernmental meetings per month.

- So far, the average number of county locations participating in the regular intergovernmental meetings has been six; the average number of participants at each location, two. Attendance at special seminars and the in-service training courses is often 10 to 20 at each site. Although a comparison with the costs of face-to-face meetings was not part of this study, a rough comparison suggests that when the value of people's time is taken into account, the MRC-TV cost per meeting,

even at today's relatively low level of programming, compares closely with the travel alternative.

- Despite the fact that visual information, such as graphic materials, maps, charts, photographs and the like, has played a minor role in MRC-TV meetings, many observers felt that the use of full two-way video is justified by the added contact and increased rapport that it appears to generate between participants. These judgments are made in relation to telephone interaction on the one hand, and face-to-face meetings on the other, since these are generally the only two interactive methods with which most people have had experience.

- The system appears to be enthusiastically accepted by most of its users; it is doing what it was intended to do; it is clearly resulting in a large amount of local government interaction that was not going on before; and it is bringing interactive training in many subject areas to county and municipal employees who did not have in-service training opportunities before.

Clearly, the Rand Report[9] represents a very positive picture of what was taking place. Was this picture "real" or "illusory", or did something significant change between April, 1974 and July, 1979? It is interesting to note here that the final phase of the Rand study of MRC-TV was never carried out. It was intended to involve analysis and measurement of major changes in intergovernmental interaction brought about by the system, test hypotheses previously formulated by Rand[6] and determine cost-effectiveness of services developed by the system.

A final study of MRC-TV was made by Columbia University[10], again funded by NSF. Apparently, what they studied was what Rand expected to do as the final phase of their work. Why a change was made is not known. Unfortunately, the Final Report on the Columbia University study was never accepted by NSF—it was considered inadequate. Nevertheless, a draft of the Final Report has been reviewed. It purports to measure the relative costs and benefits of conventional versus televised conferencing in the MRC-TV region. Some of the "tentative" conclusions from this Study were:

- The relative effectiveness of conventional and TV conferences depended on specific goals. Subjective measures of participant satisfaction generally yield better ratings for TV than for conventional conferences. However, in tests of information retention, those who attended conventional conferences scored about 10% higher than those attending TV conferences. Conventional conferences were also more effective in facilitating most forms of information exchange among participants and in inducing attitude change. When participants were asked about the extent of their follow-up contacts with fellow participants and the usefulness of the conference to their subsequent problem-solving activities, those who attended the TV sessions rated the value of the conference more highly than those who attended the conventional sessions. Taken together, these findings suggest that no single statistic is appropriate as a summary indicator of the relative effectiveness of TV and conventional conferencing. Either can be superior to the other, depending on conference goals.

- The cost analysis demonstrates that the TV system can be as much as one-third lower than conventional conferencing formats in overall conference costs, per participant (including institutional overhead, operating, and facilities costs as well as travel and time costs of participants). Three factors dominate the relative cost-effectiveness of the two conferencing media: the utilization rate of the TV conferencing facilities, the extent to which a TV conference can draw more participants than a conventional conference, and the extent to which costs of individual travel and participation are lower for TV than for conventional conference participants.

- The political, administrative, and transportation characteristics of individual metropolitan areas seem likely to be the best determinants of whether a competently managed TV conferencing capability would be likely to have significant cost advantages.

- Of primary concern is the extent of region-wide governmental activity. Because the system's utilization rate is critical, it is important that there be an extensive and continuing range of regional issues that are of sufficient concern to government officials to draw them regularly to the TV sites. Where a single government with region-wide jurisdiction does not exist to define such issues, maintain agencies to manage them, and direct its employees to coordinate activities throughout the region, there is only a limited natural constituency for a TV conferencing system.

- It is possible for a council of governments to organize individual conferences--as MRC did for this Study--but to do so on a scale that makes effective use of the TV system's capacity requires a very sizable organization of conference organizers and planners, as well as a capacity to consistently draw external funding for programming support. Though such a strategy is quite plausible for major metropolitan areas, it does impose a heavy programming burden on the

sponsoring organization. The task of achieving and maintaining cost-effective levels of utilization is likely to be much less demanding where operating government agencies have on-going programs that require regular communication among their personnel located in field offices throughout the region.

- A second factor of importance in determining the potential cost-effectiveness of a TV, relative to a conventional, conferencing system, is the extent of clustering of potential participants. Where administrative and/or political structures group large numbers of officials from various agencies into regional field office centers; or where Municipal, County, and State offices are clustered together in local government centers, access to satellite TV studios located in such centers is greatly facilitated. This is likely to strengthen the drawing power of a TV system relative to a single-site conventional conferencing system and to reduce costs to participants, two of the key variables in determining the cost advantage of a TV system.

- The New York City data suggest that the character and quality of public transportation are important in determining the cost saving potential of a TV system. Effective mass transit systems that facilitate fast movement into central city locations are likely to significantly reduce the travel and time cost advantages of a TV capability that show up very clearly among participants who used private automobiles.

- The closed-circuit interactive microwave technology appears to be a promising innovation for use in urban government. The characteristics of individual metropolitan regions should be a major consideration, however, in any strategy of diffusion.

MRC-TV - The Reality

MRC ceased operations on June 30, 1979. They ran out of money. For several years prior to this, MRC did not receive the level of funding from government contracts and grants that was anticipated. Regular dues from MRC members were insufficient to cover the deficit. As a result, starting in early 1976, MRC utilized income from TV system participating members to cover the financial gap. These were funds, which,by agreement, were to be used to meet MRC's financial lease obligations to Genesys. The shortfall in funding resulted in programming becoming ineffective and system maintenance becoming inadequate. This in turn led to increasing dissatisfaction on the part of TV system users and eventually to reduced participation. The inevitable downward spiral was never checked.

The MRC financial situation was known to all of its constituent municipalities. It created a very negative image for MRC within its own family, one which was not conducive to providing MRC with support. It certainly mitigated against growth of the system and the willingness of government groups to make continuing financial commitments. The situation also created a severe emotional and time drain on key people.

Also fundamental to MRC's poor financial situation was their inability to receive timely payments for TV system usage from the beginning of operations. MRC's management often relied on an indication of interest or a promise to pay at some undetermined later time from the participating municipalities. They unwisely did not insist on signed commitments from the participants before implementing the two-way facilities and providing program service. Thus, even the TV system income due MRC was often six months to a year late.

Figure 7 shows the MRC-TV program schedule for July,1978. Note that it had "degenerated" to providing training only. Gone were the "glory" days when intergovernmental teleconferencing was the "raison d'etre". MRC did not have the people to arrange for conferences and it did not have the people to do a decent job on performing on the government grants it did get. And, it didn't get paid anything reasonable for providing the educational programming that it offered over the TV network. The bubble had burst.

Also, long before the end, it was clear that the proposed extension links of MRC-TV to the State Capitols would not be implemented. This resulted in loss of a potentially powerful link between State officials and those of local governments--a link which in itself might have saved the system. As it was, a number of State and Federal officials came to New York to hold meetings over the TV network. When this happened, they were usually successful--but they never happened often enough.

MRC-TV - The Fix

In January, 1978, Genesys made a desperate attempt to help the MRC situation. It initiated a series of "Recommendations for Improving MRC-TV Operations and Improving the MRC-TV Financial Position" which were presented to the MRC Board. These recommendations were apparently never considered. The following outlines these recommendations:

"MRC-TV Operations - Situation Summary"

- Technically, the system is well designed, and when it is well maintained, it operates very satisfactorily. Unfortunately, it is not always well maintained. There is only one man, and he needs help --and more help means more money. Therefore, more money available for system maintenance is a fundamental necessity for success.

- More technical help is necessary but by no means sufficient. If the system operated perfectly all of the time, it would still be unsuccessful as it is presently being managed and used. MRC is now headed by Mary Haris. Mary is a gifted, charming and capable lady, but she has no experienced back-up support. She has some young planners working for her, but they do not have the background and experience needed to help make MRC-TV viable. As a result, everything falls on Mary's shoulders, an almost impossible situation for her and perhaps for any single person, no matter how energetic and well qualified.

- With no increase in cost, MRC-TV is capable of serving 10 - 100 times as many locations as it now does. This makes it obvious that,within the context of what MRC-TV is now doing, the first priority is to provide service to more "customers", i.e. more counties or cities. The necessity for this has been known from the beginning, but the best efforts of Bob Slocum and Mary Haris have not achieved this goal.

- For years, MRC-TV has lived a hand-to-mouth existence, kept alive by a few dedicated people. Its lack of success is not related to lack of effort by those people or lack of dedication. Nor is it related to the concept of the TV system not being viable. Frankly, the TV system has never been given a real chance. It has never received serious attention by the MRC Board--the kind and level of attention which is needed and which could make MRC-TV a success.

"MRC-TV Operations - What is Proposed"

- As indicated, improvement in technical operations will require more money. Getting more money is dependent on reaching a larger constituency. To reach a larger constituency requires more help, help of a different kind than is available now to MRC. What is proposed in the fol-

	Monday	Tuesday	Wednesday	Thursday	Friday
10 AM - 12 NOON	Speed Learning (Session 8-Live)	Solid Waste Management (Seminar III - Waste Reduction - Live)	Investigative Function (Session 5-Live)	Your Money Matters (Session 1-Live)	The Management Process (Session 1-Live)
12:15-12:45 PM	Public & Business Administration (Session 35 - Tape)	Public & Business Administration (Session 38 - Tape)	Public & Business Administration (Session 39 - Tape)		Public & Business Administration (Session 40 - Tape)
1:30-3:30 PM	Adv. Arson Investigation (Session 9-Live)	Grantsmanship Seminar (Session 6)	Shorthand II (Session 9-Live)	Bus. English & Communication (Session 4-Live)	Visual Aids - The Techn. Know-How (Session 1-Live)
3:30 - 5:00 PM					
5-6 PM					
6-8 PM					
8 PM					

FIGURE 7 - MRC-TV ACTUAL PROGRAM SCHEDULE - JULY, 1978 (1 WEEK)

lowing paragraphs is based on precedent. Whether it is viable depends on the willingness of the MRC Board to implement the plan and put in the effort to make it work.

- Specifically, it is proposed that two full-time senior people be "loaned" to MRC-TV on an annual basis as a public service. These people could be made available either by one or more MRC members or from local business or industry. The precedent (at least in the San Francisco Bay Area) is business and industry doing exactly this in connection with Annual United Fund drives. During the year, new people would be lined up for the following year and brought in early (part time) for full indoctrination. One of these two people would be responsible for maximizing utilization of MRC-TV by existing members and for bringing in more participating government groups. The other person would be responsible for maximizing utilization of MRC-TV by other users, such as H.U.D., Commerce, Business, Industry, Education, etc.

- Key to expanded utilization of MRC-TV by existing members and expanding its utilization of MRC-TV by others are volunteers who would serve on program committees to arrange for and schedule programs of interest to their special interest groups and who would also help solicit new participants. Each participating county (or city) could assign volunteer committeemen to serve in different areas, i.e. finance, planning, public works, administration, etc. Each would serve a three-year term (staggered) and each committee would be made responsible for the success of the system by programming serving their special interests and needs. This would bring in many more people, all senior and well qualified,and all identifying with the common goal of making MRC-TV successful. There is much precedence for this approach. Many Professional Societies have moved in this direction in order to be successful. For example, again in the S.F. Bay Area, about twenty years ago, the Institute of Radio Engineers had one organization, the San Francisco Section, which held monthly meetings. The attendance was often marginal. Then the "Professional Group" system was inaugurated. Now there are twenty-five groups involved, scheduling 25 monthly meetings, and instead of five volunteer people doing the work, there are more like 125 volunteers. It is not necessary or even desirable to use full-time staff to do the planning or arranging of professional meetings. The staff should provide support and service so that the profession-

als can accomplish their objectives.

"MRC-TV Financing - Sources of Funds"

- This is an area where action by MRC Board members (at their level) is essential. What is needed now is an "angel" or "angels" to provide transition funds so that the actions recommended above can be given time to succeed and MRC-TV can then become self-supporting. MRC-TV is a system unique in the world. It is located in the heart of the most densely populated part of the country. Its potential to benefit the community--in many different ways--is very large. Surely, in the area in which MRC-TV operates there are Foundations, Corporations and individuals that might be made to see the benefit of "bridge" financing for this great resource. None of the MRC-TV staff know these sources or people or know how to approach them. But, the MRC Board surely does. The amount of money "given" by organizations and people in the MRC-TV area as gifts or grants every year is in the tens of millions--and it is given for every imaginable kind of thing. Why not MRC-TV?

There is another major way that MRC could have been saved,which again was ignored by the MRC Board. It relates to "commitment". As an example, when the Stanford ITV Network was being planned, part of the plan evolved a concept of forming an "Association for Continuing Education" (ACE). ACE was to be a non-profit educational consortium of all Stanford Network participating organizations which would program the Stanford ITV Network with programs of interest to a broad spectrum of employees (not just engineers and scientists) during the hours that Stanford University did not offer graduate engineering courses. ACE was formed with enthusiasm but went bankrupt 2 - 3 years later,another apparently good idea which didn't work out in practice. But, the reasons for ACE's initial failure were analyzed and attributed to the policy of ACE charging a fee for each student participating in an ACE course, along with lack of identification with and support for the ACE concept within each participating organization. The "fix" was to get the participating organizations to make annual financial commitments, scaled to organization size, which in total were sufficient to insure ACE's operational viability. The quid pro quo for such commitments was to allow unlimited student participation in each course without additional fees. In addition, each organization appointed an ACE coordinator who was responsible for promoting ACE within the organization so that the organization would derive maximum benefit from its commitment. ACE turned around immediately and has been growing and operating with a surplus for years.

There are a number of essential truisms which can be learned (again) from the MRC experience. First and foremost, having a good concept is necessary for success, but it alone is not sufficient. How the concept is implemented is critical. If success is not achieved the first time around, the concept must be re-evaluated. Positive results from re-evaluation must lead to analysis of causes for failure by competent people who are experienced but who also care. And, this analysis must be conducted in the context of experiences elsewhere--both positive and negative. Even if the concept is perfect and the implementation plan is well conceived, good management and good leadership are essential as is sound financial planning. Finally, a policy setting group (Board) must be established which is composed of people with both competence and dedication. In the case of MRC-TV, the concept was excellent. But, in every other respect, the points outlined above were either ignored or violated--and failure was inevitable.

References

1. "Utilization of Broadcast Television for Government In-Service Training: A Feasibility Study", Robert F. Wilcox, Public Affairs Research Institute, San Diego State College, November, 1968.

2. "Telecommunications in Urban Environment", Eugene Rostow Committee Report, 1969.

3. "Telecommunications in Urban Government", Dordick, Chester, Firstman, Bretz, Rand Report RM-6069-RC, July, 1969.

4. "Communications Technology for Urban Improvement", Committee on Telecommunications, National Academy of Engineering, Report to HUD under Contract H-1221, June, 1971.

5. "MRC-TV: A Two-Way Audio-Visual Communications Network", Morris, Martin-Vegue, Farrer, Tallmadge, Microwave Systems News, December/January, 1975.

6. "Method of Evaluation for the MRC Telecommunication System", Alesch, Sumner, Rand Report R-1000-MRC, May, 1972 (NSF Grant).

7. "Intergovernmental Communication in the New York-New Jersey-Connecticut Metropolitan Region", Alesch, Rand Report R-977-MRC, May, 1972 (NSF Grant).

8. "Communication and Government: A Regional Report", John C. McGee, MRC, May, 1972.

9. "Two-Way TV Teleconferencing for Government: The MRC-TV System", Bretz, Dougharty, Rand Report R-1989-MRC, April, 1974 (NSF Grant).

10. "Improving Government Communication in the Metropolis: When Does Interactive Television Pay?", Alpert, Heginbotham, Bureau of Applied Research, Columbia University, August, 1976 (NSF Grant).

11. "The Tri-State Link", Howard Polskin, Videography, January, 1977.

The Teleconferencing Resource Book: A Guide to Applications and Planning
Lorne A. Parker and Christine H. Olgren (eds.)
Elsevier Science Publishers B.V. (North-Holland)
© Center for Interactive Programs, University of Wisconsin-Extension, 1984

THE IMPLEMENTATION OF TELECONFERENCING:
SOME LESSONS LEARNED

Ian Young[1]
Manager, EIU Informatics

Jim Birrell[1]
Consultant, EIU Informatics

and

Alan Derbyshire
Principal Consultant
Communications Studies & Planning Ltd.

1. INTRODUCTION

The accelerating development of teleconference systems has caught the imagination
of forward looking management, as well as att-
racting the attention of cost-conscious accountants and divisional managers. They need to
review questions such as whether to install
private systems or use public services, what
level of communication they need to provide:
audio only, audio and graphics, slow-scan or
full motion video. In this respect, the 1982
WEX conference, along with previous WEX conferences, will provide food for thought, argument
and invective for the person charged with
initiating and sustaining teleconferencing in a
variety of organisations.

But concern with system design, the technical advantage of product A over product B, is
only a part of the total story. Teleconferencing is about people as well as technology,
and a teleconference system, however technically sophisticated it might be, can only be a
success if it establishes a substantial and
recurrent user base.

This paper deals with the people side of
teleconferencing and in particular with the
management of the introduction of change to
working practices. The social psychology of
change management has been understood in
principle for some time: the need to involve
shop floor workers in consultative design; the
avoidance of threat and uncertainty; the
maintenance of earning power and benefits. All
these concepts have been established from the
introduction of machinery and automation in
blue-collar occupations, but here we are trying
something different; the introduction of new
technology to the working lives of the managers
and professionals who normally determine the
job development of subordinates. This situation has a certain intrinsic circularity, within which it is not known if the principles for
change management drawn from the factory floor

can be legitimately applied.

Rather than argue from a wholly theoretical viewpoint, we consider the best way to
learn the 'does' and 'don'ts' of introducing
teleconferencing would be to draw on the experience of organisations which had some years of
learning what can go right and wrong. Case
histories of success stories are told often
and with pride and conviction. But without
wishing to denigrate those successes, they
are, by and large, less informative than
instances where problems have been found and
overcome. The principle of learning by mistakes is sound, provided that the mistakes are
made at someone else's expense.

In this paper, we have been fortunate to
be able to draw on the experience of two
organisations whose activities with teleconferencing over the past five years have
been monitored and actively participated in
by the authors. One organisation is a large
commercial chemical products manufacturer
(I.C.I.), the other is the administrative arm
of the British Government as represented in
several departments.

Both examples concern group-to-group audio
conferencing, hence the length of track record
to draw on and both are UK-based activities,
but the experience and lessons should travel
well across the Atlantic and transplant to
forms of videoconferencing as well as audio-
only conferencing.

2. TELECONFERENCING IN I.C.I.

2.1 The Company

Imperial Chemical Industries PLC is the
largest chemical company registered in the
United Kingdom. It operates on a world-wide
basis with large subsidiaries in the United

States, Canada, Australia, Continental Western
Europe and with a significant presence in many
other countries. Of its $12bn total sales in
1980, more than half was due to overseas operations. In the UK it functions with a relatively autonomous Divisional structure, based
on specific product sectors. There are approximately 90 locations but many of these are in
close geographical proximity. Its operation
in Western Europe is very much linked to those
in the United Kingdom and are administratively
controlled from headquarters in Belgium.

2.2 Historical Background to Teleconference Activities

The earliest experiments with teleconferencing in I.C.I. began in the early 1960s using
simple speakerphones on private telephone
circuits. Although their use was found to be
successful on two large projects, there was
no continuity of use established and the operation lapsed. Interest revived in 1975 when
specially built equipment was developed and
installed between fixed points and with limited
interconnection.

The aim of introducing teleconferencing
was to add to the spectrum of communications,
particularly between Divisional Headquarters
and their remote works, and between the Corporate Headquarters in London and the separately located Divisional Headquarters. By 1981
the number of terminal locations had grown to
25 studios in UK, Belgium and West Germany.
The system has now been converted to star network which is manually switched at the centre
between dedicated high quality four-wire speech
grade circuits directly wired into the studios.

2.3 Current System Design

All sutdios are equipped with Neve Audio
Teleconference Units which operate unswitched
open microphones and loudspeakers with 5-7
hertz frequency shift incorporated in the send
amplifier to give 6db volume improvement if
required and greater stability at all times.
The equipments are fitted with send and receive
level setting devices. Most studios are also
equipped with Talos Telenotes to provide
scribble pad facilities.

The switching centre consists of a four-
wire switching frame fitted with straight
patching jacks for two-way conferences and 4-
way conference bridge and amplifier panels for
multi-way conferences.

2.4 Establishing a User Base

Having installed a system, the next concern is to establish a user base amongst the
personnel at linked sites. As part of a field
experiment in implementation, on the initial
introduction of the first teleconference links
various methods were tried to attract potential
users to the service. It was found that publicity in the form of posters and items in a
company newsletter created an awareness of the
service but did not attract any spontaneous
requests to use the facility. Demonstrations
of the service were held on an occasional basis
and were not tried as a systematic attempt to
gain real users. Such information as was
available from the demonstrations indicated
that they did not provide a very effective
means of soliciting users.

A more effective method, but one which is
highly intensive of time, was found to be
holding individual interviews with potential
users to establish their current pattern of
meetings and to identify forthcoming meetings
which could reasonably be held by the teleconference system. The selection of suitable
meetings involves the use of an implicit or
explicit model of the characteristics of a
meeting which can be successfully moved from
an existing face-to-face format to a proposed
teleconference format. The model used in this
instance was one developed at the Communications
Studies Group at the University of London
between 1970 and 1975. It is an inherently
conservative model, rejecting meetings as
suitable for specific forms of teleconference
on consideration of physical characteristics
(e.g. numbers of participants and locations),
the need to handle graphic material of various
forms and the behavioural effectiveness of
the medium for handling the meeting content.
More detailed information about the components
of the model can be gathered from other sources
(ref. 2).

By discussing the patterns and contents of
an individual's meetings, a trained implementor
can rapidly identify instances where a teleconference could have been held successfully
and with minimum disruption to the normal operation of the meeting. This identification can
then be followed up by projecting ahead to the
next such meeting and gaining a commitment
from the manager to try the teleconference
system for this meeting. The concentration
on a specific meeting means that generalised
anxieties can be assuaged and the commitment
to the implementor makes a change of heart
more difficult later on.

At I.C.I. the implementor was only able to
go as far as the first step, discussing and

identifying suitable meetings and then with only a small sample of potential users. None-the-less the technique did show some success, with half the sample expressing an interest in trying teleconferencing and a small number following through within the timeframe of the study to actually initiate a teleconference. A larger timescale, providing the prospect of going back to potential users to remind them of their expressed intentions, would undoubtedly produce a more marked response. The pattern of direct approaches to likely users was repeated at later sites through onto the systems.

The rate of takeoff of the service at I.C.I. was initially slow, until such time as a bedrock of regular users had been established and the value of the system had been spread by word of mouth. Five years on, the teleconference studio utilisation level is over 800 users per quarter, representing an estimated 300 teleconferences per quarter (some of the teleconferences involve more than two sites). At this level of utilisation, the system is far from reaching its full capacity, though some key points in the network are used quite heavily.

2.5 Operation of Service

To encourage use the location studios have, wherever possible, been equipped fairly lavishly with a good degree of comfort and accoustic treatment. But, in addition to equipping the studios and the centre, an infrastructure is necessary to translate the engineering to an ongoing service.

The most obvious is a nominated person at each location who is responsible for all aspects. This person in turn nominates a teleconference supervisor who is the interface to the users and actually makes and confirms the arrangements. These named people, and their deputies, need to be known not only on their own site, but throughout the system. The key post is the teleconference supervisor. By the very nature of his job the manager responsible for telecommunications on the site has always been made responsible for teleconference and usually the telephone or telex supervisor has been the most convenient person to be appointed teleconference supervisor. The risk is that their other duties sometimes clash.

A centre booking service is also necessary so that the demands can be passed to the engineers in a rational manner so that the teleconference calls can be set up as required. This in turn entails an availability over the period the teleconference service is offered of a person to carry out the switching. Whether

this should be done by an engineer or operator could depend on the specific organisation concerned but in this case the advantage lay in this being an engineer's function.

On receipt of a teleconference booking, the site teleconference supervisor obtains as much detail as possible and then checks first that her studio is available at the preferred time, that the distant studio or studios are also available at that time and makes a provisional reservation on them. The booking is then placed with the centre booking service who confirms the reservation with all the studios involved. Each afternoon a composite teleprinter (teletype) message referring to all the bookings for the next day is passed to the engineers and all locations that are to be involved. This acts as a double check and outstations are advised to query if any of these steps are not carried out accurately.

Depending on demands for calls, the following day the engineers make out a switch schedule which is primarily for morning and afternoon with the occasional need to change the set-up halfway through to accommodate successive calls.

"On demand" teleconferences are set up on telephoned instructions provided they do not clash with prebooked calls. Fifteen minutes before the time of the booking, each teleconference supervisor goes to their respective studio, checks the equipment and confirms connection. When the conferees arrive, if it is their first teleconference, a short instructional session is given and then they are left to get on with it. Regular users, of course, soon get into the pattern.

2.6 Problem Areas

Having considered the implementation and the operation of the teleconference service in I.C.I., we can now go on to look at areas of potential and actual problems, and to address the title of the paper - What Lessons can be Learned?

Initial Attitudes. When a new means of operating is being planned to be introduced, there are bound to be a range of attitudes in anticipation. Given the frequently pessimistic nature of humanity, it is not surprising that amongst these attitudes, the negative views expressed tend to outnumber the positive ones. In the implementation interviews mentioned earlier, note was taken of the attitudes expressed by the potential users.

The negative attitudes tended to focus on

the loss of social contact that normally came with attendance at meetings, including a sense of presence, participation in site meetings and keeping the grapevine communication channel open. Other opinions expressed doubts on the capability of the teleconference system to handle the types of meetings they took part in. The doubts were about either the handling of text materials used during the meetings, or on the content of the meeting itself (e.g. if some persuasion were involved).

Only a small number of people spontaneously saw the positive aspects of teleconferencing, such as the ability to involve a wider range of people in meetings, or holding meetings where previously there was no cost-effective way of maintaining an adequate hive of communication.

The imbalance of opinions, and the nature of them, suggest our first set of lessons:

- The importance of prior information on the nature of meetings in the organisation to aid the selection of a teleconference system capable of handling most of those meetings.

- Good advance publicity on the types of meetings that can be handled effectively over the system.

- Publicity to include the use of teleconference facilities to hold new types of meetings.

- Less emphasis on the replacement of existing meetings.

- The establishment of principles that will permit social contact to be maintained (e.g. a mix of tele-conference and trips).

Technical Performance. The main concern of engineers with the implementation of tele-conferencing is that the equipment should per-form well and be highly reliable. The short and long-term experience at I.C.I. was that this was achieved, and overall there have been few complaints from users about technical quality. Certain additional features of the system have not, though, been so well received. The Telenote electronic scribble-pad for example, was initially in a configuration which prohibited speech transmission while it was being used. This was unacceptable to anyone who wanted to give a verbal commentary to their drawing. Even since reconfiguration, the Telenote has been found to be tempermental and overall has not been a valuable aid to meetings.

Other attempts to address the issue of graphics in audioconferences have included a trial of slow-scan video, but so far this has not produced the quality of image required by the users. In particular, it has not been able to give the required resolution over practical working segments of large engineering drawings.

Trials are being planned for compressed digital video transmission in 1983. With the network now existing, the upgrading of equip-ment becomes highly expensive once the multi-ples of the number of studios is applied. With 25 studios, replacing the Telenote with some-thing better is estimated to cost over $100,000 for the whole network, whilst introducing even basic video would cost around $1 million on studio costs alone.

Lessons.

- Technical Performance needs to be adequate but not necessarily brilliant.

- The need for and use of ancillary equipment for documents and graphics should be identified at the start and implemented to meet those needs.

- Changes can be highly expensive once a large working network is installed. Small-scale pilot trials with good monitoring are the best way of making progress.

Operational Difficulties. The basic requirement in providing a troublefree tele-conference service is to motivate everybody concerned as to the importance of their part in the operation, so that they never slip up. This has not been achieved yet on this system and the authors would doubt if it has been on any comparable system. The problem with teleconferencing is that a lot of users are reluctant at best and any small slip up can result in lost customers.

To a large extent this can depend on the Company culture as indeed much of the success of teleconference depends on this factor but where the individual is allowed a reasonable amount of choice the selling job is that much harder.

The elements that need to be carefully watched include booking procedures which must be watertight. For a normal conference where the conference room is double booked, the un-lucky meeting can usually be fixed up somewhere else or postponed until the room is available. Not so with teleconference on a multi-system. The cause of this is usually changes from the original booking which have not been followed through correctly, but with anything up to six changes being recorded against a single con-ference, a slip-up becomes very possible.

Another aspect of booking is duration. It is very easy to overrun on a successful tele-conference which can make the next one from any of the studios involved very much unsuccessful. The only answer is that where there is a succeeding conference the first one must be cut down without fear of favour. Easier said than done, however, when the participants are the management board of that group.

The unavoidable absence of the teleconference supervisor and deputy needs to be covered. Important last minute conferences often seem to coincide with this eventuality and whilst a third person may have been instructed in the procedures, they have then been called upon so rarely that at a critical moment they can slip up.

In the actual conduct of the teleconference itself, it has been found that meetings are successful when the participants already know each other and when there is a chairman located at each of the sites involved. Best of all are the regular progress meetings called at the same time each week, fortnight or month. There the roles and personalities are established, people are familiar with both the booking procedures and the operation of the equipment.

The occasional user is more likely to make a mistake in the initial set-up, and although he may have been introduced to the equipment, may well have forgotten details of operation.

The final consideration of operational difficulties is the location of the teleconference room. In general in I.C.I., the rooms have been well located, with easy access usually within the same building as most of the user population. One example though illustrated the potential dangers. One conference room was located in a separate building external to the main offices and standing in very pleasant grounds. There were no major problems with this location throughout a winter of use. However, in the summer the participants found that they were sometimes competing to hear and be heard above the noise of petrol-driven grass mowers that regularly kept the picturesque lawns in trim.

Lessons.

● Creation of a simple and fool-proof teleconference room booking service.

● Establish the role of teleconference room supervisor with high priority given to that role and adequate coverage.

● Booked times for teleconferences must be respected even by senior grade staff

or else the service will be considered unreliable and not used.

● Rapid establishment of regular user groups is desirable.

● Advice and protocols for running teleconference and operating equipment must be clear and concise.

● Teleconference rooms should be easily accessible to the user population and should be located away from areas with risk of high background noise.

Resourcing Problems. The introduction and operation of teleconferencing in an organisation is highly time-consuming. The experience at I.C.I. was that it went beyond expectations of the time needed from ancillary staff to run the service effectively. Development time and trouble-shooting at the start of an operation added to the manpower cost, and extention of the service to embrace video in some form would be a further demand on time.

Lessons.

● Time spent before installing a service, in identifying user needs and developing poerational protocols may be off-set against costs of trouble-shooting after installation.

● Make sufficient allowance of time for people to service the operation of a teleconference system.

3. TELECONFERENCING IN THE CIVIL SERVICE

3.1 Background and History of Teleconferencing

The administrative arm of government, known in the UK as the Civil Service, has the job of implementing government policies across a wide range of issues. Organised into ministries and departments with only occasional reference to each other, the Civil Service might be seen as a group of independent (but not autonomous) companies under the direction of a central Head Office/Holding company - the government.

Historically, these ministries and departments have themselves been located close to the seat of government in Westminster, so that their parliamentary heads, the Ministers and Secretaries of State, could play a dual role of running the department and appearing in the House of Commons. Close proximity means that the permanent and professional staff of the Civil Service can rapidly brief the Ministers on contentious issues raised in parliament for discussion.

On the other hand, there started to be a reaction in the 1960s to the over-centralisation of government in Central London. The concern, which has been experienced in many other countries, was that central government was unaware of the problems of the regions, because senior Civil Service personnel all lived in the suburbs of London. At the same time, traditional labour-intensive industries based in the regions were running down, creating pockets of unemployment. The regions wanted their share of the employment offered by the Civil Service.

At the end of the 1960s, following the report of a commission, a major programme was started of relocation away from London. This raised the concern of how communications could be maintained at an adequate level between devolved sites and necessary residual London offices without involving senior grade Civil Servants in repeated travel to and from London. Telecommunications appeared to hold the solution to this problem, with a specific place for teleconferencing. It was partly as a result of this requirement for a solution that the initial research into the effectiveness of teleconferencing was commissioned from the Communications Studies Group at University College, London[2].

Following from this research contract and from some technical development work, an audio-conferencing system was developed called the Remote Meeting Table (RMT) which was installed for trial into specific offices from 1972 onwards. The prototype RMT was developed commercially by Plessey as the Remote Conference Table (RCT), with the programme of implementation continued during the 1970s. A description of the unit and the service is given in the following sections.

It has already been indicated that the Civil Service is a number of independent departments. In consequence, this paper does not attempt to deal with the operation of tele-conference to the Civil Services as a whole, but instead draws on a number of linked studies at three departments: The Scottish Office, Customs and Excise and HMSO - the government publishing arm.

2.2 System Design

The basic system discussed is the RMT, a group to group audio conferencing system that operates over a private 4-wire network. The RMT extends normal group to group audio conferencing by having an individual loudspeaker and indicator light for each participant at the far end of the link, so that conferees can rapidly identify who is talking at the remote site at any time. The communications channel is always open in both directions, but voice activated switching is used to direct the speech to a specific and uniquely identified loudspeaker at the remote end of the link. The system is designed to take a maximum of six conferees at each site.

RMT rooms in the Civil Service department are comfortably though not lavishly furnished. They contain no capability for the transmission of graphics materials inside of the conference rooms, though some studios have facsimile machines nearby.

There is no RMT network as such, with individual departments   making their own arrangements to have either standard fixed linkages between sites or some limited degree of switching capability.

3.3 Implementation

The process of introducing teleconferencing to potential users in the Civil Service rested heavily on promotional material including glossy brochures. For the most part people were attracted to try the service by word of mouth from colleagues who were already established users, but this process necessarily requires someone to be adventurous and try the service. Breaking into a pattern of travel and meetings can be extremely slow unless there is a strong motivation to do so. In each of the departments studied, the alternative to teleconferencing was a tedious and not very enticing journey, usually by road or rail, which may well have encouraged the more frequent traveller to experiment with the new service of teleconferencing.

The actual operation of the teleconference facility varies from location to location. However, in all cases there is one contact point for both booking the teleconference rooms and establishing the right connection at the right time.

3.4 Problem Areas

The experience of users of the RMT in the Civil Service was gathered by the use of log-sheets recording both usage and problems, and direct interviews with users in the departments studied.

Technical Performance. As was found in I.C.I., users are very tolerant of sub-optimal performance of the system once a pattern of use has been established. Most of the users interviewed had at some time experienced some technical problem, but only one or two were discouraged from using the system again.

Amongst some of the users, the experience of the fixed RMT audioconference system had led them to experiment with other forms of teleconferencing, notably the use of desk-top loudspeaking telephones, to hold 'instant' teleconferences. The quality of speech on many of these loudspeaking telephones reinforces the view that speech quality needs to be adequate for the job but wide variations are acceptable if not desirable. Once a body of regular users of teleconferencing has been established, then minor technical problems with the teleconference system are not likely to be critical to the success of the service.

Lessons.

- Technical performance may be more critical when establishing a service.

- Established users will tolerate minor technical problems and poor speech quality.

- There needs to be a mixed strategy for teleconferencing, providing different qualities of conference with different levels of access (e.g. studio-based and desk-top systems).

Operational Difficulties. The main problems concerning the operation of the teleconference systems have been associated with booking procedures. The prime users of the system are not always aware of the procedures and delegate the booking to secretaries or more junior staff. While this works well for the most part, it does incorporate some degree of risk, and could cause problems when the staff who usually do the booking are not available for any reason. A simple instruction guide for all staff would be useful.

There are reports of previously booked teleconferences being dropped at the last minute to make way for a rapidly called teleconference involving more senior staff in the organisations. Similarly, some users have complained that they have had to cut short their teleconference because senior staff wanted to start their teleconference at short notice. This form of priority for senior staff tends to undermine confidence in the booking procedure and confidence in the operational reliability of the system.

A third problem concerning booking is the bunching of meetings. A third of teleconference bookings were found to start between 2.00

and 3.00 pm. This effectively reduces the capacity of the system to only one meeting in the afternoon. Greater utilisation would be achieved if users could be encouraged to start their teleconferences at a wider range of times.

A different form of operational problem concerns the operation of the meeting itself. Although the RMT is designed to take a maximum of 6 participants at each studio, the average number of people involved in teleconferences over a two-year period studied was 3 per studio. After two years of operation, 25% of meetings involved only one person at one of the studios. The design of the RMT is not particularly suited to smaller meetings and there is a specific problem when only one person is located at one of the studios. The strongly uni-directional microphones used to reduce the risk of howl-around on an open audio system require participants to sit close to their own microphone. With only one person at one end, there tends to be far more discussion between participants at the other end than is found in most teleconferences. Unfortunately, they are less conscientious about sitting and speaking in a manner which will allow them to be heard by the solitary participant.

The problem is partly one of procedures within the teleconference, but also has relevance for both technical design (making the system more flexible to meet the demands of different sized groups) and a broader teleconference strategy (providing alternative and better suited teleconference capabilities to meet a variety of needs).

Lessons.

- Teleconference room booking procedures should be simple and highly visible to all potential users.

- Prior bookings should be respected by senior staff if more junior staff are to feel confident in using the system.

- Users should be encouraged to hold teleconferences at differing times to permit greater utilisation of studios.

- Teleconference systems should be seen as part of a strategy of overlapping systems or services.

User Base. Probably the main problem identified from the study of RMT usage is that the users are not drawn evenly from the organisation. They are heavily biased toward the highest grades in the Civil Service, even though it is a traditional pyramid shaped organisation, and travel to meetings is greatest amongst the middle to lower grades of staff. Several factors were identified as contributing to this imbalance, including the loss of confidence through senior staff over-riding prior bookings:

- Lower level staff tended to be involved in more meetings, with participants from multiple locations, which are not suited to a point-to-point teleconference system.

- The teleconference studios tended to be placed close to the offices of the most senior staff. They were then seen as being a high status facility, which was off-putting for lower level staff.

- There was little awareness amongst the lower level staff of the value of their own time, and little encouragement to save costs by not spending time travelling.

- Although awareness of the facility was high, the lack of a thorough introduction programme meant that only a small proportion of lower level staff had tried the system.

Lessons.

● Planning a teleconference system requires first-hand information about the communication requirements of all the intended user population, and not just the opinions of the senior executives.

● The location of teleconference studios may carry a message about its status that can deter one section of the intended user population.

● An introduction programme is required to encourage a wide user base and to educate awareness of the value of time.

4. CONCLUSIONS

The case studies described have been used to draw their own conclusions. All that remains here is to restate briefly the main points emerging:

● Good preliminary work to identify user needs is of major importance in the selection and design of a teleconference system, and can

reduce later costs associated with under-utilisation or major re-design work.

● Technical performance is not critical after a good base of repeat users has been established.

● The operational infrastructure is crucially important to gaining a rapid take-up of teleconferencing.

● The location of teleconference studios should be readily accessible to all potential users.

● Time spent in a thorough introduction and education programme will be recouped by higher and broader utilisation in the organisation.

● Education should include procedures for running teleconferences and information on how to make the best use of the teleconference facility.

[1] The work referred to in this paper was conducted while Ian Young and Jim Birrell were members of Communications Studies & Planning Ltd.'s Applications Studies Division. The Applications Studies Division has now moved to The Economist Intelligence Unit to form EIU Informatics.

[2] For a complete review of work carried out by the Communications Studies Group see, Short J., Williams E., and Christie B., Social Psychology of Telecommunications, N.Y.: WILEY, 1976

The Teleconferencing Resource Book: A Guide to Applications and Planning
Lorne A. Parker and Christine H. Olgren (eds.)
Elsevier Science Publishers B.V. (North-Holland)
© Center for Interactive Programs, University of Wisconsin-Extension, 1984

IMPLEMENTATION FACTORS IN FREEZE FRAME TELECONFERENCING

Robert Curwin
National Sales Manager
NEC America, Inc.
Broadcast Equipment Division
Elk Grove Village, Illinois 60007

Introduction

Teleconferencing takes many forms. The one that is most familiar to all of us is the telephone which when expanded into a conference call becomes the simplest form of audio only teleconferencing. At the other end of the spectrum, audio is supplemented by the addition of full motion video conferencing requiring a 6MHz bandwidth which is only practical over short distances. Today, full motion video systems utilizing full bandwidth video are usually limited to intra-city systems. Full motion video over great distances is accomplished by the implementation of a bandwidth compression codec so that full motion with some limitations may be transmitted in bandwidths as low as 1.5 megabits, which is approximately 1/30 of full video bandwidth. There are movements within this industry to reduce this bandwidth and it no doubt will happen, but at the present time a 1.5 megabit channel is required for each direction of the transmission. The cost of this transmission is extremely high which has in many cases forced users to look for alternatives. This alternative is freeze frame or captured frame video conferencing.

The technology that leads to freeze frame conferencing was derived from the concept of storing a frame of video in a solid state memory and then transmitting it over narrow band channels. Very typically these channels are represented by one or more direct dial telephone lines. With freeze frame video, the use of wider bandwidths results in faster transmission times with the quality of the picture being a constant. Table I below shows the transmission time of a single frame of video

TABLE I

4.8 Kbit	50 seconds
9.6 Kbit	25 seconds
19.2 Kbit	13 seconds
56 Kbit	4.5 seconds
1.5 Megabit	0.3 seconds

Transmission Bandwidth vs Transmission Speed.

over various bandwidth facilities with the highest possible picture quality. A separate line is required for audio so that the minimum transmission requirement is two dial up telephone circuits.

The cost differential in the transmission of full motion signals versus freeze frame signals is readily apparent. This paper describing the implementation of a conference room is based upon a freeze frame configuration.

Video Teleconferencing Benefits

When it has been proposed that video teleconferencing in some format might be an asset to corporate communications, it is first necessary to determine exactly what the company's needs are and will be in the near future. The very obvious consideration is the amount of travel currently being experienced by company personnel both on a regular and irregular basis. There are, for our purposes, two types of travel. One type of travel involves the movement of company personnel into the field to call on other organizations or to trade shows and conventions. The second type of travel, considering the breakdown we are looking for here, involves travel to other facilities within the organization or to subsidiaries of the same company. It is this latter type of travel where teleconferencing can be a valuable tool. In this category of travel within the company we can further break-down the reasons that people travel. A certain number of meetings are held on a predictable repeated basis such as monthly or bi-monthly. It is these meetings that may be most easily converted to a teleconferencing format. Another type of meeting is the problem solving meeting which occurs on a random basis with little or no advance notice. These also lend themselves to teleconferencing as a solution.

Within every organization there are a certain number of people that enjoy traveling. These generally are those people who do not have a heavy travel schedule and look upon travel as somewhat of an adventure and possibly even a reward or a chance to get away from the dull routine of the office. One of the important considerations of teleconferencing that must be brought out is that it will not replace all travel. One prime example that we like to cite is the case where a monthly meeting was reduced to a quarterly meeting, with teleconferencing used to supplement the information transfer given up by the two missing meetings. In this instance the two missing meetings could easily be replaced by three or even four teleconferencing meetings and still experience an increase in productivity and decrease in meeting costs. Not to be overlooked is the fact that additional unscheduled participants can be brought into teleconferencing meetings at a moments notice.

System Hardware Choices

When it comes to making a determination as to the hardware required, the two most obvious questions are; (1) How many participants will typically be in a meeting, and (2) what will be the format of the meeting. Quite often these questions will have multiple answers, therefore, the hardware and the conference room need to be designed for the worse case. The question that you need to ask yourself is, if you were at the other end looking into that conference room through a television monitor, what must you see.

First of all, it has been predetermined that we are dealing here with freeze frame television or as it is sometimes called, captured frame television. It is not appropriate with this medium to try and conduct a meeting totally built around people pictures keeping in mind that the pictures viewed on the monitors do not move. Pictures of people's faces taken ten to thirty seconds earlier are not particularly interesting. The key to successful freeze frame video conferencing is graphics. If you think about a typical meeting, even a face to face meeting you are more than likely supporting what you are seeing with graphics in one form or another. This may be 2 x 2 slides, viewgraphs or overheads, a flip chart, a blackboard or some other device to communicate the written message. People pictures even if they are still, do add a certain amount of presence to a meeting, however, you should consider that the faces are there to support and add credibility to the graphic presentation. As a minimum requirement then, we must have a camera which is used for graphics. We would recommend that a second camera be included in this minimum requirement which could be used for people pictures or it could be used for larger graphics as well. In a typical configuration we would suggest that the graphics camera be located overhead, per-

haps in the ceiling of the conference room and be used for picking up documents or objects no larger than perhaps 24" x 30". Another alternative for a document camera is to place it on a document stand which in appearance approximates a photo enlarger. This camera on the document stand and looking down on a flat surface is useful with documents up to letter size. Typically a document stand is located off of the conference table in another part of the room.

The second camera may be mounted on a pan and tilt device so that it can be positioned to look at either the people in the room or wall mounted graphics, a blackboard, or a flip chart. This minimum system gives us the ability to hold an effective freeze frame teleconference although there is still considerable room for expansion and improvement.

Figure I below shows a typical system layout for a two camera system. The size and shape of the conference table in this system is somewhat arbitrary and depends upon the size of the room. In this design it is presumed

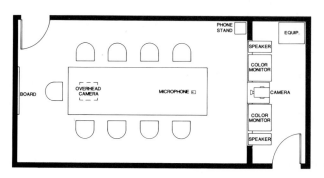

Figure I

Two Camera System Layout

that the table is a catalog item and not custom made. In keeping with the cost of this minimum system we are resisting the urge to use any sort of a microprocessor control system. Cameras are positioned remotely, but there is no memory system for positioning. The controls

TABLE II

Tranceiver	$ 18,700.00
Remote Control	1,500.00
Modem	2,400.00
Cameras and Lenses	12,000.00
Monitors	1,900.00
Audio System	2,300.00
Miscellaneous	1,100.00
Total	$ 39,900.00

Cost Breakdown of Minimum Two Camera System.

which will be push button are best kept to a minimum and designed to provide the basic system functions without any degree of elaborateness which is sometimes exercised in systems of this sort. Table II shows the projected cost of a system such as shown in Figure I. This does not include any of the transmission cost which, of course, must be considered as it has a direct relationship to the cost of the hardware. There is quite expectedly a tendancy to spend more on hardware if there is a willingness or a need to spend more on transmission.

With consideration given to the minimum hardware system just described we can expand to a more elaborate system by adding one or two cameras. If we add one camera it would be presumed that this one camera would be equipped with a fixed focal length lens and be dedicated to people pictures. This frees up the camera previously mentioned as being on a pan and tilt available for any number of graphic displays which cannot be handled by the overhead camera. The fourth camera is sometimes utilized to provide additional graphic capability should this be deemed important. Keep in mind that although the system cost goes up as cameras are added, it also reduces the amount of zooming and positioning that is required during the course of the meeting. Figure II shows a more elaborate four camera system. In addition to

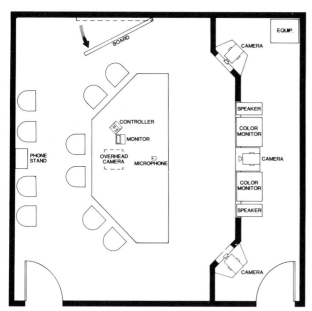

Figure II

Four Camera System Layout

adding a camera, a more elaborate system may have such options included such as the NEC pointer generator or the Interand or Optel graphic systems. These systems provide a real time effect to what is in actuality a freeze

frame system. Another common option that is typically added to an expanded capability system is a device which will allow the recording and storage of images. This traditionally has been an analog video disc, but is now more often accomplished using home computer type digital technology. This allows the storage and recall of upwards of fifty or seventy frames of video during the course of a meeting. The following Table III shows the total projected cost of an expanded system as described here.

TABLE III

Tranceiver	$ 24,200.00
Remote Control	1,500.00
Modem	2,400.00
Cameras and Lenses	28,000.00
Monitors	1,900.00
Audio System	2,300.00
Miscellaneous	2,100.00
Total	$ 62,400.00

Cost Breakdown of Four Camera System.

Room Considerations

There are many considerations to be given to good room design. It has, however, been our experience that there is a tendency in this industry to overdesign video conference rooms. There are unfortunately a number of so-called experts who have sprung up in the last two years who feel that a video conference room must approach an audio recording studio or a broadcast television facility in quality. NEC's feelings in regard to this are best expressed by one of our customers who in taking a common sense approach stated "I already have a recording studio, what I need is a good video conference room." Most organizations begin with an existing conference room which may already be sufficient in terms of lighting, acoustics and overall dimensions to serve as a good video conference facility. Even if the room is not an ideal size, it is quite often possible to work with the dimensions that you have even at a slight compromise rather than expend great amounts of money to make construction changes. We find that in terms of dimensions an absolute minimum room size is probably somewhere in the 12 x 20 foot range. Ideally we would like to see a room approximately 20 x 30 feet but this is not usually realized.

Many organizations use internal partioning that may be acoustically poor. An example would be the movable metal partitions that are quite often found in office buildings. Acoustically these represent the poorest form of

material that we would like to see in a conference room.

Most office conference rooms today presently have some form of florescent lighting
which usually can be used with some modification to the fixtures or possibly by adding to
or relocating the existing fixtures.

At NEC we recently constructed a video
conference room starting with an existing room
that measured 12 x 30 feet. This room as it
was initially configured included a low grade
commercial carpet and walls which were either
wallpapered or wood panelled. It was subject
to noise from passing aircraft and from our
own warehouse within the building, but still
it represented the best room we had to use as
a starting point, with the consideration that
we wanted to hold the cost to a reasonable
level and not build an isolated room in the
center of the building. Using the concept of
a separate room for equipment we constructed a
wall reducing the room size to approximately
12 x 25 feet. This gave us approximately a
5 x 12 foot room in the back to mount monitors
and install other equipment. In this particular instance the room deviated from a normal
video conference arrangement in that it was to
double as a demonstration facility and we,
therefore, included a shelf on the front wall.
We elected to use 25" monitors at the front of
the room which were installed so that the center point on the monitor was at the eye level
of a person standing. The reason we put the
monitors this high was that we did not plan to
use any sort of specially constructed conference table and this height would minimize the
problem of heads being in the way. This wall
shown in Figure III was then covered with wallboard with shelves constructed in back to sup-

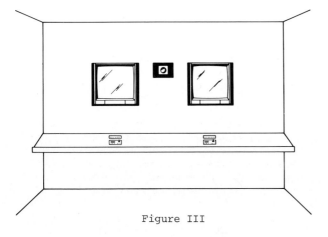

Figure III

Conference Room Monitor Wall

port the monitors in the wall. We also provided a window for a camera mounted on a pan
and tilt to observe the people in the room or

any back of the room graphics. This gave us
the greatest flexibility for this room which
was to be a two camera facility. An investigation of suitable acoustic related wall treatments suggested that the best solution was an
acoustic wallboard as manufactured by a number
of companies. We chose to use Conwed Silent 95
wallboard over the other brands mainly because
it was the choice of our particular contractor.
We feel that other brands with the same specifications would be equally suited for this purpose. Characteristics of this particular wallboard are as shown in Table IV.

TABLE IV

Substrate	Fiberglass
Fabric Cover	Non-Woven
	Polyester
Thickness	1 1/8"
Width	30"
Height	9'
NRC	90 - 1.00
NIC'f	20
SAC	1.04

Acoustical Wall System Specifications

One of the considerations for the monitor
wall was that we might want to change the monitor placement after the first year or two of
operation. There are several reasons that one
might want to do this not the least of which is
the possible introduction of new models of
monitors or perhaps the conversion to a suitable large screen projector. To have this
flexibility and at the same time to add color
to the room we elected to carpet this wall.
Carpeting provided the additional advantage of
being able to upholster the monitor and camera
openings in the wall. The upholstering provided a lot more tolerance in cutting the opening size and made it easy to put the components
into the wall.

This room, when construction began, had
six 4 tube florescent fixtures. These same
six fixtures were relocated into a more compact area to fit over the conference table that
would be purchased. We realistically considered that we might have to add two fixtures
following the completion of the room. These
fixtures were reequipped with eggcrate type
baffles to direct the light downward preventing it from reflecting into the camera. Since
it was our plan to use CCD solid state color
cameras in this room, lighting was a little
less critical than it might normally be since
this camera is more sensitive than conventional tube type cameras. One consideration that
must be taken into account relative to lighting is that color cameras operate best when

all of the lighting in a room is one color temperature. We were careful to use all florescent fixtures and although they would have been
desirable at times, avoided the use of incandescent spotlights. The fixtures were wired so
that half of the florescent tubes could be
turned on at a time giving us a way of dimming
the lights without affecting color temperature.
As it was our plan to add a marker board at one
end of the room we positioned florescent lights
over that board to provide lighting for it.

The acoustical paneling for the most part
solved the external noise problem, however, the
aforementioned warehouse noise input was
through the ceiling tiles in the conference
room. These rather inexpensive tiles were replaced with a higher quality and heavier tile
and 6" of fiberglass insulation was laid on top
of the ceiling after the tiles were in place.
This combination of ceiling treatment, wall
treatment and new carpeting on both the one
wall and floor along with acoustically lined
drapes on the two windows went a long way
toward dampening and/or removing parasitic
noises.

We selected a conference table twelve
feet long since as was previously stated it was
desired to confine our expenditures to catalog
items. This table is the very common rectangu-
lar shaped table that is found in most office
supply catalogs. The coloring on the table was
a medium oak to avoid the large dark expanse
often seen with the more traditional walnut
colored conference tables. The chairs were
selected to blend with the other materials in
the room but with care to select chairs that
were void of chrome arms or legs. Chrome legs
would have been acceptable but it was to our
advantage to insure that the cameras were not
able to see any chrome parts, thus avoiding reflections in the pictures. In determining the
number of chairs we allowed 30" of table space
per person as an absolute minimum with 36" more
desirable. The table itself was located eight
feet from the camera/monitor wall insuring that
all participants would be within the viewing
angle of the lens. The two cameras were located as we previously stated in the monitor wall
and in the ceiling which is in line with the
example we gave earlier for our minimum system.

The audio system utilized the very small
pressure zone microphone which in turn require-
ed a microphone preamplifier to interface with
the conference telephone. In this case the
conference telephone was the Northern Telecom
Conference 2000 except that we used neither
the microphone or speaker that comes with this
system. We chose to use column speakers which
were positioned beside the monitors at the
front of the room effectively making the voices
come out of the monitors.

To insure user acceptance of the teleconferencing system it is necessary to identify
one individual at each location as the room
coordinator. This individual will hopefully be
very receptive and enthusiastic toward the concept of teleconferencing. It will be the responsibility of this individual to insure that
there is continuity in room usage and that
those who do use the room are confortable with
the technology. In many cases companies have
assigned the room coordinator to operate the
equipment during the meeting. This is not
always feasible, however, because of confiden-
tialities and so we feel it is of the utmost
importance to design and configure the room to
be user friendly. One consideration that must
be given to a new installation is the training
of the users by the vendor to not only use the
hardware but to become conversant with video
terminology and the limitations and advantages
of video as well. It would be desirable if
preexisting graphics could be used without any
consideration given to the new video media but
this is not always possible.

Conclusion

You may get the impression from talking
to many people that there is a definite right
and wrong way to build a freeze frame video
conference room. Our experience is that there
is no right or wrong way. You may build a room
with as few or as many cameras as you wish,
with or without extensive room acoustic and
lighting treatment and with or without the many
video accessory items that are available today.
The right way is whatever works for you. There
is another general feeling that says that
freeze frame conference rooms, fully equipped
should not cost as much as full motion conference rooms exclusive of the two codecs. This,
we believe, is true and is brought into practice by keeping the costs of the cameras and
accessories in line with the overall budget
that led to freeze frame conferencing to begin
with. Even so these rooms can later be converted to full motion with little obsolescence.

To supplement the equipment costs that you
have already seen we would like to point out
that all of the contractor work that was accomplished on the room described herein cost less
that $8,000.00. This included the acoustic and
lighting work and the subdivision of the room
into two rooms.

The old rule "keep it simple" certainly
does apply in the case of freeze frame video
conferencing.

The Teleconferencing Resource Book: A Guide to Applications and Planning
Lorne A. Parker and Christine H. Olgren (eds.)
Elsevier Science Publishers B.V. (North-Holland)
© Center for Interactive Programs, University of Wisconsin-Extension, 1984

256

ESTIMATING TELECONFERENCING TRAVEL SUBSTITUTION
POTENTIAL IN LARGE BUSINESS ORGANIZATIONS:
A FOUR COMPANY REVIEW

Richard C. Harkness
Peter G. Burke

SATELLITE BUSINESS SYSTEMS
8283 Greensboro Drive
McLean, Virginia   22091

INTRODUCTION

A methodology was developed at SBS in 1978/1979 to estimate the future demand for teleconferencing (TC) within any large organization with multiple locations, and to estimate the economic benefits and costs asssociated with that level of demand. This methodology is based on the premises that a sizable portion of teleconferencing demand will result from teleconferences which substitute for business trips made to attend meetings, and that by analyzing existing travel and meeting patterns it is possible to estimate this important component of future demand. It has been used jointly by SBS and several large U.S. corporations.

The methodology involves a travel survey, two computer models, and a plan for analyzing the survey data base and model outputs. It is used to address the following planning questions:

1) What portion of an organization's travel and meetings could be replaced by teleconferencing? In other words, what is the potential demand or usage?

2) To what extent could the benefit of displaced travel expenses offset the cost of teleconferencing?

3) Which company locations should have teleconferencing facilities if the objective is to maximize travel displacement?

4) How many teleconferencing rooms and associated transmission channels will be needed to handle demand for teleconferencing?

In addition, the survey and models yield numerous statistics on travel patterns, travel costs and business meetings that have never

before been available to an organization's management.

Total teleconferencing demand will probably be much greater than what arises from travel substitution because teleconferencing will lead to a general increase in the amount of communication.

The methodology is likewise conservative in its accounting of benefits to be derived from teleconferencing. Only the displaced travel expenses and travel time are quantified. It is generally felt that teleconferencing will improve communications, thus leading to better decisions and increased productivity and that these benefits will outweigh direct travel savings. However, productivity implications are difficult to assess and are not the focus of this methodology.

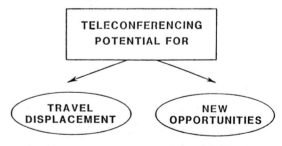

FIGURE 1

Figure 1 lists a wide range of teleconferencing benefits. The approach taken probably underestimates both teleconferencing demand and benefits, perhaps by several hundred percent. Any underestimate of demand will not cause serious problems because either demand can be managed downward by an organization's internal policies and user charges, or additional TC rooms can be added to handle unexpectedly high demand. What is important is that the methodology establish whether demand at least meets the minimal level needed to justify a network in the first place. Of course, the risk taken with a conservative approach is that a network which would, in reality, be justified by value-added benefits might fall short of being justified by the techniques described here.

Since this methodology was expensive to develop and requires each user organization to conduct a large travel survey, easier and less expensive alternatives were examined before this approach was decided upon. These include analysis of completed travel expense reports, interviews with managers and travelers, and extrapolation of findings from European studies. All were found deficient and it was finally decided to start from scratch and develop a methodology specifically for the purpose at hand. The result goes well beyond any approach that has been taken in the past toward the quantification of TC demand in terms of statistical accuracy, comprehensiveness, and relevance.

This paper discusses the methodology process and will convey some of the nonproprietary findings gathered in four surveys of large companies, as well as similarities and dissimilarities across all four surveys.

THE MODELING PROCESS

The basic approach taken to estimate TC demand is to gather, by questionnaire, detailed data on a random sample of business trips and, by examining them one at a time, to determine what fraction of the total could be substituted by teleconferences. The fraction substituted in the sample is then assumed to hold in the universe of all trips. In this way, the results of analyzing the sample are extrapolated to annual totals for trips substituted, teleconferencing demand, TC costs and travel savings. The model is run for each individual firm and several hundred to several thousand travel questionnaires are used.

This general modeling technique is similar to the "model split" models used by transporta-

tion planners to estimate the demand for automobile travel versus public transit. Both require a large random sample of individual trips, and both simulate behavior at the individual trip level before compiling totals. To our knowledge, this is the first time a model split model has been applied to the teleconferencing versus travel tradeoff.

A computer model called the TC Demand Model was developed to perform the necessary data processing. When given a TC network (a set of company locations with TC), the Demand Model estimates the amount of teleconferencing demand in terms of hours per year in that network. An APL program was written to process "value-added" data not used by the Demand Model but considered important in developing a business case. This program also derived standard statistical summaries of all the questions in the survey.

The cost of teleconferencing is computed by a separate model called the TC Cost Model. By making multiple runs of the models, it is possible to roughly optimize the set of locations chosen for TC and to determine the benefit/cost ratio associated with any particular location in the TC network.

The Demand Model works by examining each individual trip, using data from the questionnaire, and determining whether the trip could be partly or totally substituted by the hypothesized teleconferencing network.

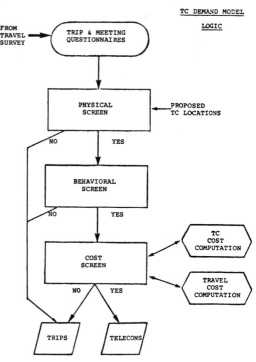

FIGURE 2

This determination is made by testing the trip and its meetings against three screens in sequence. The screens are illustrated by the flow chart in Figure 2. Only if a meeting passes all screens does the model consider it substitutable by teleconferencing.

Since not all company locations will have on-premise teleconferencing rooms, the model considers the possibility that meeting attendees will travel to the closest TC location. The cost of this "feeder" trip is calculated and considered as part of the cost of the teleconference.

The following statements can be made about trips which pass all screens and which the model designates as substituted by teleconferences:

1) The trip was made within the Continental U.S. for the purpose of attending meetings.

2) Teleconferencing was physically available to all the meeting attendees, and TC rooms were large enough to handle the number of attendees.

3) Teleconferencing was a suitable substitute from a behavioral or communications effectiveness viewpoint.

4) Teleconferencing was less expensive than travel.

This is not the same as saying that teleconferencing would necessarily have been chosen had it been available to those who filled out the questionnaires. Clearly other factors such as personal attitudes toward travel and company policy would have influenced their choice between TC and travel. For this reason, the model is not intended to be a predictive tool. It does, however, yield a reasonable planning estimate of the demand that might arise from travel displacement.

After all computations are complete, the model produces a series of reports which include not only the key demand and benefit totals but also travel matrices, meeting and trip statistics, results of interim calculations, etc. This information is combined with statistical data derived from the APL analysis to develop the business case.

THE TRAVEL SURVEY

In January of 1979, the Travel and

Meeting Characteristics Survey was developed. The development process included many reviews and several tests on SBS travellers. Keypunch formatting was also applied and tested.

The questionnaire itself has two parts. Part one is a tan booklet with instructions and three sections:

- General information
- Travel information
- Non-business meeting activities

The second part of the questionnaire is the business Meeting Record Sheets. One meeting record sheet is completed for each meeting which occurs on the trip. Unreported meetings and the number of unreported hours and minutes of those meetings are recorded as the last question in the tan booklet. Three meeting record sheets are provided in each questionnaire.

A self-addressed envelope with a preprinted return address and a cover letter rounded out the questionnaire package. The return envelope in all cases has an internal company mail return address, and a UPS address just in case. The cover letter is authored by a recognized figurehead within the organization. Both the return address and the author of the cover letter are carefully chosen to maximize response and minimize biases to the results. The cover letter as well as the questionnaire assure "all answers will remain confidential" and the responses are "not intended to be an audit of activities."

The formal field test of the Travel and Meeting Characteristics Survey was conducted in February, 1979. Questionnaires were distributed in our Stamford field office and the SBS corporate headquarters in McLean, Virginia. The intent of the test was primarily to test the survey form, keypunch format, instructions, and the questions themselves. There was some concern about the length of the questionnaire so a short form was devised which omitted several "value-added" questions. Half of the test group received long forms and half received the short form. The response rate was not significantly different in the test, but we decided to provide both versions for future surveys. Two of the four surveys used a 50/50 split and the other two used long forms only. (In all cases, there was no significant difference in response rate.)

Modifications were made to all parts of the questionnaire and survey procedures following the internal SBS test. Instructions were shortened and clarified, keypunch instructions rewritten, and administrative tasks modified.

The questionnaire package was attached to airline tickets, private aircraft reservations, and, in a few cases, railroad tickets. While the methodology suggested a random sample distribution, it was decided to attach questionnaires to all tickets.

IMPLEMENTATION

Two surveys were conducted in 1979, one in 1980, and one in 1981. Each survey differed in size and companies motivations but the overall objectives were the same. "How can we improve communication between and among our remote office locations?" In all cases, increased productivity of professionals and managers, shorter decision cycles, and access to key individuals were considerations.

Modifications were made to the questionnaire form by each company. Functional areas were defined and interviews were conducted to construct a representative list of meeting types for the meeting record sheets. Additional questions were also added and administration and distribution procedures reviewed. Questions critical to the Demand Modeling process were never changed and additional "value-added" information was obtained from new questions.

Distribution and administration varied from one company to another, but every attempt was made to attach a questionnaire to each ticket distributed during the survey period, and historic records were used to verify anticipated volumes. Call back procedures were implemented to increase the return rates, but coercion was not used.

Questionnaires were returned to the companies' administrators for processing before keypunching. After keypunching, cards or tape were forwarded to SBS for analysis. Some of the companies chose to analyze the data in parallel with SBS using their own statistical software packages.

A total of nine thousand three hundred (9,300) questionnaires were distributed with an average response rate of about 50 percent. Response rates varied significantly, however, from a high of 71 percent to a low of 30 percent. Diligent efforts of administrators accounted for very high response rates in two of the surveys. Four thousand one hundred eight (4,108) validated questionnaires and over five thousand (5,000) meeting record sheets in total were processed.

THE RESULTS

Compiled here is a summary of findings which we feel are representative of large, regionally dispersed business organizations. (Specific information about each of the companies travel and meeting characteristics is proprietary. The individual data bases created by each company's survey remain confidential. Much of this proprietary information would be of little interest to other organizations in any case.)

First, let us address the planning questions (see Introduction). It can be assumed that an organization will not invest time and resources in a teleconferencing study unless it feels there is a need to improve communications between distant locations. Often companies have specific locations in mind and many have one or more applications identified. What we have tried to do with the travel survey approach is provide enough hard data to confirm the organization's "gut feel" and suggest the appropriate implementation approach. The data base and model have been used to develop realistic, defensible business cases. Actions taken are then supportable by data supplied by the actual intended users. The risk in this approach is, of course, the possibility that a teleconferencing system is not justifiable. A survey is, however, a much cheaper way of finding that out than a failed pilot program.

The actual findings reported here come from the Demand Model Process and an APL DI program analysis of the data. Each questionnaire provided up to two hundred and eighty-four (284) separate fields of data (if all three meeting record sheets were used).

Tables 1 through 3 as well as Figure 3 convey some of the statistics collected about trips and meetings from the Demand Model.

There are three points of special interest. First, although the average meeting duration in the several firms examined was over 3 hours, these same firms are finding that teleconferences last only about one and one-half (1-1/2) hours. With certain assumptions, this corroborates and roughly quantifies the oft expressed observation that teleconferences are shorter than the meetings they replace.

Second, we found that visual materials of one sort or another are used in almost all meetings, thus substantiating the requirements to provide appropriate scanners and displays for them in the teleconferencing room. The six types of graphics grouped at the top of

TABLE 1
MEETING DURATION PROFILE*

Duration Hours	% Of Total Meetings
0 - .25	10.1
.26 - .50	3.2
.51 - .75	3.4
.76 - 1.0	2.6
1 - 1.5	9.3
1.5 - 2.0	6.8
2 - 2.5	9.5
2.5 - 3.0	6.3
3 - 4	11.6
4 - 5	10.7
5 - 6	7.4
6 - 7	6.4
7 - 8	4.3
8 - 10	3.1
10 - 12	1.6
12 - 14	0.9
14 - 16	1.1
16 - 18	.3
18 - 20	.4
over - 20	1.2

Average: 3.8 hrs.
Median: 3 hrs.

* Sample size: 1,112 meetings.

TABLE 2
MEETING STATISTICS*

Average Duration	3.8 Hrs.
Average Number Of Attendees	5.9
Average Number Of Traveling Attendees	3.0
Percent At Company Locations	40%
Percent Of Meetings Where All Attendees Are Company Employees	42%
Travel Cost Per Meeting Hour	$180 ⎤ Total For All
Hours of Lost Time Per Meeting Hour	5.7 ⎦
Travel Cost Per Meeting hour	$ 53 ⎤ For One
Hours of Lost Time Per Meeting Hour	1.6 ⎦

*Based on a sample of 1020 meetings in one large company.

TABLE 3
TRIP STATISTICS*

Total Number Of Trips In Sample	100%
Trip To One Location Only	85%
Average Trip Distance	900 miles one way
Average Trip Cost	$380
Average Cost of Trip With Meetings	$300 + 8.9 Lost Working Hrs.
Average # of Meetings on Trip with Meetings	1.5

*Sample size 763 trips based on a survey done in mid-1979 in a large manufacturing company.

Figure 3 are those which require resolutions higher than provided by standard NTSC television. Their frequency is sufficient to indicate the need for a high-resolution graphics subsystem in any teleconferencing room where widespread use is desired. One to two hundred lines per inch resolution has been found appropriate in this respect.

FINDINGS: GRAPHICS

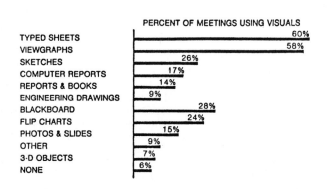

PERCENT OF MEETINGS USING VISUALS

TYPED SHEETS	60%
VIEWGRAPHS	58%
SKETCHES	26%
COMPUTER REPORTS	17%
REPORTS & BOOKS	14%
ENGINEERING DRAWINGS	9%
BLACKBOARD	28%
FLIP CHARTS	24%
PHOTOS & SLIDES	15%
OTHER	9%
3-D OBJECTS	7%
NONE	6%

BASED ON COMPANY "B" TRAVEL SURVEY   SAMPLE: 1070 MEETINGS, 1979

FIGURE 3

A final point is the finding that approximately 9 business day hours are lost in the typical business trip. The importance of lost time as a displaceable cost was anticipated and the questionnaire was carefully designed to capture the parameter accurately. The exact departure and return times were recorded and the computer model subtracted the hours between 5 p.m. and 8 a.m. as well as the time spent in meetings. The remainder was called "lost time." The corporations surveyed each assigned a per hour value to this time. The final result was that for every dollar of direct travel expense, there was $.50 to one dollar in lost time.

The results of the modeling itself are mostly proprietary. It will, however, be of interest to researchers to note that SBS used the same behavioral algorithm used by the Communications Studies Group (CSG) for their classic research in Europe[1]. Figure 4 compares the results for one of the U.S. firms SBS analyzed with that obtained earlier by CSG from analyzing a cross section of European firms. Both pie charts show the percentage of meetings which could, according to the CSG algorithm, be substituted by teleconferencing with no detrimental outcome on the effectiveness of the meetings. The charts also indicate the long-term potential for teleconferencing when it becomes widely available.

BEHAVIORAL ELIGIBILITY OF MEETINGS

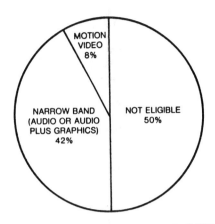

BRITISH RESEARCH BY CSP, LONDON

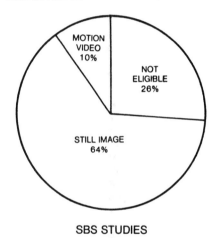

SBS STUDIES

FIGURE 4

In order to get a reasonable estimate of the percentage of types that might be substituted near term, the SBS model had, of course, to consider other factors: principally access to TC facilities and costs. Typically, we examined scenarios where the largest 5 to 20 corporate locations in the U.S. were equipped with a private (intracompany) teleconferencing network. The computer runs showed that until on-premise teleconferencing facilities are ubiquitous and fully interconnectable throughout the U.S. business community, physical access will remain the primary limit to the percentage of trips addressable by teleconferencing. Cost was found to be an important consideration with full motion teleconferencing, but less so with still frame. Although recent advantages in video CODEC's which reduce transmission speeds to 1.5 Mbps or below naturally make motion easier to

cost-justify, the analysis still underlines the need for corporations to have hybrid facilities where users can choose either motion or still-frame depending on the meeting type and willingness to pay.

Several questions on the Travel Questionnaire probed for value added benefits. For example, in one firm 64 percent of those surveyed felt more contact with distant parties would be beneficial to their jobs whereas only 9 percent wanted to travel more often. Everyone is familiar with how long it often takes to arrange a meeting once the need to do so is felt. We probed this and found that the median delay between the time need was felt and the meeting occurred was 14 days. The average delay was 47 days. These delays suggest something we might call "information float." It certainly has some impact on decision cycles and productivity. Teleconferencing can reduce this delay considerably, not principally because it avoids travel time but because the time each participant must block out on his calendar is an hour or two rather than a day or two. The probability of being able to schedule the meeting sooner, rather than later, rises accordingly.

The APL analysis was used extensively to identify similarities and differences in the travel and meeting characteristics of the four companies. While some of these findings stand alone, others need to be looked at in the context with other answers.

SIMILARITIES

o   Attitude toward travel and contract with distant parties

   -   Fifty-eight percent (58%) of the respondents felt more contact with distant parties would be beneficial, but most felt their current amount of business travel was "about right". (See Chart 1 and 2.)

o   Percent of trips with meetings

   -   Meetings were held on over seventy percent (70%) of the trips taken by the respondents. Over forty percent (40%) had two or more meetings. (See Figure 5.)

**PERSONAL ATTITUDE TOWARD
CONTACT WITH DISTANT PARTIES**

- MORE CONTACT WOULD BE BENEFICAL   58%

- CURRENT LEVEL APPROPRIATE   40%

- LESS CONTACT WOULD BE BENEFICAL   2%

CHART 1

**PERSONAL ATTITUDE TOWARD
AMOUNT OF BUSINESS TRAVEL**

- TOO MUCH   12%

- ABOUT RIGHT   73%

- TOO LITTLE   15%

CHART 2

MEETINGS PER TRIP

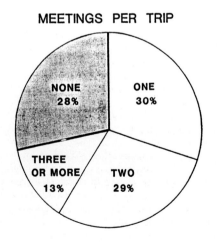

FIGURE 5

o   Types of meetings on trips

-    One of the most significant similarities across all four surveys was the types of meetings held (Chart 3). Not only did these three types head the list in all four surveys, the Demand Model Behavior Screen ranked them highest in teleconferencing substitutability. In all four companies, these three meeting types comprised between fifty (50) and sixty (60) percent of all meetings held on trips. Some Formal Presentations did have a low suitability rating due to the conflict and personal performance review nature of the meeting.

**TYPE OF MEETINGS
PEOPLE TRAVEL TO ATTEND**

- INFORMAL WORKING SESSION/DISCUSSION

- INFORMATION EXCHANGE

- FORMAL PRESENTATIONS

CHART 3

o   Importance of meetings on trips

-    An attempt was made to measure relative importance of trips and each meeting on the trip. As one might expect, most respondents felt that the first meeting on the trip was important enough to justify the trip and could not have been delayed (Chart 4). We were somewhat surprised, however, at the importance registered for second and third meetings on the same trip. The impact of delayed meetings on conducting business is further raised when you consider that the average delay to arrange a meeting for which attendees need to travel is fourteen (14) days.

IMPORTANCE OF INDIVIDUAL MEETINGS

- **IMPORTANT ENOUGH TO JUSTIFY TRIP**
    - **FIRST MEETING ON TRIP   80%**
    - **SECOND MEETING ON TRIP   53%**
    - **THIRD MEETING ON TRIP   43%**

- **COULD NOT HAVE DELAYED MEETING**
    - **FIRST MEETING ON TRIP   86%**
    - **SECOND MEETING ON TRIP   70%**
    - **THIRD MEETING ON TRIP   63%**

CHART 4

o   Characteristics of meetings on trips

-   Characteristics of the meetings further highlight the impact on productivity of the traveling business person (Chart 5). Most of these meetings were not regularly scheduled and two-thirds of them were not scheduled by the person traveling. For these and other reasons, some people who might like or need to attend were not able to attend. Were teleconferencing available at these locations, not only would teleconferenceable meetings be able to occur sooner, but key individuals would be more accessible.

MEETING CHARACTERISTICS

- **80% WERE NOT REGULARLY SCHEDULED**

- **TWO THIRDS WERE NOT SCHEDULED BY THE PERSON TRAVELING**

- **IN MANY CASES OTHER PEOPLE WOULD ATTEND IF THEY DID NOT HAVE TO TRAVEL**

CHART 5

DIFFERENCES

The nature of a business tends to structure its operations. Manufacturers operate very differently than insurance companies, for instance. For that reason, many of the findings in the four surveys were dissimilar across all companies.

o   Respondents characteristics

-   Three of the four companies had about 50/50 split on managers versus non-managers traveling. The fourth (a financial institution) had a much higher percent of managerial level travellers.

o   Functional areas traveling

-   The functional areas within an organization which travel are very dependent on the nature of the company's business. Identification of the right functional areas and appropriate teleconferenceable applications is key to a successful teleconferencing implementation.

o   Factors that limit travel

-   Limiting factors to travel varied by company also, but respondents in 1979 were less cost conscious than those in the 1980 and 1981 economic situation.

o   Locations of meetings

-   Predictably, meeting locations varied by company. This information is, of course, critical to planning an implementation effort on a case-by-case basis.

o   Meeting visuals

-   Visuals used during meetings were different at each company, but they are also of critical importance in determining teleconferencing suitability. They, in cojunction with other variables like number of attendees, dictate the support systems required to handle the meetings. The less an individual has to modify his current meeting behavior, the more likely he is to accept the innovative meeting options TC provides.

o   Other activities on the trips

-    Non-business meeting activities varied by type and amount across all companies. Giving and receiving training ranked hightest overall, but "inspection of machinery, structures or other things which cannot easily be brought into a conference room" ranked highest in one organization. Some of these activities, of course, might be addressed with the proper TC application design.

SUMMARY RECOMMENDATIONS

The information provided in this paper should be helpful in developing a teleconferencing business case for any large geographically dispersed organization. Some amount of primary research is however required, and one should focus on determining the right applications at the right locations. Historically, systems that support engineering applications have been very successful, but that would not help you if you are developing a system for a bank.

It is possible, by conducting a travel survey and using modeling such as that described, to quantify the travel substitution potential of teleconferencing with greater credibility than more casual methods yield. The basis for a business case can thus be established. Unless the model can be calibrated against emperical evidence, its absolute accuracy is, of course, unknown.

The Demand Model has proven most useful in running sensitivity tests on the multitude of variables from the Survey. It provides suitability evaluation of meetings, locations, and user groups for teleconferencing. When used in conjunction with the TC Cost Model, it also provides financial data for planning.

The findings listed here can reduce the amount of research required in evaluating teleconferencing potential in an organization. We suggest you focus on the areas listed in the results under Differences. While a travel survey might be warranted for a company which plans a major multi-node network, indepth interviews by qualified researchers may be adequate to supplement secondary data from this and other research findings. Cost and sophistication of the teleconferencing design required to satisfy the intended application(s) can dictate the amount of justification research required.

REFERENCES:

1   Pye, Roger; 'Teleconferencing: Is video valuable or is audio adequate?' Telecommunications Policy, June 1977.

The Teleconferencing Resource Book: A Guide to Applications and Planning
Lorne A. Parker and Christine H. Olgren (eds.)
Elsevier Science Publishers B.V. (North-Holland)
©Center for Interactive Programs, University of Wisconsin-Extension, 1984

VIDEOCONFERENCING IN AMERICAN BUSINESS:
PERCEPTIONS OF BENEFIT BY USERS OF INTRA-COMPANY SYSTEMS

(A REPORT ON WORK IN PROGRESS[1])

Kathleen J. Hansell
and
David Green

Satellite Business Systems
McLean, Virginia

METHODOLOGY

Although there is a great deal of interest in teleconferencing, there is very little real understanding of its value. A great deal of speculation on the communication tool's ability to displace travel provided the grist for theorizing in the 1970's. Major efforts considered the question and postulated various substitution ratios.

The results of these efforts have been inconclusive and contradictory.[2]

More recently, observers have begun to consider "justifications" other than travel substitution. Jeff Charles suggests four justification approaches in addition to travel cost avoidance which are used by organizations to support a decision for teleconferencing: the intuitive leap, timeliness of decisions, improved work coordination, and extending the range of communications available to an organization.[3]

The field has remained one primarily of theory and experimentation. But there is growth in the number of systems installed in the last few years. There is also growth in numbers of users and their collective experience. This critical mass of real experience allows one to begin to understand, for the first time, actual users' perceptions of the medium. In the latter half of 1981, Satellite Business Systems undertook such a study, with the authors as principal investigators.

The major purpose of the study was to understand the kinds of benefits perceived by corporate users of videoconferencing. To accomplish this purpose, a survey methodology was formulated with the assistance of Dr. Karen File of National Analysts, Booz-Allen & Hamilton. The population was defined as domestic business organizations using an on-premise captured-frame[4] or motion videoconferencing system which was in place at least six months during 1981. Further definition limited the population to those with "daily" (a minimum of 20 hours per month) use for business meetings. Although arguments can be made that there is much to be learned from new or limited implementations, an equally valid argument is that some amount of experience is necessary before one begins to observe and categorize change; hence, the six-month and 20-hour usage criteria.[5] Videoconferencing systems internal to AT&T operating companies were excluded. Public agencies and organizations whose uses of videoconferencing were primarily other than business meetings (e.g., testing, training, sales demonstrations) were also excluded.

More than sixty leads from earlier studies, equipment vendors, consultants, and the popular press were checked. Evidence of an inverse correlation was frequently found: Organizations with the most publicity about their videoconferencing efforts were often those doing the least actual videoconferencing. Potential candidates were qualified against the criteria outlined above, and sixteen organizations were initially found which were thought to fit the population definition. Closer examination disqualified six. Examination of scheduling records showed four which did not meet the usage level criteria, one with primary usage in specialized industrial testing rather than business meetings, and one with recently discontinued operations and a usage level that could not be verified. The remaining ten organizations met the criteria and responded affirmatively to an invitation to participate in the study. (See Company Profiles.)

Assistance in formulating a survey instrument was sought from a variety of people knowledgeable in videoconferencing and survey research. Two areas of concern were particularly heeded. The first (obvious from the beginning and repeatedly confirmed during the qualification process) was that the perceptions of videoconferencing use and effects vary greatly with the respondent's position in the organization relative to the medium. For this reason, three distinct groups within the using organizations were identified for survey: the decision maker who authorized the expenditure for the facility; the manager of the videoconferencing function; and a random sample of actual users.

The second concern was that "benefit" to one may be "detriment" to another. Examination of the benefits issue raised some interesting examples. The literature and conference circuit have repeatedly touted faster decision making, involving more people, greater task orientation, and travel expense reduction as benefits of videoconferencing. In a test of a preliminary survey instrument, a user pointed out that faster decision making with videoconferencing was often detrimental to the quality of the decision. He cited a greater tendency to "shoot from the hip" and make hasty decisions because of perceived time constraints. Increasing the number of meeting participants may not have a positive effect (more is not necessarily better). Greater task orientation can have the potential of negatively affecting user comfort and satisfaction. A decrease in travel, while beneficial to the organization's budget, may be detrimental to the personal visibility of an employee. And so on.

The instrument was therefore designed to present neutrally items which could be affected by videoconferencing. For example, instead of asking whether a respondent had observed faster decision making, he was asked to comment whether or not he had noted any change due to videoconferencing in the time required to make decisions. If so, he was

asked to characterize the change as an increase or decrease and to rate the change as beneficial or detrimental. The user was asked to assess benefit or detriment to himself personally. The decision maker and function manager were asked to comment from an organizational perspective. Decision makers and function managers were queried both on expectations prior to installation and on outcomes observed. Users were queried on outcomes only; additionally they were asked about their experiences with videoconferencing in a series of open-ended and closed-ended questions.

Function managers and decision makers were interviewed in person (except one who was interviewed via telephone) between October 1981 and February 1982 using a standard interview guide. The guide for function managers was the same as that for decision makers except for the addition of items describing the facility and its management.

Part of the interview with the function manager involved a look at the videoconferencing facility and usage records. Facilities varied from "bare-bone" multi-purpose rooms to elegant executive conference suites. All of the participating organizations kept scheduling records or logs which listed at least the name of the person who called a videoconferenced meeting. These records varied from annotations on a secretary's calendar to a computer-assisted central scheduling system with detailed printouts. In all but two cases, the records were centrally compiled.

It was from these scheduling records that the researchers drew a random sample of users. The technique involved counting the number of videoconferenced meetings (excluding demonstrations, training sessions, photo sessions, maintenance, and testing), dividing by 30 to determine "n," and sampling every "n^{th}" name in the chronological record. There was no attempt to stratify the sample by kind of meeting, functional area of the respondent, or his position within the organization. More frequent users had a greater likelihood of being selected, a weighting which paralleled the study's selection of experienced user organizations.

The survey instrument used in the personal interviews with function managers and decision makers was reviewed with the National Opinion Research Center, University of Chicago. With NORC's assistance, it was modified to a questionnaire suitable for written response from users. Two significant changes were made in addition to creating a more

COMPANY PROFILES

Company A is a multi-billion dollar financial services organization. Due to overcrowding at its headquarters, the company recently relocated a large number of its data processing personnel to a new facility. To enable good communications between programmers in the new location and their users at headquarters, Company A installed sophisticated motion videoconference rooms in each location late in 1980.

Company B is a Fortune 50 energy and resources company operating worldwide. The organization installed a captured-frame link between research and engineering facilities in the East and on the Gulf Coast early in 1981.

Company C operates telephone companies in a number of areas and manufactures communications and electrical products. Communication between headquarters and operating companies was perceived as a significant issue in the late 1970's resulting in the installation of a six-node captured-frame network in 1979.

Company D is a multi-billion dollar aerospace organization. To improve communications between a headquarters engineering center and a manufacturing facility approximately 500 miles away, the company installed a captured-frame videoconferencing system in early 1981.

Company E is a Fortune 100 heavy equipment manufacturer concentrated in the midwest. Primarily to control travel costs between its headquarters and a large manufacturing facility in an adjacent state, the company leased two captured-frame units for each facility in early 1981.

Company F is a Fortune 50 information industry organization. Following a captured-frame experiment between engineering design and manufacturing facilities in the mid-70's, Company F began installing a sophisticated captured-frame network now approaching twenty locations. The network is used by all business functions.

Company G is a group of insurance companies underwriting a wide range of policies. As with Company A, this organization separated its data processing department from its headquarters. It established a sophisticated motion videoconferencing system in late 1980 to link the two groups.

Company H is a rapidly growing group of high technology companies dispersed around the U.S. A five-node captured-frame network was installed in early 1981, at the request of the CEO, to enable regular communications among senior executives.

Company I is a multi-billion dollar consumer goods company. Its product research is well known. To enhance communications between research scientists and engineers located at several sites throughout a metropolitan area, the organization installed a four-node motion network in 1980.

Company J makes computing and aerospace products. What started as a test of satellite transmission characteristics grew into a communications tool for engineers designing and manufacturing computer systems. The system is captured frame.

attractive format. One change provided for collection of demographic information from the user. The second strengthened the manner in which observations of changes due to videoconferencing were reported. The revised questionnaire was sent during February 1982 to 219 user names [6] with a cover letter describing the study and seeking input.

User responses will be statistically analyzed by NORC. Reports subsequent to this one will detail those findings.

ORGANIZATIONAL PERSPECTIVES

As of this writing (March 1982) the study is not yet complete. It is possible, however, to analyze responses of function managers and decision makers in order to understand better their perceptions of organizational expectations and outcomes relating to videoconferencing.

Several caveats are in order. The first involves closeness to actual use. Generally the function managers were videoconferencing users themselves, although some of this use was of a demonstration nature. Most seemed to be in close touch with other users in their organizations. They tended to base their responses on personal observation or the results of their own efforts to gather feedback from users. Decision makers were farther removed. Many had moved on to other positions within their organizations.

The second caveat stems from the fact that function managers and decision makers have direct vested interests in their video-conferencing systems. In the case of the decision maker, a multi-thousand dollar or million dollar decision in favor of videoconferencing was made based on an expectation of usage and some return on investment. The function manager's interest is similar. He is judged within his organization and by the videoconferencing community largely on the success or failure of his system, measured in usage statistics and subjective user satisfaction. This is particularly true of present function managers who were involved in the design and implementation of their systems. Assessments of videoconferencing's accomplishments, therefore, must be accepted in context of the close ties between system success and respondents' responsibilities for installation and management.

A third caveat stems from the compilation of expectations. Decision makers and function managers were asked to remember what expectations they held at a point just prior to the start of their operational systems. They were then asked to match these expectations to outcomes observed today. Compiling expectations after the fact is difficult at best-- memory fades. There is a tendency for remembrances of expectations to be reinforced by actual experience. There is also a tendency for memories to be influenced by reports from the literature and conference circuit of what ought to be the case; function managers and decision makers were familiar with the popular press and conference activities on video-conferencing.

The remainder of the report reviews some of the significant findings from the survey of videoconferencing function managers [7] and decision makers.[8] The reader is cautioned against drawing final conclusions based on this small segment of respondents and is urged to obtain a copy of a subsequent report which will describe findings from the user portion of the study.

Clusters of change reported by function managers and decision makers were found in five areas: travel, amount of communications, resource accessibility, visibility, and meeting effectiveness.

Travel

Two of the travel-related questions posed to function managers were: "Did you expect to see a change in the amount of time spent in travel?" and, "Did you expect to see a change in travel expense?" All function managers indicated that they had indeed held expectations of a decrease in travel time and a reduction of travel expense. Decision makers concurred, with the exception of one who said he had not expected any change in travel expense as a result of videoconferencing.

When asked whether they had actually observed any changes in travel time, affirmative responses predominated. All but one function manager and one decision maker noted a decrease in time spent in travel. Related questions pertaining to time away from home and time away from the office showed similar responses. The decreases were all rated as beneficial to the organization.

Respondents had no formal measurement system to calculate time saved. Most organizations assumed that all or a portion (one organization used 70 percent) of the videoconferenced meetings would have been held by traveling to a face-to-face meeting had videoconferencing not been an option. In some cases, calculations had been made of estimated travel time savings. Travel times between a particular set of locations were multiplied by the number of participants who video-conferenced from those locations, in an attempt to quantify the value of time savings to the organization.

As for travel expense, the majority of the function managers and decision makers stated they had observed a decrease as a result of videoconferencing. Again, very few of the organizations had attempted to measure these decreases. Only one pointed to a decrease in actual travel expense budget accounts. It was unclear what influence videoconferencing may have had in minimizing increases in travel budgets.

The evidence pointing to a decrease in travel expense was most often reported to be personal observation based on numbers of people using videoconferencing. In a few organizations, travel expenses between loca-tions were estimated and applied against the number of participants in videoconferences between those same locations.

Assuming that the number of video-conferencing participants equals the number of people who would have traveled has long been discounted in the research community. Responses in two other areas of this study partially illustrate why this direct substitution assumption is invalid. Respondents were asked if they had observed any change related to videoconferencing in meeting frequency or numbers of meeting participants. The majority noted an increase in both the frequency of meetings and in the number of people involved in a meeting, decision makers more often citing the latter increase than function managers. Clearly, looking at numbers of video-conferencing participants can give an inflated view of presumed travel displacement. A few of the organizations recognized this and adjusted estimates downward.

Others felt simply that attempting to quantify travel dollars saved was a futile activity. The term "value-added" was used by one function manager to explain how his organization viewed its videoconferencing system and why calculating travel expense was not a valid exercise for his organization.

This is not to say that organizations didn't initially cost justify a videoconferencing system on projected savings in travel dollars and travel time. Many did. One organization based its business case for installing video-conferencing entirely on projections of travel time to be saved. While justifications on projections are common, conclusive data after installation is lacking. It is impossible to quantify precisely what cost savings are accruing to organizations who use video-conferencing as a replacement for some amount of travel.

Amount of Communications

One of the outcomes most sought by organizations through videoconferencing is an increase in the amount of communications among managers and professionals within the organization. This expectation was cited by eight of the ten function manager respondents. The two who reported no expectation of change were from the organizations that had imple-mented videoconferencing concurrently with user relocations. Their expectations were to maintain the same level of communication via videoconferencing as had occurred before the relocations. Observed results were similar to expectations. Two function managers with short distances between locations reported no change observed in the amount of com-munications with other parts of the organization. All others responded that they had seen an increase in the amount of communications. The majority viewed that change as beneficial, and no one categorized the increase as detrimental to the organization.

Increased communications with dispersed parts of the organization was found to be a strong factor in organizational decisions to implement videoconferencing. One respondent said that his company's prime objective was "to get people to talk with each other." Another function manager had been instructed to "get my scientists a better way to exchange ideas." The foresight of the two organizations that relocated service departments came from a concern for amount of communications. Both organizations realized that distance had the potential of reducing necessary contact. Videoconferencing was seen as a way to maintain the amount of communications that existed when distance was not a factor.

Respondents indicated an increase in frequency of meetings and numbers of people involved in a meeting. These may have contributed to an overall increase in amount of communications. It is interesting to note that a decrease in meeting length was also frequently observed. The effect this had on amount of communications is unknown.

Resource Accessibility

Resource accessibility in videoconferencing is widely discussed in the literature; videoconferencing obviously allows for increased proximity to one's colleagues and files.

All but one function manager anticipated an increase in accessibility to people resources while meetings are in progress. Many of the systems allowed for such accessibility. In addition to being located in easily-reached company locations, many rooms had a built-in conference telephone capability. A call could be placed from the videoconferencing room, allowing all conferees to participate in an audio discussion with a resource person.

All but two function managers observed an increased ability to reach resource people during the course of a meeting. The increase was viewed as beneficial to the organization. The decision makers had not anticipated any change in resource accessibility. All reported, however, that they had observed such an increase.

Visibility

Function managers were asked whether they had anticipated any change in their personal visibility as a result of their activities in videoconferencing. All but two modestly answered no. Once the system became operational, though, the majority discovered that their visibility within the organization had indeed increased. The decision makers likewise indicated an increase.

Function managers noted that they received a number of requests for assistance in planning and executing meetings and that generally they had greater exposure to higher management. One noted his recent promotion as evidence of visibility. All would have agreed with one decision maker's observation that the increased visibility from videoconferencing "cuts both ways." While none of the respondents rated the visibility as detrimental, several chose a neutral position rather than clearly labeling their visibility as a benefit.

Another item in the survey checked on corporate visibility. Respondents were asked whether or not they had anticipated any change in their company's image of corporate leadership as perceived by the business community. Only a few indicated expectations of an increase. But in fact only a few reported that they had not observed a change. Increases noted in the business community's perceptions of their company's corporate leadership were thought to have occurred because of an increase in publicity--publicity that frequently presents erroneous information but is nevertheless very favorable to videoconferencing and the organization in particular. Most organizations had been besieged by researchers and reporters. Many had received invitations for speaking engagements on the topic of their organization's efforts in videoconferencing. Requests for visitations and demonstrations from representatives from other companies were also cited. Indeed, some of the organizations' scheduling records looked like a "Who's Who" in corporate America.

The increase in their organizations' leadership image was viewed as beneficial and desirable. One function manager summed it up by saying that his organization wanted to be regarded as a "modern business using modern techniques."

Although one might postulate a connection between leadership image and competitive advantage, respondents didn't indicate such was the case. While the majority had expected an increase in their company's competitive advantage, the majority reported that they had observed no change due to videoconferencing.

Meeting Effectiveness

Many of the organizations studied had informally surveyed users on whether or not videoconferencing had been an effective way to conduct a meeting. Function managers frequently mentioned evaluation or feedback forms or user comments as evidence when asked about meeting effectiveness. The majority of the function managers indicated that they had anticipated an increase in meeting effectiveness and that they had in fact observed such an increase. None of the decision makers reported expectations of a change, but all reported they had observed an increase in meeting effectiveness related to videoconferencing.

What makes for an effective meeting? Respondents cited several factors: more structure, more preparation, more task orientation, less time.

More structure, or more organization, was mentioned by a number of respondents. They reported that agendas were used more frequently in videoconferenced meetings and that materials were more likely to be prepared and distributed in advance. More preparation, particularly of visual aids, was felt to contribute to meeting effectiveness. Whether or not all participants benefit from the increased structure of videoconferences was unclear. One respondent suggested that "highly structured people do better" in videoconferences than less structured people.

Closely allied with an increased structure is an increase in the task orientation of participants. Only half of the respondents had anticipated a change, but a majority reported they had seen greater task orientation in videoconference meetings. Participants "get down to brass tacks more quickly," said one; others cited less idle chatter, less "side talk," more involvement, more attention.

A decrease in meeting length was reported by a majority of respondents. The function manager of one organization said that 95 percent of the videoconferenced meetings were scheduled for 45 minutes, the length of its standard scheduling block, and 5 percent scheduled for multiple blocks. Another commented that because a specified amount of time is allocated, there is a greater commitment to accomplishing the task within that boundary. (A number of systems had clocks prominently mounted in the rooms, a constant reminder to meeting participants of the time remaining.) Whether increased task orientation causes shorter meetings or whether other factors such as scheduling practices limit the time available and thus create "more pressure to get it done" and hence more task orientation is unclear.

USER PERSPECTIVES

This study is the first look at the impacts of videoconferencing in American business. This report describes the rationale and methodology of the research. It further describes perceptions of the system managers and those who made the decision to implement the application.

The primary focus of the research is on the perceptions of actual users. A report detailing those findings will be issued in May.

References

1. This preliminary report was prepared in March 1982.

2. A decade of these studies is summarized in Kenneth Kraemer's "Telecommunications: Transportation Substitution and Energy Conservation," Part I, Telecommunications Policy, March 1982, pp. 33-59.

3. Jeff Charles, "Approaches to Teleconferencing Justification: Towards a General Model," Telecommunications Policy, December 1981, p. 296.

4. Alternate terminology includes "still frame," "freeze frame," or "slow scan."

5. Another argument points out that there is much to be learned from failures. In the course of investigating videoconferencing, examples were found where systems had been unsuccessful and discontinued. Although a study of failures may shed some light on what "makes" or "breaks" videoconferencing implementations, such an investigation is not within the scope of this study. Rather, this study looks at experienced organizations with a current minimum usage level to ascertain effects of videoconferencing observed by users.

6. Ideally, the sampling technique would have produced a total of 300 names (30 from each of ten organizations). Questionnaires were mailed to 219 people because of rounding, removal of duplicate names, movement of a few users who left the participating organization between the time the names were selected and the questionnaires prepared for addressing, and the inability in three instances to draw a full 30 names because of small user populations.

7. Eight of the function managers had been in the job at least six months. In the two organizations where this was not the case, an individual who had been closely associated with the videoconferencing function over a six-month period was interviewed. Each of the two people in this category had been heavily involved in installation and operation of the system until the recent appointment of a function manager. Their responses, together with those of the eight experienced managers, were used to tabulate what is referred to as "function manager" response.

8. Pinpointing the decision maker was often difficult, and tracking him down for an interview proved a formidable task. Results shown here are based on responses from three very senior and clearly identifiable decision makers.

The Teleconferencing Resource Book: A Guide to Applications and Planning
Lorne A. Parker and Christine H. Olgren (eds.)
Elsevier Science Publishers B.V. (North-Holland)
© Center for Interactive Programs, University of Wisconsin-Extension, 1984

THE UTILIZATION OF VIDEO CONFERENCING:
A PRELIMINARY REPORT OF THE TELEDECISION PROJECT*

William H. Dutton
Janet Fulk
Charles Steinfield
Annenberg School of Communications
University of Southern California

Several trends suggest that the use of video telecommunication technologies is likely to expand significantly over the next decade. Travel costs are increasing in relation to telecommunication costs. In addition, the technical quality and feasibility of video telecommunications is continuing to improve with advances in communication satellites, video compression techniques, and more experience with systems now in use. Already, several large organizations are embarking on the development of in-house systems. Other organizations are increasing their use of publicly available systems.

The increased use of video teleconferencing has reawakened concern over the implications of new communications technologies for the organizations that employ them. For example, cost studies have called into question the efficiency of the telecommunications-transportation trade-off (Elton, 1978; Harkness, 1973; Kraemer, 1981; Nilles et al., 1976). Other research indicates that for many communication tasks, video telecommunication technologies are no more effective and sometimes less effective than audio-only communications, yet far less economical (Johansen et al., 1979; Pye and Williams, 1977; Short et al., 1976). Concern has also been expressed regarding the effects of video teleconferencing on the relative influence of various participants in organizational decisionmaking (Kohl, Newman, & Tomey, 1975). For example, will the utilization of video systems tend to increase the influence of central management in organizations with geographically dispersed offices? Or, will the technology facilitate decentralization by allowing greater participation in central management decisions by managers in distant facilities?

Interestingly, such important organizational issues have generated much speculation, but little research, on video teleconferencing users and uses. Little is really known even now about who utilizes video teleconferencing systems for what purposes and with what effects. The purpose of the pilot study reported here was to provide some preliminary information regarding users and uses of one publicly available system--AT&T's picturephone meeting service (PMS). PMS facilities were chosen because they were one of the few major, full-motion, two-way video conferencing systems in use at the time.

A Survey of PMS Users

In May of 1981, questionnaire packets were sent to the PMS room attendants in Los Angeles, Sacramento, and San Francisco. Each packet included a project cover letter, a self-administered questionnaire, and a return envelope. AT&T managers also attached a cover letter explaining the independence of this project from both Pacific Telephone and AT&T. PMS room attendants were asked to distribute one questionnaire to each meeting participant at the end of each meeting over a three week period ending on June 10, 1981, the date at which the PMS marketing trial period ended. Because our project personnel were not on site, questionnaires were numbered sequentially to permit both an estimate of response rates and the grouping of questionnaires by conference. Questionnaires were completed anonymously, except that those respondents interested in obtaining a report of findings voluntarily provided their name and address. This administration procedure yielded a response rate of 53%. Follow-ups to nonrespondents were impossible given the anonymity of the questionnaire and of each PMS customer. Despite this, the response rate is higher than many mail surveys with extensive follow-ups.

The Users of Video Conferencing

A developing stereotype is that advanced communication media are "elite technologies," basically used by the largest and wealthiest private corporations (Laudon, 1977; Schiller, 1981). However, survey responses suggest more variation among the users than is suggested by the elite image.

First, 50% of the respondents were from public and/or nonprofit organizations. This finding is explained in part by our sampling from a major government seat (Sacramento), and the fact that some governmental uses such as Board Hearings result in relatively more people being involved in a given meeting (only 39% of the meetings were conducted by public/nonprofit organizations). Nevertheless, the public sector presence is more substantial than conventional stereotypes suggest.

Moreover, there is a diversity of users within each sector, as shown in Table 1. Almost half of the public sector users are from health services industries. And, nearly one-third of the private sector respondents are from the telecommunications industry. Beyond these two industry concentrations of users, the range of industries seems to be the most characteristic feature of the users, particularly in the private sector. Included are profit-making and not-for-profit organizations, as well as voluntary associations. Also represented are manufacturing and service subsectors, and some industries not typically considered "high technology."

Table 1
Industries Represented in Sample

	Percent
Government	
Health services	23(16)
Transportation/motor vehicle	10(7)
Energy regulation	6(4)
Water resources	3(2)
Board of control	3(2)
Law enforcement	1(1)
General services	1(1)
Private industries	
Telecommunications	16(11)
Insurance	7(5)
Professional associations	6(4)
R&D defense related	4(3)
Education	3(2)
Manufacturing/distribution of assorted products	4(3)
High technology conglomerates	3(2)
Management consulting	1(1)
Health	1(1)
Retailing	1(1)
Financial	1(1)
Building products & manufacturing	1(1)
Engineering & construction	1(1)
Publishing	1(1)
TOTAL	100(69)

Second, video conferencing usage was not limited to only large corporations, although larger organizations were heavily represented. Figure 1 shows the percentage of respondents employed by organizations of varying sizes. Almost one-third (31%) of the respondents represented organizations employing fewer than 100 people. Still, over half of the respondents are employed by organizations of over 1,000 people. Respondents were also asked to rate the size of their organization relative to other organizations in their primary industry: 53% were "very large," 18% were "large," 12% were "medium" and 18% were "small." The fact that almost three fourths (71%) rated their organization as either "large" or "very large" in comparison with other organizations in their primary industry suggests disproportionate use of public video conferencing facilities by larger organizations. This trend toward larger users may actually be understated by our data, for two reasons. Larger organizations are more frequent purchasers of "in-house" systems, which reduce the need for using public facilities. Also, during the trial, the hourly charge for the use of video conferencing was kept below its actual cost. Because the accessibility of video conferencing is likely to be sensitive to rate structures, use by smaller organizations might be relatively high during this trial period.

The geography of video communications is more in line with conventional stereotypes, and is consistent with expectations based on organization size. Over 80% of the respondents indicated that the facilities and offices operated by their organization were dispersed beyond any single local area.

Just as stereotypes suggest video communication to be an elite media of the Fortune 500 company, it is viewed as an elite media within organizations--a tool of top executives, top managerial officials, and their staffs. Again, our survey responses suggest more diversity than this image implies. Middle (35%) and lower (16%) level managers and staff comprised over half of the users. Still, a disproportionate number of respondents came from top executive and (15%) upper management (21%) ranks.

Video conference users span a variety of departments and divisions within organizations. As shown in Table 2, marketing/sales, general management, and program/project management were the most frequent users, with a wide array of other departments employing this new communications tool (Table 2). Nor is the technology limited only to larger organizational units. Small (26%) medium (47%) and large (24%) organizational units were all well represented. Only 3% of the respondents reported their department as "very large."

The Uses of Video Conferencing

Following Pye et al. (1973) and Noll

Figure 1
Size of Organizations Using Video
Teleconferencing[a]

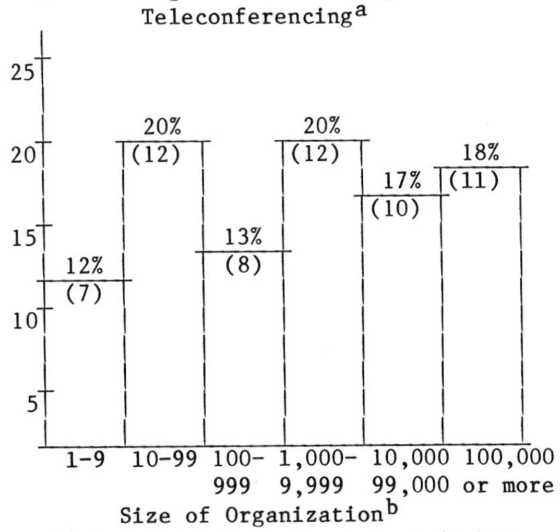

[a]Respondents were asked: Approximately how many people are employed by your company/agency?
[b]Mean = 28,438, N = 60

Table 2
Divisions/Departments of Respondents[a]

	Percent (N)
Marketing/sales	12(8)
General management	12(8)
Program/project	10(7)
Legal	8(5)
Personnel	6(4)
Finance	5(3)
Research	5(3)
Design/development	5(3)
Planning	3(2)
Accounting	3(2)
Production	3(2)
Other[b]	30(20)
TOTAL	100(67)

[a]Respondents were asked: "In what division or department do you work?"

[b]Other responses included: Computer, standards, energy conservation, instruction, procurement, mass transportation, clerical, engineering, training, editorial

(1977), respondents were asked to rate the amount of time that was devoted to each of several communication activities during the teleconference. The primary purpose of these self-reports was less to obtain accurate time estimates than to determine which types of communication activities tended to dominate the conferences, and, conversely, which activities rarely occurred. Obviously, the data obtained can provide only a broad brush sketch, since different individuals might perceive the same interchange differently from each other and differently from an outside observer.

Early studies of interactive video conferencing suggest that the media is best suited for information exchange activities, in contrast to communication activities involving interpersonal conflict, bargaining, or negotiation (Champness, 1973; Noll, 1977). As shown in Table 3 the survey results confirm the prevalence of information exchange activities. However, the meetings involved significant amounts of a variety of other communication activities: presentations, progress checks, discussions of documents, budget or plan reviews, coordination, brainstorming, project planning, advice giving, and problem solving. This diversity of communication

activities suggests that current video conferencing technology is flexible and able to satisfy a variety of communication needs. Less frequently reported activities include interviewing candidates, negotiating, delegating tasks, instructing, conversing socially, persuading, and resolving disagreements. However, the latter three activities were reported as receiving at least minimal attention by nearly half of the respondents.

The literature on teleconferencing suggests that many people prefer to handle negotiations or persuasion in face-to-face groups (Champness, 1973). Noll (1977) calls these "high personality involved" tasks. In some respects, the responses of users reinforce this view. People do not believe that much time was spent on such communication tasks as persuading and negotiating.

The communication activities in Table 3 have been arranged according to the broad communication tasks they appear to support: surveillance, socialization, or control (Lasswell, 1948), or entertainment (Schramm, 1971). Clearly, the activities within a category are not exhaustive. Superimposition of this broad frame, however, highlights the extant focus in telecommunications research

Table 3
Time Devoted to Communication Tasks and Activities[a]

Communication Tasks and Activity[b]	1 No Time Spent % (N)	2 Little Time Spent % (N)	3 Moderate Time Spent % (N)	4 Much Time Spent % (N)
Task A. Surveillance (sharing knowledge of environment)				
Exchanging information	2(1)	6(4)	38(26)	54(37)
Exchanging opinions	8(5)	18(12)	36(24)	39(26)
Giving presentations or reports	21(14)	22(15)	29(20)	28(19)
Brainstorming/generating ideas	43(29)	21(14)	27(18)	9(6)
Giving advice or assistance	34(23)	39(26)	18(12)	9(6)
Task B. Socialization				
Instructing or teaching	73(48)	20(13)	8(5)	0(0)
Interviewing candidates for jobs, promotions, transfers	95(62)	0(0)	2(1)	3(2)
Task C. Entertainment				
Social conversation	54(37)	44(30)	2(1)	0(0)
Task D. Consensus (gain consensus, persuade, control)				
Monitoring performance or checking progress	39(26)	21(14)	21(14)	18(12)
Discussing documents	28(19)	37(25)	25(17)	9(6)
Appraising or reviewing plans or budgets	41(27)	23(15)	21(14)	18(12)
Coordinating activities	38(26)	32(22)	15(10)	15(10)
Planning a future event, project, or activity	39(26)	33(22)	21(14)	8(5)
Solving a problem	37(25)	37(25)	19(13)	6(4)
Resolving a disagreement	52(35)	21(14)	19(13)	8(5)
Persuading	51(34)	32(21)	14(9)	3(2)
Delegating tasks	77(51)	15(10)	6(4)	2(1)
Negotiating an agreement	75(50)	21(14)	5(3)	0(0)

[a]Respondents were asked: "During the video conference you just completed, how much time was devoted to each of the following kinds of communications activities?"

[b]Task categories are based in those of Lasswell (1948) and Schramm (1971).

[c]The mean was calculated using a scale where 1 equaled "No time spent" and 4 equaled "Much time spent."

on two categories in particular: surveillance and consensus. The survey results suggest that interactive video was, indeed, primarily a tool for surveillance and consensus building. Some form of surveillance activity was reported by nearly all respondents; and over half of the respondents perceived at least some time being spent on each surveillance activity. While not as pervasive as surveillance, consensus activities also tended to be a common fact of video conferencing.

The dominance of surveillance, the prevalence of more subtle as opposed to more blatant consensus building activities (discussing a document versus persuading), and the relative lack of socialization activities might reflect the relative status of conference participants across nodes. As Table 4 shows, most interactive video links involved horizontal rather than vertical (superior-subordinate) communication. Most respondents (62%) were linked with individuals from other companies and agencies. The next most frequent channel involved individuals at one's own level within the respondent's company or agency (45%). Diagonal links were reported by 31% of the respondents (18% to other superiors and 13% to other subordinates). Less than 20% of the respondents reported vertical communication links (13% to their immediate superiors and 5% to their immediate subordinates). This may reflect the fact

that superiors and subordinates are more likely to be physically proximate and will thus engage primarily in face-to-face interactions. It also suggests the role that video conferencing may play as an integrating mechanism and as a means to reduce vertical information flow and consequent managerial overload. Direct horizontal or diagonal contact among units whose activities require coordination obviates the necessity of referring many joint problems upward until they reach a common superior for coordination and control (cf. Galbraith, 1977).

Table 4
Other Participants in the Video Teleconference[a]

	Present at other end % (N)	Not Present at other end % (N)	Not applicable % (N)	Total % (N)
Individuals in other companies/agencies	62(38)	15(9)	23(14)	100(61)
Individuals at your level within the company/agency	45(29)	24(15)	31(20)	100(64)
Other superiors in your company/agency (besides immediate superior)	18(11)	34(21)	48(29)	100(61)
Your immediate superior(s)	13(8)	38(23)	49(30)	100(61)
Individuals at lower levels in your company who are not your immediate subordinates	13(8)	40(24)	47(28)	100(60)
Your immediate subordinates	5(3)	41(24)	54(32)	100(59)

[a]Respondents were asked: "Who did you video conference with (who was at the other end)?"

The image of video conferencing as a regular tool available to organizational elites is also contradicted by the survey's data on respondents' experience with the medium. Most users (61%) had only used video conferencing once or twice over the last six months, and for less than two hours (Figure 2). Although most respondents had some limited experience with video conferencing (only 15% had never video conferenced), the tendency was toward infrequent usage (only 5% had employed it more frequently than once per month). In part, this may reflect the fact that we surveyed publicly available facilities--convenient in-house systems might be more routinely used where frequent usage is required.

The routine use of video conferencing by _any_ part of the organization appeared to be

Figure **2**

Average Length of Video Teleconferences[a][b]

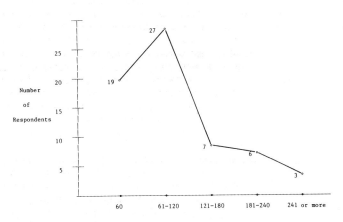

[a]Respondents were asked: On the average, how long do these video conferences last?

[b]Mean = 112 minutes

tors are employing this medium for organizational and inter-organizational communications. Use is not confined to the largest corporations, but ranges across organizations of varying sizes with varying degrees of geographical dispersion. Individual participants also cannot be stereotyped as the organizational elite. Not only are policymakers and marketing personnel using video conferencing, but also individuals from a host of different departments and across all levels of management. This is not to say that video conferencing is a "citizen technology" that is accessible to all economic sectors. Neither extreme characterizes the uses of this technology.

Video conferencing is most commonly used for information and opinion exchange, what is generally classified as surveillance. These tasks are fairly pervasive, however, and apply to many situations that might involve subtle efforts to persuade and negotiate with peers. Also, a variety of consensus-building activities that involve group decisionmaking were reported by respondents. Although support for the reluctance to use teleconferencing for highly interpersonal activities like negotiating and persuading is found here, the diversity of communication activities engaged in over this medium suggests caution in attempting to characterize communication in a typical video conference.

In fact, characterization of a typical video conference is generally difficult. Linkages may be horizontal, diagonal or vertical, and nodes may be within or across organizations. Few respondents were frequent users, and most organizations were not regular video conferencing users. The most consistent characterization is that the conferences included many individuals who would not otherwise have attended the meetings. Given the group composition, interactive nature of the communication, and the decisionmaking relevance of many of the communication activities, the impacts of video telecommunications may well extend beyond the meeting rooms to broader decisionmaking processes and outcomes of modern organizations.

more the exception than the rule. Although 23% of the respondents reported that the medium was used regularly by a few departments, none reported regular use by many departments; the majority reported that video conferencing usage was rare (28%) or only occasional (26%). Again, however, this may reflect sampling from users of a publicly available facility. Interestingly, 20% of the video conferencing user sample were unaware of the extent to which their company employs this new medium.

Finally, video conferencing generally is viewed as a mechanism for substituting telecommunications for travel, allowing people to teleconference instead of travelling for face-to-face meetings. One other effect may be to increment meetings by including people in meetings who would not otherwise have attended due to travel restrictions. The survey data support this notion. Over 60% of the participants reported that they would not have attended the meeting had it not been teleconferenced.

Summary

Although conclusions are limited by the brief timeframe of this survey and the preliminary stage of the analysis, several trends appear strong enough to merit attention. First, video conference users represent a wider array of organizations and departments than expected. Organizations from a variety of industries in both government and private sec-

Footnote

*This report is part of the Teledecision Project, supported by a grant from the Information Science and Technology Division of the National Science Foundation. We are grateful to American Telephone and Telegraph Company and Pacific Telephone and Telegraph Company for their cooperation and advice in the conduct of this study. The views expressed here are solely those of the authors

and do not necessarily reflect the views of
the National Science Foundation, AT&T, or
Pacific Telephone and Telegraph Company
officials. Further information about the
project may be obtained by writing Dr.
William H. Dutton, Annenberg School of
Communications, University of Southern
California, Los Angeles, CA 90007.

References

Champness, B. G. (1972) "The Assessment of
Users Reactions to Confravision: II Analysis
and Conclusions," Communications Studies
Group, University College, London, England,
rep. E/73250/CH.

Elton, Martin C. J. (1978) "Videoconferencing:
Some Reflections on Recent Research." Alternate
Media Center, School of Arts, New York Univer-
sity (mimeo).

Galbraith, Jay (1977) Organization Design.
Menlo Park, CA: Addison-Wesley Publishing
Company.

Harkness, Richard Chandler (1973) Telecommunications Substitutes for Travel. Ann Arbor,
Michigan: Xerox University Micro films.

Johansen, Robert, Vallee, Jacques, and
Spangler, Kathleen (1979) Electronic
Meetings: Technical Alternatives and Social
Choices. Menlo Park, CA: Addison-Wesley
Publishing Company.

Kohl, Day, P. G. Newman, and J. F. Tomey (1975)
"Facilitating Organization Decentralization
Through Teleconferencing," IEEEE Transactions
on Communications COM 23(10): pp. 1098-1104.

Kraemer, Kenneth L. (1981) Telecommunications-
Transportation: Substitution and Energy
Productivity," Telecommunications Policy,
forthcoming.

Lasswell, Harold D. (1948) "The Structure and
Function of Communications in a Society,"
The Communication of Ideas, edited by Lyman
Bryson. New York: Institute for Religious
and Social Studies.

Laudon, Kenneth C. (1977) Communications Technology and Democratic Participation. New
York: Praeger.

Nilles, Jack M. and Carlson, F. Roy, Jr.,
Gray, Paul, Hanneman, Gerhard, J. (1976)
The Telecommunications-Transportation Trade-
off. New York: John Wiley and Sons.

Noll, A. Michael (1977) "Teleconferencing
Communications Activities," Proceedings of
the Institute of Electrical and Electronics
Engineers, Inc., pp. 8-14.

Pye, Roger, Champness, Brian, Collin, Hugh,
and Connell, Stephen (1973) "The Description
and Classification of Meetings." London:
Communications Studies Group, University
College London, Ref: P/73160/PY.

Pye, Roger and Williams, Ederyn (1977) "Teleconferencing: Is Video Valuable or is Audio
Adequate." Telecommunications Policy (June):
230-240.

Schiller, Herbert I. (1981) Who Knows:
Information in the Age of the Fortune 500.
Norwood, N.J.: Ablex Publishing Company.

Schramm, Wilbur (1971) "The Nature of Commu-
nication Between Humans," pp. 3-53 in Wilbur
Schramm and Donald Roberts, editors. The
Process and Effects of Mass Communication,
revised edition. Urbana, IL: University of
Illinois Press.

Short, John, Williams, Ederly, and Christie,
Bruce (1976) The Social Psychology of Tele-
communications. New York: John Wiley and
Sons.

The Teleconferencing Resource Book: A Guide to Applications and Planning
Lorne A. Parker and Christine H. Olgren (eds.)
Elsevier Science Publishers B.V. (North-Holland)
© Center for Interactive Programs, University of Wisconsin-Extension, 1984

TELECONFERENCING AND LONG-TERM MEETING:
IMPROVING GROUP DECISION-MAKING[1]

J.A. Birrell
Consultant

and

Ian Young
Manager
EIU Informatics
The Economist Intelligence Unit
London

ABSTRACT

Much of the emphasis of teleconference system design has, in the past, concentrated upon the need for users to function at the same level as would be the case in 'round-the-table' meetings. This, until now, has often resulted in system designers aiming to provide an environment which will allow as close an emulation of the face-to-face meeting as is feasible considering cost and technical advancement. However, recent work by Social Psychologists has cast doubt on the assumption that 'normal round-the-table' meetings provide the best group decision-making procedures. In fact, it now seems evident that face-to-face group decision-making suffers serious defects resulting in poorer decision-making performance than had previously been realised.

Stemming from a concern for the implications of these findings, research was commissioned by the Defense Advanced Research Projects Agency of the Department of Defense to investigate the possibilities for the implementation of video-teleconferencing for high level crisis decision-making. Thus, Communications Studies and Planning Ltd. is presently involved in carrying out laboratory-based experimental research charged with the brief of investigating methods of using the electronic intervention implicit in teleconferencing to enhance the decision-making abilities of groups of individuals. This paper presents a selective review of the social psychology of group decision-making as relevant to teleconferencing and goes on to outline some of the findings of the first year's research. Particular attention is paid to the issue of teleconferencing over long periods of time involving an investigation of groups which were required to work together (remotely) for periods of up to one week.

1. INTRODUCTION

The understanding of group processes involved in crisis decision-making has, for some time, held some importance to many in both business and government. Decision-making in a crisis often relies as much on the processes used as on the abilities of the decision-makers, yet it is only recently that social psychologists have turned their attention to the problems which are inherent in group decision-making.

A good decision is difficult to define, but it is the aim of decision-making groups to approach the problem in hand logically and to produce a solution of high quality in a short time. This is an ideal; reality is usually a compromise, but how much should we compromise?

Crisis decision-making is of particular interest to the Departments of Defense and the Defense Advanced Research Projects Agency (DARPA) has been looking at novel ways of improving decision-making as carried out by group and individuals. The Applications Studies Division of Communications Studies & Planning Ltd.[2] (CS&P) was commissioned to carry out an explorative study of teleconferencing as one way of enhancing group decision-making. The work commissioned was to go beyond the normally accepted advantages of teleconferencing and to look for points in the social processes of the meeting where benefits could be felt.

In general, the advantages of using groups in decision-making or problem-solving over that of individual decision-makers are as follows:

● The increased experience collectively held by the individuals in the group.

● The greater range of intellectual abilities present in the group.

● The use of a more analytic procedure of open discussion and debate between individuals to solve problems.

The group's greater pool of resources should therefore add to the diversity of perspective and opinion during discussion, making for a higher quality decision. Consequently, the larger the group, the greater should be its problem-solving abilities. This, however, is manifestly untrue and many would argue that the intellectual and problem-solving abilities of the committee are inferior to that of an individual.

Understanding why this is relies upon knowledge of the processes involved in group decision-making and of the social psychological influences acting upon individuals within the group. It now seems evident that the advantages of group decision-making are offset by disadvantages stemming from:

● Difficulties in moving individuals to come to agreement on dialectical or logical grounds.

and

● The stifling of the intellectual abilities of some group members by social psychological pressures.

These disadvantages can have such powerful effects that groups of four or more individuals may have their problem-solving capabilities substantially reduced.

The aim of this explorative study was to use laboratory based techniques to investigate group decision-making held over a multipoint video teleconference system. Special emphasis is being put on those meeting processes which have been shown to be susceptible to adverse social pressures and to monitor whether teleconferencing can offer advantages over the 'round-the-table' meetings.

2. GROUP DECISION-MAKING, SOCIAL PRESSURE AND TELECONFERENCING

Moving through the decision-making process, the discussion group has been shown to pass through identifiable stages (Osborne, 1957[3]; Kelly, 1974[4]; Gardner and Edwards, 1975[5]). While the precise number of stages is still a point of discussion between schools of thought, two stages are common and prevalent in the literature. Stage1, the generative stage, is concerned with the production of ideas, actions and possible outcomes to the meeting. It is this stage which is the focus of attention of "brain-storming": a technique for max-

imising the number of ideas, especially the novel, by inhibiting criticism and showing equal consideration in development of all ideas produced (Osborne, 1957[3]). Stage 2 is concerned with the evaluation of ideas, to exclude the weaker solutions and to develop the stronger with the aim of finally bringing the group to a mutually agreed decision (see Figure 1).

FIGURE 1   SCHEMATIC REPRESENTATION OF THE DECISION-MAKING PROCESS

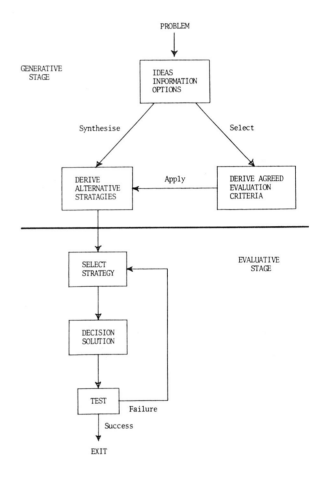

Ideally, group decision-making should be carried out with the aim of capitalising on the abilities and experience contained within the group. Moving smoothly and efficiently through the discussion process, the group should give due consideration to all possible alternatives, rejecting on the grounds of logic those which are not likely to be productive and further considering the rest until the decision is made as to which, in the whole group's opinion, is the best solution to the problem. Sadly, this is not what is found in reality. Decisions are often made by groups on the basis of inadequate discussion of the problem (although much discussion per se may have taken place), and the resulting solution or decision may have been the outcome of something far removed from logic. General influences upon group decision-making have been shown to be firmly rooted in the social interaction processes of the group. These processes serve other purposes within the group and very often result in reduced decision-making effectiveness.

Classic studies of conformity in groups were conducted by Sherif, 1975[6] and Asch, 1952[7]. Sherif showed that pressures to conform within a group are such that people can distort quite simple judgments to satisfy the group norm. Asch went on to show that when judging lengths of lines, a real subject would 'go along with' a group of confederate subjects when they were all giving what seemed to be the wrong response.

For the individual who conforms, one of the major reasons he does so is his dependence on the majority:

- dependence for information (informational influence)

and

- dependence for approval (normative influence).

These ideas arose out of Deutsch and Gerard 1975[8] and are well accepted in the current literature (Shaw, 1976[9]). Cartwright and Zander (1968[10]) have noted that a group's influence on an individual increases as attraction to the group increases. It would seem, then, that the less dependent a member is on a group, the less he will conform.

Pressures to conform in groups may be overt, but they do not always need to be imposed; frequently they come from within the individual members themselves. Thus, although a group may meet with the intention of obtaining a greater diversity of opinion on an issue, they may slip into what Janis (1972[11]) has termed 'Groupthink'. By this Janis means the tendency for a group of people to rapidly agree on a decision with little serious discussion and no personal animosity.

Janis' (1972[11]) "Victims of Groupthink" contained both theoretical analyses of group processes, and some devasting historical analyses of what he called 'foreign-policy decisions and fiascoes', which resulted from decisions taken by groups containing extremely high levels of intellect and knowledge. Groupthink can be described as a way in which members of a group tend to think alike, without properly considering alternatives to expressed opinions. When group members experience strong feelings of loyalty and solidarity, the desire for unanimity may override the motivation to

logically and realistically evaluate courses of action. He used the term Groupthink to refer to the "deterioration of mental efficiency, reality testing and moral judgment that results from in-group pressures". Everyone who feels that he privately disagrees, thinks he is the only one who disagrees, and that it would not be worth bringing up the matter.

The research finding by Stoner (1961[12]) that groups of people make more risky decisions than individuals challenged conventional wisdom about the cautiousness of committees. It was probably for this reason of disbelief, rather than the important implications of the finding, that the 'risky shift phenomenon' became one of the most abundantly researched topics in social psychology in the 1960s.

A number of explanations were offered for the findings, the most important being Brown's (1965[13]) value norm hypothesis. This said that there exists in our culture a value for taking risks -- that we overtly or covertly admire the daring, courageous entrepreneur type, and that when we find ourselves less risky than some others in a group, we move towards the valued ideal.

Stoner later in 1968[14] showed with a different set of questionnaire items that he could achieve a cautious shift in groups. As Brown put it "to be virtuous, in any of an indefinite number of situations, is to be different from the mean, in the right direction, and to the right degree" (1974[15]). There are certain decisions where to be virtuous is to be cautious.

Thus, the research with the choice dilemmas questionnaire is not a consistent shift to risk tendency, but rather a tendency for group discussion to enhance the initially dominant point of view.

Two explanations of this behaviour have received experimental support:

- Social comparison theory

and

- The Theory of Informational Influence.

Social comparison theory proposes that exposure to others' positions stimulates a person to adjust his response to maintain desirable self-perception. Informational influence theory says that during discussion, arguments are generated which predominantly favour the initial alternative; these can include some persuasive points the typical person has not yet considered. So what people learn from the discussion is mostly in the direction supporting the majority's initial preference.

There are two types of supporting evidence. First, increased information per se leads to polarization of opinion (Sears, 1969[16]; Burnstein and Vinokur, 1973[17]). Second, analysis of discussion content (Vinkur and Burnstein, 1974[18]; Bishop and Myers, 1974[19]) showed directly that the arguments expressed do mostly favour the initially dominant direction, and that these arguments were independently rated as more persuasive than arguments favouring the alternative.

The processes just described are primarily concerned with uniformity, conformity and cohesiveness in groups; particularly of a kind that may be detrimental to the generative stage of decision-making. However, Group Problem Solving is also under the influence of divisive forces, especially at stages where ideas and opinions have to be evaluated. As soon as disagreements are generated, the possibility exists that coalitions may form and the literature is clear on the possible consequences of that. Within-subgroup conformity may increase, and between-subgroup divisiveness may also increase. It is unlikely that the formation of antagonistic coalitions will enhance decision-making, as it is not conducive to proper consideration of all alternatives, but rather will result in bargaining about the particular alternatives favoured by the competing subgroups.

Coalition theory suggests that coalitions would be expected to occur only in mixed-motive situations, where there is an element both of conflict and coordination (Gamson, 1964[20]). They especially form where two or more persons can achieve greater rewards through joint action than can either acting alone. The second stage of decision-making may thus be ripe for coalition formation. An individual may not be able to persuade others of the rightness of his solution; in coalition with another, he has more chance of so doing and has validation for his idea.

All the discussion so far has been based on research conducted on groups meeting 'around the table' and has shown that this is not necessarily an effective way of reaching high quality decisions in a short time. This paper though deals with the conduct of comparable meetings by route of teleconferencing.

The study of social interaction via electronic media was pioneered by the Communications Studies Group during the early to mid-1970s. The work carried out by the CSG combined rigorous laboratory experiments with field trials producing several recurrent findings. This research used the general experimental paradigm of two people involved in a discussion held in one of three ways: face-to-face, in separate rooms linked by two-way audio

and by closed-circuit television. The two latter conditions were used as simulations of teleconferencing. The major recurrent findings from this and other work are:

- Teleconferencing reduces the interpersonal aspects of discussion, focussing attention on the issues. Meetings are thus far more task oriented than people anticipated.

- Participants are less dogmatic and more compromising in teleconferencing resulting in more opinion change in meetings and less coalition formation.

- Teleconferences are also shorter and more businesslike.

- No measurable loss could be found in the communication process, information exchange and ideas generation were unaffected by the medium.

- When coalitions did occur they were more likely to form between the groups in group-to-group conferencing than within groups.

- Participants, while just as able, were less confident in making judgments about others over an electronic medium.

- Any loss of visual non-verbal cues was amply compensated for in other ways.

- In discussions involving negotiation and bargaining, the side with the strongest case was more likely to win when the meeting was held over a teleconferencing system.

- The effects of teleconferencing are generally stronger for audio-only.

The concern of the research project reported here was to achieve a marriage which would use the established properties of teleconferences to overcome the known difficulties of face-to-face meetings.

3. THE STUDY: EQUIPMENT AND METHODOLOGY

At a time of national crisis, it would be hoped that decision-making would be based on analysis of the arguments and facts rather than a member's charisma or social psychological pressures. Based upon the evidence summarised above, it was hypothesised that teleconferencing, in crisis decision-making, may have value in reducing the influence of

282   J.A. Birrell and I. Young

social psychological pressures on individuals
and pave the way for higher quality solutions
to problems.

The research procedure adopted set out to
use realistic teleconference meetings and used
a combination of middle/senior managment
personnel recruited from central London
businesses and graduate students as participants. This procedure involved the participants in discussions of a problem task with
the brief to come to a mutually agreed decision
within a specified time-frame. The measurement tools used were direct observation, paper
and pencil testing and open-ended debriefing
interviews. Participants met the teleconference controllers and were shown to conference
rooms individually. They would discuss the
task freely over the video conference system
and come to their decisions. Afterwards, they
would be interviewed collectively about the
meeting, teleconferencing and related inhibitions and frustrations experienced. The major
emphasis of the study was explorative rather
than analytic and as a consequence focussed
upon identifying ways in which the decision
process may be enhanced. One way of doing this
was to use the decision-making experience of
the participants to point to those aspects of
teleconferencing which are inhibitory and dis-
advantageous. This was done during the open-
ended interview.

The teleconference facility used for the
project was a 4-point full motion video system
constructed in five non-identical rooms: one
for each of the participants and a teleconference controller's suite, see Figure 2. The
equipment installed in each room was identical
consisting of monitors, camera and loudspeakers
housed in a free-standing unit and a separate
floor standing microphone (Figures 3 and 4).
An open audio system was used in preference
to a voice activated switch system to gain
advantages of conversational fluency so
appreciated in group discussions. This was
achieved successfully by using unidirectional
microphones and a moderate amount of room
treatment.

The video system was arranged as a 2x2
array around a centrally positioned camera and
used 12" monochrome television monitors. The
configuration of monitors from room to room
was consistent in the sense that each viewer
saw all four participants (including himself)
in the same arrangement. This arrangement of
monitors was felt to be the best possible for
the following reasons:

  • The distance between monitors is kept
    to a minimum so that all participants
    can be held in the field of view
    without undue head movement nor placing
    the monitor screens too far away.

  • Correspondingly, the reduced dis-
    tance between the camera and the
    monitors reduces the number of
    "side of head" shots transmitted.

  • The centrally located camera can
    transmit cues of axis of inter-
    action.

FIGURE 2   LAYOUT OF TELECONFERENCE ROOMS AND ADJACENT AREAS

(not to scale)

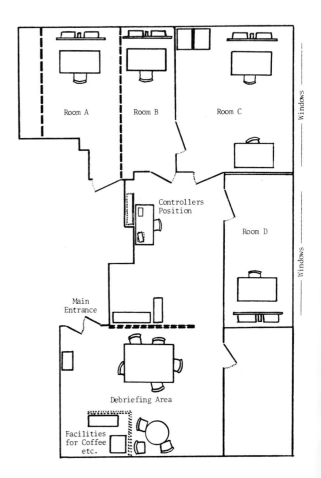

Often in conversation, the talker directs
part of his speech to a particular listener.
This link between talker and listener; the axis
of interaction, will be constantly changing
during conversation, but the participants
know where these axes lie at any time by the
direction of looking. In teleconferencing,
this can be a major problem for free-flowing
discussion as evidenced by the increase in
name tagging which can be observed. To reduce
this problem advantage was taken of the fact
that, with the camera central between four
monitors and positioned level with the participant's head, a different aspect would be
shown for each of the monitors looked at.
Thus a viewer looking at the person displayed
on the bottom left monitor would be seen to

look down and to the left. Full advantage
was taken of this aspect by reversing the scan
of video monitors. The direction of looking
shown by the monitor images, when reversed,
is consistent with the direction they are
actually looking in (see Figure 5).

FIGURE 3   PLAN VIEW OF PARTICIPANTS TELECONFERENCE EQUIPMENT

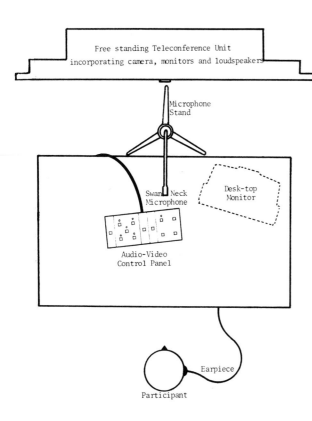

It was hypothesised that this consistency of direction of looking would compen-
ate for the loss of actual eye-gaze cues and
would not result in name tagging or loss of
conversational fluency.

4. MAIN FINDINGS OF THE FIRST YEAR'S WORK

There is a wide and deep body of research
knowledge on teleconferencing but remarkably
little has been concerned with multipoint video
conferencing, and what there has been has focussed on educational applications. Therefore,
this research in many ways breaks totally new
ground. It was not known for example whether
research findings from two-person meetings
would scale-up to four-person meetings. In
addition, the scope of this research, both
to generate and test ideas for improving
group effectiveness, necessitated an approach
which emphasised breadth rather than depth of

research detail.

FIGURE 4   FRONT VIEW OF FREE STANDING TELECONFERENCE UNIT

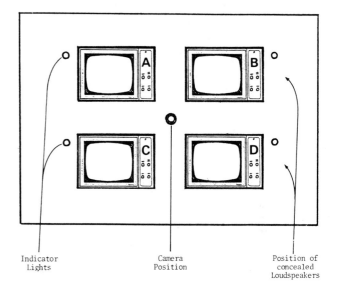

Indicator Lights   Camera Position   Position of concealed Loudspeakers

FIGURE 5   EFFECT OF REVERSED SCANNING ON DIRECTION OF LOOKING

NORMAL SCANNING   REVERSED SCANNING

As a consequence, the main findings from
the year-long project are not drawn, in general,
from tight research designs and rigorous
statistical analyses. They are indicators
used by the research team to formulate new
ideas and directions for subsequent phases of
the research project. In total though the
findings are based on over 1,000 man-hours
of use of the teleconference system by a broad
cross-section of business and professional
personnel in meetings lasting from 40 minutes
to a week. These are:

- A flat matrix array of video monitors,
  with a single camera per station
  (scanning in a reverse direction) is

perfectly adequate to permit full and
effective teleconferencing. It
permits the ready identification of
who is talking to whom in a meeting
without the need for name tagging
all comments.

- The lack of precise 'eye-contact'
between participants, partly associated with the use of only one camera per
station, while noted by some users
at the start of teleconference
sessions, was not found to interfere
with the progress of the discussion.
The only comments made on this point
after meetings were by some users who
found it useful to be able to look at
another person in the meeing without
risking the occurrence of anxiety-
provoking mutual-gaze. (Nicols &
Champness, 1974[21]).

- The reverse scanning of the image
cameras was not noticed by participants unless the fact was pointed out
to them. It resulted in a much more
comfortable meeting.

- A self-view monitor is valuable for
keeping meeting participants in camera
shot, though it need not necessarily
be available all the time.

- An open-audio teleconference sound
system was found to provide adequate
speech levels in teleconference rooms
with only modest treatment for sound
absorbency. It resulted in much
freer discussion than could have been
possible with a voice- or manually-
switched conference system.

- The use of separate loudspeakers
associated with video monitor images
provided sufficient orientation
cues to identify the main speaker in
a discussion, even when there was
no prior familiarity with the normal
speaking voice of fellow participants.
The inclusion of additional signalling
lights as cues to the current main
speaker were found not only to be
unnecessary, but also intrusive and
distracting.

- A video monitor presenting text and
graphics information was preferred at
a closer viewing distance than the
monitors carrying the images of the
other participants.

- Users found that their meetings held
over the teleconference system, compared to their normal experience of
face-to-face meetings, were at least

as satisfying, more task oriented,
and less time consuming. The individual level of contribution was
not thought to be any different from
normal.

- When users were provided with just
audio teleconferencing and no
video image, they were less favourable
toward the teleconference. Video
appears to increase user satisfaction,
and other participants were felt to
be more cooperative and friendly in
the video conference meetings compared with the audio-only. This
added value from the visual channel in
teleconferencing is at variance with
previous research which saw little
value in using video.

- The teleconferencing equipment itself
was seen as helpful to the progress
of the meeting, rather than distracting.

- The perceived benefits of the teleconference were found to be greater
in those tasks involving competition
and conflict than in the more cooperative tasks. This is again at
variance with the traditional wisdom
of teleconference research, which
considers conflictual meetings to be
unsuitable for teleconferencing.

- Feedback from other members of the
group during discussions was
reported to be heightened in the
teleconferenced meeting compared
with face-to-face.

- Groups undertaking cooperative tasks
produced better outcomes when other
group members were experienced as
being more friendly, and when one
person was felt to have greater
control over the meeting.

- Attempts to replicate teleconference
research on attitude change in two-
person communication, showed that
substantial shifts of attitude could
be achieved in the short term of the
discusssion, but these rarely produced
sustained opinion changes by individual members. Nevertheless, the
presence of moderate opinion-holders
in a discussion did tend to produce
more sustained attitude change than
otherwise. Furthermore, agreements
reached were not necessarily by a
simple process of averaging the individual opinions of the group members.
They could involve movement of
opinion from, as well as towards, a

moderate position. Teleconferencing was found not to assist isolated group members (i.e. those holding opinions different to the other three) in persuasive discussions. They were usually forced to come in line with the majority view.

The study of teleconferencing over long periods of time was included as part of this program of research, when its importance and relevance were realised from comments made by users during the early phase of the research. Some aspects of long-term teleconferencing could not be effectively simulated, particularly those involved with actual geographic separation, since the location of all tele-conference nodes in one building meant that it was not possible to stop participants from meeting face-to-face sometimes during the long-term sessions without recourse to complex and artificial scheduling.

The investigation of longer term meetings looked at both single meetings of longer dur-ation (ranging from 40 minutes up to 4 hours) and sustained teleconferencing (involving a number of meetings over a virtually continuous period of up to one working week). The objects of investigation were:

● group dynamics

● user fatigue

● medium and person perception.

The first sustained teleconference involved four company staff who, in the course of their normal work, were engaged in a series of meetings. At times these meetings would need all four staff, but a large number of the meetings would comprise only a part of the group. In normal practice such a project would require arranging meetings in the company conference room, followed by periods of working alone or working in subgroups when some conven-ient work space was available.

The staff were occupied over five days. They used the conference stations on the understanding that all interaction between group members, including graphics and docu-ments transfer, would be via the teleconference facility or one of the operators. This arran-gement of interlinked offices would provide an opportunity for both investigating behaviour and identifying the problems which will occur in long term sustained teleconferencing. The bonus here is that we are not involved in staged or simulated meetings, but are using staff members in normal work.

The pattern of the work included:

● project meetings - all group members

● subgroup meetings - involving 2 or 3 staff

● individual tasks.

Project meetings were held at the start and finish of the contract and at periods during the week as arranged by the project leader. Subgroup meetings usually concerned the performance of some specific group task, but also included short meetings for the exchange of information or the airing of a problem. Many such meetings occurred during the week.

Group Tasks consisted of:

● Information exchange

● Generation of work plans

● Problem solving

● Individual task allocation.

Other sustained teleconferences held included a one-day meeting of company engineers working to a pre-arranged agenda and a series of experimental studies. The experimental studies involved graduate stu-dents in teleconferencing continuously over 4 days. Five such sessions were run and all discussions were continuously monitored by the two experimenters and tape recording. Discussion tasks over the week covered a wide range of meeting aspects including coop-erative problem-solving, role-play and information exchange. The major findings of the sustained teleconference programme were:

● Teleconferenced meetings and dis-cussions involving the same people for periods between several hours and a working week show first that it is possible to sustain this type of interaction for longer than any previous research has shown.

● During teleconferences sustained over four days, meeting effectiveness and efficiency were maintained and there was no measurable fall-off in atten-tion or vigilance.

● This type of prolonged meeting is more tiring than would be expected from a comparable face-to-face pattern of work meetings, but it is not possible to determine whether this fatigue is due to the teleconference itself or to the higher work rate and more focussed activities found in tele-conferences. The fatigue was not identified in any of the tasks nor on the specific measures of attention,

it was only reported by participants
at the conclusion of their activity
(i.e. on relaxation). It is not
yet determined how long this sustained level of activity could be
maintained.

● The video channel of communication is
particularly valuable at the start of
prolonged teleconferencing. Strategies for establishing an effective
work group from participants not
previously acquainted, were found
to be easy and effective when the
video channel was available, but
strained and awkward when it was
absent.

● There was a general pattern of
adaptation to teleconferencing
apparent during series of discussions
by the same groups held over 4 days.
The video images and communication
pattern were reported as becoming
increasingly more natural.

● Aggressive behaviour was reported
to increase, and friendly behaviour
to decrease amongst group members
after the second and third consecutive day of teleconferencing.
This may be a symptom of stress, or
could, more positively, indicate that
the group is passing through well-
documented phases of development.

5. CONCLUSION

In summing-up the work reported here, it
should be restated that the findings were
drawn from a study that emphasised innovative
and exploratory research rather than scientific testing of precisely defined hypotheses.
All of the comments and results should really
be tested under more rigorous conditions
before they are accepted into the forelore
of teleconferencing.

There was though one very loose hypothesis
being tested throughout the project; that
teleconferencing could be used to positively
overcome the known problems associated with
group decision-making. The indicators are
that it can be used in this way. The view
that teleconferencing is always an inferior
but cheaper substitute for some other way of
doing business or gaining education should
not be allowed to continue unchallenged.

The research programme reported here
subjected one form of teleconference to wide-
ranging testing by a broad variety of users
and over a number of different tasks, inclu-

ding recurrent sessions over a period of
several days. Through all this it was usually
felt by users to have made a positive contribution to the outcome of the discussion and
to their own contribution to that discussion.
We need from here to go on to look at the creative use of teleconferencing within the set of
office automation tools. We need to be
designing teleconference systems which will
be an aid to the meeting process, at times
cutting across conventional meeting habits
where they are unproductive. Too often teleconference design has been motivated by the
desire to replicate the face-to-face meeting.
We should be considering more deeply whether
the face-to-face model is really so very
valid.

(1)   Our gratitude to Mr. B. Champness,
Drs. M. Hyland and F. Reid for their contribution to this work, and to all those who
participated in teleconference sessions.

(2)   The work reported in this paper was
conducted while Mr. Birrell and Mr. Young were
members of the Applications Studies Division
of Communications Studies & Planning Ltd.,
London.

(3)   Osborne, A.F. (1957) "Applied Imagination"
(revised edition) New York: Scribner.

(4)   Kelly, J. (1974) "Organizational
Behaviour: an existential-systems approach"
Irwin: Illinois.

(5)   Gardner, P.C., and Edwards, W. (1975)
"Public values: multi-attribute utility measurement for social decision making" In M.F.
Kaplan and S. Schwartz (eds.) Human Judgement
and Decision Making NY: Academic Press

(6)   Sherif, M. (1935) "A study of some social
factors in perception" Archives of Psychology,
No. 187.

(7)   Asch, S.E. (1952) "Social Psychology"
Englewood Cliffs, N.J.: Prentice Hall

(8) Deutsch, M., and Gerard, H.B.A. (1955) "A study of normative and informational social influences upon individual judgement" Journal of Abnormal and Social Psychology, 51, 629-636.

(9) Shaw, M.E. (1976) "Group dynamics: the Psychology of small group behaviour" (2nd edition) McGraw Hill.

(10) Cartwright, D., and Zander, A. (eds.) (1968) "Group dynamics: research and theory" (3rd edition) New York: Harper and Row.

(11) Janis, I.L. (1972) "Victoms of groupthink" Houghton Mifflin: Boston.

(12) Stoner, J.A.F. (1961) "A comparison of individual and group decisions involving risk" Unpub. master's thesis, Massachusetts Institute of Technology.

(13) Brown, R. (1965) "Social Psychology" New York, Free Press.

(14) Stoner, J.A.F. (1968) "Risk and cautious shifts in group decisions: the influence of widely held values" Journal of Experimental Social Psychology, 4, 442-459.

(15) Brown, R. (1974) "Further comments on the risky shift" American Psychologist, 29, 468-470.

(16) Sears, D.O. (1969) "Political Behaviour" in Handbook of Social Psychology, Vol. 5, G. Lindzey and E. Aronson (eds.) Reading, Mass: Addison-Wesley.

(17) Burnstein, E., and Vinokur, A. (1973) "Testing two theories about group-induced shifts in individual choice" Journal of Experimental Social Psychology, 9, 123-37

(18) Vinokur, A., and Burnstein, E. (1974) "The effects of partially shared persuasive arguments on group induced shifts: a group problem solving approach" Journal of Personality and Social Psychology, 29, 305-15.

(19) Bishop, G.D., and Myers, D.G. (1974) "Informational influence in group discussion" Organizational behaviour and human performance, 12, 92-104.

(20) Gamson, W.A. (1964) "Experimental studies of coalition formation" In L. Berkowitz (ed.) Advances in Experimental Social Psychology Vol. 1. New York: Academic Press.

(21) Nichols, K.A., and Champness, B.G. (1971) "Eye Gaze and the GSR" Journal of Experimental Social Psychology, Vol. 7(6), 623-626.

2500-8X2T138-82

The Teleconferencing Resource Book: A Guide to Applications and Planning
Lorne A. Parker and Christine H. Olgren (eds.)
Elsevier Science Publishers B.V. (North-Holland)
© Center for Interactive Programs, University of Wisconsin-Extension, 1984

288

SPEAKER INTERACTION: VIDEO TELECONFERENCES VERSUS
FACE-TO-FACE MEETINGS

Karen M. Cohen
Bell Telephone Laboratories
Holmdel, New Jersey   07733

ABSTRACT

Human factors research was conducted
to compare speaker interaction patterns in face-to-face meetings with
those in video teleconferences.
Groups of subjects completed two
problem-solving tasks using either a
face-to-face (FTF) medium of communication or the new system to be used
in the service offering of
PICTUREPHONE® Meeting Service (PMS).
Upon completion of each task, subjects rated the perceived ease of
communication and the social dynamics
of the meeting. In addition, observational records of speaker interaction patterns were obtained. Analyses of these data reveal ways in
which FTF meetings and PMS video
teleconferences differ. The observational records of communication
behavior parallel the subjective
responses concerning the perceived
ease of communication during the discussion tasks. These data indicate
that many more speaker "turns" and
simultaneous speech events occur in
FTF meetings than in PMS conferences,
though the social dynamics during the
tasks are comparably rated. Hence,
these results question the appropriateness of FTF communication as the
"ideal" model for video teleconferencing. However, they also demonstrate that the offering of PMS is
perceived to provide positive social
dynamics comparable to those found in
FTF meetings. This new version of
PMS, in contrast to the previous
market trial version, also eliminates
the negative effects associated with
interruptions and simultaneous speech
events previously reported for video
teleconferences held in the presence
of transmission delay. In addition,
this research provides data on subjects' perceptions of the effectiveness of different communication media
for various conferencing tasks.

1. Overview

Recent technological advances in
the telecommunications industry have
provided the capability for groups of
individuals in distant locations to
speak with and see one another
through the use of video teleconferencing systems, such as PICTURE-
PHONE Meeting Service (PMS). A frequently espoused goal of such video
teleconferences is to emulate FTF
meetings among conference participants. Recently, this goal or
"ideal" has come under scrutiny since
human factors research and anecdoctal
evidence from PMS market trial users
have indicated that video teleconferences are often considered more
structured, more efficient, and more
productive than (FTF) meetings[1].
These reports have raised speculation
about the extent to which actual communication patterns differ as a function of the medium of communication.
While there is considerable literature documenting differences in perceived effectiveness and acceptability of video teleconferencing systems
compared with FTF meetings and other
media[2,3], there is a lack of objective data on speaking behavior among
participants in video teleconferences, compared with FTF meetings.

In order to compare video
teleconferences and FTF meetings from
both objective and subjective perspectives, human factors research was
conducted in which groups of subjects
completed two discussion tasks using
either an FTF or PMS medium of communication. The PMS facility
employed the newly designed, voice-
gated audio circuit (VG) that will be
used in the PMS offering[4]. Upon completion of each task, subjects rated
the ease of communication, the social

dynamics of the conference, and their satisfaction with the task solution. Observational records of speaker interaction patterns during the conferences were also obtained. In addition, subjects rated the perceived effectiveness of different communication media for various communication tasks.

In general, FTF meetings and PMS conferences differed on some subjective and objective measures of communication ease, while the social dynamics for the two meeting types were similarly perceived. The results indicated that video teleconferences with the new PMS system were perceived to have fewer interruptions among speakers than did FTF meetings. Though such interruptions are a natural part of interactive communication and may or may not be an asset in FTF meetings, such events have been previously associated with perceived audio communication difficulties in video teleconferences in the presence of transmission delay[5]. The objective records of speaker interaction patterns indicated more speaker "turns" or alternations (both natural and disruptive) and simultaneous speech events in the FTF meetings than with the PMS facility. PMS sessions were, however, similar to FTF meetings on measures of positive social dynamics (e.g., task-oriented, cooperative, democratic). The ratings on the perceived effectiveness of different communication media for different conference tasks indicated that video teleconferencing systems like PMS are viewed as an electronic communication medium that helps bridge the gap between FTF meetings and regular telephony, especially for communication tasks in which personal contact or "social presence" is deemed important.

2. Background

2.1 Communication Media and Tasks

In the recent literature on the social psychology of telecommunications, there has been considerable attention devoted to the relation between different communication media (e.g., face-to-face, video teleconferencing, audio teleconferencing, computer conferencing, regular telephony) and various communication tasks (information exchange, problem-solving, bargaining, meeting someone). Research has addressed the social psychological dimensions by which communication media differ as well as the effects of various media on meeting outcomes and social interactions[2]. There have also been attempts to categorize and describe FTF meetings in order to identify those types of FTF meetings for which a teleconference could be substituted.[6]

The most obvious difference between FTF meetings and those held by regular telephony or audio teleconferencing is the presence or absence of visual cues during social interaction. Though telecommunication systems like PMS provide conferees with many of the visual, nonverbal cues associated with social interaction (e.g., gestures, head-nods, body posturing), there remains a difference between PMS and in-person meetings in the actual physical (and possibly, psychological) proximity and the ability to gauge eye contact. These nonverbal cues are particularly relevant in signaling speaker acceptance and turn-taking. They are also important for tasks in which emotional expression and interpersonal relations are salient. In such tasks, the substitution of a video teleconferencing medium for an FTF meeting is likely to be less appropriate and satisfactory than for those whose main focus is the factual exchange of information.

Researchers with the Communication Studies Group in London[3] have suggested that communication media differ in their "social presence" or the "degree of salience of the other person in the interaction and the consequent salience of the interpersonal relationships" (p. 65) afforded by a given communication medium. This statement implies that those tasks that are the most sensitive to the conveyance of emotional expressions and interpersonal relations (e.g., persuasion, meeting someone) should be most impacted by the choice of communication medium because of the difference in social presence that various media provide. Assessment of the social presence of different media has been based primarily upon factor analyses of rating scale data along different dimensions (e.g., formal-informal, sociable-unsociable, active-passive).

The effects of different media on task outcomes and social interactions have also been addressed empirically by social psychologists[2]. In addition, both empirical and anecdoctal evidence have been reported that indicate video meetings are often considered more efficient, effective, and productive than FTF meetings[1]. This result appears due in part to a reduction in the amount of extraneous conversation during video meetings, compared with in-person meetings. In this case, the lesser social presence of the video teleconferencing medium, compared with FTF communication, likely contributes to this finding, since video teleconferencing systems are often considered more formal and impersonal than the FTF mode of communication.

In the evaluation reported in this paper, participants were asked to rate the effectiveness of different communication media for various communication tasks. Analyses of these data reflect the users' perception of the similarity and dissimilarity in the social presence afforded by the different media (i.e., the range in ratings for the different media for tasks which vary in their demands for emotional expression and social interaction).

2.2 Speaker Interaction Patterns

The way in which speakers interact during conversations and meetings has long been the subject of psychological observation and investigation[7,8]. Of particular interest to this evaluation are the communication signals and rules for speaker turn-taking during social interactions[9,10]. These signals and rules may be verbal or nonverbal cues that facilitate the orderly exchange of information among speakers. Interruptions or simultaneous speech events occur when one party in a conversation fails to send or interpret properly the cues for turn-taking and/or refuses to yield the floor to another speaker. Though such events occur frequently in interpersonal interactions, they are usually quickly resolved since such behavior is considered impolite in our culture and often impedes accomplishment of meeting objectives.

Attempts to analyze interactive speech patterns during conversations have employed a variety of techniques, ranging from observers' recording of speaker exchanges to computer analysis of the on- and off-patterns of speech signals. Until recently, most of these computer techniques were only applied to dyadic in-person meetings or telephone conversations that lasted less than 10 minutes. Automatic analysis of such on-off patterns of speech was reported by Brady[11] with particular focus on the identification of talkspurts, pauses, simultaneous talking, and mutual silence.

While subjective ratings of the frequency of such events is readily obtainable through rating scale questions, an objective estimate of the occurrence of such events was also desirable. However, methods of extensive computer analysis of on-off patterns of speech have not yet been refined for the conditions studied in this evaluation (e.g., meeting duration, number of conferees, etc.). Consequently, a method of obtaining observational records of speaker interaction patterns was developed. In general, observers in both PMS room locations during a conference (or two observers in the same room at opposite ends of the conference table for a FTF meeting) maintained a running log of speaker turn-taking, frequently noting specific comments and/or the precise time of the speaker exchange. Particular attention was given to whenever a simultaneous (or nearly simultaneous) speech event occurred and whether the event involved speakers from the same or different locations (or sides of the table in FTF meetings). When the records from both observers were merged, the number of such events for each 20-minute task during a meeting could be computed. If there were frequent occurrences of simultaneous (or nearly simultaneous) speech events between PMS locations that were heard by only one of the locations, these events would, in effect, be considered unsuccessful attempts to interrupt a distant speaker and could lead to user dissatisfaction with the conference. Simultaneous speech events that were heard by only one observer during FTF meetings could also occur since observers were seated at opposite ends and sides of the conference room (see Fig. 1).

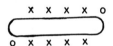

Fig. 1. FTF meeting arrangement.
(O=observer, X=participant)

It was thus possible for "side" conversations across the conference table or on the same side of the conference table to be heard by only one observer (and some of the participants). The data obtained with this method were used to assess the effects of variation in communication media on speaker behavior and were also related to users' subjective ratings of speaker interaction patterns and group dynamics.

3. Methodology

3.1 Subjects

Fifty-eight Bell Laboratories members of technical staff participated in this study. Subjects in this experiment had little or no experience with video teleconferencing facilities.

3.2 Study Design

The comparison of in-person meetings and video teleconferences was conducted by having different groups of subjects conduct two conferencing tasks either across a conference table or across a video link. Hence, type of communication medium was a between-subject factor, with each subject experiencing only one of the two alternative communication media. Conferencing task was a within-subject factor, with each subject participating in a consensus task and a role-playing task with the same communication medium. There were four different groups of subjects (or sessions) using each of these media, with a total of 28 different subjects using each medium. There were 6-8 participants in each meeting, evenly distributed to the extent possible between rooms or across sides of the conference table.

3.3 Facilities and Equipment

Subjects in this study were seated around an oval table in either of two PMS rooms in Holmdel, New Jersey[4]. Subjects using PMS facilities faced two monitors, one of which displayed the incoming picture from the distant PMS room, while the other displayed the outgoing picture. Three cameras, mounted above the monitors, were used, with each camera focused on two adjacent positions around the conference table. The video from these cameras was voice-switched so that a particular outgoing picture from a room was a function of who was, or had last been, talking in that room. The subjects wore lavelier microphones on neck crooks. Incoming audio came from two loudspeakers mounted in the ceiling.

Participants in the FTF sessions were seated on both sides of the oval conference table in one of the PMS rooms. They communicated with their fellow-conferees without the use of the video cameras or microphones.

Audio and video transmission delay of 705 ms was used in the PMS portion of this study to simulate full round-trip satellite and picture processing delay. The delay for audio and video signals was synchronized. Prior research has indicated that, with the introduction of audio transmission delay, the number and length of double talks or simultaneous speaking events increase[12,13] and that with audio and video transmission delay using the previous market trial version of PMS, difficulties with audio communication are produced[5]. These results are not surprising given that transmission delay disrupts the pace of normal conversation, makes the appropriate timing of interruptions more difficult, and impedes the smooth resolution of simultaneous speech events. Such events, then, are likely to affect the ease by which conferees interact with distant speakers and their ability to engage in natural, interactive communication as found in FTF meetings. This, in turn, may influence users´ appraisal of the social dynamics of their PMS meetings.

3.4 Conferencing Tasks

Two interactive discussion tasks were devised for this study. The first task, National Issues, required subjects to select and put into priority order the six most important issues from a list of 13 suggested topics for a successful presidential campaign. After individual rankings were completed, participants were required to reach group consensus on a final ranking. The second task, Factory Relocation, was a role-playing task adapted by the author from a human relations exercise[14]. Participants were asked to make a recommendation to an electronics company´s board regarding a major relocation of its factory from the Northeast to the Southeast. The participants or "managers" discussed the advantages and disadvantages of the move. Though positions on the move were assigned, subjects were free to change their positions as a function of the discussion. A recommendation to the company board, based upon majority vote, was required to complete the task.

3.5 Procedure

At the beginning of each session, the experimenter explained to the participants that the purpose of the study was to obtain their reactions to the technical aspects and social dynamics of video teleconferencing. Subjects in the FTF sessions were told that their reactions to the social dynamics of meetings would be used as comparative data with PMS sessions.

The discussion period for each task was about twenty minutes. For each task one participant was selected by the experimenter to be the group leader. The location of the group leader was varied so that if the group leader came from one location/side of the table for the National Issues task, the group leader for the Factory Relocation task came from the other location/side of the table. The first task was National Issues, followed by Factory Relocation for each session. Following the conclusion of each task, participants completed questionnaires concerning the technical quality of PMS facilities (for those in the PMS sessions only) and the social dynamics of the conferences. Subjects were also asked to rate the perceived effectiveness of various communication media for different communication tasks.

4. Results

Analyses of variance were performed on the rating data with each subject as the unit of observation. Most of the rating scale questions were of an absolute, not relative, form and were based upon a 9-point scale with every other point labelled according to the following semantic differential scale: not at all, slightly, moderately, quite, and extremely. Questions that referred to "different/same" locations for the video teleconferencing sessions were compared with "different/same" sides of the table for FTF sessions. Such comparisons preserved the spatial orientation among conferees (a conferee in front of or to the side of a fellow-conferee), regardless of communication medium.

4.1 Ease of Communication

The questions addressed here reflect the perceived ability to share ideas and communicate easily using a given communication medium. Subjects were specifically asked to rate the perceived frequency of interruptions and simultaneous speech events that occurred at their own location/side of the table and between locations/sides of the table. They were also asked to assess the ease of gaining the floor and getting their points across during the discussion.

The results indicated that subjects perceived that their ability to gain the floor was comparable to their usual participation at meetings, regardless of conferencing medium. Subjects using the PMS facilities reported that the audio

system had minimal effect on their ability to gain the floor. Hence, the PMS facility was considered somewhat of a "transparent" communication medium in terms of gaining the floor. Participants in the FTF sessions were asked to identify the factors that most affected their ability to gain the floor. Awareness of group dynamics was the most frequent response, followed by personal characteristics, knowledge and interest in the topic, and stewardship of the chairperson.

Subjects were also asked how frequently they were interrupted by fellow-conferees in the same room/side of the table and by others from different rooms/sides of the table. The data are shown in Table I.

Table I

Mean Perceived Frequency of Interruptions
(1 = not at all, 9 = extremely often)

Medium:	Same Rm./ Side of Table Mean (Task 1, 2)	Diff. Rm./ Side of Table Mean (Task 1, 2)
PMS	2.68 (2.54, 2.82)	3.18 (3.07, 3.29)
FTF	3.34 (3.39, 3.29)	4.13 (4.07, 4.18)

As is evident from the data, there were fewer perceived interruptions within a room and between rooms for the PMS sessions than for the same sides or different sides of the table for the FTF sessions. These findings suggest that the PMS facility produces more orderly turn-taking and fewer speaker exchanges that are viewed as interruptions than does the FTF medium of communication. It should be noted that for both communication media there were more perceived interruptions between locations/across sides of the table than within a location/on same side of the table.

Subjects were also asked to rate how often they perceived that two people from either the same location/side of the table or from different locations/sides of the table tried to talk at the same time. These data reflect subjects' perceptions of simultaneous (or nearly simultaneous) speech events. The data are shown in Table II.

Table II

Mean Perceived Frequency
of Simultaneous Speech Events
(1= Not at all, 9 = Extremely often)

Medium:	Same Rm./ Side of Table	Diff. Rm./ Side of Table
	Mean (Task 1, 2)	Mean (Task 1, 2)
PMS	2.63 (2.54, 2.71)	3.46 (3.50, 3.43)
FTF	3.86 (4.11, 3.61)	3.50 (3.50, 3.50)

These results indicate that the PMS sessions have fewer perceived simultaneous speech events within a room than FTF sessions have on the same side of the table. However, the perceived number of such events across rooms/sides of the table is equivalent.

Subjects were asked if they experienced any difficulty in getting their point across during the discussion. One-eighth of the subjects using PMS or FTF communication media reported difficulties in getting their points across. Personal attributes such as shyness or autocratic fellow-conferees were considered the primary sources of their difficulties. Difficulties with interruptions and simultaneous speech due to the audio system and/or transmission delay were not major complaints from subjects using this new PMS facility.

In summary, subjects using the PMS facility perceived fewer interruptions and simultaneous speech events within the same room than did subjects using FTF facilities perceive on the same side of the table. There were also fewer perceived interruptions across rooms for the PMS sessions than across different sides of the table for FTF sessions. The number of perceived simultaneous speech events across rooms for PMS and across sides of the table for FTF were equivalent. Though interruptions and simultaneous speech events are natural components of interactive communication during FTF meetings, they may affect, either positively or negatively, the ease by which task objectives are accomplished and may impact users' assessment of the social dynamics of video teleconferences.

4.2 Social Dynamics

These questions are concerned with conferees' perceptions of the group discussion and general meeting atmosphere. Subjects were asked to rate the conferences along several psychological dimensions of interpersonal communication[15] as well as the information flow during the meetings.

The mean ratings for the psychological dimensions of interpersonal communication are shown in Table III.

There were no significant differences in the ratings for these dimensions as a function of communication medium. Since these dimensions have produced some significant differences between FTF meetings and other video teleconferencing facilities in other research by the author, the lack of difference found here between PMS and FTF is probably not due to insensitivity of the measurement tool. Thus, for these psychological dimensions of social interaction, there is no evidence of a difference between this new PMS facility and FTF communication.

Table III

Subjective Ratings for Psychological
Dimensions of
Interpersonal Communication
(1=Not at all, 3=Slightly,
5=Moderately, 7=Quite, 9=Extremely)

Medium of Communication

	PMS Mean (Task 1,2)	FTF Mean (Task 1,2)
Dimensions		
Task-Oriented	6.68 (6.7, 6.7)	6.86 (7.0, 6.4)
Cooperative	6.73 (7.0, 6.4)	6.72 (7.0, 6.4)
Intensity of Involvement	6.18 (6.0, 6.4)	6.80 (6.6, 7.0)
Democratic Participation	6.43 (6.6, 6.2)	6.88 (7.2, 6.6)

Participants rated their fellow-conferees as quite attentive, regardless of the medium of communication. There was a tendency for subjects in PMS sessions to rate conferees in their own location slightly more attentive than those in the distant location, while subjects in FTF sessions tended to rate those across the table as slightly more attentive than those on the same side of the table. This pattern suggests a potential difference in perceived group dynamics as a function of the presence of a physical barrier between conferees. It is difficult to predict whether extended interactions using these media would accentuate or attenuate these differences in perceived social dynamics.

About two-thirds of the subjects using each communication medium rated the locations/sides of the table equivalent in both the quality and quantity of ideas generated. Choice of a future person with whom to confer was equally divided between rooms (i.e., own location vs. other location) for PMS sessions. However, those meeting FTF more strongly pre-ferred a future partner from across the table, rather than from the same side of the table. This finding suggests that the "social distance" among conferees in the same room may vary as a function of the ease of establishing eye contact and the presence of a physical barrier.

Subjects rated the FTF sessions significantly more enjoyable than the PMS sessions, though the magnitude of the difference is small (7.00 vs. 6.32, respectively). There was no significant difference in subjects' ratings of their satisfaction with the group solution to the conferencing tasks as a function of conferencing medium (PMS = 6.34 and FTF = 6.14).

4.3 Records of Speaker Behavior

As described previously, records of speaker behavior were made that permitted the frequency of speaker exchanges and simultaneous (or nearly simultaneous) speech events to be determined. The data are shown in Table IV. The total speaker exchanges reflect the number of speaker turns in addition to the number of simultaneous speech events (SSE). An SSE could occur between speakers in opposite rooms (PMS) or sides of the conference table (FTF). An SSE could also occur between speakers in the same room (PMS) or on the same side of the table (FTF). All of these events could have been heard by one observer or both observers.

Table IV

Mean Number of Observed
Speaker Interactions*

	Medium of Communication	
	PMS Mean (Task 1, Task 2)	FTF Mean (Task 1, Task 2)
Speaker exchanges plus simultaneous speech events (SSE)	99.6 (126.5, 72.8)	190.3 (273.0, 107.5)
SSE only	16.4 (19.5, 13.3)	28.4 (40.5, 16.3)
Total SSE Between Rooms/ Sides of Table	14.0 (15.3, 12.8)	18.6 (24.3, 13.0)
Heard by both rooms/sides of table	8.1 (9.3, 7.0)	9.5 (11.8, 7.3)
Heard by one room/side of table	5.9 (6.0, 5.8)	9.1 (12.5, 5.8)
Total SSE Within a Room/ Side of Table	2.4 (4.3, 0.5)	9.8 (16.3, 3.3)
Heard by both rooms/sides of table	0.9 (1.3, 0.3)	4.0 (7.3, 0.8)
Heard by one room/side of table	1.6 (3.0, 0.3)	5.8 (9.0, 2.5)

* The mean reflects the average of the two, 20-minute discussion tasks.

These objective records of speaker interaction patterns reveal a major difference in the number of speaker exchanges (alternations) and simultaneous speech events in FTF sessions and PMS sessions. There were almost twice as many such events for the FTF meetings, compared with the PMS conferences. There was also anecdoctal evidence that participants in FTF sessions were not as close to task solution as their PMS counterparts after 15 minutes of discussion and frequently had to be reminded of the 20-minute time limitation to complete the task. Since the conferencing tasks and the number of conferees were constant across communication media, these findings may provide evidence for why video teleconferences are often considered more efficient and productive than FTF meetings. They also confirm the anecdoctal reports that more extraneous "side" conversations are likely to occur in FTF meetings than in video teleconferences.

These data also parallel the previously reported subjective responses from users regarding the number of perceived interruptions and simultaneous speech events, with their being more for FTF than for PMS. It should be noted that a simultaneous speech event heard by only one room may be considered an unsuccessful attempt to interrupt a speaker. The lower frequency of such events for the PMS sessions is matched by a lower frequency of attempted interruptions (Total SSE). The percentage of successful interruptions (SSEs heard by both rooms or sides of table/total SSEs) is the same for both media (55% and 48% for PMS and FTF, respectively), and this may underlie the equal subjective ratings on group dynamics previously discussed (e.g., cooperativeness). It should also be noted that the number of simultaneous speech events shown in the table reflects each of two twenty-minute, problem-solving tasks. Hence, the total number of events experienced by each group of subjects was, on the average, twice the mean value indicated in the table.

These data on speaker interaction patterns provide evidence that FTF meetings are more interactive and apparently less polite than video teleconferences. PMS meetings are apparently more orderly than FTF meetings, a finding that may positively impact the ability of conferees to conduct their business effectively across a video link.

4.4 Effectiveness of Communication Media and Tasks

Subjects were asked to rate the perceived effectiveness of different communication media (FTF, PMS, and regular telephony) for various communication tasks. Only those subjects who used the PMS facilities were asked to rate PMS. The data are shown in Table V.

Table V

How effectively can different kinds of discussion tasks be handled by different communication media?
(1=Not at all, 3=Slightly, 5=Moderately, 7=Quite, 9=Extremely)

Mean Effectiveness Ratings

	PMS Ss N = 22			FTF Ss N = 25	
Medium:	FTF	PMS	RTP*	FTF	RTP*
Task:					
Giving/Receiving Info	8.4	7.2	5.5	8.0	5.8
Generating ideas	8.6	7.0	4.7	8.3	4.8
Asking questions	8.6	7.3	5.9	8.0	6.3
Persuasion	8.8	7.0	4.3	8.5	4.0
Exchanging opinions	8.7	7.3	5.3	8.1	5.4
Bargaining	8.6	6.3	4.2	8.3	4.4
Problem-solving	8.7	7.0	4.3	8.3	4.7
Getting to know someone	8.8	6.1	3.1	8.6	3.5

* RTP = regular telephony

The results indicate that the three communication media were rated significantly different in their perceived effectiveness. PMS, however, was rated closer to FTF than to regular telephony for most of the discussion tasks. Those tasks that require less personal contact or social presence (e.g., asking questions, giving and receiving information, and exchanging opinions) had the smallest range in ratings across the different media, while those tasks that required more personal contact (e.g., getting to know someone, persuasion, and bargaining) had the greatest range. This finding supports the previous suggestion that perceived social presence may be more important for some tasks than others. The data further indicate that a video teleconferencing medium such as PMS is perceived to help bridge the gap between FTF meetings and regular telephony for many different types of communication tasks.

5. Conclusions

The results from this human factors research indicate ways in which conferences using the PMS offering are similar and different from conferences held in-person. These results challenge whether FTF meetings should be considered the "ideal" model for video teleconferences since the level of interactions and simultaneous speech events differs dramatically between these types of communication media. Rather, the goal for video teleconferencing systems should accentuate the positive social dynamics of FTF meetings such as cooperativeness, task-orientation, and enjoyment. In addition, although interruptions and simultaneous speech events may be typical of FTF meetings, they seem to contribute to user dissatisfaction with some video teleconferencing systems[5]. However, the new Bell system design appears to eliminate the problems associated with interruptions and simultaneous speech events. Both the objective and subjective data presented here indicate that the video teleconferencing system that is implemented in the offering of PMS accomplishes these goals.

REFERENCES

[1] Williams, E. Holloway, S. The
     Evaluation of Teleconferencing:
     Report of a Questionnaire Study
     of Users' Attitudes to the Bell
     Canada Conference Television
     System, Communications Studies
     Group, London, England, Paper
     P/74247/WL (discussed in refer-
     ence 2).

[2] Johansen,   R.,   Vallee,   J.,
     Spangler, K. Electronic Meet-
     ings: Technical Alternatives
     and Social Changes. Reading,
     Mass.: Addison-Wesley Publish-
     ing Co., 1979.

[3] Short, J., Williams, E., Chris-
     tie, B. The Social Psychology
     of Telecommunications.   New
     York: John Wiley & Sons, 1976.

[4] Menist, D. B., Wright, B. A.
     PICTUREPHONE Meeting Service:
     The System. Paper presented at
     the University of Wisconsin
     Conference on Teleconferencing
     and Interactive Media, May
     1982.

[5] Wolf, C. G. Video Telecon-
     ferencing: Delay and Transmis-
     sion Considerations.   Paper
     presented at the University of
     Wisconsin   Conference   on
     Teleconferencing and Interac-
     tive Media, May 1982.

[6] Brecht, M. A Study of Meeting
     and Conference Behavior, Techn-
     ical Report. Baltimore: The
     Johns   Hopkins   University,
     Psychology Dept., July, 1979.

[7] Argyle, M. Social Interaction.
     London: Methuen, 1969.

[8] Duncan, S., Fiske, D. Face-
     to-Face Interaction. Hills-
     dale, N. J.: Lawrence Erlbaum
     Associates, 1977.

[9] Kendon, A. Some Functions of
     Gaze-Direction in Social
     Interaction. Acta Psycholo-
     gica, 1967, 26, 22-63.

[10] Duncan, S. Some Signals and
     Rules for Taking Speaker-Turns
     in Conversations. Journal of
     Personality and Social Psychol-
     ogy, 1972, 23 (2), 283-292.

[11] Brady, P. A   Statistical
     Analysis of On-Off Patterns in
     16 Conversations. The Bell
     System Technical Journal, Janu-
     ary 1968, 73-91.

[12] Brady, P. Effects of Transmis-
     sion Delay on Conversational
     Behavior on Echo-Free Telephone
     Circuits. The Bell System
     Technical Journal, January
     1971, 115-134.

[13] Krauss, R. , Bricker, P.
     Effects of Transmission Delay
     and Access Delay on the Effi-
     ciency of Verbal Communication.
     Journal of Acoustical Society
     of America, 1967, 41 (2), 286-
     292.

[14] Pfeiffer, J. W., Jones, J. E.
     A Handbook of Structured
     Experiences for Human Relations
     Training, Vol. I-VII. La
     Jolla, Ca.: University Associ-
     ates Publishers, 1979.

[15] Wish, M., Kaplan, S. J. Toward
     an implicit theory of interper-
     sonal communication. Sociome-
     try, 1977, 40 (3), 234-246.

The Teleconferencing Resource Book: A Guide to Applications and Planning
Lorne A. Parker and Christine H. Olgren (eds.)
Elsevier Science Publishers B.V. (North-Holland)
© Center for Interactive Programs, University of Wisconsin-Extension, 1984

LEARNING FROM EXPERIENCE:
SOME USER REACTIONS TO TELECONFERENCING

Vicki Smith Sherman
Member of Technical Staff
Bell Laboratories
Holmdel, New Jersey

Abstract

In the past, there has been a recurring pattern to teleconferencing usage in new systems. Initially, the system was heavily used; as the novelty wore off, usage dropped off.

The in-house AT&T/BTL Audiographics Teleconferencing Trial started in February, 1980. Over the past 18 months, the Trial has shown a pattern of steadily increasing usage. As of September 30, 603 teleconferences have been held. Over 1000 people have used the service (some as many as thirty-five times). Considering that no promotion has been done and that this is strictly a word-of-mouth service, this amount of usage demonstrated a high level of user need.

One of the Trial's primary objectives was to study teleconferencing usage. Three aspects of usage data will be presented: usage statistics, preTrial travel data, and user characteristics. PreTrial travel data was collected through a survey of all technical supervisors at Trial locations regarding their travel patterns between Trial locations over a three-month period. User characteristics have been compiled from a premeeting questionnaire which studied teleconferencing-related experience, project meeting history, and travel history, as well as from Trial data.

Another of the Trial's primary objectives was to examine meeting management issues. What are some of the problems users encounter when they teleconference? How does their behavior change with increasing teleconferencing experience? These data were collected from oral and written questionnaires, focussed group interviews, and teleconferencing observations.

Introduction

For eighteen months, Bell Laboratories (BTL) has been conducting an in-house audiographics teleconferencing trial. Much has been learned about teleconferencing. Perhaps the most important finding has been the heavy usage the service has received. Between February 21, 1980 and September 30, 1981, 603 teleconferences were held. Usage data of three types will be presented: usage statistics, preTrial travel data, and user characteristics.

Another objective of the Trial was to study meeting management issues. What are some of the problems people encounter when they teleconference, and how do they solve them? The data used to answer these questions primarily are from questionnaire and interview questions.

Background Information

The Trial includes seven locations, five of which are BTL locations (Denver, Colorado; Holmdel, New Jersey; Indian Hill, Illinois; Indianapolis, Indiana; and South Plainfield, New Jersey), and two AT&T locations (Basking Ridge, New Jersey and 195 Broadway, New York). Holmdel and Indianapolis conducted the first teleconference on February 21, 1980.

These rooms have essentially the same equipment. All Trial rooms are maintained by room attendants (except for South Plainfield) and are scheduled through a centralized reservation system. Meetings are set up by the room attendants and bridged if necessary. Questionnaires are filled out by all participants at the end of each teleconference.

Overall Usage

As of September 30, 1981, 603 teleconferences have been held. Figure 1 shows the number of teleconferences held per month. The pattern that emerges is one of growing usage. Meeting usage grew slowly but steadily through December, 1980. In January, 1981, the Trial allowed simultaneous two-point meetings if the rooms were available. People responded to the availability, and again usage grew rapidly. "Slack" periods seemed to correspond to likely vacation periods.

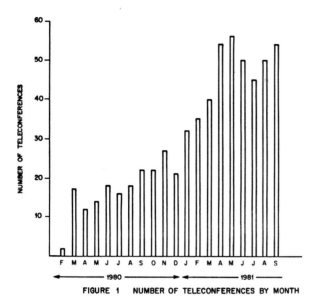

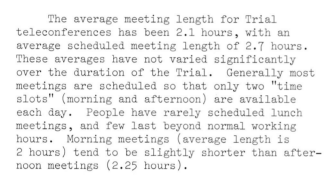

FIGURE 1   NUMBER OF TELECONFERENCES BY MONTH

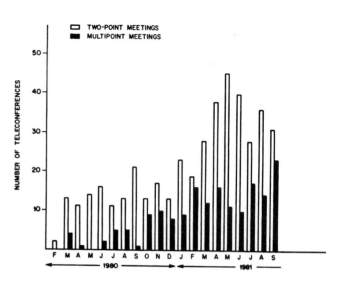

FIGURE 2   TWO-POINT VERSUS MULTIPOINT TELECONFERENCES BY MONTH

The average meeting length for Trial teleconferences has been 2.1 hours, with an average scheduled meeting length of 2.7 hours. These averages have not varied significantly over the duration of the Trial. Generally most meetings are scheduled so that only two "time slots" (morning and afternoon) are available each day. People have rarely scheduled lunch meetings, and few last beyond normal working hours. Morning meetings (average length is 2 hours) tend to be slightly shorter than after-noon meetings (2.25 hours).

Two-point Versus Multipoint Usage

Since the beginning of the Trial, two-point meetings have been more common than multi-point meetings. However, the number of multi-point meetings has been increasing (see Figure 2). There are now (or have been) several "regular" multipoint meetings. The first was a seminar series. Twice a month a presentation is presented with 12 locations participating. All of these "regular" meetings are audio-only (no graphics equipment).

However, just as the number of multipoint meetings is increasing, the number of two-point meetings is also increasing. Figure 2 shows the percentage of multipoint meetings as compared to two-point meetings. Although the percentage of multipoint meetings fluctuates, there does appear to be more multipoint meetings than during the first eight months of the Trial. In October, November, and December, one group teleconferenced more frequently than once a week with three or four locations. They

stopped teleconferencing in January when their task force ended. This kind of meeting activity leads to large fluctuations in the percentage of multipoint meetings.

Reservations Patterns

Another way to look at usage and demand is to examine reservations patterns. Initially (through June, 1980), teleconferences could be scheduled less than a week in advance. However, by December, 1980, people were forced to make reservations over two weeks in advance on the average. Usage was growing throughout the year, and reservations also were being made more and more in advance. This leveled off somewhat in 1981.

Figure 3 gives a more detailed look at the number of working days in advance that tele-conferences were scheduled (as of September 30, 1981). Sixty-four percent of all reservations were scheduled within two weeks of the tele-conference. The overall mean is 12 working days in advance. However, the range demon-strates that many meetings have been scheduled a long time before they are to occur, probably before an agenda or even ideas about an agenda have been formed.

There are two factors which could have contributed to this pattern. During the summer, people began to schedule blocks of regular teleconferences (usually once per week, but sometimes less frequently) over extended periods of time. September was the first month to have teleconferences that had been scheduled

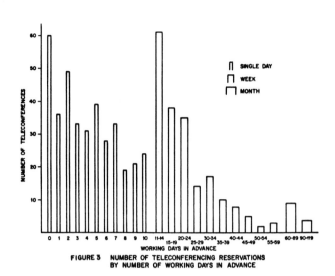

FIGURE 3 NUMBER OF TELECONFERENCING RESERVATIONS
BY NUMBER OF WORKING DAYS IN ADVANCE

more than 20 working days in advance (18% of September meetings were scheduled more than 20 days in advance). Since then, between 7 and 26 percent of the meetings have been scheduled more than 20 working days in advance every month. Most of these were made as part of a block of regular meetings. This suggests that teleconferences are considered to be more than a last minute option when one cannot get airplane tickets. Users choose to teleconference these meetings, rather than meet face-to-face.

Another factor contributing to the growing number of days reservations are made in advance is the growing usage itself. There is a fixed number of rooms and potential meeting times. Afternoons are the most frequently used time of day. In general, teleconferencing is light on Monday and Tuesday, and heavy towards the end of the week. This would lead to making reservations further in advance for popular days and times.

Number of Users

The number of users has been steadily increasing. On the average, 60 new people and 50 experienced people have an audiographics teleconference each month. As of September 30, 1981, 1192 people have used the audiographics teleconferencing service. As was mentioned previously, the number of audio-only teleconferences began to affect overall teleconferencing usage to a noticeable extent in November, 1980.

Forty-one percent of the people who have ever used the service have used it more than

once. Clearly there is variety in the patterns for repeat usage. On the average, people teleconference once per month, although a few people teleconference more frequently. This pattern indicates that people begin to teleconference as a way of doing business.

This usage is somewhat unusual when it is realized that virtually no publicity has been given to the Trial. Initially, a few supervisors were approached and told about the Trial service. However, almost all of the users have heard about the service by word of mouth.

Travel Survey

Travel substitution has been suggested as a reasonable way for buyers of teleconferencing systems to cost justify the purchase. It is also necessary to have a method for determining how many and which locations need teleconferencing facilities. To investigate the travel patterns of potential teleconferencers between Trial locations, a travel survey was conducted. Questionnaires were sent to all technical supervision to determine the number of trips made between the AT&T and BTL locations to be included in the Trial. The results are of interest now (two years later) in terms of actual behavior comparisons.

Looking at the number of trips that supervisory groups made, it seemed that Holmdel people traveled the most to all Trial locations, with Basking Ridge people traveling a close second. 195 Broadway and Indianapolis, on the other hand, seldom traveled to any other Trial locations.

It is interesting to compare the number of trips with the frequency of teleconferences between the Trial locations. Although Holmdel-Basking Ridge trips involved the most people, it was not a frequent teleconference (only 5 teleconferences of 603 have been Holmdel-Basking Ridge only). Holmdel-Denver trips involved few people, yet they comprised a large percentage of the teleconferences (99 of 603 teleconferences have been Holmdel-Denver teleconferences).

This suggests that people prefer to travel short distances and teleconference longer distances, even if the further distances are to a more "attractive" place. Returning weekly to a distant location, no matter how attractive, is something some people clearly want to avoid. Travel selection seems to be important to teleconferencers. The ability to decide for which meetings to travel is important to people who use teleconferencing.

User Background

During the first part of the Trial, a pre-meeting questionnaire was administered to first-time users. It was designed to learn about users' past experience related to teleconferencing, their expected teleconferencing graphics use, and their project meeting history. These were mailed out through December, 1980; 113 were returned.

Relevant Past Experience

There were several aspects of a user's past experience that could affect their attitude toward, and their knowledge of, teleconferencing. The audio equipment provided in the Trial is a voice-switched system, similar to 4A Speakerphones. Use of the Gemini® 100 Electronic Blackboard and mailroom facsimile service gives some familiarity with the graphics equipment provided in the Trial. All of these have been experienced by the first-time users to varying degrees, demonstrating some familiarity with teleconferencing-related equipment.

Expected Equipment Use

Users have had many forms of equipment and graphics present in their past meetings on their current project. The table below lists the types used in previous project meetings.

Type of Graphics	Percent of Users
Handouts	94
Viewgraphs	89
Blackboard	84
Slides	33
Objects displayed	21
Movies, film	11
Tape recorder	9

Based on this type of usage, one would expect the facsimile equipment to be the most often requested piece of graphics equipment. However, first-time users expected to use primarily the electronic blackboard (89%). Facsimile was only expected to be used by 57% of the users.

Project Meeting Experiences

Sixty-nine percent of the first-time users

had previously attended meetings on the project. Of these people, 85% claimed that the locations involved in their upcoming teleconference had also been involved in previous meetings. This suggests that people may have some familiarity with the people at remote locations. However, a surprisingly large percentage (31%) had never attended a project meeting before their first teleconference. This may have been a "new" project assignment, or it may be that people who would not be sent to another location for a meeting might participate in a teleconference from their own location. This would tend to reduce familiarity with other teleconference participants.

Users who had had project meetings prior to their first teleconference had attended them with varying degrees of frequency:

30%	Several per month
35%	About once a month
35%	Less than once a month

Most people had attended meetings once a month or less which corresponds well to the average number of teleconferences per month data collected in the Trial (1.0-1.3 meetings per month on the average).

About two-thirds of the first-time users had traveled to another project meeting (12% had traveled often; 45% had traveled several times; and 10% had traveled once). This suggests that teleconferencing may substitute for travel for some people, but some people who attend teleconferences may never travel to attend project meetings.

Of the people who travel, 75% sometimes conduct other business in addition to business related to their project. Forty-eight percent sometimes combine their project trips with trips to friends or relatives. This again suggests that people want to select which trips to take, rather than have to travel or teleconference all of them.

Meeting Management Issues

During the early stages of the Trial, it became apparent that users had different expectations about how to conduct a teleconference. Some people clearly were more prepared, and had more successful meetings. The problems that people described in debriefing sessions and on written questionnaires seem to cluster around the same general issues. In order to learn more about these problems, 20 people were interviewed in focus group sessions. It is important to remember that the discussion of the following issues occurred in a session that focused on problems in meeting management; these problems need to be viewed from the perspective of the high level of user need and teleconferencing usage.

First Time Use

When people first walk into one of the Trial rooms, they often claim to have had an uneasy feeling. Generally what happens during the first couple of teleconferences is that conferees "play" with the equipment. They use the equipment enough to either learn how it works, or to give up and ignore it. Sometimes they summon help if they cannot figure out how the equipment works. They tend to feel self conscious when the slow scan television sends their picture. Often comments following first teleconferences suggest that conferees were distracted from the purpose of their meeting by the equipment.

This problem is exacerbated by the fact that many first-time users do not know what equipment will be available to them before they walk into the teleconference room. Some of them have had teleconferencing-related experience before (they have used a speakerphone, or sent facsimiles via the mailroom). This may lead them to make incorrect assumptions about what the equipment can do, or what will be available.

Part of the "first time user" problem is related to the general problem that any new experience can be and usually is awkward. As one conferee stated, "Initially it was a little awkward, but I think you can overcome it fairly quickly." In this sense, teleconferencing is no different from any other first time experience. Often people hesitate to speak the first time, and the chairperson (or someone else) needs to pointedly draw them into participating.

Speaking and Listening

Most teleconferencing communication is auditory. Consequently, problems conferees have while speaking and listening are very important as they can significantly influence the amount accomplished at a teleconference.

A comment made by several conferees concerned speaking to the other location. "It feels like no one is out there." The speaker, seeking some reaction may ask if anyone has any questions. If no one answers the silence is often uncomfortable, resulting in speakers repeating themselves over and over to make sure people understand. The problem is further compounded by the lack of body language and other forms of feedback. This is particularly true if no one else is in the same room with the speaker.

The problems of the speaker are two-sided. Not only does it feel like no one is listening; unidentified people may also be listening. One person commented that the slow scan television helped one to know that people were listening. "It is kind of reassuring to know they are there ... It is better than nothing." The privacy issue is a tough one because clearly anyone who wants only to listen may do so. People may not purposefully be hiding their existence; it may be that they came in late without announcing their arrival.

People who are listening also experience problems. Sometimes it is hard to distinguish a pause from the end of a speaker's comment. Sometimes it is clear whether the speaker is pausing or is finished speaking from the context or the voice inflection, but often it is not. This makes it difficult to decide when to start speaking. This is much more of a problem than in a face-to-face meeting where it is often quickly apparent that the previous speaker has finished.

Another listening problem is deciding who said what at the remote location. This is particularly true for groups that have not met face-to-face, or have done so infrequently. It is hard to differentiate voices, particularly without a face or name to associate with it. Meeting people face-to-face before teleconferencing helps as do telephone conversations. Over time, a person's voice becomes recognizable in its own right.

Clearly speaker identification is a more severe problem in a multipoint meeting. At least in a two-point meeting, the location of the voice is known. When the location is not clear, the identification of a voice is even harder.

Teleconferencing, however, offers some advantages over face-to-face meetings with regard to listening. For example, the "microphone off" button permits a location to discuss privately among themselves so that they can reach a consensus without the other location(s) hearing their discussion.

These speaking and listening problems are difficult to solve completely. Asking for additional comments, or addressing people at particular locations can help. The chairperson can facilitate this process by making sure people who want to comment are given the chance.

Speakers prefacing their comments or questions
with their name (and/or location) also helps.

Attention and Efficiency

Conferees differed in their ability to
attend to the ongoing teleconference. Most felt
that their attention wandered less easily than
in a face-to-face meeting. One factor that
contributed to this was the physical isolation
of the room. The room was dedicated for teleconferencing use. Consequently, there were
less telephone calls to conferees, and, since
the conference room door was kept closed, less
distractions were present from "outside."

Conferees also felt they could critique
faster in a teleconference. The teleconference
(when well run by the chairperson) generally
did not deviate significantly from the agenda.
This resulted in a more efficient handling of
the topics on the agenda.

Conferees are also more likely to say what
they think in a teleconference. They feel less
constrained by a voice than by actual physical
presence. While conferees want to know who is
at the other location, and what rank people are,
they are not constrained in terms of what they
will say. They will try to tailor the information to the type of information that each
type of person would want to hear (in terms of
level of detail, etc.).

Accomplishment and Ease

Most conferees claim that they generally
accomplish what they expected to accomplish,
which they claim would have happened in a face-
to-face meeting. This feeling of accomplishing
"about the same" in a teleconference seems to
be the opinion of the large majority.

Most conferees comment that they are
comfortable when teleconferencing and find it
very convenient. A few even claim that it is
fun! However, initially there are also some
comments that teleconferencing is something
that you get used to, that you learn how to do
with experience.

Many debriefings included comments about
the "benefits" of teleconferencing, particularly when people were commenting on accomplishments. In general, the comments made
about teleconferencing benefits centered around
one major theme: travel. Many of their
comments reflected the increased efficiency
and their sense of using their time more effectively. Conferees clearly chose to

teleconference rather than fly to hold it face-
to-face. This is not to say they teleconferenced all meetings. From their comments it was
also clear that they would fly for some of them.
Travel "selection" seems to describe the situation. As one conferee commented, "You commute
whenever you want (TC.9-5)." People want to
control their life, and choose when to meet by
teleconferencing, and when to fly to meet face-
to-face.

Discussion

The data from the Trial clearly show that
there is a need for teleconferencing facilities
at most Trial locations. With very little
publicity, very high usage has occurred. It is
also clear that simple measures of the number
of trips, or the number of people traveling
between locations cannot predict teleconference
volume. However, locations with communities of
interest will teleconference more than locations
with none.

Usage continues to grow in the Trial. As
more locations are added, the network of
communities of interest will become more complicated, and the number of multipoint meetings
should continue to increase. A method for
predicting teleconferencing traffic needs to be
developed. It is clear that travel is related
to the amount of teleconferencing, but it is
not a simple relationship, and it is not the
only variable. Much still needs to be learned
before a model of teleconferencing traffic can
be developed.

Users tend to teleconference for many
reasons. They teleconference because they want
to, not because they feel they have to, or
because there is no other alternative. Teleconferencing becomes a way of conducting a
meeting, an alternative to consider when discussing when and where to have a meeting. Teleconferencing can also facilitate spur of the
moment meetings that may increase productivity/
communication among project members. People
whose input may be valuable, or who need to be
aware of what is happening in meetings may teleconference, even though they would not travel
for a meeting.

Participating in a teleconference is not
the same as in a face-to-face meeting. There
are "new" problems in this means of conducting
a meeting. The first time someone teleconferences may be an awkward, even "scary" experience. However, conferees get used to it, and
integrate it into their means of doing business.
Problems such as speaker identification, knowing
when to start speaking, and insuring participation by all conferees become less difficult

as conferees gain experience and devlop protocols. Conferees clearly feel at ease, and feel
that they generally accomplish what they desired to accomplish. They cite the benefits of
teleconferencing frequently when discussing
their increased efficiency by saving travel and
meeting time (many conferees find teleconferenced meetings to be shorter than face-to-face
meetings even though they accomplish the same
things).

Clearly teleconferencing is becoming a
popular way of doing business for some applications for some people. As more people gain
experience with teleconferencing, more can be
learned about its merits and shortcomings.
Understanding the variety of uses and situations
encountered can help in the design of a successful service.

The Teleconferencing Resource Book: A Guide to Applications and Planning
Lorne A. Parker and Christine H. Olgren (eds.)
Elsevier Science Publishers B.V. (North-Holland)
© Center for Interactive Programs, University of Wisconsin-Extension, 1984

ONCE YOU'VE GOT A COMMUNICATIONS SYSTEM, HOW TO KEEP PEOPLE USING IT
OR
KEEPING THE SHOW ON THE ROAD

Edward O. Hugdahl
Professor of Music and Chairman
University of Wisconsin Extension Music Department
Madison, Wisconsin

Distance learning! Being on the cutting edge of today's electronic explosion. The maneuvering to get the system installed. After interminable delays the equipment finally arrived and it all became operational ...at last! The bugs were ironed out. Then the dazzle and excitement of those first ventures with the unseen learner. A sheaf of press clippings and glowing feedback. It's heady wine.

Once the novelty has worn off with presenters and participants, sustaining interest, expanding into new endeavors and developing programming to become self-sustaining, plague educators with communication resources at their disposal.

Geritol Is For Everyone

These days the term of "burn-out" surfaces in educational circles everywhere. Whether it's related to static salary scales, unrealistic increased income expectations, expanded student and course loads, or decreased federal and state funding, conversations with adult educators at coffee breaks reveal disenchantments. All of this is coupled with the challenge of avoiding ruts and routines as programming falls into yearly cycles.

Weary-worn adult educators can slip into the doldrums and with a ho-hum attitude xerox last year's schedule for an easy re-run. It's one way to placate one's superiors by complying with expectations to use that communications equipment that arrived with so much difficulty. Communications systems do have staggering numbers of time slots in the course of a week, let alone in a semester or a year. Administrators do get restless when these systems are not being used.

The constant presence of a communications system can become old hat. No longer is there the tantalizing intrigue of beguiling gadgetry ...a love affair with electronics that grew out of a proposal period in the early grant writing days. The system is in. It's there. It's installed. It's operational.

Ho-hum attitudes rapidly catch up with one. Yellowed lecture notes take longer to catch up with an instructor in face-to-face learning. But in distance learning it's faster. In working in adult education the pressure is to produce or these students just opt to stay home.

Regardless where one is in the educational scene a shot or two of "motivational Geritol" can be for everyone these days. Students need constant motivation. But so do teachers and adult educators as well. Combatting the doldrums and the ho-hum syndrome is especially crucial to those engaged in distance learning: from the technical back-up staff to adult educators and instructors.

Effective distance learning requires tremendous teamwork. The backup crews working behind the scenes are as equally crucial for successful distance learning as those in the spotlight at the microphone. No longer is it a solo act with an instructor at a lectern in a classroom as in face-to-face learning. The communications engineers keeping abreast with the latest technical developments and pushing a system to its limits deserve more applause than those at the microphone. Without a dedicated and competent technical staff there just is no system to use. Clerical staff and operating personnel play key roles in a smoothly functioning system. These people deserve recognition and appreciation. They need to know they play key roles in this team effort in education. They, too, need Geritol.

Adult educators often become discouraged as some of the most carefully researched and designed offerings self-destruct. It takes greater effort to maintain an interest in students whom one does not see. The weekend workshops in retreat settings provide dynamic interpersonal relationships that adult educators revel in and can see tangible evidence of their efforts materializing. It takes more effort to sense student appreciation in distance learning. Behind rosters of participants scattered hundreds of miles apart in a distance learning offering are people. And each and everyone of them is important. It takes carefully designed evaluations as well as reading between the lines to sense why a student responded to a distance learning offering. With all of these frustrations adult educators, too, need Geritol.

Instructors who use distance learning on a regular basis need reassurance that their efforts are meeting a need. There is a loneliness sitting in front of a microphone when most of the time there are students populating a classroom. It does take extra effort to teach in the distance learning mode. It does take extra effort to communicate one's excitement about the subject matter with these unseen students. Energies expended for distance learning are exhausting. Teachers, too, need Geritol.

Wherever one is in distance learning and adult education a malaise of indifference is all too contagious as yearly cycles relentlessly reappear in seemingly shorter and shorter time spans. It is imperative that everyone so fortunate to have distance learning potential available retain an excitement about the potential that this medium has for reaching people. At all times being in the people business must be paramount in one's thinking. There must be imagination. There must be vision. There must be a sense of adventure. Through the marvels of the electronic age reaching people and serving their needs is possible as never before in the history of civilization. Those are staggering responsibilities. Perhaps the limitations are those of our abilities to utilize most creatively the resources that we don't really know how to exploit most fully. Yes, Virginia, there is Geritol...for everyone!

Climbing Mount Everest: Expanding Into New Clientele Groups

In those first ventures in distance learning one cautiously dips one's toes into the water fearing unseen drop-offs and treacherous currents. The unknown always activates hesitancies, until the dipping assures wading, before that dive into deep water. Then it's, come on in the water is fine!

But just paddling around and not getting into open water makes those distance learning efforts just a dip in the old swimming hole. The potential for programming in the distance learning mode is as vast as the ocean. Not dreaming beyond the unlimited horizon is just walking along the beach. Being content with an offering or two is only swimming in shallow water. There's a lure of adventure in distance learning that can make Olympian swimmers. But to become an Olympian, swimmers must be in the water, not in a deck chair philosophizing about developing endurance. In distance learning adult educators must be in the water aggressively swimming beyond the horizon.

To expand horizons everyone, regardless of their subject matter area, must view their subject area as having many sub-groups in it.

Figure I

Clientele Groups in the Field of Music For While The UW Extension Music Department Used the Distance Learning Mode 1973-1982

1. Private Teachers of Piano
2. Private Teachers of Spinet Organ
3. Public and Parochial School Music Educators Teaching at Various Levels and In Specialized Areas
   a. Elementary School
      (1) General Music
      (2) Choral Music
      (3) Mainstreaming Students with Developmental Disabilities
   b. Junior High School
      (1) Choral Music
      (2) Instrumental Music
         (a) Band
         (b) Orchestra
   c. Senior High School
      (1) Choral Music
      (2) Instrumental Music
         (a) Band
         (b) Orchestra
         (c) Jazz ensembles
      (3) General Music
   d. Music Administrators of School Systems
4. Church Music
   a. Choir Directors
      (1) Children's Choirs
      (2) Youth Choirs
      (3) Adult Choirs
   b. Organists
      (1) Beginning
      (2) Intermediate
   c. Clergy and Lay Leaders
5. Teachers of Pre-school and early childhood education

The University of Wisconsin Extension Music Department has been involved in programming in the distance learning mode for ten years. During this time period there has been programming designed for a number of sub-groups within the field of music. Figure I shows the groups for which there has been programming during these first ten years.

Such offerings did not all begin at once. Over a period of time new emphases were gradually added as confidence and experience in using this mode developed. Other factors influencing adding offerings were (1) appropriate contacts within the sub-group for programming guidance; (2) available mailing lists; and (3) adequate promotional support.

Yet, Figure I is only the beginning for the field of music. There are other sub-groups within music that have not yet been programmed. Whenever there is staff time to pursue them, whenever there is a viable potential to insure a successful response, whenever there can be appropriate planning groups for designing offerings, then efforts will aggressively move forward to expand. For example, just a few fields yet untouched are: (1) music merchandising in both small and large operations with varying specializations; (2) private violin teachers; (3) music therapists; and (4) composers of both serious and popular music. And the list can go on.

What is so crucial for the adult educator is forever to be unsatisfied in seeking additional clientele groups to serve. This does not mean spreading oneself so thin that only a superficial effort results. On the contrary, a solid, continuing program must be established with each clientele group. This is key to sustaining a program.

In education today the greatest potential for unlimited growth and development is with the adult learner. It's an awesome responsibility for the adult educator. Everywhere there are growing numbers of adult learners needing professional continuing education to stay abreast with the knowledge explosion as well as adults seeking general knowledge and skills for self-fulfillment and life enrichment. The distance learning mode can bring resources to people wherever they are, even in remote regions. No longer do people need to be deprived because of their place of residence. The world's resources can be brought to one's own backyard. It's staggering the impact distance learning can have on people and society.

Adult educators must climb every mountain searching high and low for new clientele groups to be served. And Everest is attainable in the distance learning mode.

Waging A Two Front War

A never ending battle in adult education is to lure people into continuing education experiences. When professional associations and state regulatory agencies require periodic courses and classes for renewal of certification, offerings serving such clientele groups have built in enrollment guarantees.

The non-credit offerings which provide self-enrichment but are purely voluntary are some of the greatest challenges to the adult educator, yet they can be the most exciting and rewarding. Perhaps some of the most imaginative and creative efforts in education today can be in the non-credit area. Being especially sensitive to needs and interests of a particular clientele group and bringing the most authoritative expertise to it are key to having dreams become realities with viable

enrollments to support the offering.

Nevertheless it is a perennial challenge to motivate adults to participate in continuing education offerings. Complacency afflicts all too many. Exciting people about learning and the joys it can bring to one's personal and professional life is the mission of adult education.

The distance learning mode is relatively new in adult education. Face-to-face learning is what everyone knows and expects. In the thinking of many presence in a classroom is equal to learning, even though in actuality no learning might take place.

For adult educators using the distance learning mode extra efforts are needed to persuade clientele that there are many kinds of offerings which can be equally effective on an interactive aural communications system. People must be persuaded that so much learning in a student-teacher relationship is through the aural sense with the visual sense merely reinforcing concepts gained through the aural sense. Key to learning is the interaction between student and teacher . With an interactive aural communications system the basic dialogue between student and teacher is possible. But we have only begun showing adults that distance learning can be a great resource and should be used.

In Figure II data on participants in distance learning offerings in music in Fall 1981 indicate that even after ten years of programming for music clientele groups there are people experiencing distance learning for the first time. This underscores that adult educators using distance learning must view their efforts on a long term basis. Minimal response in early years is to be expected in any new endeavor, especially one so radical in basic learning modes. Being patient and persevering is absolutely essential. Once one ventures into distance learning one must constantly be presenting offerings so that over a period of time a following for this mode of learning is developed. The most important message to convey to one's clientele groups is that the distance learning mode is here to stay and that each year the clientele groups can look to this source as a constant and tremendous resource for meeting professional and continuing education needs.

The high percentage of participants willing to use the distance learning mode for their future continuing education needs in Figure II is evidence that the potential of this medium is being understood. Of course, positive and productive experiences in distance learning are the best advertising one can obtain.

Figure II

Non-Credit Offerings of the UW Extension Music Department in the Distance Learning Mode in Fall 1981

	First Experience With ETN	Previous Experiences With ETN	Willing To Use ETN Again	Not Willing To Use ETN Again
BEGINNING CHURCH ORGANISTS: BASICS FOR EFFECTIVE SERVICE PLAYING (Sept. 14 & 21) (Church Organists)	71%	29%	100%	0%
IN TUNE WITH GENERAL MUSIC (Sept. 22 & 29) (Elementary School Music Teachers)	33%	67%	100%	0%
SMALL CHURCH CHOIRS: A BONANZA OF MATERIALS FOR ALL CONDITIONS OF CHOIRS (Sept. 28 & Oct. 5) (Church Choir Directors)	38%	62%	95%	5%
MAKING SUNDAY MORNING CLICK (Oct. 6, 13 & 20) (Clergy and Lay Leaders)	46%	54%	96%	4%
MODELS FOR GETTING RESULTS IN MUSIC (Oct. 15 & 22) (High School Band and Choir Directors)	38%	62%	100%	0%
BOWING SKILLS AND STYLES FOR ORCHESTRAL PERFORMANCE (Oct. 27, Nov. 3 & 10) (School Orchestra Directors)	71%	29%	100%	0%
CHILDREN'S CHOIRS IN CHURCHES (Oct. 28, Nov. 4) (Church Children's Choir Directors)	54%	46%	98%	2%
JAZZ ALIVE: TEACHING JAZZ AND IMPROVISATION ALONG WITH THE STANDARD REPERTOIRE (Sept. 14 & 21) (Piano Teachers)	20%	80%	100%	0%
GETTING DOWN TO THE NITTY GRITTY OF PIANO TEACHING (Sept. 25 and Oct. 2) (Piano teachers)	18%	82%	100%	0%
BETWIXT AND BETWEEN: SAVING THE INTERMEDIATE STUDENT (Oct. 21 & 28) (Piano Teachers)	8%	92%	100%	0%
JUST ORDINARY KIDS CAN DO IT TOO: TEACHING AND LEARNING TO PLAY BY EAR (Dec. 1, 8, 15) (Piano Teachers)	28%	72%	100%	0%

In Figure III the response to the non-credit offerings indicates a positive response in terms of enrollment in the offering as well as the motivational efforts to gain this response. The percentage of counties in which the participants were located underscores the extensive impact, statewide as in the case of a statewide system in Wisconsin, distance learning can have.

As adult educators venture into distance learning, patience and perserverance are key to establishing a continuing and sustained program.

Putting Pizazz Into Programming

Luring adults into non-credit continuing education experiences is a never-ending challenge. Being especially sensitive to the needs of various clienteles and designing offerings to meet these needs are key ingredients in adult education

Having advisory committees is a helpful technique in gaining a sense of grass roots interests and desires. Sometimes these advisory committees can be very helpful. At other times one has to temper their thinking with one's own knowledge, experience and sensitivity to the field. One must be very careful with advisory committees so that they understand their role is one of being advisory only; otherwise, one might get into difficult positions with advisory committees issuing directives for programming which might not be successful.

In addition to advisory committees it is most healthy to be able to communicate with colleagues in one's subject matter so that ideas can be mutually evaluated for their merits. Then, of course, ultimately one must have enough confidence in one's own thinking coupled with knowledge of the field and its needs to make final decisions.

Sometimes celebrities teaching in a given field are key factors in luring people into continuing education experiences. At other times it is a crucial topic that is most persuasive.

What one communicates in announcements and brochures advertising an offering is also crucial to developing enrollments which can sustain the offering.

Titles which capture attention in all the mail and announcements which cross a person's desk these days are so important. If the title of the offering causes a second look, then it is the program description which must convince a prospective participant that this offering will meet a continuing education need. This

is the hard sell. It might be the topic and the scope of the content. It might be a celebrated authority who is presenting the material. It might be a combination of factors.

One of the key points of persuasion in distance learning, especially when presenting celebrities, is that all participants have equal access to the instructor, merely by depressing a bar at the base of a microphone. So frequently in face-to-face learning situations one does not have such easy access especially in large lecture hall situations.

In Appendix I there is selected copy used to promote non-credit programming for Fall 1981 of the U.W. Extension Music Department. At all times quality and substance in content are crucial in adult education. Programs built on substance inevitably will survive. The intent of this copy in Appendix I was to capture attention, stimulate interest and motivate response by enrolling in the offering.

It takes time, effort, imagination and just plain hard work to develop promotional copy. But this is time well invested.

Because the data in Figure II substantiates there are many participants experiencing distance learning for the first time, it is important to include brief descriptions of what distance learning is in all publications. An example of this type of copy is in Appendix II.

In bibliographic reference number 11 there is a detailed accounting of the concert pianists of the world as well as celebrities in the field of piano pedagogy who have presented sessions for private piano teachers in Wisconsin. This chronology also indicates the locations from throughout the United States from which these presentations were made. Not one of these people set foot in the state of Wisconsin, but because of the tremendous flexibility of distance learning were able to have statewide impact from wherever they were in the country.

The first effort to bring European resources to Wisconsin was in Spring 1982. At that time, via satellite connection, the Director of the Royal School of Church Music, Dr. Lionel Dakers, lectured to church leaders all over Wisconsin while he sat in his office in Addington Palace in Croyden, England, just outside of London.

Drawing upon resources of the country and the world is now possible in distance learning. Utilizing such resources can bring rich rewards to one's programming for various clientele groups.

In Appendix III there is a chronology of the programming efforts of the U.W. Extension Music Department in the distance learning mode for the various clientele groups served to date by category listing. Such a listing indicates the various emphases pursued with each grouping.

An inter-active aural communications system need not just be limited to aural presentations. Because the series, "Bowing Skills and Style for Orchestral Playing," needed full motion video to be effective, a video tape was developed and disseminated to each site where there were students. The video tape was designed in segments with the musical examples reproduced in accompanying print materials. The format was that segments of the video tape were played (and re-played when necessary) and then there was interaction on the aural interactive communications system.

In Appendix I there is description of an offering entitled, "Small Church Choirs: A Bonanza of Materials for All Conditions of Choirs." Music materials are a great need for church choir directors. Frequently they travel great distances to larger cities for music reading sessions presented by music dealers and other institutions. But it is possible to use the distance learning mode most effectively for this. Publishers supplied music which was sent to each participant. The music was recorded during the summer. During the fall distance learning session participants could follow with their music as the recordings were played. Then there was an analysis of the music given by the instructors along with suggestions for rehearsing and performing this music.

There are unlimited ways in using the distance learning mode. Adult educators should constantly force themselves to use this great resource more creatively and imaginatively.

YES!! It Works!: All of the Above

Sceptics abound everywhere. Can it all work? Can it all be effective? A resounding, YES! With all of the above necessary!

The data in Figure IV gives evidence from the offerings for Fall 1981 of the U.W. Extension Music Department. Based on an unusually high rate of return on the participants in each offering it is evident that in almost all offerings for this particular time period that the vast majority of participants rated the offerings in the top 20%

This evidence coupled with the willingness to use the distance learning mode for additional continuing education needs underscores that the distance learning mode of instruction is accepted and can be successful. (See Figure II)

For enthusiasts of the distance learning mode this reconfirms the basic conviction that the distance learning mode is viable for many types of offerings. For the newcomer to distance learning this should give encouragement that distance learning can be most effective.

Once a program in distance learning is established, it is imperative that the momentum be maintained and that there be vitality in one's offerings every year. This is not easy, but it is crucial in keeping the show on the road.

Summary

Enthusiasm for distance learning is key for any adult educator using this mode of instruction. The possibilities of what it can do are staggering and perhaps even bewildering. An excitement for learning and desires to share one's enjoyment and satisfactions with a given discipline are basic ingredients for every adult educator. Yet, everyone involved in distance learning, from technical and administrative staff to adult educators and instructors need constant encouragement and reinforcement.

Having vision and viewing the world with unlimited horizons are essential to realize the full potential of distance learning for a given discipline. Adult educators must view their subject matter from the viewpoint of having many sub-groups. With such a view one can expand program offerings and broaden into new clientele groupings.

Distance learning is a new mode of learning. One must have much patience and great perseverance as well as having a strong committment to a yearly program of offerings over an adequate length of time. In this way clientele groups will be persuaded to look to distance learning as another option in meeting their professional and continuing education needs.

Developing a dynamic program in distance learning takes time, effort and commitment. Putting some pizazz into one's programming can aid in persuading people into this mode of learning. With an imaginative and creative view of distance learning one can turn this resource into programs which continue and are sustained over the years.

It is becomming most evident that distance learning can be effective and that adults are willing to look to distance learning for their professional and continuing education needs.

Now it's up to adult educators to fully utilize distance learning most imaginatively and creatively, not being hesitant to adapt and use additional technological resources as

Figure III

Offerings of the UW Extension Music Department in the Distance Learning Mode in Fall 1981

	Number Enrolled	Percent of Counties In Wisconsin In Which There Were Participants Enrolled
BEGINNING CHURCH ORGANISTS: BASICS FOR EFFECTIVE SERVICE PLAYING (September 14 & 21) (Church Organists)	108	63%
IN TUNE WITH GENERAL MUSIC (Sept. 22 & 29) (Elementary School Music Teachers)	107	70%
SMALL CHURCH CHOIRS: A BONANZA OF MATERIALS FOR ALL CONDITIONS OF CHOIRS (Sept. 28 & Oct. 5) (Church Choir Directors)	133	64%
MAKING SUNDAY MORNING CLICK (Oct. 6, 13 & 20) (Clergy and Lay Leaders)	68	43%
MODELS FOR GETTING RESULTS IN MUSIC (Oct. 15 & 22) (High School Band and Choir Directors)	41	30%
BOWING SKILLS AND STYLES FOR ORCHESTRAL PERFORMANCE (Oct. 27, Nov. 3, & 10) (School Orchestra Directors)	13	13%
CHILDREN'S CHOIRS IN CHURCHES (Oct. 28, Nov. 4) (Church Children's Choir Directors)	75	45%
JAZZ ALIVE: TEACHING JAZZ AND IMPROVISATION ALONG WITH THE STANDARD REPERTOIRE (Sept. 14 & 21) (Piano Teachers)	64	34%
GETTING DOWN TO THE NITTY GRITTY OF PIANO TEACHING (Sept. 25 & Oct. 2) (Piano teachers)	99	43%
BETWIXT AND BETWEEN: SAVING THE INTERMEDIATE STUDENT (Oct. 21 & 28) (Piano Teachers)	101	53%
JUST ORDINARY KIDS CAN DO IT TOO: TEACHING AND LEARNING TO PLAY BY EAR (Dec. 1, 8 & 15) (Piano Teachers)	57	40%

Figure IV

Offerings of the UW Extension Music Department in the Distance Learning Mode in Fall 1981

	Overall Evaluation of the Offering On A Scale Of 0 (low) to 9 (High)										Percentage of Enrollment Responding
	0	1	2	3	4	5	6	7	8	9	
BEGINNING CHURCH ORGANISTS: BASICS FOR EFFECTIVE SERVICE PLAYING (September 14 & 21) (Church Organists)	-	-	-	-	-	3%	3%	29%	36%	29%	31%
IN TUNE WITH GENERAL MUSIC (Sept. 22 & 29) (Elementary School Music Teachers)	-	-	-	-	-	4%	6%	15%	27%	48%	51%
SMALL CHURCH CHOIRS: A BONANZA OF MATERIALS FOR ALL CONDITIONS OF CHOIRS (Sept. 28 & Oct. 5) (Church Choir Directors)	-	-	-	-	-	12%	6%	19%	41%	22%	32%
MAKING SUNDAY MORNING CLICK (Oct. 6, 13 & 20) (Clergy and Lay Leaders)	-	-	-	-	-	10%	16%	22%	42%	10%	35%
MODELS FOR GETTING RESULTS IN MUSIC (Oct. 15 & 22) (High School Band and Choir Directors)	-	-	-	-	-	8%	6%	33%	33%	-	31%
BOWING SKILLS AND STYLES FOR ORCHESTRAL PERFORMANCE (Oct. 27, Nov. 3, & 10) (School Orchestra Directors)	-	-	-	-	-	-	-	33%	33%	33%	54%
CHILDREN'S CHOIRS IN CHURCHES (Oct. 28, Nov. 4) (Church Children's Choir Directors)	-	-	-	-	3%	9%	9%	47%	20%	12%	49%
JAZZ ALIVE: TEACHING JAZZ AND IMPROVISATION ALONG WITH THE STANDARD REPERTOIRE (Sept. 14 & 21) (Piano Teachers)	-	-	-	-	-	-	6%	3%	16%	75%	55%
GETTING DOWN TO THE NITTY GRITTY OF PIANO TEACHING (Sept. 25 & Oct. 2) (Piano teachers)	-	-	2%	-	4%	2%	14%	14%	20%	44%	39%
BETWIXT AND BETWEEN: SAVING THE INTERMEDIATE STUDENT (Oct. 21 & 28) (Piano Teachers)	-	-	-	2%	-	4%	8%	-	12%	74%	37%
JUST ORDINARY KIDS CAN DO IT TOO: TEACHING AND LEARNING TO PLAY BY EAR (Dec. 1, 8 & 15) (Piano Teachers)	-	-	-	-	-	5%	5%	20%	20%	50%	44%

they become practical and available. That's
the way to keep the show on the road!

Bibliographic References on Music in the
Learning Mode

Hugdahl, Edward O., "Bringing In-Service
    Education to Your Backyard," Wisconsin School
    Musician, February 1975, Volume 44, Number 3,
    pages 10-11.

_____, "A Communications System
    Serving the Continuing Education Needs of the
    Private Piano Teacher," American Music
    Teacher, February-March 1976, Volume 25,
    Number 4, pages 27-29.

_____, "Utilizing Teleconferencing
    for Music Education," The Status of the
    Telephone in Education, compiled by Lorne A.
    Parker and Betsy Riccomini, Madison: Division
    of Educational Communications, University of
    Wisconsin-Extension, 1976 pages 100-107.

_____, "Wisconsin's Wired for
    Organists," Hurdy-Gurdy, January-February
    1976, Volume 4, Number 1, pages 26-27.

_____, "Instituting Extended and
    Short Term Offerings in Music Education for
    Teleconferencing," The Telephone in Education
    Book II, compiled by Lorne A. Parker and
    Betsy Riccomini, Madison: Division of
    Educational Communications, University of
    Wisconsin-Extension, 1977, pages 83-98.

_____, "Teleconferencing to
    Enrich Collegiate Music Education Curricula,"
    Dialogue in Instrumental Education, Winter
    1977, Volume 1, Number 1, pages 8-13.

_____, "Seeing the Whites of His
    Eyes," Wisconsin School Musician, February
    1978, Volume 47, Number 3, pages 8-10.

_____, "Distance Learning in Music
    Through The Teleconferencing Principle: A
    Six Year Experience in Wisconsin On A State-
    Wide Basis." (Biennial National Convention
    of the Music Educators National Conference--
    Chicago, Illinois, April 1978) ERIC ED 168
    534

_____, "Impact of Technology on
    Hymnody," The American Organist, Volume 14,
    Number 1, January 1980, pages 30-31.

_____, "Impact of Teleconferencing
    on the Church and Church Music," Teleconfer-
    encing and Interactive Media, compiled by
    Lorne A. Parker and Christine H. Olgren.
    Madison: Center for Interactive Programs,
    University of Wisconsin-Extension, 1980,
    pp. 107-119.

_____, "Continuing Education for
    Private Piano Teachers: A Breakthrough For
    the '80s." (The Second National Conference
    of Piano Pedagogy - University of Illinois
    at Champaign-Urbana, October 1980) ERIC ED
    198 811

Raccoli, Susan, "Technology Makes the Workshop
    Scene," Clavier, March 1974, Volume 13,
    Number 3, pages 31-33.

Weerts, Richard. "Calling On the Voice of
    Experience," Music Educators Journal,
    Volume 64, Number 1, September 1977, pages
    82-84.

APPENDIX I

Program Descriptions from Selected Offerings of the UW Extension Music Department in Distance
Learning Mode in Fall 1981

FOR PRIVATE PIANO TEACHERS

GETTING DOWN TO THE NITTY GRITTY OF PIANO TEACHING

      Whether you've just started teaching piano or you've been at it for awhile,
this series is for you. And if you're just even thinking about taking on a few
students in the neighborhood now that your children are in school, this series is
for you. There's a lot more to piano teaching than a cute child playing the little
Bach "Minuet" with dangling feet not yet touching the floor or playing the "Minute
Waltz" in forty-five seconds. Have you heard of "Moon Rock?"
      During these two sessions you will be able to draw upon the collective years of
experience of three highly successful teachers who have students trooping through
their studios every day. Two are full time-private teachers making their entire
living from their teaching. One teacher is part-time and balances this career with
the responsibilities of being a wife and mother of four.
      There's always a new wrinkle that can give you more mileage to your teaching,
whether you're overwhelmed by it all or you have things well under control. These
sessions focus on the very nitty-gritty of being a private piano teacher.

Session I
IT'S MORE THAN A PIANO IN A LIVING ROOM
Getting started and keeping it running...smoothly
Piano teaching is a three legged stool: student, teacher and parent
Inside scoop on motivation from the inside out
Private teaching: It's a profession! It's a career!

Session II
THERE'S A LOT GOING FOR YOU
Work on it as a business: fees, billings, tax deductions, health insurance, tax
     sheltered annuities
Know what you're all about by getting your act together
The personality quotient
Questions you've always wanted to ask but have been afraid

JAZZ ALIVE!: TEACHING JAZZ AND IMPROVISATION ALONG WITH THE STANDARD REPERTOIRE

      It's not oil and water! A comprehensive curriculum can include standard
repertoire along with experiences in the jazz idiom. Such approaches have
combatted pernicious dropout rates with all ages. Teaching the "basics" so
crucial to establishing a firm piano foundation need not be sacrificed.
      For the classically trained teacher this series shows jazz and pop music can
be effectively integrated in developmental and sequential ways. Teachers no longer
need feel ill-at-ease in not having grown up with a jazz component in their backgrounds.
      Improvisation is so essential for the well-grounded pianist. It is possible to
use the jazz idiom effectively for these objectives at the elementary and intermediate
levels.

Session No. I
JUST HOW TO GET STARTED
     Some basic skills and how to use them; presenting the multitude of jazz
     and pop teaching literature using easy, elementary techniques with an
     authentic style; basic requirements of jazz piano; how to begin studying
     these skills.

Session No. II
HEY, IMPROVISATION IS FUN! I CAN DO IT!
     Getting over the initial hurdles; training the ear for improvisation; taking
     scales and modes and using them to enhance an understanding and inflection
     of jazz rhythm.

APPENDIX I - Continued
FOR CHURCH MUSICIANS AND CHURCH LEADERS

SMALL CHURCH CHOIRS: A BONANZA OF MATERIALS FOR ALL CONDITIONS OF CHOIRS

"We don't have many men in our choir, what can we sing?" "If our sopranos
are weak, is there something for us?" "During hunting season, we've got problems!"
"What about those 'lean' Sundays after Christmas and Easter?" "Get-away weekends'
can cripple us!"

Such questions haunt choir directors as they dig through bins of choral music
whenever they can get to a large music store. Church choirs are so crucial in the
worship life of a parish. No matter how large or small the choir the quest for
materials should never be abandoned. New, fresh materials are always needed.

With church choir directors there's a never-ending search for materials...
materials that appeal, challenge, motivate and fit into the needs of the Church
Year. And materials that will provide for all of the expected and unexpected
contingencies.

In this series, it will be a bonanza of materials for the choir director.
Instead of going to a choral reading session in Minneapolis or Chicago, the choral
reading session comes to you via ETN wherever you are in Wisconsin...right in your
own backyard! A packet of choral music from six major publishers will be sent to
each registrant to use as each work is presented in recording and discussion. Each
participant retains this packet of music for consideration for the church's future
musical needs. An exciting variety of materials appropriate for all denominational
needs has been selected.

Another feature is that each of the six publishers will present a brief overview of his choral catalogue before his publications are presented.

MAKING SUNDAY MORNING CLICK: PLANNING SKILLS FOR REFINING WORSHIP

Throughout Wisconsin each week churches of all denominations in their own
particular heritage and tradition gather the faithful for worship. Effective
weekly worship in a parish no matter its size is subject to a vast array of
factors. Making Sunday morning click is the bottom line for church leaders--
both clerical and musical. Church leaders can make Sunday morning click by
refining planning skills for worship.

By drawing upon dynamic leaders in worship from a variety of denominational
traditions church leaders in Wisconsin can gain insights that can bring new
dimensions of spirituality in the lives of a parish. The emphasis in this series
is on practicality for a given parish situation. No matter where one is located
in Wisconsin because of the fantastic flexibility of the Educational Telephone
Network (ETN) these authorities are brought into your own back yards. They
become immediately accessible for dialogue and discussion.

For everyone in church leadership coping with resistance and obtaining
involvement in the worship life of a parish are basic concerns. There are always
"hot" and difficult decisions. But pitfalls can be avoided. Developing a team
spirit between clergy and musicians is a reality and music can be an integral and
purposeful part of corporate worship.

Session No. I
A VISION OF WORSHIP
        Raising a consciousness with a corporate group of worshippers in a
given community for an understanding of worship.
        The emerging role of the involvement of an entire congregation in
worship and obtaining this involvement.
        Coping with resistance and developing educational challenges.

Session No. II
A DIVERSITY OF GIFTS
        Drawing upon the total resources of a worshipping community.
        Organizing planning sessions for worship.
        Developing a repertoire of congregational hymns.
        Planning music so it is an integral part of worship.
        Aids and resources for the planning process.

APPENDIX I - Continued

Session No. III
IMPLEMENTATION--MORE THAN MECHANICS
    Lay involvement and lay leaders.
    Models for a planning process.
    Avoiding pitfalls.
    Getting feed-back.
    Instituting an on-going system of evaluation.
    Being practical and specific

FOR MUSIC EDUCATORS

MODELS FOR GETTING RESULTS IN MUSIC

All music educators desire more instructional mileage from the bands and choruses or orchestras under their batons. Primarily the taxpaying public sees these organizations in performing roles in concerts, parades and festivals. These are times of increased accountability, budgetary retrenchment and back-to-basics movements. Recently, some dramatic and innovative approaches have been developed to get more educational results from school offerings which have performance as a logical and expected result.

In this series you will have an opportunity to analyze the features and principles of the Wisconsin Comprehensive Musicianship Project and learn about the enthusiasm it is creating. This is not a gimick. It's not buying another "method book." This is an approach every music educator can adopt and adapt at no cost. You'll be able to discuss the experiences being obtained through this process. Excitement spreads because of the results this approach is obtaining with music educators who are applying these principles to their bands and choruses and orchestras.

IN TUNE WITH GENERAL MUSIC

Is your school a "Singing School?" Much has changed since the early days of music instruction in our country, but singing is still considered the heart of a good school music program. That's because singing is such a natural way for children to make music and such a satisfying way to learn!

Our kindergarteners arrive with five years' experience cooing, humming, and chanting. If we nurture that innate love of singing, they will leave the elementary grades with a uniquely personal link to music which will last a lifetime.

Session No. I
THE EARLY GRADES--OFF TO A GOOD START
    This session will focus on early singing experiences which encourage
    self expression and musical growth. Ways of helping the inaccurate singer
    gain pitch control will be demonstrated along with tips for choosing and
    introducing new songs.

Session No. II
OLDER AND WISER--SINGERS STILL
    The ability to sing in harmony develops gradually as children experience
    part-singing in a variety of ways. Appealing examples of ostinatos, echo songs,
    partner songs, rounds, and descants will be shared. Other topics include
    harmonizing by ear, assembly sings, and the elementary school chorus.

APPENDIX II

Description of Distance Learning Used in Promotional Brochures of the UW Extension Music Department

A FORUM VIA ETN

Where you reside in the state is no longer a barrier to revitalizing and enriching yourself in professional continuing education. The UW-Extension Educational Telephone Network (ETN) makes it convenient to up-grade yourself professionally by drawing upon the experiences and insights of nationally acclaimed specialists in piano pedagogy and to personally question and discuss with them and your fellow teachers throughout the state these essential aspects in your teaching. And it's as close as your county court-house or UW campus--or wherever there is an ETN listening location.

The Educational Telephone Network (ETN) is located in every county in Wisconsin (see location listings) where there is a room especially equipped with high fidelity communications equipment. At any time during the Forum you may ask a question and be heard by the forum leader and all the other participants. What results is a huge state-wide classroom composed of segments linked together by an electronic system with two-way communications between teacher and each participant. Since this system is carried over telephone lines--that's why it's called the Educational Telephone Network--it is possible to have persons give presentations from anywhere in the country.

FROM TWO PARTS OF THE COUNTRY

ETN now brings you--from other parts of the country--some of the most insightful thinking today in piano teaching. On this series Guy Duckworth and his students will be in his studio on the campus of the University of Colorado at Boulder. Louise Bianchi will originate from the piano preparatory department at Southern Methodist University in Dallas, Texas.

No one travels, but via telephone lines and through the wonders of electronics you can discuss ideas and problems with these superb teachers. An exciting educational opportunity for private teachers in Wisconsin!

APPENDIX III

Offerings of the UW Extension Music Department in the Distance Learning Mode 1973-1982 by
Clientele Groupings

CLIENTELE GROUP   TOPIC

Private Teachers of Piano

Spring 1973	First Piano Teacher Forum on Piano Pedagogy
Spring 1974	Second Piano Teacher Forum on Piano Pedagogy
Spring 1975	Third Piano Teacher Forum on Piano Pedagogy
Spring 1976	Fourth Piano Teacher Forum on Piano Pedagogy
Fall 1976	Symposium on 1977 Piano Contest Music
Fall 1976	Preparation for Certification for Private Piano Teachers
Spring 1977	Preparation for Certification for Private Piano Teachers
Spring 1977	Fifth Piano Teacher Forum on Piano Pedagogy
Fall 1977	Symposium on 1978 Wisconsin School Music Association Solo Piano Contest Music
Fall 1977	Preparation for Certification for Private Piano Teachers - Aural
Fall 1977	Symposium on 1978 Wisconsin School Music Association Piano Duet Contest Music
Fall 1977	Preparation for Certification for Private Piano Teachers - Written
Spring 1978	Sixth Piano Teacher Forum on Piano Pedagogy
Fall 1978	Piano Pedagogy: The Dynamics of Teaching
Fall 1978	The Teaching of Piano Music: A Study in Depth
Fall 1978	Lead, Follow or Get Out of the Way: Teaching High School Accompanists
Spring 1979	Seventh Piano Teachers Forum on Piano Pedagogy
Fall 1979	From Presentation to Performance
Fall 1979	Music Reading: Antidote to Frustration and Open Sesame to Achievement
Fall 1979	Transfer Students and Teaching
Spring 1980	Eighth Piano Teachers Forum on Piano Pedagogy
Fall 1980	First Steps in Teaching Pop Music: Motivating and Retraining the Prospective Dropout
Fall 1980	In the Muddle of the Middle: The changeling Student
Fall 1980	Preparation for Certification for Private Piano Teachers; An Overview of Music Theory
Fall 1980	Symposium on the 1981 Wisconsin School Music Association Solo Piano Contest Music
Spring 1981	Ninth Piano Teachers Forum on Piano Pedagogy
Fall 1981	Jazz Alive!: Teaching Jazz and Improvisation Along with the Standard Repertoire
Fall 1981	Getting Down to the Nitty Gritty of Piano Teaching
Fall 1981	Betwixt and Between: Saving the Intermediate Student
Fall 1981	Just Ordinary Kids Can Do It Too: Teaching and Learning to Play By Ear
Spring 1982	Never Too Late To Learn
Spring 1982	Tenth Piano Teachers Forum on Piano Pedagogy

Private Teachers of Spinet Organ

Fall 1973	Seminars on Teaching The Spinet Organ
Fall 1974	Seminars on Teaching The Spinet Organ
Fall 1975	Seminars on Teaching The Spinet Organ
Fall 1976	Seminar on Popular Music
Spring 1981	Organ-izing Your Studio: Materials That Motivate
Spring 1982	Pizazz in Playing Pop Organ: Teaching Easy Steps For Instant Stylings and Alternate Harmonizations

Public and Parochial School Music Educators - Elementary School

Fall 1974	Music and the Exceptional Child
Fall 1976	Putting Music Into The Mainstream of the Elementary Classroom
Spring 1977	Mainstreaming in Music
Fall 1977	Turning Kids on: A Positive Approach
Fall 1978	Learning to Read Music Can Be Painless

APPENDIX III – Continued

Spring 1979	The Special Learner and Music
Fall 1979	Instructional TV: More Mileage from "Music"
Spring 1980	Making Magical Moments in Music Classes
Fall 1980	Creativity?...On Monday Morning?...On Friday Afternoon
Fall 1981	In Tune With General Music
Spring 1982	Getting Familiar With An Elementary Series and Making Better Use of Yours

Public and Parochial School Music Educators – Junior High School

Fall 1975	Ideas for General Music
Spring 1976	Boys Can and Want to Sing
Fall 1976	Designing Music Courses for Non-Performers in Junior and Senior High Schools
Spring 1977	New Techniques for the Autoharp in the Classroom
Spring 1978	Electronic Music: Tool for Creativity
Spring 1979	Getting Your Act Together in General Music
Spring 1980	Dealing with "Uncertain Singer" Problems Through Careful Selection of Music
Spring 1981	Boys Can and Want to Sing!
Spring 1981	Group Teaching: Can It Solve Your Problems
Spring 1982	The Scheduling Hassle: There Are Solutions

Public and Parochial School Music Educators – Senior High School

Fall 1975	The Wind Ensemble in the High School Music Program
Spring 1976	Choral Concepts
Fall 1976	Improvisation: Teaching It In Group Settings
Fall 1976	Teaching Musical Content in Performing Groups
Spring 1977	Assessing Musical Learning
Spring 1977	The Cello: Techniques and Repertoire for High School Cellists
Fall 1977	The 1978 Class B High School Band Contest Music
Fall 1977	The 1978 Class A High School Mixed Chorus Contest Music
Fall 1977	Lowering Student Dropout Rates In Music
Fall 1977	The 1978 Class A High School Band Contest Music
Spring 1978	The 1978 Class B High School Mixed Chorus Contest Music
Spring 1978	Misconceptions in Oboe Playing
Spring 1978	The Violin: Its Practice and Performance for the Intermediate and Advanced Player
Fall 1978	Music Interpretation and Score Analysis of Band Literature – Series I (Class A)
Spring 1979	Teaching Jazz Ensembles in High School
Spring 1979	Music Interpretation and Score Analysis of Band Literature – Series II (Class B)
Fall 1979	A Comprehensive Approach to Teaching String Improvisation and the Golden Strings Concept for Schools
Fall 1980	Bowing Skills and Style for Orchestral Performance
Spring 1981	Teaching Beginning Improvisation in Elementary and Secondary School Music
Fall 1981	Models for Getting Results in Music
Fall 1981	Bowing Skills and Style for Orchestral Performance

Music Administrators of School Systems

Fall 1975	Music Administration and Supervision
Spring 1976	Careers and Career Education in Music
Spring 1976	Issues for College Music Education Majors
Spring 1977	Improving Instruction Through Effective Supervision
Spring 1977	Issues for Collegiate Music Education Majors
Spring 1978	Planning and Evaluating the Music Curriculum
Fall 1979	Bridging the Gap Between Theory and Practice: Implications of Psychology for Music Teaching

APPENDIX III - Continued

Directors of Church Choirs

Fall 1977	Sunday School Music and Children's Choirs in Churches
Spring 1978	Refining Folk Music in Churches
Fall 1978	Sunday School Music and Children's Choirs in Churches
Fall 1979	Small Church Choirs: Don't Give Up
Fall 1979	Children's Choirs in Churches
Fall 1980	Children's Choirs in Churches
Fall 1980	Small Church Choirs: Don't Give UP!
Spring 1981	Solid Solutions for Recruiting Singers
Fall 1981	Small Church Choirs: A Bonanza of Materials for All Conditions of Choirs
Fall 1981	Children's Choirs in Churches

Organists in Churches

Spring 1980	Small Church Organists: Lots Can Be Done!
Fall 1980	Small Church Organists: Lots Can Be Done!
Fall 1981	Beginning Church Organists: Basics for Effective Service Playing
Spring 1982	Church Organists: Expanding Repertoire and Service Playing Techniques

Clergy and Lay Leaders

Spring 1979	Cultural Roots in the Hymnody of the Church
Fall 1979	Hymns: Dump Them Or Sing Them!
Spring 1980	Psalmody: Roots, Traditions and Use In Today's Church
Fall 1980	Tradition and Worship in Today's Church
Fall 1980	Creative Use of the Hymnal
Fall 1981	Making Sunday Morning Click: Planning Skills for Refining Worship
Spring 1982	Sunday Morning Slump: Enlivening Hymn Singing
Spring 1982	Some Aspects of Church Music Today

Teachers of Pre-School and Early Childhood Education

Fall 1979	Music In Early Childhood Education
Fall 1980	Music In Early Childhood Education

Semester Long Courses for Graduate Credit in Music Education (Credit from University of Wisconsin-Whitewater)

Spring Semester 1978	Folk Music in the Kodaly Curriculum
Spring Semester 1979	History and Philosophy of the Kodaly Concept
Spring Semester 1980	Folk Music in the Kodaly Curriculum
Spring Semester 1981	History and Philosophy of the Kodaly Concept
Summer Semester 1981	Principles of Kodaly for Choral Directors
Spring Semester 1982	Folk Music in the Kodaly Curriculum

The Teleconferencing Resource Book: A Guide to Applications and Planning
Lorne A. Parker and Christine H. Olgren (eds.)
Elsevier Science Publishers B.V. (North-Holland)
© Center for Interactive Programs, University of Wisconsin-Extension, 1984

TELECONFERENCING: A TOOL FOR KNOWLEDGE
DIFFUSION IN EDUCATION

Donna D. Mitroff, Ph.D
Program Development Associate
WQED-TV
Pittsburgh, Pennsylvania
and
R. T. Eichelberger, Ph.D.
Department of Educational Research
University of Pittsburgh
Pittsburgh, Pennsylvania

Background

In the Fall of 1979, the Pennsylvania Department of Education, under the leadership of a newly-appointed State Secretary of Education, unveiled a state-wide plan for school improvement. The School Improvement Plan (SIP) was essentially a major state-wide effort to promote educational renewal and reform by involving school personnel, community members, and higher education institutions in a cooperative, long-range planning process. The long-range planning process included: a.) needs assessment; b.) program planning, identification, and development; c.) implementation; and d.) program evaluation. Under the SIP these activities would ultimately occur at the local school district and building level.

Considering that the state of Pennsylvania contains over 500 school districts in a state which ranges from large, complex urban settings to small, stable rural areas, the magnitude of the effort was staggering. In order to make the SIP implementation process more manageable the Pennsylvania Department of Education (PDE) planning team designed an implementation scheme which phased in districts and activities in successive waves. "Wave I" consisted of 78 school districts; "Wave II" of about 130 school districts, and so on.

As each wave embarked on the process, PDE staff would provide technical assistance and help coordinate assistance from higher education and educational R&D resources.

Even though the number of districts selected for Wave I implementation was proportionally small (78 out of 505), the entire education community in the state of Pennsylvania was aware of and concerned about the implications of the SIP effort.

In the Fall of 1979, the PDE issued a request for project proposals that would assist in carrying out the goals of the school im-provement plan. The Educational Services Department of WQED-TV, the public broadcasting station in Pittsburgh, PA, responded with a proposal in which the advanced technology of the Pennsylvania Public Television Network (a network of seven public broadcasting stations whose combined signals can blanket the entire state) would be coordinated to conduct two state-wide interactive School Improvement teleconferences. The actual teleconferences would be the high point of a multi-stage process designed to provide knowledge about and affect attitudes toward the SIP on the part of school personnel.

This paper will describe the multi-stage process and the teleconference formats. It will also report the outcomes of the project evaluation. It is the position of this paper that there is a great future in applying television technology to the diffusion/dissemination needs of education. However, in order to realize that potential, applications of the technology must be based on theory which guides the design of the application and leads to objectives which are evaluated.

Theoretical Framework for the Project

A large scale educational change effort such as the Pennsylvania School Improvement Plan requires many supportive functions in order for it to move from plan to practice. This fact was acknowledged by Pennsylvania's Secretary of Education himself in the following comments to a group of educators and researchers: "Transfusing schools with new and creative practices requires a carefully engineered approach - one that considers not only the efficiency of the proposed innovation and the effectiveness of the strategies employed, but also the impact of the change process on the people involved over a long period of time. Building relationships and understanding roles and responsibilities between state agencies, local school districts and the individual school principal is a key ingredient in the dissemination of innovations."[1]

As a strategy for change, the SIP would be expected to proceed through two basic change stages: Initiation and Implementation. Each of the two major stages has been characterized by sub-stages which have been described by Zaltman, Duncan and Holbek[2] according to the following model:

1.) Initiation
   a.) Knowledge-awareness
   b.) Attitude formation
   c.) Decision

2.) Implementation
   a.) Initial implementation
   b.) Continued-sustained implementation

At its inception in the Fall of 1979, the SIP entered the first of the two major stages - Initiation. During an Initiation stage, appropriate change tactics need to be selected and applied which will enable potential users or clients to become familiar and comfortable with the object of the change effort, e.g., the School Improvement Plan. Once familiar and comfortable with the SIP, they can form attitudes which will lead to informed decisions about participation in the process.

The change tactic most appropriate to the Initiation stage is Information/Linkage. Information/linkage tactics are needed at this stage to stimulate, motivate, and fuel the change effort.

The School Improvement Teleconferences project was based on the premise that television technology is a potent vehicle for information/linkage during the critical initial stage of any change effort. The power of television can be used to mobilize and energize a target audience around the goals of the change effort. The use of television technology can help to provide a "common language" with which the educational community can communicate about the effort. The use of television can provide a set of "shared images" on the meaning of particular change efforts. And the use of television can, because of its pervasiveness in society, facilitate reaching a broad range of stakeholders.

The term "television technology" is being used in place of just television or teleconferencing. Before proceeding, a definition of that term is in order. Whereas, television refers to the viewer in front of the television set, television technology refers to a more complicated situation which includes interactive capability between persons gathered in front of television sets, yet separated by many

miles. It refers to active rather than passive viewing. Active viewing occurs because viewers have been prepared for and are involved in the material which is coming over the screen. In addition to being prepared for the programming, they are expected to participate in some sort of follow-up activity after the screen image disappears. The willingness of the viewer to become engaged in this process is enhanced by the fact that in the case of television technology, the viewer in the group setting interacts with professional peers. Therefore, social interaction and the support that comes of sharing an experience with others is added to the television technology experience.

Description of Project Activities

The rationale and theoretical framework guided the team in the design of project activities. As indicated above, the on-air teleconferences were actually the high point of a multi-stage process which included both pre- and post-conference activities utilizing print as well as media.

Extensive collaboration and planning between the Department of Education and public television staff resulted in:

1.) an extensive publicity campaign.
2.) a set of pre-conference print materials which provided conceptual background.
3.) pre- and post-conference questionnaires.
4.) the organization of both school-site groups and studio groups for viewing and participating in the broadcast.
5.) two live 1½-hour teleconference broadcasts in which seven public television stations across the state were interconnected and interactive.
6.) the packaging of the resultant video and print into a set of workshop modules for use by Department of Education staff in subsequent training.

Each of these components is described briefly below.

Publicity Campaign. In order to assure a high level of involvement in the teleconference, an extensive publicity campaign was conducted in advance of the broadcast dates. The campaign served two main purposes: to inform the target audience about the event; and, to build some conceptual background among target audience members (i.e., provide "advance organizers" on program content). To meet these goals a series of one page "ads" was developed for the biweekly newsletter published by the State

Department of Education. The ads were carried in three issues preceding the broadcast dates. In addition to the newsletter ads, informative posters (18 x 20 blow-ups of the ad) were developed and distributed to the schools.

Pre-conference Print Materials. A multi-fold brochure was created for each teleconference. Each brochure provided an overview of the teleconference content and some key concepts. The brochure was intended as part of the overall information dissemination vehicle. Its content was, in fact, included in the evaluation. The brochure consisted of the following sections: a.) details on the broadcasts such as dates, times, on-air participants; b.) teleconference agendas; c.) background on the teleconference theme; d.) key concepts related to each theme; and e.) a list of questions frequently asked on the theme by teachers and other school personnel.

Obviously, the process of developing the brochure helped to focus the development of the on-air content. As frequently asked questions were culled from interviews and included on the brochure, it was obvious that those questions needed to be directly addressed by the on-air presenters. The brochures were mailed to reach the schools one week before the tele-conferences.

Pre- and Post-conference Questionnaires. The questionnaires were designed to determine the demographics of teleconference participants and determine the qualitative impact of the teleconference experience on the participants.

Both pre- and post-questionnaires were mailed to the schools before the teleconference broadcast. More detail on the evaluation procedures and outcomes will be included in the next section.

Viewing Groups. The target audiences for the teleconferences were organized into "viewing groups." The rationale for "viewing groups," as stated above, is that the critical social interactions and building of common language and shared images is facilitated by having viewers together in fairly small (10-15 person) groups.

Packets containing general information on how to participate in the teleconference, how to set up viewing groups, and how to administer and compile pre- and post-evaluations were distributed to school principals prior to the broadcasts.

Teleconference Broadcasts. The two live 1½ hour teleconferences originated from the studios of WQED in Pittsburgh. The other six stations of the Pennsylvania Public Television Network were interconnected by microwave link and staffed with panels of local educators. In addition to the two-way interactive between the originating station, WQED, and the satellites, there was a telephone call-in number so that participants in the school viewing groups, administrative offices, etc. could phone questions in during the broadcast.

The teleconference format included pre-produced "trigger" films, interviews, panel presentation, studio reactions, and call-in segments.

The content of each teleconference was determined by the PDE task force in collaboration with project staff.

Post-teleconference Workshop Modules. The video and print from the teleconferences were re-configured into modules of 15-20 minute videocassettes with supplementary print components. Each module concentrated on a different aspect of the SIP effort. These modules were developed for use by PDE technical assistance staff for orientation and training of subsequent "waves" into the SIP process.

The foregoing discussion of the project's components has indicated that this was a highly coordinated campaign of carefully selected tactics. As indicated at the start of this report, this project attempted to be guided by theory. In the case of selecting and orchestrating the tactics which would be employed to support the effort, we utilized an evaluation framework provided in the work of Zaltman, Florio, and Sikorski.[3]

Zaltman et al. have found that in most instances, the selection of tactics to introduce and implement change in education is a "casual process." Typically, change planners do not know enough about "what innovations or types of innovations disseminated in what manner to what kinds of schools result in educational renewal and reform." (p. 183).

In order to lift our effort from "casual" to at least the level of "reflective" the work of Zaltman et al. was used to a.) consider available change tactics from which to select; and b.) to analyze the strength of the selected tactics along dimensions relevant to this particular application.

Categories of Change Tactics. The two categories which are most important during the initiation of a change effort are information/ linkage and user involvement. According to Berman, Greenwood, and McLaughlin and reported in the Rand Change Agent Study (1975):[4] "Initiation is a period when the LEA staff conceives and plans possible innovations, looks for resources, and decides which projects to select for support. The kind of support and commitment that the project receives then, casts a long shadow over the implementation process that follows."

"Information/linkage tactics" in general include direct mail, mass media, demonstrations, field agents, telephone, training manuals, meetings, etc. "User involvement tactics" are actions intended to obtain commitment from those who must implement change by having them participate in the decision-making and trial activities that preceded change. The user involvement tactics included in this project were involvement in open communication, involvement in approval, and in endorsement (leadership).

According to Zaltman, et al., each of the tactics can be characterized on a number of dimensions. Eight of those dimensions are relevant to this project. The eight key dimensions along with a brief definition of each are presented below.

The matrix chart on the following page combines the tactics which were included in the project (with the specific examples of each one) and the dimensions of the tactics. It shows how each of the tactics utilized in the project rates on each of the dimensions. The chart shows that the combined use of the change tactics results in an array in which one or more tactics rate high on every dimension.

Evaluation Methods & Techniques

The first teleconference, May 1978, was targeted at those school districts (78 out of a total of 505 districts) which had been designated by the State Department of Education as "Wave I" for the implementation of the statewide effort. Therefore, although the teleconference was publicized and broadcast over open air, the distribution of the pre-conference materials, the viewing groups, and the evaluations were carried out within the 78 Wave I districts.

The second teleconference, October 1978, was targeted at the "Wave II" school districts. The distribution of pre-conference materials, the viewing groups, the telephone hookups, and the evaluations were limited to the 125-130 Wave II districts.

The teleconference evaluations focused on assessing the impact of the project on each of the sub-stages of the initiation phase, i.e., knowledge-awareness, attitude formation, decision making. Also, the evaluations were designed as exploratory investigations to be used formatively, i.e., in planning future telecon-

RELEVANT DIMENSIONS OF INFORMATION/LINKAGE TACTICS

Dimensions	Definitions
Stability	The extent to which the tactic, once implemented, is likely to operate in the manner predicted.
Personal Contact	The extent of in-person contact involved.
Feedback/Interaction	The extent of interaction possible between the message sender and the message receiver.
Action Implications	The extent to which the tactic implies a course of action.
Potential Coverage	How large an audience can be reached with the tactic.
Immediacy	How much time elapses from the initiation of the tactic and the beginning of its impact on the target system?
Ease of Use	Can the user access the information easily or does it require special skills?
Imagery	The extent to which the information is presented in concrete rather than abstract form.

D.D. Mitroff and R.T. Eichelberger

DIMENSIONS OF CHANGE TACTICS

TACTICS	Stability	Personal Contact	Feedback/ Interaction	Action Implications	Potential Coverage	Immediacy	Ease of Use	Imagery
Information/Linkage								
use of direct mail (pre-conference publicity and Educator Brochure)	H	L	L	M	H	H	H	L
use of mass media (statewide interconnect teleconference broadcast)	H	L	L	M	H	H	M	H
use of telephone (viewer call-in)	M	M	H	L-M	L	H	H	L
training modules (video and print package)	H	L	L	H	M-H	M	M	M-H
meetings (viewing groups, Advisory Committee)	M	H	H	M	L	M	M	H
User Involvement								
involve in approval (Advisory Committee, conference participants)	L	M-H	M-H	M-H	L-M	L-M	L	M-H
endorsements (leadership) (Advisory Committee, conference participants)	L-M	H	L-M	L-M	L-M	M	M	L
discussion and planning (viewer call-in, viewing groups, local studio participation)	L	H	M	M	H	H	M	H

L - Low   M - Medium   H - High

ference experiences. Furthermore, the results contribute to our understanding of the use of teleconferencing for knowledge diffusion.

Evaluation Procedures. The evaluations both utilized a pre- and post-questionnaire design. The instruments consisted of two major parts:

1.) Determining the demographics of the teleconference's viewers.

2.) Determining the qualitative impact of the teleconference experience on participants.

The teleconference experience was considered to include the pre-conference print component, participation in the viewing groups, and the interactive teleconference broadcast. The pre-questionnaire requested information as to the respondent's feelings about the SIP, and their willingness to participate in and promote it, and their knowledge of concepts/terms relevant to the Plan.

For the first teleconference it was decided to sample teachers and other personnel. Each principal was sent four pre- and four post-questionnaires, and asked to complete one and distribute the other three to teachers representing a cross-section of teachers at that school. About one third of the questionnaires were returned. Approximately 400 persons returned both the pre- and post-questionnaires. About 100 returned only the post-questionnaires. About one half of the Wave I districts were represented in the returns. One hundred and eighty-six (186) of the respondents were teachers, 80 were principals, and 125 were classified as "other"--a classification including administrators, board members, parents, etc.

For the second teleconference pre- and post-teleconference questionnaires were sent to principals in about 4,000 schools. Both questionnaires were completed and returned by 388 participants: 208 teachers, 79 principals, 26 administrators, 19 superintendants, and 56 "other."

Results

In each evaluation the viewer responses were initially analyzed as to the characteristics and attitudes of persons responding to the pre-teleconference questionnaire, the conditions under which the teleconference was viewed, and the impact of the teleconference on their knowledge, attitudes and willingness to participate in the SIP (decision making).

In this report, results from the two eval-

uations are summarized with responses compared and contrasted. Recommendations and implications based on the data and associated experiences comprise the final section of the paper.

Initial Attitudes Toward SIP

In both evaluations initial attitudes toward SIP and toward local districts' involvement with it ranged from neutral to positive. For example, on the item which asked about "...the likelihood that SIP will improve the education of pupils" responses were that SIP "probably would" (255 of 391; and 280 of 382). The evaluation results for this item are reported by each teleconference in Table I.

The educators who viewed each teleconference also agreed on their need for more information about SIP, and on what they expected from the teleconferences (See Tables II & III).

Teleconference Viewing Conditions

Nearly everyone viewed the teleconferences in their schools (381 of 391 and 359 of 379). The size of the groups that watched the teleconferences seemed to vary somewhat. In May, most respondents watched with more than 10 people, while in October, more viewed it with "under 4" people (Table IV).

In the larger viewing groups there was a tendency for people to feel that the program was more meaningful (217 "yes," 119 "no"). Perhaps the Wave I districts which participated in the first teleconference made a greater effort to emphasize the activity since it would have immediate application. To reinforce this view, more people in the earlier viewing groups called in questions or comments (217 in May, 26 in October).

Given these differences, activities of the local districts should be investigated. Evidence indicates that viewing conditions impact on the program's effects. Recommended conditions should probably be a part of planning new teleconferences.

Impact of Teleconferences on Knowledge, Attitudes, and Willingness to Participate in SIP

Results from the two teleconferences were very consistent. Respondents were asked to rate their current knowledge of selected terms or concepts central to SIP on both the pre- and post-questionnaires. The five-point rating

TABLE I - SIP Will Help to Improve Education of Pupils

Attitudes	Teleconf. I	Teleconf. II	TOTAL
Definitely will not	3	5	8
Probably will not	75	48	123
Probably will	255	280	535
Definitely will	48	49	97
TOTAL	381	382	763

TABLE II - Presently Know All That You Need to Know About SIP?

Attitudes	Teleconf. I	Teleconf. II	TOTAL
No	371	280	651
Yes	15	36	51
Not Sure	5	71	76
TOTAL	391	387	778

TABLE III - What Did Respondents Expect From SIP Teleconferences? (Each selected two choices)

Choices	Teleconf. I	Teleconf. II	TOTAL
What PDE has in mind	198	179	377
What SIP means to our district	149	163	312
What benefits from SIP	119	178	297
What my role will be in SIP	98	–	98
What support is available	–	110	110
TOTAL	564	630	1,194

TABLE IV - Number of People in Viewing Group

Size of Group	Teleconf. I	Teleconf. II
Less than 4	62	189
4 - 6	65	60
7 - 10	87	40
Over 10	177	89
TOTAL	391	378

scale ranged from "know nothing about it" to "understand it thoroughly." On the concepts of "Building Level and Action Plan" and "Registration" there was some change in rated knowledge in both evaluations. Initially, most respondents were in the lowest three categories, while after the teleconference most of the responses were in the upper three categories (See Tables V & VI). These results indicate that the viewers believe that they did learn something about SIP from the teleconference. Respondents to the first teleconference indicated that they had learned something new (337 "yes" and 47 "no"), while those in October were about evenly split on this question (187 "yes" and 184 "no").

The impact of the teleconferences on attitudes and associated willingness to do things to implement SIP were not very strong and tended to be inconsistent. Both groups indicated that there was little or no attitude change, although more indicated a positive change than indicated a negative change (Table VII).

The impact of the teleconferences on the decisions that the educators would make in participating in SIP-related activities were assessed in both teleconferences by three questions included on both the pre- and post-questionnaires. They were asked to rate their willingness to:

1.) participate in long range planning groups.

2.) do whatever is necessary to implement school improvement.

3.) promote school improvement among colleagues.

There was some tendency for respondents to indicate less willingness to participate in SIP-related activities after the teleconference, although the vast majority did indicate that it was "moderately likely" or "very likely" that they would do all of the activities specified in the three items. Results are reported in Tables VIII & IX.

Implications and Recommendations

The primary purpose of this evaluation was to describe the impact of the PDE teleconferences on the School Improvement Plan on education personnel in participating school districts, and to assess the teleconference procedure as part of the change effort. There was evidence that the procedure had some value in communicating basic information about SIP.

Differences between the conditions under which each teleconference was viewed, variations in program content, and differences in respondents' responses need to be looked into further to identify the most effective combination. The value of the interactive aspects of the program is one area that deserves more attention. As these questions are investigated, it is important that differences among the subgroups such as teachers, administrators, board members, supervisors, etc., be taken into account. In these analyses, teachers were the primary respondents and tended to be the focus of the programming.

There was some indication that the educators tended to be fairly open and positive about the PDE's School Improvement Plan, but are slow to make any changes in their views. Since this area is of such importance to large scale school change efforts, future knowledge dissemination efforts which attempt to improve education should include more comprehensive efforts to assess its impact. This effort, while providing some information about the impact of the teleconferences, tends to raise more questions than provide answers.

Given the experiences of these two teleconferences, it is clear how massive an effort is required. All activities associated with conducting a purposeful teleconference implementation need to be more clearly specified, and more procedures for monitoring the implementation are needed. For example, such issues as when all associated materials are to reach the target audience and who is responsible for actually sending them and who is responsible for receiving them need to be specified at the beginning. A phone number to call if mailings are not received, or are incorrect, should also be a part of the planning.

The types of activities and materials to be integrated with teleconferencing needs further consideration. Pre-conference materials and follow-up workshops might be another combination which would have positive effects.

Should participants become more involved in the interactive aspects of the program? If so, how could this best be accomplished? What are the relative costs and benefits of alternative ways to accomplish the purposes of the teleconferences?

Importance of the Study

Over the fifty years that television has been a "presence" in our lives, it has had a profound effect on nearly all aspects of society; and yet, we have been slow to harness the medium and create some intended consequences in education. The findings of this study are just a beginning. The findings have clear implications for the application of television technology to the diffusion/dissemination needs of education. In this era of increasing travel costs and decreasing travel budgets it will be necessary for educators to look to and understand controlled applications of the expanding telecommunications field. This study and the findings yet to come from this project provide knowledge, data, and methodologies on which to build.

References

1. Scanlon, Robert G. (August, 1973): "Building Relationships For The Dissemination of Innovations." Publication No. BR. Research for Better Schools, Inc., Philadelphia. Presented at CEDaR Communications Group Workshop, Denver, Colorado, July 30 – August 1, 1973.

2. Zaltman, Gerald, Robert Duncan and Jonny Holbek (1973): "Innovations and Organizations." New York: Wiley.

3. Zaltman, Gerald, David H. Florio and Linda A. Sikorski, (1977). "Dynamic Educational Change," Models, Strategies, Tactics, and Management. New York: Free Press.

4. Berman, Paul, Peter W. Greenwood, Milbrey Wallin McLaughlin, John Pincus, (April 1975): "Federal Programs Supporting Educational Change," Vol. V: Executive Summary. Prepared for the U. S. Office of Education, Department of Health, Education, and Welfare.

TABLE V - Knowledge of Term, "Registration"

Attitude	Teleconference I		Teleconference II	
	Pre	Post	Pre	Post
Know nothing	204	45	163	116
Almost nothing	74	82	69	53
Something	86	184	107	141
A good bit	18	58	41	49
Understand it	8	22	6	11

TABLE VI - Knowledge of "Building Level" and "Action Plan"

Attitude	Teleconference I		Teleconference II	
	Pre	Post	Pre	Post
Know nothing	186	30	134	74
Almost nothing	74	57	71	66
Something	94	213	119	155
A good bit	30	72	54	66
Thoroughly	7	18	8	9

TABLE VII - Attitude Toward SIP Change

Attitude	Teleconf. I	Teleconf. II	TOTAL
Less Positive	38	28	66
No Change	213	256	469
More Positive	136	91	227
TOTAL	387	375	762

TABLE VIII - Willingness of Teleconference I Respondents to Participate in SIP Activities

Attitude	Long-Range Planning		Implement SIP		Promote SIP Activities	
	Pre	Post	Pre	Post	Pre	Post
Not Very Likely	44	58	8	17	15	33
Moderately Likely	195	173	172	159	179	171
Very Likely	151	156	208	212	192	183

TABLE IX - Willingness of Teleconference II Respondents to Participate in SIP Activities

Attitudes	Long-Range Planning		Implement SIP		Promote SIP Activities	
	Pre	Post	Pre	Post	Pre	Post
Would Not Participate	6	22	4	13	13	20
Might Participate	119	105	90	90	106	87
Would Participate	176	174	197	195	186	179
Would Volunteer	85	70	94	72	78	72

The Teleconferencing Resource Book: A Guide to Applications and Planning
Lorne A. Parker and Christine H. Olgren (eds.)
Elsevier Science Publishers B.V. (North-Holland)
© Center for Interactive Programs, University of Wisconsin-Extension, 1984

SATELLITE TELECONFERENCING IN CONTINUING EDUCATION:

WHAT LIES AHEAD?

by
Randall G. Bretz, Ph.D.
University of Nebraska-Lincoln

Higher education in the United States is facing a number of changes in the decade of the 1980's. Among the changes are a predicted decline in traditional enrollments, an increase in the demand for professional continuing education, and a growth in the amount and exchange of information. There are two ways that the higher education system is meeting these challenges. One is retrenchment. This calls for a stabilization or decrease in the size of the institution to compensate for the decline in the traditional college age enrollment. The second is to change directions to meet the needs of new clientele. To provide increased information to clientele seeking professional continuing education and to compensate for declining enrollments on campus, many institutions are experimenting with satellite teleconferencing. This has the advantages of sharing resources among many institutions, quick replies to the need for information, broader dissemination of information, and meeting the needs of our adult population.

Satellite Teleconference History

The beginning of satellite teleconferencing can be traced to the middle 1970's when the National Aeronautics and Space Administration (NASA) fostered experimentation in this area with the Applied Technology Satellite (ATS-6) which was utilized for a number of educational teleconferences. Another significant development in the area of satellite communication was the implementation of a national satellite network by the Public Broadcasting System (PBS) which now connects more than 160 ETV and PTV stations throughout the country. (Bartlet, 1982) While primarily for program distribution, this network can also be used for teleconferencing. More recently the Public Service Satellite Consortium (PSSC) initiated the Campus Conference Network (CCN) which proposes to tie together institutions of higher education in large cities around the country. (PSSC, 1983) Still another development in higher education's use of satellite teleconferencing was the creation of the National University Teleconference Network (NUTN) which brought together more than 65 institutions for the purpose of sharing educational programming. (NUTN, 1982) This

organization, created in 1982, has already coordinated two successful satellite teleconferences and one audio teleconference for its member institutions.

The technology for satellite program delivery is in place and ready to be used. A number of satellites are in orbits which allow them to relay signals from almost any origination point in the country to an unlimited number of reception points. It is the utilization or techniques for satellite teleconferencing in continuing education which still need development. As is often the case, the technology is available and needs only the development of programming to become of greater benefit to education. Most satellite teleconferences to date have made use of a basic presentation format placing the content expert(s) at one location and transmitting them via satellite to participants at a number of sites. The viewers then interact with the presentors by return audio connection, usually telephone. (Robertson, 1980) The creative faculty talent and program planning talent which exists on university campuses has largely been untapped.

Description of Study

In 1982 a Delphi study was conducted to investigate the technical, utilization, policy, and procedure questions related to satellite teleconferencing in continuing education. The study considered what technology would be needed to conduct satellite teleconferencing and related activities, what formats could be developed to foster and improve utilization of teleconferencing, and what policies and procedures needed development to govern and guide the operation of teleconferences.

The Delphi technique was the methodology used to project the future of satellite teleconferencing in continuing education. This technique is characterized as a method of structuring a group communication process to allow a group of individuals to deal with a complex problem. (Linstone, 1975) A review of Delphi studies suggests that either three or four questionnaire rounds be conducted to obtain the results. For the purposes of this study four rounds were conducted.

To select the participants for the study, contact was made with the Public Service Satellite Consortium in Washington and Denver, the Educational Telecommunications Utilization Division of the National University Continuing Education Association, the Joint Council on Educational Telecommunications, and the Public Broadcasting System. Those identified were invited to participate. Twenty-eight persons representing these groups or agencies participated in the study.

The study included four questionnaire rounds. The first round asked for comments or statements from panel members in the areas of utilization, technology, policies, and procedures related to satellite teleconferencing. Rounds two through four listed each of these statements by category and asked the participants to suggest the probability, using a 1 to 5 scale, of each statement occurring within the 1982-1992 decade. Ninety-seven statements were tested in this study.

Results of Study

The results of the study are presented here in two tables. Table One presents 20 statements in the four categories which achieved a high degree of consensus as defined by having a standard deviation of .600 or less. Table Two lists those statements which resulted in high disagreement among the panel members. This disagreement was defined as those statements which had a standard deviation among the probability scores of .900 or greater.

Because of the nature of Delphi studies and the statistical results which can have a statement reaching consensus, yet a relatively low probability, the discussion of the results presented here focuses primarily on those areas which reached consensus and had probable or very probable ratings. The discussion includes the areas of administration, a teleconference profile, teleconference delivery, and those areas which are not likely to happen based upon a high level of disagreement among panel members.

Administration: Members of the panel were not certain that satellite teleconferences in higher education would be administered by the continuing education unit on their campuses. In most institutions it is this unit which has the budgetary and personnel capabilities to develop and coordinate teleconferences. It was suggested by the panel that there would be an institutional coordinator appointed to work with teleconferences. This person would most likely be responsible for the combined technical and educational needs called for in satellite teleconferences. It is interesting to note that in the establishment of the NUTN, these two theore-

tical statements have already been put into practice. A single person at each of the NUTN member campuses serves as teleconference coordinator. These coordinators are within the continuing education units at member institutions.

Other aspects of administration noted by the panel were that joint arrangements would be made to handle a number of problems or potential problems. This might include registration, evaluation, and examination procedures. It was also suggested that there would be priority systems and procedures established for scheduling conflicts including the use of satellites during high demand periods. And as more institutions have receive facilities, administrative issues such as privacy and copyright will be raised. Some technical developments have been made to scramble signals in answer to the need for privacy, but this is a costly process.

TABLE 1

List of Consensus Statements by Degree of Consensus

Technology

1. A mixture of technologies will be used in teleconferencing. (ie: video disc, satellites, computer, teletext)
4.93*   .262 **

2. There will be increased trans-border (international) flow of continuing education.
4.00   .471

3. Domestic satellites will be placed in orbits with 2 degree separation.
3.32   .476

4. Satellite ground systems will be less expensive.
4.50   .509

5. Transmit (some) and receive (many) sites will be established in virtually every major academic institution, industrial firm, government agency, and various "public" sites throughout the country and some foreign countries.
4.26   .526

6. Direct Broadcast Satellites (DBS) will be available.
4.32   .548

* Denotes probability of statement
** Denotes standard deviation of answers

TABLE 1 Continued

Utilization

7. Utilization will expand as cost and availability of facilities make teleconferencing more
attractive for meetings and educational
programs.
        4.86   .356

8. There will be increased numbers of short
courses for professional up-dating.
        4.68   .476

9. Educational programming will be delivered to
learning centers via satellite.
        4.32   .476

10. Written materials will be specifically
designed to complement video programs.
        4.36   .559

11. Interinstitutional liason will be a problem
to successful implementation and utilization.
        4.36   .559

12. Programs will be delivered directly to
professionals at their places of work.
        4.46   .576

Policies and Procedures

13. The need to protect privacy and confidentiality will require policy development.
        4.07   .539

14. The need for copyright protection will
result in new policy.
        4.00   .544

15. University policy will place responsibility
for satellite tele-conferences with the continuing education administrative unit.
        3.43   .573

16. Participating institutions will appoint and
train a primary coordinator to work with internal and external parties using teleconferencing.

        4.11   .416

17. The economy of using satellite transmission
for conferences will cause professional associations to utilize satellite teleconferences for
their meetings.
        4.25   .441

18. Teleconferencing systems will be developed
with sufficient flexibility to maximize potential utilization.
        4.04   .508

19. The cost of continuing education teleconferences will be offset by user fees.
        4.25   .518

20. Registration, evaluation, and examination
procedures will be developed by institutions and
consortia of institutions.
        4.39   .567

Teleconferencing Profile: One of the
criticisms of satellite teleconferencing in
continuing education to date has been that
utilization of educational techniques and practices has not been adequately integrated with
the technology to provide the best educational
product. Some of the statements in the study
which were agreed upon by the panel speak to
this need. First, it was agreed that satellite
teleconferences will actually include a mixture
of technologies including satellite transmissions, computers, audio networks, facsimile,
videodisc, and others. Diagram One depicts a
future teleconferencing system which shows a
number of media providing for interaction and
information sharing among the participants.
This diagram suggests that there will be video
and audio interaction between and among the
various sites through satellite, audio, computer
conferencing, telephone bridges, and the exchange of documents and other print materials
via facsimile.

The results of the study also suggest that
satellite teleconferences will involve written
materials developed specifically for the teleconferences. This may be a partial answer to
the privacy problem noted earlier. A teleconference designed so that it relies on printed
charts, graphs, and other materials will be of
little value to those who do not have the
printed material in hand.

Teleconference Delivery: It was generally
agreed by members of the panel that there would
be satellite receive facilities constructed or
at least available at many institutions of
higher education across the country by 1992.
This suggests that colleges and universities
will be receive sites for satellite teleconferences in the future much the same as public
broadcasting stations are today. Members of the
panel also suggested that satellite teleconferences should provide service to remote areas of
the country as already demonstrated by experiments using ATS-1, ATS-6, and CTS satellites.

Direct Broadcast Satellites (DBS) are also
expected to be in service by 1992 according to
the panel. The FCC has cleared the way for this
new generation of satellites which will be
providing service in the next two or three
years. DBS may bring about a new type of
teleconferencing. Rather than providing educational conferences on campuses, members of the
panel suggested that professional continuing

education would be provided via DBS to places of work, homes, and community learning centers.

Areas of Disagreement: The discussion to this point has focused on those statements in the study which reached consensus and had a relatively high rating on the probability scale. There is also value in comparing some of the statements which had a comparatively high degree of disagreement among the panel members.

Members of the panel could not agree on the role faculty or staff members would have in regard to satellite teleconferencing. There were statements suggesting that faculty would become guiders of learning, and would administer learning rather than be responsible for learning. These statements had a high level of disagreement among panel members. This suggests that the roles of faculty and staff members as well as institutions of higher education are still to be defined.

It seems uncertain whether or not the FCC will set aside frequencies or transponders for use solely by education. This has been the practice with television, radio, and some microwave bands, but the panel could not agree whether or not this would be the case with satellite channels. This suggests that education will compete in the market place along with business and industry for satellite time.

Members of the panel also could not agree on the statement which suggested that there would be one or more central agencies coordinating and scheduling technical arrangements for satellite teleconferences. This may indicate that there will be many agencies providing this service, or that it is too soon to determine whether this type of activity will be accomplished centrally.

Finally, panel members could not agree to the statement suggesting that the fear of technology would inhibit its use. This may mean that they were beginning to see greater acceptance and use of technology in education, or it simply could mean that they did not know whether this would be a factor in satellite teleconferencing.

TABLE 2

Statements with High Disagreement

1. One or more central agencies will coordinate technical arrangements for most teleconferences.
    3.57   1.103

2. Faculty and staff at institutions of higher education will become administrators of learning rather than operating places of learning.
    2.86   1.008

3. Faculty members will become guiders of learning activity rather than dispensers of information.
    3.46   .999

4. Competition from the business community for use of teleconferencing sites will keep costs sufficiently high that many educational applications will be discouraged.
    3.18   .945

5. Fear of technology will greatly inhibit its use.
    3.00   .943

6. More than half of the homes in the U.S. will have on-site computer systems.
    3.43   .920

7. Potential users will be screened to sift non-appropriate from appropriate users of the technology.
    3.93   .916

8. Program scheduling will have a high degree of flexibility to meet needs of learners.
    3.61   .916

9. Federal statutes will set aside frequencies or in some other way reserve transponder space for educational utilizatin. (similar to PTV, educational radio, ITFS)
    3.00   .903

Conclusion

In conclusion, satellite teleconferences in continuing education can be expected to involve a mixture of media under the coordination of one person at each of the participating institutions. It can be expected that institutions of higher education will jointly develop working agreements to handle registration, evaluation, and examination procedures as well as the development and dissemination of print materials to augment teleconferences. Privacy and copyright are issues which will need careful consideration and policy development. The technology and educational methodologies will probably receive attention as the state of the art of delivering continuing education via satellite is developed.

Broader dissemination of educational materials can be expected through the use of satellites as residents of remote regions acquire the necessary equipment to view teleconferences. This will in turn call for the development of interinstitutional, interstate, and international policies to handle the technical and administrative questions raised by satellite transmissions which cross geographical boundaries.

The role that faculty and staff members play in satellite teleconferencing is still to be defined. The same is true of the method of coordinating the technical aspects of teleconferencing. It is uncertain whether or not there will be a central agency or agencies to coordinate transmissions.

This study has offered answers and suggestions related to the use of satellite teleconferencing in continuing education, however, there remain many areas within this complex combination of technology and educational program development and delivery which still need to be considered or developed. Institutions of higher education are already joining forces to coordinate and offer satellite teleconferences through such consortia as the PSSC and NUTN. This study makes some suggestions about how those conferences might be administered and delivered, and lists some areas which need further study.

Bibliography

1. Bartlett, Roger. Chief Engineer for Nebraska ETV Operations. Interview, July 15, 1982.

2. "The Campus Conference Network." Concept paper circulated by the Public Service Satellite Consortium, 1982.

3. Conference call of 45 colleges and universities to discuss the establishment of the National University Teleconference Network, June 29, 1982.

4. Robertson, James. Satellite Teleconferencing, June-November 1980. Corporation for Public Broadcasting Technical Report 8011, 1980: 4-6.

5. Linstone, Harold A. and Murray Turoff. The Delphi Method Techniques and Applications. Reading, Massachusetts: Addison-Wesley, 1975.

MULTI-UNIVERSITY TELECONFERENCING NETWORK

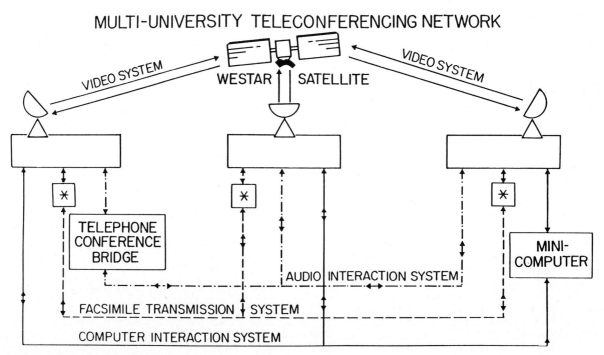

* FACSIMILE MACHINE

The Teleconferencing Resource Book: A Guide to Applications and Planning
Lorne A. Parker and Christine H. Olgren (eds.)
Elsevier Science Publishers B.V. (North-Holland)
© Center for Interactive Programs, University of Wisconsin-Extension, 1984

TRAINING TELECONFERENCE USERS: HOW TO TACKLE IT

Marcia Baird
Associate Director
Instructional Communications Systems
University of Wisconsin-Extension
Madison, Wisconsin

Mavis Monson
Instructional Design Coordinator
Center for Interactive Programs
University of Wisconsin-Extension
Madison, Wisconsin

*Training new teleconference
users means more than writing
a single how-to instruction
booklet and more than an
hour's introduction to your
new teleconferencing system.
It means setting up situations
and developing materials that
ease fears and build skills. It
means training that never ends.
Here's how we meet the challenge.*

Unless you're unusually gifted and can sell teleconferencing know-how to whomever you talk to within an organization, what you need is a blueprint for training teleconference users. You need a blueprint whether you're responsible for that training, or advising others who will fill that role. The more and varied training ideas and materials you have, the easier it will be to speed up acceptance of teleconferencing and stimulate everyday use.

We've done plenty of teleconferencing and training. Our national organization trains nearly 2,000 people each year in the art and science of teleconferencing -- audio-only, audiographics, freeze-frame and videoconferencing. Our state organization has one of the longest teleconferencing track records anywhere, stretching back 17 years to 1965. Between the two of us, we've spent 14 years in the field, churning out brochures, handbooks, books, videocassettes, workshops, seminars.

Each year our Wisconsin assignment involves working with about 150 individuals who are preparing to lead statewide teleconference training sessions or meetings. To do this successfully, we have a variety of training techniques. They're designed to deal with a wide spectrum of users, from on-the-run people who say,

"Teleconferencing is no different than my face-to-face meetings" to those who crave any teleconference training and say, "I want to know everything!"

You don't have to meet all potential users face-to-face to discover what training help they want; you only have to meet their needs. The most direct approach to determine those needs, of course, is to ask a small sampling. Many of our ideas spring from talking with new users and veterans about their needs -- over the phone, in the elevator, or during one-on-one interviews. It's also important to study the research in teleconferencing and related fields. Finally, it's important to try to walk a little in the other person's shoes. We have done -- and continue to do -- teleconferencing ourselves. We always try to remember how it was for us, and try to conjure up the fear and insecurity that's often associated with that first teleconference when we work with new users.

That leads to another bit of advice: don't draw all your training materials from the same well. After interviewing users and sifting through a variety of research, for instance, we boiled the important bits of information on teleconferencing program design into a concise, three-fold brochure. Designing for Interactive Teleconferencing is a hip-pocket guide, an appetizer, that's given to every new user. We've also found you need to balance talking about teleconferencing with doing it. A 3-hour hands-on training workshop allows users to discover for themselves what teleconferencing is all about. After talking with users, we also established a "Brown Bag Series on Interactive Media." Each monthly seminar focuses on a different topic, ranging from developments in audio station equipment to applying listening skills to teleconferencing. Lesson: variety is the spice of life.

We're concerned about the individual--their attitudes towards teleconferencing, their perceptions and abilities. And, we're not alone. Tom Hoff, Director of Sales Promotion and Telemarketing for Hoffmann-LaRoche, says, "You're putting a person on a new stage, and you can't afford to let that person fail. You want to do everything you can to help that person to start with a win." He couldn't say more clearly what any trainer must do--not just to introduce teleconferencing to Hoffmann-LaRoche employees, but to introduce it into any organization in the hopes of it being routinely used.

Teleconferencing is here to stay: a recent Center for Interactive Programs' study found teleconferencing incorporated as a regular communications tool in 62 companies in business and industry, 55 colleges, universities and medical groups, 19 governmental agencies and 11 other organizations.[1] Today there are a variety of quality teleconferencing products as well as audio and videoconference suppliers to meet this growing demand. The equipment and services, however, are only half the battle in implementation. It's the job of getting them into regular use within an organization that's the most difficult.

Historically, there has been a casual approach to training the teleconference user -- by both organizations and equipment suppliers. The assumption for many years seemed to be that there was little to learn about teleconferencing aside from turning a few knobs and switches. Johansen, McNulty and McNeal, for example, found that although 80 percent of the 50 systems they contacted recognized training as a factor in success, most efforts towards training were limited to a users' manual.[2]

The apparent simplicity of teleconferencing has been and is deceptive. Some executives and faculty can't be convinced that any training in teleconferencing is necessary. Some view audio teleconferencing, for instance, as simply an extension of a telephone conversation. If two people can easily talk to each other and communicate, they reason, what's the big difference when 4, 8 or 16 or more gather on the

partyline? Audiographic systems have been likened to being as easy as writing on a class-room chalkboard.

Training Framework

If historically, there's been a casual approach to training, then what's the argument for more in-depth training of new or potential users? Why use dollars and personnel to design and implement training programs? How can training influence user acceptance and commitment to teleconferencing?

Based on our experience, and data from organizations who have focused resources in this area, we're convinced that a well-designed training program provides an opportunity for individuals to deal with the changes in attitudes and behavior that must go hand-in-hand with the adoption of technological innovation like teleconferencing. And in dealing with these changes--to become more successful users, faster.

Most of the attitudes hindering full acceptance of a new system center around a common, albeit difficult-to-acknowledge human reaction to change: fear. Fear of the unknown, fear of failure, of inadequacy, fear of loss of control or power. In addition, there may be risks in giving up old and comfortable habits--habits both psychological and social. Psychological habits which guide and protect individuals from loss of face: "We've always done it this way." Social habits which protect comfortable relationships with other people: "I always talk to him in person." Negative attitudes may also result from a lack of understanding of the innovation, stemming from distorted information fueled by grapevine rumor.

If individuals can gain control of the new system by mastering successful teleconferencing technique, and if the individual early on recognizes and seizes the power afforded by this form of communication, that individual becomes an open advocate of the new system. If on the other hand, the potential or new user forms an unrealistic opinion as to the benefits of the change or feels threatened by the demands made for behavior change, then resistance may occur.

Resistance is not necessarily bad per se. Resistance--that conduct which serves to maintain the status quo despite pressure to change--can be used constructively. Resistance tells us something about that individual or group--its resources, attitudes towards change, value system and the way it relates to the rest of the environment. Resistance may even be nothing more than a way to gain time to make the adjustments perceived necessary to cope.

It's within a training session-- if the session is open and supportive-- that resistance can be dealt with in a constructive way. Training programs can help clear the air by allowing people to vent their feelings, gain information, and develop the teleconferencing skills they need to bolster their self-confidence.

Successful teleconferencing is not a spectator sport. It demands not only a favorable attitude towards this way of communicating but actual skills - a relearning of some of the communication behaviors that work well in a face-to-face setting but need to be adapted for this new environment.

Obviously, the type of training format suggested above--a group workshop--cannot be the only approach. It is unrealistic to expect that every individual will have enough motivation, or perhaps time, to attend an in-depth session. And, certainly, <u>requiring</u> users to attend such a session before allowing them to use the system would not be recommended policy! Thus, a variety of training resources need to be developed.

How those materials are designed-- the general approach, the "look," the amount and type of information, etc., will depend on the types of users, their level of experience, and the type of system being used. If you are hunting for a standard training formula for these materials, keep one thing in mind: there is no rulebook.

There is, however, a general strategy or blueprint that has proven effective for us in developing an overall training package. We base this strategy on a list of innovation attributes developed by the well-known diffusion theorist, Everett Rogers.*In his <u>Communication of</u>

Training sessions allow people to develop skill in using the technology and to master successful teleconferencing technique.

*We recognize that these categories are only a part of the entire implementation/diffusion process, and may be arbitrary. In Rogers' recent publication <u>Communication Networks</u>, he frankly criticizes the linear models of communication in which there is a Source-Message-Channel-Receiver model in a one-way mode. He develops what he calls a "convergence" model in which communication is examined as a complete cycle. We agree. Training is a two-way process, not something that is done to someone--all the while reinvention is going on as people interact with the system and with each other.

Innovations, Rogers suggests five
attributes of an innovation which
relate to rate of adoption.

Of those attributes which relate
positively to a faster rate of adoption,
Rogers lists:

Relative advantage - the degree to
which an innovation is perceived
as being better than the idea it
supersedes.

Compatibility - the degree to
which an innovation is perceived
as consistent with the existing
values, past experiences and
needs of the receivers.

Trialability - the degree to which
an innovation may be experimented
with on a limited basis.

Visibility - the degree to which
the result of an innovation are
visible to others.

The single quality which is negatively
related to adoption rate is:

Complexity - the degree to which
an innovation is perceived as
relatively difficult to understand
and use.

User perception is a key issue in
training strategy. If an individual
perceives an innovation as complex--
difficult to use--that perception becomes a self-fulfilling prophecy.
Psychological studies point out the
many ways that selective perception
and retention prevent a person from
seeing that the status quo is inadequate.
For various reasons, a person may not
"see" problems, nor by the same token,
"see" the solutions.

Our strategy, then, is two-pronged.
One aspect is the use of Roger's
attributes in defining what it is we
are dealing with; the second is to
identify what individual perceptions
may be interfering with a person's
ability to move forward in accepting
the innovation. It is a diagnostic
rather than didactic approach.

The Five Components

Let us now expand upon how we
carry home the blueprint Rogers has
given us. The five components of all
our training materials and sessions
include:

Identify rewards (relative advantage)

Know if teleconferencing fits (compatibility)

Make it easy (complexity)

Try it out (trialability)

Spread the word (visibility)

One. Identify the rewards.

Show new and potential users what
benefits teleconferencing contains for
them. Address benefits here from the
user's perspective; not from that of
the organization.

It's a fact of life many folks
have preconceptions about teleconferencing, fueled by a lack of understanding, negative past experiences,
fears, or the grapevine. Each of these
can lead new users to some very quick,
but inaccurate conclusions about new
teleconferencing situations.

It's also a fact that inexperienced users often can't -- or don't --
distinguish between unskilled use of a
teleconferencing system and the system
itself. In a face-to-face situation
if we hear a poor speaker, for instance
we generally evaluate the situation and
say "That was a poor speaker." Not,
that was a poor meeting. In teleconferencing we can also have good and
poor speakers. More often than not in
teleconferencing situations, however,

a poor speaker leads to a negative reaction to teleconferencing and new users write off the entire medium.

The secret here is: get the facts out about teleconferencing. Present these facts about what teleconferencing is -- and is not -- so people can form a realistic opinion about how it's going to affect them.

Some will recognize the opportunity that teleconferencing offers immediately: the opportunity for increased communication, increased coordination, increased power, etc. You may need to help others -- through training -- become aware of the rewards. Have some case studies ready, for instance, that address cost savings because it's a repeated reward theme. A key part of one state agency's training efforts focuses on cost comparisons of the same meeting held face-to-face, held via operator assisted audio conferencing and held via Meet-Me dial-in audio conferencing. They don't just tell them, they show the savings an agency can begin making immediately without giving up necessary meetings. And, the more specific the examples are to the user group, the better.

Productivity is another buzz word when we talk about teleconferencing and its advantages. You don't lose eight hours of job time and energy, for instance, traveling to a two-hour meeting. Give some examples or case studies about how teleconferencing allows groups to do tasks quicker, allows more input and adds to the quality of life. We often cite the example of 3,500 state plumbers who participated in teleconference training to learn new solid waste codes passed by the state legislature. Training was completed in less than two months. The sponsoring state agency estimated it would otherwise have taken 14 months to complete the compulsory training.

In addition to the facts, your training programs must also address the whole area of past experience. What kinds of teleconferencing have your new users participated in? Make sure you know how they felt about that experience. You need to know the good experiences and more important, the bad.

Teleconferencing is also a high learning product. It demands or generates change, and that's scary for all of us. But teleconferencing won't survive within an organization unless there is some change. People must change their communication behavior. If they don't, they're not going to make it in teleconferencing. Brainstorm with a group of new users. Ask them what frightens them the most about their upcoming teleconferencing experience. Go from there. Like any topic, teleconferencing can be understood and used effectively with a little work.

Two. Know if teleconferencing fits.

The fact that people understand what teleconferencing is -- and isn't -- doesn't necessarily mean they will use it. Teleconferencing must "fit" particular subjects, clientele and styles. The closer it fits to one's ideas about how people can learn, one's past experience and needs, the easier it will be adopted.

Values. Teleconferencing may be compatible with existing values or incompatible. For example, many people believe that face-to-face is the only effective way to meet or to learn. They use criteria based on the characteristics of face-to-face meetings; teleconferencing comes up second best in their scheme.* Another value, perceived as threatened by adoption of teleconferencing, is travel. This issue may or may not be out in the open. Other values may be more subtle.

In dealing with the area of values, we present teleconferencing as a "tool" for accomplishing communication tasks rather than as a replacement for face-to-face. There is a delicate balance between overselling teleconferencing on the one hand and not taking an upbeat-enough approach on the other. We present the "facts" as we know them

---

We have discussed the advantages of a teleconference under "Identify the Rewards" and would refer you to that section.

from the research but then go on to
show by successful case studies how
creativity in program design can do
much to make or break a teleconferencing application.

A good example of overcoming the
"facts" of research is a recent airline
negotiation session. Research indicates
that teleconferencing is a "weak"
medium to use for negotiation. However,
a contract negotiation was successfully handled by United Airlines using
a combination of video tapes for pre-
education and a one-way video, two-way
audio teleconference among 9 locations
to reach a successful agreement.

People need to see successful
applications in terms of case studies,
not just a listing of what teleconferencing can do. We need to resist
the attempt to make teleconferencing
"just like" face-to-face, however.
(Many system designers have lost sight
of the fact that it is not the same.
Not necessarily inferior, but not the
same.) If we attempt to do this,
through technology or design, individuals may be lulled into a false
sense of security.

For example, in an attempt to make
teaching appear to be compatible with
previous face-to-face courses, a
videoconferencing system was designed
by a governmental agency with the
objective of taking care of any
contingency through the technical
design.

A one-way video, two-way audio
system was put into operation and the
instructors given the advice that
they shouldn't change any of their
typical classroom behaviors. The
result was less than successful.
Instructors broke all rules for
effective presentations and interaction.
Individuals were frustrated in their
attempts to participate. Visuals were
poorly prepared and not timed so that
appropriate camera work could capture
them efficiently.

Past experience. People have had
past experiences--not all of them
positive! We need to recognize that
everyone has a data bank of informa-
tion--an image of what teleconferencing
is and what it can do. You need to
begin where people are. A workshop
or a one-on-one consultation can help

to get out negative feelings; past
less-than-successful experiences can
be built upon.

Old ideas are the main tools
which new ideas are assessed, thus
transfer of information from one type
of teleconferencing to another may
retard acceptance because of the need
for "relearning." For example, in
introducing a meet-me type bridging
system to Wisconsin users familiar
with the capabilities of the dedicated
network (which allows more than one
individual to talk at a time), there
was a need to "relearn." Users had
to adjust to the one-way at-a-time
capabilities of the meet-me system.

Needs. The closer the users'
needs are to the characteristics of
teleconferencing, the quicker and more
widespread the adoption will be. A
good example is the heavy use of
teleconferencing by health care
professionals. Here's a group that
needs continuous updating of skills,
access to current information, ful-
fillment of continuing education
requirements, and convenient times and
places to get that instruction so they
don't need to leave their practices
for long periods of time. Teleconferencing is a natural fit.

Training sessions can be helpful
in finding out the mission of users,
both in broad terms and in terms of
content and tasks. Try to identify
those needs that are felt; then help
push towards areas users could not see
before, in an incremental approach to
the innovation.

The whole area of compatibility is
a delicate one. It's very easy for
those who are in the implementation/
training area to be seen by potential
users as having a vested interest in
the use of the system. And trainers
are guilty of promising too much too
soon. It is critical for those in
the role of working with new users to
gain teleconferencing experience for
themselves and actively use the systems
they are professing.

In addition, it is helpful if
testimonials and applications come
from the mouths of peers--those
individuals to whom new users relate to
on a day-to-day basis. Our training
materials reflect this philosophy. The

sophistication of the publications, the approach and the actual content material is designed to "look" like them, incorporating as many actual quotes from successful peer programmers as possible.

There is also a critical need to be aware of the value of feedback in terms of making the system compatible. This is the time when "reinvention" can occur. Small changes can help individuals "own" the technology-- rather than feeling as if they are constrained by it. Stay open to what people are really saying when they make subtle suggestions or complaints.

Three. Make teleconferencing easy.

Our goal here is to make tele- conferencing easy to understand and use. To reduce perceived complexity, make sure your training materials address complexity at two levels-- technical and program design.

Crusade against overstuffing. We sometimes have the habit of providing new users with more information than they really need to do teleconferencing. Too many instructions. Too many how- tos. If you overstuff, you run the risk of the user ignoring the important bits of information.

A generous choice of training materials about teleconferencing should be available to users on a selective basis. Together with other activities, these materials ease fears and build skills.

We've condensed instructions on how to work audio station equipment at each teleconference site, for instance, into a prerecorded tape cartridge that's less than one minute in length. It's played over the system at the start of each teleconference for instructors and participants alike. Details on the functions of all the knobs and switches and how to perform operational tests on the equipment are included in print form with each unit. Finally, nuts-and-bolts technical information, complete with schematics, is included in a technical brochure that's available upon request.

Users need to have a basic understanding of the technology they're working with. The interactive capability on a two-wire teleconferencing system, for instance, is different from the interactive capability on a four-wire system. Yes, we want the teleconferencing technology to be as transparent as possible, but as experienced users know, technology can choose very inappropriate times to fail. New users can lose control of a training session or meeting very quickly if they do not understand the technology. This dependency underscores the importance of teleconferencing design. In teleconference training workshops, we frankly discuss the possibility of "Murphy" cropping up during any teleconference situation and what a meeting leader and teleconferencing service can do in those situations.

You can't make full use of any teleconferencing medium until you understand program or meeting design. Our secret of good teleconferencing program design is based on four concepts: humanizing, participation, message style and feedback.

An initial step at reducing the complexity of program design is through our hip pocket brochure called Designing for Interactive Teleconferencing. It's strong on basic how-to tips for implementing the four concepts. For those whose appetite has already been whetted, we add depth with a book called Bridging the Distance. Its 65 pages expand on specific techniques for humanizing the experience, soliciting participation, selecting the most effective message style and getting and receiving feedback. It's also laced

with how-to quotes from teleconferencing users. The reason for the quotes is simple: teleconferencing users follow other teleconferencing users. The quotes in themselves have helped instruct many readers.

Our training efforts don't just appeal to readers. For the visually-oriented and on-the-run people, we've designed a series of videocassettes. A tape on how to design a videoconference that's strong on participation, for instance, runs 20 minutes instead of 100 pages.

No Place for Lone Rangers, another videocassette, shows how one meeting leader, with the help of some know-how in good program design, improves his teleconferences. Twelve Interactive Techniques for Teleconferencing is a set of four videotapes that focus on practical tips and guidelines for presenting information and getting participation. They're light, but packed with information on such interactive techniques as tandem teaching, reactor panel, buzz groups, case study and role play. Print materials are given to viewers to serve as a valuable reference tool back at the office.

Complexity is also reduced through hands-on workshops. Some are a combination of face-to-face and teleconference sessions within our Madison headquarters. Others are completely held via teleconference to meet the training needs of out-of-state instructors or far-flung statewide faculty.

To introduce potential users to the call-in Meet-Me system, we had participants call in to the training session from their office telephones. The process became the content. Participants, from first-hand experience, were able to compare and contrast Meet-Me vs. other systems and thus gained a deeper understanding of the new service.

Four. Try It out.

The best way to learn about teleconferencing is by doing it and experimenting with it.

By simulating teleconferencing between a number of locations, workshop participants gain hands-on experience with the equipment, learn new communication behaviors and establish a baseline for their own programs.

For example, can the new idea be tried on the installment plan, a piece at a time, before deciding whether or not to accept it? If so, that innovation represents less risk to the person who is considering it and will generally be adopted more quickly.

The "installment plan" theory immediately suggests that a pilot test or experimental approach to introducing teleconferencing would be ideal. Two schools of thought exist on this issue, however. Proponents of pilot tests say they are a way for people to get their feet wet before committing and thus relieve some of the pressure, reducing the resistance which comes from an imposed change.

Those who disagree with the pilot test approach say there is a danger in labeling a teleconferencing system as "experimental" within an organization. They say this approach is ladden with pitfalls. Opinion leaders,

in particular, don't have the time to spend on "probables." They want solid ground to stand on. If they are going to endorse an idea, they need assurance that if they commit time and energy to the project, that it will be a part of the organization in the future. There is merit in this argument, especially at the level of organizational commitment to teleconferencing.

And if we look at the level of direct user training, we can see that some commitment is also necessary. There is, in adoption of new technology, a "time-for-learning" factor. A certain time period (which varies from individual to individual) is needed for people to feel comfortable with the system and begin to consciously and unconsciously adapt their communication behaviors to it. That time-for-learning factor may be one or two teleconferences--perhaps more. A one-time meeting may not be enough for some individuals-- and the chance for acceptance may be lost.

Probably a balance between full commitment and casual experimentation is called for. At the level of user acceptance, there needs to be a certain amount of testing-of-the-waters combined with a firm commitment on the part of individuals to engage in some follow-up programming.

Simulations and games can provide a chance to try the system in a mistake-free setting, and as such can provide a beginning step. It is an approach we frequently use. Other groups have found it of value as well. One marketer of videoconferencing systems uses the game approach dually--to introduce the concept of video-conferencing as well as to train individuals in good technique. This hands-on simulation has as its objectives the enabling of individuals to:

> participate more comfortably in future videoconferencing

> make use of techniques that increase the effectiveness of meetings

> and ultimately use videocon-ferencing successfully for business meetings"

These objectives are carried out using a two-point videoconferencing system for solving a business problem set.

Our objectives, although focused on audio and freeze-frame videoconfer-encing, are similar for what we call the TELETECHNIQUES workshop. We do not have dual marketing/training ob-jective -- the Wisconsin systems are a "given" for our users, but we are giving individuals a mistake free setting in which to operate. During the TELETECHNIQUES workshop, we simu-late a number of teleconferencing lo-cations, with groups at each site. Using various studios here at Old Radio Hall, we can duplicate the experience of physical separation and provide each individual with an environment close to that found at our other state locations. We present the sessions using methods of teaching and learning adapted for teleconferencing--in the hope that our instructors will consider using them. The hands-on experience with the equip-ment, the setting and the interaction provide modeling on which they can base their own behavior.

Post-simulation discussion pro-vides another chance for learning. By analyzing strengths and weaknesses ("I could model my agenda after this one," or "I wouldn't need to spend that much time on humanizing," etc.) there is a chance to see what can be done, what works and what doesn't and time for a little mind-stretching in terms of future program directions.

The orienting of new groups to technology through the technology is not a new one. It is simply that the awareness of using this technique to introduce teleconferencing has become more in vogue. Many of the commercial bridging services have seized upon "call in" get-acquainted-with-telecon-ferencing sessions. Generally these are no-charge chances to try out a meet-me type audio bridging system. Increasingly, they provide more than just the "process" but some "content" as well.

If the chance for trying a system just isn't possible, the experience should be as closely paralleled as possible. Either a full motion video-conference or a freeze-frame video-conference may be previewed from video or audio tape respectively-at least to give a flavor of what those experiences might be like.

Five. Spread the word.

Recognize that the easier it is for others to see the results of tele-conferencing, the more likely telecon-ferencing will spread. Teleconferen-cing needs to make itself highly visible by using a variety of promo-tional devices. Fortunately for all of us there has recently been a tremendous explosion in the visibility of teleconferencing, not just in trade journals, but in the daily press and electronic media.

Here are some guidelines and suggestions for getting the word out. We consider all as pieces of the training effort.

Publish and route. Publish your-self, and encourage users to publish first person "I did it!" articles in professional and trade journals, newsletters and mass media. Or, make sure you have a strong public relations staff. Circulate copies of published success stories to new and potential users.

Establish a user's group. We've found our informal "Brown Bag Series on Interactive Media" a good source of inspiration, information and moral support. It allows those who have particular goals in teleconferencing to cross paths with others who are going different places.

Extend invitations. Throughout the year we extend an open invitation to department chairpersons and agency directors to conduct workshops for them, or with them, on teleconferencing delivery systems. We invite ourselves to be included on district and state-wide meetings. And we invite users and potential users to hold monthly staff meetings in our facilities so we can introduce them to new teleconferencing services or equipment.

Develop up-to-date application pieces. We publish a newsprint each semester that in addition to promoting upcoming teleconferences carries stories on new and interesting applications of teleconferencing to training and instruction. The tabloid is circulated statewide to past participants, departments, administrators and to the general public. This inexpensive tabloid gives high visibility to teleconferencing and recognition to innovative users.

Schedule yourself on conferences. Professional and organizational conferences are an excellent way to export the results of teleconferencing to other colleagues. Whenever possible presentations should be made by colleagues -- those already involved in using teleconferencing, those who have high acceptance and those who can stir up excitement.

Involve interested users. Involve interested users from the start -- whether it be on a teleconferencing advisory committee, testing out a new piece of equipment or making decisions about the placement of teleconferencing rooms.

Ongoing Support

and

Evaluation

Teleconferencing systems, once set in motion, don't continue to function without the ongoing support of the organization--in terms of the hardware and the software. Just as individuals need to be committed to some degree to making the system work, organizations need to provide the commitment and resources necessary to keep teleconferencing functioning smoothly. What are some of the follow-ups necessary?

Reinforcement for one. There is a period of time after the user knows how to use a system properly before it's entirely comfortable. Reinforcement, as simple as a phone call after the first teleconference, is a way to encourage that individual as well as gain feedback on reactions to teleconferencing. It's a way to identify small problems before they become big ones.

Dependability is another factor. Staff help when it's necessary--to provide easy access to teleconferencing rooms and resource materials and to keep red-tape and paperwork to a minimum. Staff can also be a sounding-board for innovative ideas and go-betweens for resolving problems, be they technical or programmatic. Tele-conference users also need dependable service: equipment that's set up, that works and that undergoes routine maintenance. High reliability is essential and appropriate equipment and facilities must be available to make this possible.

Communication on a continual basis is a third factor. There is a continual need for publicity and training. There will be the core group but there is usually a high influx of new people. Involve the new and old users. Seek continual input from them to keep tabs on what they want and need. This may be through informal means, through periodic surveys or ad hoc committees.

And reach out to them as well with newsletter items, brown bag seminars, meetings and open houses.

Evaluation is a fourth factor. It is important to build in evaluation mechanisms not only to determine strengths and weaknesses of teleconference sessions, but also strengths and weaknesses of the teleconferencing system itself. Usage statistics, observations and participant surveys are common evaluation measures.

Final Thoughts

Our consistent experience in training teleconference users is that a single teleconferencing how-to booklet, no matter how well designed, cannot carry an entire training program. Multi-level print materials and videocassettes, hands-on workshops which actively engage the user, and continuing personal contact are critical to maintain a high level of interest, success and visibility.

In addition to a generous choice of training options, users also need the chance to control the time and pacing of their training.

The concepts and ideas we've presented here are flexible, practical and easy to put to use. They've proven effective in developing teleconferencing confidence and know-how among hundreds of users.

But, we're still restless--continually on the lookout for better ways to respond to changing training needs of individuals and organizations. Any ideas anyone?

REFERENCES

1. Lorne Parker and Christine Olgren, "CIP Releases New Teleconferencing Study," TELCOMS Interactive Telecommunications Newsletter, Vol. V, No. 1 (January, 1982), 1.

2. Robert Johansen, Maureen NcNulty, and Barbara NcNeal, Electronic Education: Using Teleconferencing in Postsecondary Organizations (Menlo Park, California: Institute for the Future, 1978), p. 74.

Bibliography

Bennis, Warren G.; Beene, Kenneth D.; and Chin, Robert, editors. The Planning of Change. New York: Holt, Rinehart and Winston, 1966.

Elam, Phillip G. "Change: How Users React." Computer World, December 15, 1980, pp. 9-16.

Galitz, Wilbert O. Human Factors In Office Automation. Life Office Management Association, Inc. (LOMA), 100 Colony Square, Atlanta, Georgia, 1980.

Johansen, Robert; Vallee, Jacques; and Spangler, Kathleen. Electronic Meetings: Technical Alternatives and Social Choices. Reading, Massachusetts: Addison-Wesley Publishing Company, 1979.

Rogers, Everett M., and Kincaid, D. Lawrence. Communication Networks: Toward a New Paradigm for Research. New York: The Free Press (a division of MacMillan Publishing), 1981.

Rogers, Everett M. with Shoemaker, F. Floyd. Communication of Innovations. New York: The Free Press (a division of MacMillan Publishing), 1971.

Zaltman, Gerald, and Duncan, Robert. Strategies for Planned Change. New York: John Wiley & Sons, 1977.

The Teleconferencing Resource Book: A Guide to Applications and Planning
Lorne A. Parker and Christine H. Olgren (eds.)
Elsevier Science Publishers B.V. (North-Holland)
©Center for Interactive Programs, University of Wisconsin-Extension, 1984

COACHING TELECONFERENCE USERS

Jim Boudle
Designer, Tele-Comfort[TM] Training
Resource Management Consultants
Derry Professional Park
Derry, New Hampshire 03038

ABSTRACT

Today's teleconference users and potential users may be overwhelmed by the variety of teleconferencing options. These options - audio conferencing, audio plus graphics including (slow scan) freeze frame video, full motion two way dedicated video and ad hoc video teleconferencing require preparation, presentation skills, participation, practice and patience by the user.

Coaching or training teleconference users by instruction, demonstration, critique and practice helps them develop confidence and competence using their system.

This paper presents an overview of key factors to consider when coaching teleconference users in each of the aforementioned options. Professional coaching involves detecting ineffective communication mannerisms, demonstrating an effective communication skill, and reinforcing corrective verbal and non-verbal (when applicable) behaviors by the user. We do not advocate "acting" when telecommunicating but do believe teleconference users can be coached to more effectively use this powerful, dynamic business communication tool. Effective coaching is observable, measurable and most important self-reinforcing to users. It builds and maintains communications competence.

INTRODUCTION

The integration of computer and communications technology has made available a powerful business communication tool - teleconferencing. The power of teleconferencing is in the direct, immediate, simultaneous and interactive communications among three or more individuals at two or more distant locations.

The options in the concept of moving meetings to people, instead of people to meetings, include: (1) audio only conferencing; (2) audio plus graphics, including (slow scan) freeze frame video; (3) dedicated full motion (point to point) two way video and audio teleconferencing; (4) ad hoc full motion (point to multi point) one way video, two way audio (a fifth option, computer conferencing, is not considered here). Each system option has a special set of characteristics, path of information flow and corresponding user requirements.

Electronic meetings or teleconferencing, require a change in the traditional face to face approach to conducting meetings. To be more effective using teleconferencing requires preparation, presentation, operational and interactive communication skills adapted to the medium.

Organizations considering implementing or utilizing a teleconference option face a continuous challenge of developing their most valuable assets -- their human resources. The users are the key success factors in teleconferencing. Such new and unfamiliar ways of business communications demand that users gain the necessary knowledge of the concept of teleconferencing, understand system capabilities and constraints, develop an attitude of openness to change and be encouraged to acquire new communications skills to effectively teleconference.

This paper, written from a consultant/trainer perspective, describes the process of coaching as one approach towards developing skilled users and promoting a higher rate of system utilization.

Corporate trainers, consultants, users and others involved with the task of promoting user satisfaction, system implementation and increasing utilization rates can use this paper as the framework to assist them toward those ends.

UNSKILLED AND UNPREPARED USERS

Unskilled users increase the risk of misapplications of teleconferencing, technical difficulties and ineffective communications resulting in user dissatisfaction. In turn, user dissatsifaction can and often does lead to resistance to continued use and future system implementation delays. Unprepared users waste valuable time, annoy seasoned users and project a less than professional image. Such does not have to occur. Organizations that commit financial resources and demonstrate support to users by providing coaching opportunities would likely benefit by more effective use, appropriate applications and a high utilization of its system. Greater likelihood of achieving the competitive and economic gains of the system can be expected.

COACHING INVOLVES:

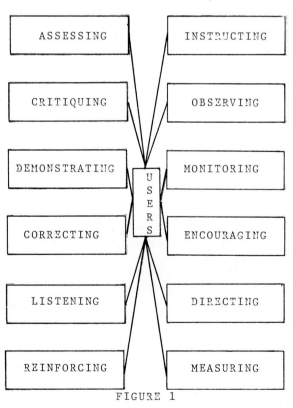

ASSESSING | INSTRUCTING

CRITIQUING | OBSERVING

DEMONSTRATING | MONITORING

USERS

CORRECTING | ENCOURAGING

LISTENING | DIRECTING

REINFORCING | MEASURING

FIGURE 1

THE PROCESS OF COACHING

Coaching is described in this article as one element of an overall training program for teleconference users. Training as a process involves gaining increased knowledge, assessing attitudes and developing new skills (using coaching) in a given area (teleconferencing). Coaching is a skill intensive form of training teleconference users (privately or in groups) by instruction, demonstration, critique and practice. The goals are to build confident and competent users - users who can communicate their message clearly, completely and concisely using this medium. Coaching teleconference users can:

● minimize intimidation, frustration and eventual non-use of tele- conferencing as a communications tool

● improve the communication effectiveness of each of the tele- conference users

● help to interface users with the unfamiliar electronic communica- tion medium, increasing their com- fort

● provide personal guidance and training for success in conducting more productive meetings

● increase users' knowledge of the concept of teleconferencing as well as the capabilities and constraints of their particular system

● develop telecompetence

● develop in users, effective pre- paration, operational, presentation and interactive skills

● facilitate an organization's imple- mentation process

● enhance users personal communi- cation skills

COACHING: A MULTI-ROLE PROCESS

In the process of developing effective communication skills in teleconference users, the coach has many roles. He or she acts as a catalyst, change facilitator, monitor and helper to reinforce corrected verbal and non-verbal behaviors. In addition, the coach has the responsibility of meeting the users realistic expectations and needs. A coach detects ineffective communication mannerisms, skillfully and positively confronts the user, and demonstrates the effective communication skill to be developed.

REQUIREMENTS OF AN EFFECTIVE COACH

In addition to possessing knowledge and skill in conducting and participating in electronic meetings, the coach must understand the culture, requirements and objectives of the organization and the users. The coach must be "people sensitive" and skill focused. He or she must possess a surplus of human qualities of persistence and persuasiveness. This means acquiring a knowledge of how adults learn including a basic understanding of behavioral psychology and communication sciences. Coaches must understand the process of change and be skilled in facilitating change. It is also suggested that a coach be an experienced teleconference user with the ability to provide a model for effective communications. Needless to say, information alone is insufficient.

MULTI-ROLES

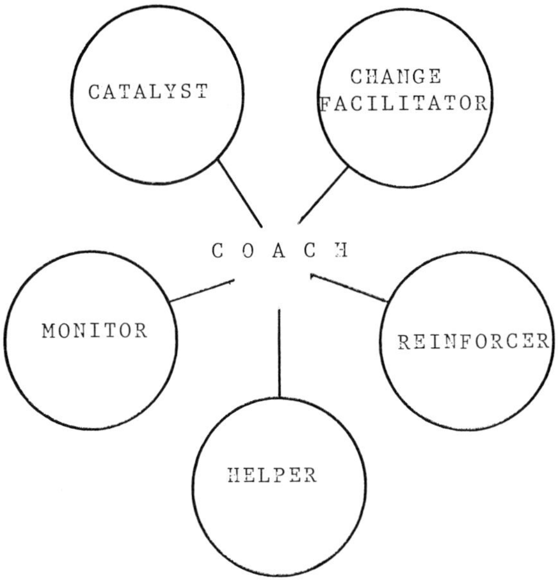

FIGURE 2

COACHING PARTICIPANTS

There are several reasons for a company to request communications coaching but most fall into one of two categories: (1) for purposes of intervention, and (2) as a preventive measure.

INTERVENTION

Intervention requests for coaching usually involve current users who are experimenting with teleconferencing and are frustrated in their attempts to meet expectations. The following example will help to illustrate:

Recently a group of managers at a medium size financial company expressed disappointment and confusion in their efforts to use a new audio teleconferencing system. Expectations were high that the system would save travel and waiting time in dealing with immediate issues and would facilitate a more rapid decision making process. Such expectations were replaced with frustration because of a lack of skill and knowledge in preparation, protocol and meeting requirements to use the new medium effectively. They sought consultation and coaching was recommended as a means of dealing with the problem. This request was clearly user reactive in nature.

PREVENTION

Prevention requests include potential users who are involved in the planning and implementation process and seek support on how to best utilize their system. This pro-active approach in preparing users is most recommended. It is during the implementation phase that user concerns and needs can be clearly identified and flexible coaching programs designed.

Uses of coaching for problem prevention and system enhancement include:

- corporate trainers who seek consultation and training services to prepare their participants for teleconferencing and help minimize resistance

- teleconferencing system hardware vendors who include, as part of their service contract, supportive services like intensive training or coaching. These services are of course in addition to their conventional operational skills training and demonstration of system capabilities

- first time teleconference users of ad hoc one way video teleconferencing. This very dynamic medium is most unfamiliar to users and requires communication skills such as relating to camera and participating in audio interaction with multi receive sites.

- universities which have implemented teleconferencing systems have a need to train instructors on how to adapt their materials and conduct classes using an electronic medium

- hotel chains involved in installing dedicated video systems have requested coaching of a core support staff to assist new users of electronic business meetings.

OPTIONS TO COACHING

Leaving human factors to chance is highly risky. But one may ask, is coaching the only solution to human factor concerns? Though coaching may be more ideal, scheduling conflicts, time constraints, unavailability of teleconferencing hardware may create barriers to a well designed coaching program. Given these circumstances, other options must be explored. They are:

- training departments providing user manuals on how to operate the system and conduct successful teleconferences (self-instruction)

- voluntary participation in user teleconference skill programs via teleconferencing (from the convenience of their office)

- brief, on-site orientation seminars

- formalized training by corporate training departments or outside consulting firms (knowledge based focus)

- vendor briefings on operational skills only

- experiential learning in actual live teleconferences

- by osmosis - gaining skills by observing peers (a haphazard hit-miss approach).

CHOOSING A COACHING PROGRAM

When feasible, a well designed coaching program provides users with needed support and encouragement and serves as an opportunity to enhance ones personal communication skills.

STEPS IN PLANNING

If one decides to assume the role of coach in an organization, the importance of planning the coaching session and targeting participants cannot be overemphasized. Planning includes:

- interviewing participants to identify their needs and concerns

- assessing the level of receptivity to coaching

- identifying users previous exposure to teleconferencing

- familiarity with the teleconferencing system to be used (it's dangerous to assume all systems are alike)

- designing the coaching session clearly, addressing the needs and concerns of users

- developing flexibility in scheduling

- determining ways the coaching program best fits within the organization.

Above all:

- Clarifying and defining the relationship, roles and responsibilities of the coach and the participants. Underscore the mutual goal involved - developing skills for improved effectiveness in teleconferencing.

INFORMATION/SKILL AREAS

Within the perspective of a desire to learn and make a positive attitude to changing existing communication approaches, users need basic knowledge and information to begin the skill development process. The knowledge base should include: (1) understanding the concept of teleconferencing; (2) characteristics of electronic meetings; (3) awareness of teleconferencing options; (4) communications flow and user requirements; (5) system capabilities and constraints; (6) appropriate applications for each teleconference system; (7) participants roles in the teleconference.

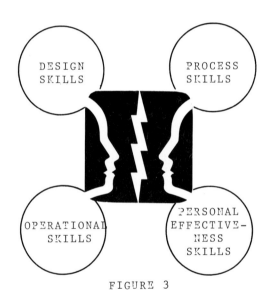

FIGURE 3

SKILLS

Skill requirements for users at the orignation site and receive sites will vary with the teleconferencing system selected. They generally fall within the following categories: (1) teleconference design skills - planning and preparation; (2) teleconference operational skills; (3) teleconference process skills - meeting management; (4) teleconference personal effectiveness skills.

The remainder of this article addresses key factors to consider in developing teleconference skills through coaching. Skill requirements common to all teleconference systems are described along with specific factors particular to each of the teleconference options.

TELECONFERENCE DESIGN SKILLS: PLANNING AND PREPARATION

Good meetings of any kind are those which successfully reach planned outcomes in the shortest period of time. Therefore, good meetings presuppose a plan and assume participant preparation. Most of us are far too familiar with the "let's play it by ear", or the "get it over with quick" approach to meetings. In most cases the result is wasted time, inefficiency and frustration in not concluding with a productive result. In face-to-face meetings, it is easy to shift focus away from business issues and on to unrelated topics. However, the electronic meeting is not conducive to such peripheral discussion and as such places planning and preparation as a prerequisite.

The electronic meeting requires a leader or meeting organizer who is capable of designing the teleconference meeting such that the meeting style and approach is appropriate and compatible with the teleconference medium being used. Meeting goals and objectives should be written in concise terms in order that outcome expectations are clear to participants. Time constraints and the reality of distance (participants interacting at different locations) require greater attention to planning the meeting content. The planner must consider how the "audience" will view the message being presented, the effect of message style and required adaptation to accommodate the teleconference medium, and the nature of the information to be presented, exchanged, and discussed. Support materials (audio visuals) must not only be prepared in advance but must be prepared in such a way as to achieve the greatest impact through the medium being utilized.

The actual format of the meeting needs to be considered in relation to the objective of maintaining the attention of participants, reinforcing retention of material and enhancing interaction. The roles and responsibilities of participants on site and at remote sites should be thought of

in advance. In addition, workable
time frames to deal with the issue at
hand need to be determined and serve
as the basis for planning the balance
between presentation and interaction
during the meeting. The agenda or
"content outline" is a product of the
planning and preparation process.

TELECONFERENCE OPERATIONAL SKILLS

This article focuses on the human
factors and skills required to increase
competence and confidence in teleconferencing. It assumes that users
have been oriented to their teleconferencing system's capabilities and
constraints and the mechanics of using
the various technical elements of
their system.

Coaching, to enhance operational
skills, focuses on the ability to integrate meeting content with available
hardware. For example, it is assumed
that the teleconference user is familiar with the equipment and mechanics
to provide close-ups, focus camera on
individuals in the teleconference room,
use facsimile and various graphics
equipment and come on line through the
audio teleconference bridge. It is the
responsibility of the "coach" to de-
emphasize the equipment and to focus
on the message to be conveyed, the use
of technical equipment to convey that
message and the timing of presented
materials in a constructive coordinated
manner.

TELECONFERENCE PROCESS SKILLS: MEETING MANAGEMENT

In the face-to-face meeting,
participants have the advantage (or
disadvantage) of conveying both verbal
and non-verbal communication and can
more easily continue the flow of interaction without leadership and
direction. Such meetings can be mis-
managed with the moderator or chairperson taking a passive position.
Electronic meetings require both good
leadership and moderator skills. A
meeting protocol needs to be established in order that issues which
prompted the meeting can be dealt with
within the time constraints of the
meetings. It is important that the
meeting protocol be conducive to

"personalizing the meeting" such that
the "tele" is de-emphasized and
"conferencing" is emphasized.

Leadership skills in managing
the teleconference require:

- Self-discipline in keeping the
  meeting focused;

- The ability to manage time;

- Skill in controlling the pace of
  meetings (i.e., new information to
  be presented at a slower rate than
  information which is being reviewed,
  changing pace for purposes of emphasis, altering pace in order to
  maintain interest, etc.);

- Ability to add variety to the teleconference meeting;

- Segmenting information into short,
  concise and yet interrelated parts;

- The ability to time visual materials
  with verbal interaction;

- Maintaining eye contact.

The skilled teleconference
moderator combines his/her planning,
leadership and presentation ability
with interpersonal skills which both
motivate and enhance interaction.
Active listening is critical to the
teleconferencing process as is the
ability to ask key questions or bring
forward key points within the discussion. Managing information flow
must be handled in such a manner as to
elicit response from remote site
participants, maintain and reinforce
interaction, and encourage/motivate
each participant to contribute to the
fullest extent possible.

Thus, the leader in a teleconferencing situation plays a key role
in assuring that desirable outcomes are
reached. Coaching individuals who
assume leadership roles will bring forward the communication skills necessary
for optimal success.

TELECONFERENCE PERSONAL EFFECTIVENESS SKILLS

As mentioned above, the teleconference moderator is a key member of
the participant group. However, he/she

cannot carry full responsibility for
the meeting outcome. Such positive
outcomes can only be the result of the
teleconferencing skills shared by the
larger user group.

Coaching teleconference users includes attention to the skills of
verbal persuasion and presentation
(i.e., controlling the rate, pitch,
volume and inflection of verbal
messages), promoting the ability to
speak concisely and clearly, encouraging enthusiastic interaction,
and minimizing the use of "non-words"
such as ahs, ums, etc. Beyond skills
in verbal communication, teleconference
users can be coached to appear alert,
responsive and in control of the
communication process by developing a
positive visual communication presence.
The use of body language as it is seen
and perceived in the video teleconferencing medium, eye contact with
viewers and symbolic communications
through appearance and gestures will
enhance or destroy both the message
and the image which is attempting to
be communicated. Distracting
mannerisms are not only exposed but are
enhanced through the video teleconferencing medium. Therefore, it is
advantageous to the presenter to recognize that such mannerisms exist in
order that they can be avoided.

SUMMARY

Skills required in planning and
preparation for teleconferencing,
operational consideration, meeting
management by the teleconference
moderator, and personal effectiveness
of teleconference participants have
been reviewed as they are generic to
all teleconference technologies.

Specific considerations for skills
development particular to each teleconference option (audio, audio plus
graphics, full motion video, and ad hoc
teleconferencing) can be summarized as
follows:

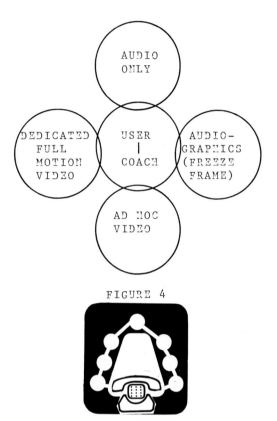

FIGURE 4

AUDIO ONLY TELECONFERENCING

This least expensive, most
accessible teleconference option is
most familiar to users, yet often mis-
understood. Many still perceive audio
teleconferencing as a simple extension
of a telephone call. It is not. The
concept of audio teleconferencing involves interactive communications among
many persons at many different and distant locations.

The critical characteristic of
audio teleconferencing is the fact
that participants are both speaking
and listening to an unseen audience.
As such, messages need to be constructed clearly and concisely and
focus needs to be on verbal description
to a higher extent than other forms of
teleconferencing. Maintaining interest
through controlled pacing of communication coupled with specific attention
to "personalizing" the communication
process are key factors to successfully using the audio teleconference
option. Self-discipline relative to
the frequency and duration of participation, variety in tone, volume,
inflection and avoidance of non-words

or excessive pauses are skills which
foster a positive teleconference experience.

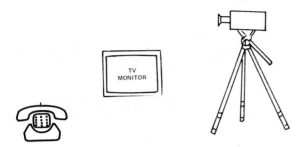

AUDIO PLUS GRAPHICS

Audio plus graphics teleconferencing adds a visual element to
real time audio only conferencing
and, depending on objectives, can result in more effective meetings.
Examples of audio plus graphics components include: electronic blackboards, electronic pens and writing
surfaces, facsimile devices, (slow
scan) freeze frame video, video
cameras, transmitters, receivers and
viewing monitors.

Skill requirements relative to
audio teleconferencing are applicable
to audio plus graphics. In addition,
the following must be considered:

● Determination of the optimal
  balance between reliance on audio
  and graphic means for presenting
  the desired message;

● The selection and preparation of
  appropriate graphics;

● Recognition of the natural tendency of participants to focus
  more on visual than verbal communications;

● Timing what is being seen with what
  is being said in real time (most
  critical in use of freeze frame
  equipment);

● The requirement of presenters of
  graphic information to "think
  graphically" when transmitting information with emphasis on what
  the participant is viewing at the
  remote site. (Example: Using
  electronic graphics requires
  attention to legibility, size and
  spacing as it is not received at

the viewing site exactly the way it is
written by the presenter.)

A simple rule of thumb in utilizing audio plus graphics technology is
to "tell users what they will be shown,
and tell them while they are being
shown, and summarize key points of what
has been told and shown to them".
Again, all skills applicable to audio
conferencing are necessary to enhance
the audio plus graphics experience.

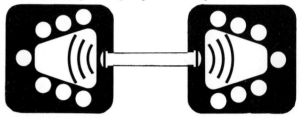

FULL MOTION DEDICATED VIDEO
TELECONFERENCING

Full motion video teleconferencing
simulates the face-to-face meeting
more closely than other forms of teleconferencing. Attempts are made to
present a conference room environment
typical of the culture most familiar to
the teleconference user. Since conscious effort is given to user needs
and environmental factors in the construction of rooms and both the placement and use of equipment, controversy
regarding the need for coaching (other
than in the use of hardware) has
emerged. From conversations with many
users comments range from "no
training is needed...Just walk in and
conduct your meeting" to "some
training would be helpful to best
learn how to integrate conventional
meeting behavior with this electronic
medium".

Video teleconferencing is the
least accessible and most expensive way
to teleconference. Transmission time
is costly while high utilization rates
are required to spread fixed costs over
a greater volume of participation.
Thus, planning the teleconference
meeting such that the pace is rapid,
time is well spent, and the combination
of verbal interaction, graphics and
video imagery utilize the capacity of
the medium is highly recommended in
order to meet cost effectiveness goals.

Most of us are not accustomed to
relating interactively with the 25"
color television monitor. We are

instead conditioned to respond in a more passive recreational way. Video teleconferencing is "face-to-face television" where one's actions and inactions may be continually viewed by remote site participants. Symbolically, by dress and body language, participants are always communicating. Facial expressions, movements and the like enhance communication or distract depending on the circumstances.

Coaching is a recommended approach to help new users communicate both efficiently and effectively and to help experienced users to improve their communication skills when using video full motion systems.

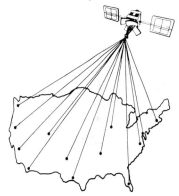

AD HOC POINT TO MULTI-POINT
TELECONFERENCING

Ad hoc teleconferencing can be divided into two categories. The first category represents the use of ad hoc teleconferencing as a means of interaction. The second use of ad hoc teleconferencing is for the production of events such as the introduction of products, special annual meetings, etc. In the first case, leadership, participation, and interaction skills described above are applicable to the ad hoc teleconferencing situation. In the second case, experts are required for the successful application of the medium for the desired production. In such cases, expert script writers, hardware vendors, producers, directors, and communication coaches must work together for the success for the success of the "special event".

This article primarily focused on skills required for teleconferencing as a means of interaction through electronic meetings. Therefore, other than to point out the fact that coaching is an important and integral

part of ad hoc teleconference, special skill requirements particular to the ad hoc teleconferencing environment and purpose will not be discussed.

THE TELE-COMFORT™ FACTOR . . .

THE TELE-COMFORT™ FACTOR

The Tele-Comfort™ Factor is the user's state of comfort, confidence and competence using the teleconference medium. Resource Management Consultant's one to two day skills development program is customized to the user's teleconference option and is designed to address their needs and concerns. By participating in the comprehensive training program, participants:

- gain a better understanding of the concept of teleconferencing and various applications of their system option

- gain knowledge of their system's capabilities and constraints

- are instructed on how to plan and prepare for electronic meetings

- are monitored while conducting simulated meetings to provide valuable skill practice

- are coached in effective communication skills using their teleconference option including presentation, interaction and, where applicable, operational skills

SUMMARY

Effectively and efficiently using teleconferencing systems as a means of interactive communication depends upon user preparation and skill. The equipment is merely the facilitator of communication's transmission while people are both the beneficiaries and the critical variables for success. Greater trust, motivation, excitement and commitment to utilizing the teleconferencing approach can be expected

from confident and competent users.
Underachievement of desired outcomes,
underutilization and disappointing
cost impact projections are often the
result of lack of human resource
planning.

Coaching can improve the level
of communications, and communications
effectiveness of each user and conference team. Participants can
benefit from skills in designing a
variety of meeting formats, can
develop an array of interactive
skills, can learn how to adapt and
integrate visual and audio material
to the electronic meeting environment. Through guided practice and
simulation exercises, meeting effectiveness can be accomplished.

Trial and error is not an alternative to a well designed coaching
program in developing user skills and
encouraging system implementation in
the business environment. Coaching
is a response to the user need for a
sense of comfort, competence and well
being in the teleconferencing environment.

REFERENCES

Boudle, Jim and McCarthy, John
Preparation - a Teleconference
Requirement, Teleconference Magazine
November-December, Volume 2, Number
2, 1982.

Gibbs, Peter The Facilitator Trainer
Training and Development Journal,
July, 1982.

Johansen, Robert, Vallee, Jacque,
and Spangler, Kathleen Electronic
Meetings: Technical Alternatives and
Social Choices, Addison-Wesley
Publishing Company, 1979.

Monson, Mavis Bridging the Distance -
An Instructional Communications
Systems, University of Wisconsin-
Extension, 1978.

Monson, Mavis A Telewriting System
for Teaching and Training - Some
Principles and Applications,
Teleconference and Interactive Media,
1981, University of Wisconsin.

Pereyra, Susan G. Conducting Business
Meetings by Teleconference,
Teleconferencing and Interactive
Media, University of Wisconsin-
Extension, 1981.

The Teleconferencing Resource Book: A Guide to Applications and Planning
Lorne A. Parker and Christine H. Olgren (eds.)
Elsevier Science Publishers B.V. (North-Holland)
© Center for Interactive Programs, University of Wisconsin-Extension, 1984

CONDUCTING BUSINESS MEETINGS BY TELECONFERENCE

Susan G. Pereyra
Director
The Darome Connection
Danbury, Connecticut

Now that escalating costs are rendering business meetings that require travel obsolete and inefficient, the teleconference has become the better, faster and more flexible method of business communication. There is an added dividend to teleconferencing: the quality of information exchange on a telephone conference is often better than in a face to face meeting because participants on a conference feel less exposed and tend to be more candid and open with their comments and opinions. Therefore, teleconferencing is not just a substitute for in-person meetings; it is, in many cases, a new and better way to conduct business.

Despite the advantages of teleconferencing, the fact remains that it is an entirely new experience for most of us. And, as with anything new, teleconferencing requires extra effort and time before we can gain familiarity and comfort with this new medium.

The purpose of this paper is to help make the transition easier for those who are planning to conduct meetings by teleconference for the first time.

As Director of the Darome Connection, the telephone conferencing service provided by Darome, Inc., I have conducted countless meetings by teleconference. In the process, I have developed a method of my own that has worked well for me and has proved simple and flexible. In fact, I never moderate a teleconference --no matter how hastily put together -- in which I do not go through the steps outlined here.

My method is, by no means, the only way to organize and conduct a meeting by teleconference. However, it does provide for what I think is the absolute rule for the most effective use of audio teleconferencing; that is, to hold attention for more than five minutes at a time, it is necessary to vary presentation style, to have a change of speakers, or best of all, to have active participation and interaction on the part of the group.

The moderator cannot rely on visual clues such as eye contact for assurance that he has been understood. Instead, he periodically polls the group for comment and response. Likewise, participants on a conference have no visual diversions to hold their attention. They will remain most attentive if they know they'll be asked to participate or contribute to the teleconference in some way.

These factors are built into my method for conducting a teleconference meeting. My objective is to provide a versatile and workable model which anyone can adapt for his own purposes while remaining confident that he is using the medium effectively.

My techniques are geared for the audio teleconference that has no visual supplements such as slides, video tapes or full motion video for support. However, there may be a written agenda or reference material in written form. The audio teleconference is the most flexible form of conferencing and requires the least amount of advanced preparation. It has the spontaneity and informality

of in-person meetings called quickly
for important information exchange.

The telephone is the only equipment required for this form of teleconference. Arrangements need to be
made with the Darome Connection to
connect all of the callers together.
The telephone company also provides
some conferencing capability with its
conference call. Occasionally, some
calling locations may have groups
participate by means of a teleconference
device attached to the telephone line.
These include the Darome Convener,
the Bell speaker phone or 50-A, or the
Northern Telecom Conference 2000 unit.

Since my experience has largely
been with teleconferences conducted
through the Darome Connection, my
methods are particularly suited to
the interactive capability of the voice-
activated Darome system. However, the
steps outlined here work well even for
face to face meetings.

I have found the secret to conducting an effective business meeting
by teleconference is in the preparation.
If I think through my objective carefully
and analyze the path of communication
flow and then select the structure and
format for each portion of the meeting,
there is very little to worry about
while I am conducting the meeting. Once
this process is reduced to a system, it
can be done in a few minutes with some
hasty notes when necessary.

Therefore, the secret to conducting
a good meeting is good preparation.
And the key to good preparation is
developing an On-Line Agenda for the
meeting. All of the following steps
for conducting a teleconference are
concerned with preparation - except
the last. Most time is devoted to
helping you prepare an On-Line Agenda.

The steps are listed below for
easy reference, and the remainder of
this paper discusses each one separately.

Reserve Conference Time

Notify Participants

What Is Your Purpose or Subject?

What Is Your Objective?

Develop an Agenda

Construct an On-Line Agenda

- Roll Call
- Welcome & Introductions
  (Announcements)
- State Purpose & Objective
- Review Agenda
- Item #1   (Select Interactive
- Item #2   or Broadcast
- Item #3   modules)
- Summary & Close of Meeting

Conducting the Meeting

RESERVE CONFERENCE TIME FOR YOUR MEETING

Make your telephone conference reservations with the Darome Connection by
calling toll-free 800-243-0991 with
the following information:

. Day and Date for the Meeting
. Time, Eastern, for the Meeting
. Estimated Duration of the Conference
. Number and Names of Participants

The Darome Connection will confirm your
conference reservation by assigning a
teleconference phone number for your
meeting. Your participants should be
given this special teleconference phone
number when they are notified of the
meeting. At the designated time, each
participant calls into the conference
from any telephone.

NOTIFY PARTICIPANTS

The following information should be
given to each participant when you
notify them of the telephone conference:

. Name of the Moderator
   (Person Conducting the Meeting)
. Day
. Date
. Time, Eastern, or in the
   Participant's Time Zone
. Teleconference Phone Number
   Issued by the Darome Connection

Optional:

. The Subject of the Conference
. The Agenda
. Preparation Required of Participant
. List of Participants
. Advance Material

Participants can be notified by telephone, by mail, mailgram or Telex.

If participants are unfamiliar with
the procedure, give complete instructions.

WHAT IS YOUR PURPOSE OR SUBJECT?

What is your reason for calling this
meeting?

It is important, for yourself particularly, to state the purpose of the
meeting simply and concisely.

Carefully expressing your purpose in
one sentence is the first step toward
structuring an effective teleconference.

For example:

"The purpose of this conference is
to discuss the entry of a competing
product ATAR into the marketplace."

If participants are given the purpose
of the meeting when they are notified,
their thinking will be more focused
for the conference.

WHAT IS YOUR OBJECTIVE?

What do you want to accomplish with this meeting.... What is the end result you want to achieve?

The objective differs from the purpose of the meeting in that it is a precise expression of what you expect the meeting to produce.

For example:

> "The objective of this meeting is to A) assess the potential threat this new product ATAR is to our present share of the market, and B) to determine what immediate action should be take to counteract this threat."

How well you define the purpose and objective for the conference will determine how effectively you structure and conduct your meeting. These two items are essential to the most productive use of conference time.

DEVELOP AN AGENDA

To accomplish your objective, what items need to be covered in your meeting, and in what order? Develop your agenda.

For example:

> " 1. Background and present knowledge of competing product ATAR - assess the threat
>
> 2. Suggested courses of action
>
> 3. Consensus on action to be taken immediately "

For an audio conference, with people on telephones and not a conference device, the maximum productive time for a meeting is between forty minutes and one hour. For this reason, it is important to limit your agenda to two or three items.

The danger in planning a teleconference is to attempt to accomplish too much in the time available. That is why the next step in preparation is the On-Line Agenda which outlines the structure of the meeting and allots realistic time segments to each agenda item.

CONSTRUCT AN ON-LINE AGENDA

The next step is to take your purpose, objective and agenda and to construct an On-Line Agenda that will be your blueprint for your teleconference meeting.

In constructing your On-Line Agenda, you will select an appropriate format or presentation style called an "agenda module" for each item you plan to cover in the meeting. Some presentation styles and meeting formats are better suited to the interactive nature of the teleconference medium than others.

A few of these have been organized into agenda modules of approximately fifteen minutes each. They are presented here for you to use in constructing an On-Line Agenda for your own meeting. By using agenda modules you can be assured of conducting your meeting in an effective teleconference format.

The On-Line Agenda will quickly become an indispensable tool for planning and running a business meeting by telephone conference.

Following is a standard outline for an On-Line Agenda that works well for almost any type of business meeting by teleconference. It can be adapted easily to fit any objective, and agenda modules can be inserted at the appropriate places.

On-Line Agenda

Roll Call
(Usually performed by Darome Connection Operator)

:00   Welcome & Introductions
(Moderator)

:01   State Purpose & Objective

:03   Review Agenda
(Explain how you plan
to proceed with the
meeting)

:05   Item #1 on Agenda
(Follow agenda module
selected)

:20   Item #2 on Agenda
(Follow agenda module
selected)

:35   Item #3 on Agenda
(Follow agenda module
selected)

:50   Summary & Close of Meeting

:55   Conference Concluded

This standard format for the On-Line Agenda is classic in its simplicity and flexibility. There isn't any meeting or program - no matter how informal or sophisticated - that cannot be accommodated by this agenda.

To use this agenda well, however, requires an understanding of the principles of good teleconferencing techniques that lie behind each component. Following is an explanation of each section of the agenda, including a discussion of appropriate agenda modules to be used:

Roll Call

The roll call may appear to be one of the more inconsequential sections of the agenda. However, it is really one of the most important in setting the meeting off to a proper and comfortable start.

On a Darome Connection conference, the Darome operator performs this essential function by addressing the moderator first and then going "around the conference" calling each person's name and asking him to respond with the name of the city and state from which he is calling.

Remember, on an audio teleconference, the only way you really know who is "in the room" is by the sound of voices. Even with a participant list in front of you, you will feel "blind" if you haven't heard the voices of all participants during the roll call.

And--the participant who may not have been acknowledged during the roll call has the disquieting feeling that he does not exist in the awareness of the group. Indeed, he does not because no one has heard his voice. If he does decide to speak during the meeting, he gives the group a jolt as if he were an intruder or an unknown eavesdropper.

Of course, there are times when a conference is so large that a roll call is out of the question. When a conference consists of a group of people at each calling location, the roll call may require a group leader at each site to respond instead of each individual.

A group has to become a group before it can be productive or creative. The roll call has this as its function. It takes the place of the handshake in a face to face meeting.

Welcome & Introductions

When a roll call is impossible, the welcome and introduction section serves as the main "humanizing" element of the conference.

After the roll call, you, as the moderator, should introduce yourself to the group if you are not already known. This is the time when you should welcome the group to the meeting and introduce any new members, guests or guest speakers - even if they've been addressed during the roll call. If you are using a Convener or other conferencing device and have others in the room with you, introduce them at this time.

When introducing anyone on line, do it in such a way that he has to respond by saying hello or where he is calling from. It is important that the other participants hear his voice.

For example:

"This is John Scott. As national sales manager for AJAX Company, I know most of you who are on line with us today.

I want to welcome all of you and I'm glad you were able to call into this important meeting.

Before we start, I want to mention that I have Tom Smith, our director of marketing, in the room with me.

Tom, it's nice to have you with us. How are you today? "

(Tom Smith replies)

"Before we begin the meeting, I have a few announcements...."

State Purpose & Objective

This section is simply a re-statement of the purpose and objective you defined earlier in your preparation for the meeting.

For example:

"Now to begin our meeting.... As you know, the purpose of this meeting is to discuss the entry of a competing product called ATAR into the marketplace.

I hope that we'll be able to accomplish two things with this meeting today.

They are as follows:

A) To assess the potential threat this new product poses to our share of the market

B) To determine an appropriate course of immediate action to counteract the threat"

Review Agenda

At this time refer to the agenda, if it was sent out with the notification and explain how you plan to proceed.

If your participants do not have the agenda in front of them, then you should list the items on the agenda in the order you will cover them.

Remember, on an audio telephone conference a certain amount of repetition is necessary for the group to be able to follow along and to perceive the direction of the meeting.

As an example:

"First we're going to assess the threat that ATAR poses by having a report from our director of marketing, Tom. And then we'll go to the group for opinions from the field.

Second, we'll discuss various alternatives or courses of action that we could take.

And finally, we'll decide on the best course of action to take immediately."

Agenda Items &"Agenda Modules"

In preparing an On-Line Agenda for a teleconference - and in conducting a meeting by teleconference - there is only one absolute rule that you should follow whenever possible:

In order to hold your listeners' attention for more than 3-5 minutes, you must have -

1. Changing speakers (voices)
2. Interaction or dialogue
3. Involvement of group members

Therefore, even your fifteen minute agenda modules need to be organized into smaller five-minute segments of time interspersed with a change of voices, interaction or involvement of some kind.

It is this basic fact - that you cannot hold someone's attention for more than 3-5 minutes at a time-that forces the telephone conference into an interactive mode. Because it has to be interactive to be effective, the teleconference done correctly is better than a meeting face to face.

To help you determine the best and most effective presentation style to use for the items on your agenda, certain models called "agenda modules" are presented here for you to use.

Each agenda module is structured to capitalize on the interactive nature of the teleconference medium and it is designed to hold the attention of participants and to involve them in meaningful ways. Since it is a good idea to break a teleconference meeting into fifteen minute segments, each

agenda module is approximately ten to fifteen minutes long. It is a good idea to assign one agenda module to each item on your agenda.

To select an appropriate agenda module, you should first analyze the direction of communication flow. In any meeting there are two paths of communication flow. The first is information out from the moderator or principal speakers to the group. The second is information in from the group to the moderator.

Any information that goes out from the moderator or speaker to the group is basically presentational in nature. For our purposes, we will call it a "broadcast" mode. From what we already know about teleconferencing, we can tell that an agenda item that necessitates "information out" or "broadcast" mode is in danger of losing the group's attention if it continues for more than five minutes without interruption.

However, there are ways to structure such material effectively and inter-actively. Agenda modules entitled "Broadcast Modules" will present some models for effective teleconference delivery of "information out" items.

There is another very important fact about teleconferencing to remember. Whenever possible, lecture material and any large amounts of information should be "packaged" in some form and sent out in advance. Videotapes, slides, or even written form is the best way to present such material. A telephone conference is an inter-active medium and should be used as such. Use the teleconference to clarify, reinforce and elucidate information sent out in advance. Obviously, when there is a critical shortage of time and information must be transmitted to everyone simul-taneously, the teleconference is the only way it can be done.

The second path of communication

flow in which information comes in from
the group to the moderator is the
"interactive" mode and is absolutely
compatible with interactive requirements
of the teleconference.

Agenda modules that suggest models
for structuring "information in" type
items are presented in "Interactive
Modules."

Broadcast
Agenda Modules

The broadcast agenda module is
designed for those items on the agenda
that are presentational in nature and
require one-way communication from
the moderator out to the group. This
path of communication flow can be
incompatible with the teleconference
medium if it is not restrained to
five minute segments and interspersed
with some form of interaction.

Following are some effective
broadcast agenda modules:

Interview .

5-10 minutes: Interview the
speaker or presenter with questions
designed to elicit the presentational
material.

5-10 minutes: Poll individuals in
group for questions or comments

Reporting.

3-5 minutes each: Ask a few
participants or speakers to prepare
a brief report on a portion of the
subject at hand. Have each, in turn,
report. Pause between each report
to comment or repeat significant
points made.

Or, more simply, parcel out portions
of the meeting to other people. This
will make conducting the meeting
easier on yourself while making it
more interesting for your listeners.

Straight Presentation.

5 minutes (10 minutes with slides or
some visual support): Deliver your
presentational material in 5 minute
segments interspersed with interaction
in any of the following forms:

10 minutes or more: Poll individuals
for comments, or better still, ask
them specific questions. (Do not
say "Are there any questions?" you
will rarely get good responses.)

10 minutes: Case study or Problem:
Focus on a case study, problem or
exercise to reinforce the presentation.

Panel Discussion.

10-15 minutes: Select a panel of
2-4 people. Pose a question or
problem and ask each to comment in
no more than 2-4 minute segments.

As moderator, interject your comments
or observations at appropriate
junctures. This works best if there
is some controversy, ambiguity or
debate.

Obviously, these agenda modules and
the requirement of five minute segments
are ideal. A moderator can often get
away with talking for as much as ten
minutes at a time. Also, constraints
and urgent situations may require that
large amounts of information be "broadcast" to many sites or individuals at
once. When necessity forces you to
stretch these rules, just be aware that
you are requiring more effort and concentration from your group and that
you may not be able to hold their
attention as effectively.

Interactive
Agenda   Modules

The interactive agenda module is designed to help you structure and organize those portions of your meeting that require getting information in from the group. This communication flow is, by nature, interactive, which makes it compatible with the teleconference medium. Therefore, the five to ten minute time segments found in the broadcast agenda modules are not so important here. In fact, if an entire meeting requires feedback and information in from the group, one of the interactive agenda modules could be used for the whole meeting.

Following are some suggested interactive agenda modules:

Roundtable Discussion.

Pose a question or ask for comment and poll each participant for his response by going "around the conference." Budget 1½ - 2 minutes for each response.

Or, if time is limited, poll four individuals for their opinions and ask the rest if anyone has something to add.

Brainstorming.

Phrase a problem or question. Set a time limit of 60-90 seconds and ask each participant to jot down as many ideas or thoughts on the question at hand as he can. The key is to write down as many as possible spontaneously while reserving judgment on each item.

Then poll each person for one of the ideas on his list that has not been mentioned by someone else. Have each participant list these items too.

Summarize and conduct a roundtable discussion on the best ideas.

Case Study or Problem Exercise.

Present a case study or sample problem in advance or during the conference. Allow a few minutes of silence on line for each person to formulate his answers on one or two questions you have asked regarding the problem.

Poll each participant for his answers or solutions or opinions. After you have obtained all of the responses, then summarize and comment on them.

You may wish to divide the conference into two groups by giving them either two different problems or the same problem with different questions.

It should be noted that it is possible to have the Darome Connection break a conference into subgroups for small-group discussion.

This module is especially effective when it provides the interactive counterpart for material that has been presented. It helps to reinforce understanding of the information presented.

Q & A.

Question and answer periods are effective when they are conducted in the manner of a roundtable discussion.

It is not advisable to say, "Are there any questions?" for two reasons: you rarely get good - or very many responses; and if you do get response, you may hear the garble of many voices at once or clipping as one voice overrides the others.

Therefore, it is best to address specific individuals for question or comment. You will get much more interaction with this method since each has to respond in some manner.

When you do have to ask a general question of the group, do so by exception. For example, "Who has <u>not</u> received the new sales kit ? Asking questions by exception will elicit the fewest simultaneous responses and will minimize the confusion of overlapping or clipping voices.

One interactive agenda module can be used for an entire meeting although changing modules every fifteen minutes will add variety and sustain interest.

A simple way to organize a good teleconference meeting is to alternate broadcast mode information with inter-active elements or interactive agenda modules. It should also be noted that involvement does not always have to be verbal. If your listeners are working on a problem or jotting down thoughts as in the brainstorming module, they are involved.

Now that the agenda modules have been explained, we can resume con-structing the sample On-Line Agenda for the teleconference meeting on the competing product ATAR. The task in-volves analyzing each agenda item to determine if the communication flow is out or in and then selecting an agenda module to use for that portion of the meeting.

Agenda Item #1 To assess the threat
    of competing product ATAR
            (Total 15 minutes)
    Background information and present
    knowledge regarding the product:

                (Information Out)

        <u>Reporting Module</u> 5 minutes
        Report from headquarters by
        Tom Smith,director of Mktg.

        Input from the Field
            (Information In)

        <u>Roundtable Module</u> 10 minutes
        Poll each participant about
        effects on sales to date

Agenda Item # 2 Suggested Courses
    of Action
            (Total 15 minutes)
    (Information Out)

        <u>Panel Discussion Module</u>
            10 minutes
        Ask three people before the
        meeting to each suggest one
        alternative and to take 3
        minutes to present and explain
        it on the conference.
            5 minutes
        Ask if anyone in group has
        something to add or suggest.

Agenda Item # 3 Consensus on
    Immediate Action To Be Taken
            (Total 15 minutes)

        (Information In)

        <u>Roundtable Module</u>
            15 minutes
        Poll each participant for
        opinion on which of the three
        suggested alternatives he
        advocates and why.

<u>Summary & Close</u>

As with all other portions of the On-Line Agenda, the "Summary and Close" is also important because it enables everyone to leave the meeting with a clear sense of what transpired and what has been accomplished.

You should review the important points covered in the meeting and summarize the conclusions reached or the decisions made. If the conclusions reached correspond closely to your objectives for the meeting, you have been successful.

Before closing, mention any next steps to be taken or the next meeting date.

Perhaps most important, thank everyone for their contributions.

As an example:

"Before closing, let me summarize -
Most of us seem to agree that the
greatest threat of ATAR to our share
of the market will occur in the
beginning when our competitor is
promoting ATAR most heavily.

Of the alternatives discussed, we
agree that the best action to take
immediately is to make a special
offer on our product to defuse the
effect of the promotional campaign
for ATAR.

We will have another conference later
this week once the details of the
special offer have been worked out.

Does anyone have any final comments
before we close the meeting? ...
Fine....

I want to thank all of you for
your valuable contributions to this
meeting. I'll be talking with
you again later this week. Have
a good day...."

This completes the steps in preparation for a meeting by teleconference.
It is obvious by now that the most
important element in conducting an
effective meeting by teleconference is
the preparation. Once the On-Line
Agenda has been developed, you have
a blueprint for the entire meeting
in front of you.

When you are familiar with this
method and have made adaptations to
fit your style or your material, you
will find that you can put an On-Line
Agenda together in a few minutes when
you have to.

CONDUCTING THE MEETING

Since the secret to conducting a
good teleconference is in the preparation,
there are only a few additional suggestions:

. Try to avoid having anyone talk
continuously for more than five
minutes without the visual support
of written material, slides or
freeze-frame video.

. Different voices will help to
hold the attention of your listeners so poll individuals for
comments and opinions or ask
different people to take over
portions of the meeting.

. As moderator, pause frequently
to allow for questions. Do not
be afraid of pauses and silences.
These give your listeners time to
think and phrase their comments.

. Have anyone who asks a question
or speaks spontaneously to
identify himself.

. Select a telephone in a quiet
place from which to conduct your
meeting. Your office or a conference room is probably the best.
Make sure you will not be interrupted by other phone calls or
be distracted by people coming
in and out of the room

. Set aside at least five minutes
before conference time to sit
quietly and collect your thoughts
alone. A teleconference requires
a slightly different type of
concentration and "one-pointedness"
than we are accustomed to. Going
directly from activity into the
conference will cause your thoughts
to be scattered and less focused.

• To be a good moderator, you have
to be a good <u>listener</u>, sensitive
to what you are hearing and to
what each person is trying to
say. For this reason, some
moderators find it distracting
to have other people in the room
with them when they are running
a teleconference. Others, however, frequently run conferences
using a speaker phone or Convener
with three or four participants
in the room with them and have
no difficulty. Experience conducting one conference will quickly
tell you whether you prefer to
be alone.

• Have the following items in
front you before starting a
conference:

On-Line Agenda
List of Participants
    Keep track during the
    meeting of those you have
    polled for responses and
    participation.
Clock or watch
    Stick closely to the time
    schedule on your agenda
Background or advance material
    sent out to participants
    for the conference

CONCLUSION

The guidelines set forth
above are for informal business
meetings that consist of several
individuals calling in on telephones.
One or two of the calling locations
may have small groups participating
by means of a speaker phone or
Darome Convener.

Conferencing with groups of people
at each location for training programs
or other special purposes requires
a special set of instructions. However,
all of these basic guidelines still
apply. It is just that group conferencing is more difficult and requires
more careful planning and coordination.

The Teleconferencing Resource Book: A Guide to Applications and Planning
Lorne A. Parker and Christine H. Olgren (eds.)
Elsevier Science Publishers B.V. (North-Holland)
©Center for Interactive Programs, University of Wisconsin-Extension, 1984

Training People to Audioconference: A Review of the Current Wisdom

Mary E. Boone and Ronald E. Bassett
The University of Texas at Austin*

Tell your colleagues in the field of teleconferencing that you're interested in "skills training" and you will probably elicit one of the following reactions: (1) a patronizing smile, (2) an extended yawn, or (3) a raised eyebrow. Many people are skeptical not only of the need for training but also of the people who allegedly provide it. The term training has, in and of itself, suffered from the actions of people who promised an oak and delivered an acorn. Unfortunately, some people who are now providing training programs for teleconferencing are doing so without the benefit of extensive knowledge or experience with interactive media--perhaps at the expense of subtracting credibility from the ranks of their more qualified and capable counterparts.

It is important for the teleconferencing field to come to grips with this issue because it ultimately affects everyone from the users to the vendors. A central question to be raised is what sources are trainers drawing from to design their programs? Research? Intuition? Experience? Literature from the field of teleconferencing? All of the above?

This paper provides a synthesis of available literature on training for audio (voice only) teleconferencing. While this focus on audio may seem limited, many of the questions addressed here apply to all forms of teleconferencing training.

Finding the "literature" for this review is a bit like going on a scavenger hunt. There are plenty of glossy brochures and "how-to" pamphlets, but very few systematic papers, journal articles, or books. This raises another question: Why is there a paucity of in-depth writing in this area? Whatever the reason(s), it is important to identify the information that is available, describe coherent patterns, and begin to formulate guidelines for teaching people how to use teleconferencing effectively.

This paper centers on several objectives; it will:

1. Identify user skills proposed in teleconferencing literature as important

2. Report the results of a study in which expert opinion served to verify the literature on user skills

3. Identify training methods

4. Examine considerations in the implementation of training programs

_____

*Mary Boone is currently an intern at the Institute for the Future in Menlo Park, CA.

5. Demonstrate the need for empirical assessment of training effects.

There are different types of training and skills associated with teleconferencing. The next section will provide definitions for training and skills as they relate to this paper. It will also give rationales both for why it is important to identify skills and for why it is important to provide training.

Definitions and Rationales

This paper is concerned with identifying critical skills for audioconferencing. Skills are actions that people perform to facilitate a smooth flow of a conference and timely completion of objectives. Such skills are needed for audioconferencing regardless of the type of meeting (for example, instruction or business).

Why identify these skills? Until behaviors are specifically targeted for change, there is little point in developing instructional programs. If we are to assess the effectiveness of our programs, we must first be sure that we have chosen the appropriate behaviors to change.

For purposes of this paper, training means providing specific information about the operation of the system and the process of communicating through it.

A rationale for providing training would include the following reasons: People often ask for it, and may avoid using the equipment without it; users may more readily adopt the process if its introduction is accompanied by effective training (Ref. 20); the system is effective to the extent that its users are effective.

Audioconferencing literature includes information on skills for effective system use. The next section of this paper will set forth the skills that are reported in this literature.

Literature Review--Skills

This review will be divided into categories. It will begin with an examination of the facilitator's role and will determine the skills necessary for preparation (prior to the meeting), conduct (during the meeting), and follow-up (after the meeting). One of the problems with this literature is the definition of the role of "facilitator." Stockbridge and Bateman (Ref. 14) suggest that there should be three separate people to fill the roles of operator, host, and chairperson. For purposes of this review, the role of facilitator will be considered as a combination of these three roles. A facilitator coordinates, manages, and structures the audioconference.

This review will also include an examination of the skills necessary for users to participate effectively in an audioconference. When conducting a training program and determining the content of that program, an instructor will have to make the determination about which skills to teach whom. It is likely that it will often be the case, however, that participants will serve as facilitators and vice-versa. This would indicate that practice in both roles is advisable during a training session.

Facilitator--Preparation for the Meeting

Most of the research we examined indicated the need for a facilitator to conduct the conference (Refs. 3,4,5,8,10,11,12,14,16). Carey (Ref. 4) provides a general definition of the function of such a person: (1) to confirm the presence of participants, (2) to control information flow, (3) to designate turns at talk, and (4) to handle inappropriate comments.

Almost all of the literature suggests that this facilitator should construct an agenda (preferably written) prior to the meeting (Refs. 3,8,11,12, 14,16). Carey qualifies this need somewhat: "Previous research has insisted that a strong moderator and written agenda are necessary for audio teleconferencing, but this research indicates that some organization is necessary to preserve continuity and to regulate the interaction but that alternative methods are viable." (Ref. 4, p. 314.)

The construction of the agenda should be based, according to these sources, on well-defined objectives. Indeed, Stockbridge and Bateman (Ref. 14) suggest that a "successful" meeting is one in which "the objectives of the chairperson are accomplished." They further discuss unstated objectives or "hidden agendas" that participants may wish to introduce, and propose that a skillful chairperson will be able to accomplish his or her own objectives while incorporating the objectives of others as well.

Much of the literature suggests sending out a notice prior to the meeting that contains the following information:

1. Name of facilitator
2. Date, time, and place of meeting
3. Agenda
4. Participant list
5. Materials corresponding to presentation

(Refs. 3,7,8,11,12,14,17).

Bolsky and Stockbridge (Ref. 3) and Monson (Ref. 8) additionally suggest sending pictures of participants ahead of time if possible. Also mentioned is the importance of attention to time zones when planning a meeting (Ref. 5). Any materials sent ahead of time should be easy to read and numbered for easy identification (Refs. 3,8,12,15,16).

Pereyra (Ref. 11) suggests setting aside time prior to the conference to collect thoughts and gain composure. She also suggests that the facilitator gather the following items prior to beginning: (1) agenda, (2) participant list (to keep track of discussion), (3) clock or watch, and (4) copies of background material that have been sent out to participants.

Other suggestions to facilitators for preparation include: arranging a face-to-face meeting between remote participants prior to the conference if possible (Refs. 3,8,14); finding out the size and composition of the group (Refs. 4,8,12); selecting a quiet room (Refs. 3,11); establishing a predetermined order for speaking (Ref. 3); assigning responsibilities to participants (Refs. 8,16); sending biographies and information about guest speakers to participants and vice versa (Ref. 16); sending instructions on how to use slides (Ref. 16); and seeking feedback during the planning stage (Ref. 16). One important consideration in the planning of a meeting is to avoid trying to accomplish too many objectives in a short time frame (Refs. 6,14,16).

Another suggestion for action prior to the meeting is to allow for a period during which people are allowed to chat informally (Refs. 3,12,14,16). Immediately following this "chat," an informal roll call should be conducted (Refs. 3,8,11,14,16). This roll call should begin the meeting.

Beginning the meeting on time is essential (Refs. 11,14,16). It is easy to justify the importance of promptness when one considers the careful planning and timing that have gone into setting up the agenda. Also, the system may be scheduled for use immediately after the conference.

Facilitator--During the Meeting

Once the meeting is under way, the consensus of the literature suggests that the facilitator should do the following:

1. State the purpose and review the agenda (Refs. 3,8,11,14,16).
2. Encourage interaction (Refs. 3,8,11,12,14).
3. Use names of participants when asking questions (Refs. 4,8,11,12,14).
4. Pause frequently for questions (Refs. 3, 11,14).
5. Have spontaneous speakers state their names (Refs. 11,14).
6. Allow for silences (Refs. 7,8,11).
7. Be a good listener (Ref. 11).
8. Keep a record of who has spoken (Refs. 8,11).
9. Avoid long presentations by one speaker (Refs. 11,14).
10. Tell people to speak briefly (Ref. 14).
11. Tell people local noise is disruptive (Refs. 6,14).
12. Use interactive formats (Refs. 8,11).
13. Use variety in formats (Refs. 8,11,12,16).
14. Encourage the use of visuals (Refs. 8, 11,12).
15. See that everyone shares the same visuals and documents (Refs. 11,14).
16. Summarize proceedings and repeat main points when closing meeting (Refs. 3,8, 11,12,14,16).

17. Address questions to specific individuals or groups rather than "any questions?" (Refs. 8,11,12,14).
18. Keep the group focused; don't stray from the agenda (Refs. 3,8,11,12,14).
19. Periodically poll the group to encourage participation (Ref. 11).
20. In closing, thank people for their contributions (Refs. 11,16).
21. Make sure remote people know how to rejoin meeting if they're disconnected (Ref. 14).
22. Indicate the order of speaking (Refs. 14,16).
23. Close the meeting on schedule (Refs. 11,14).
24. Let one's personality come through (Refs. 8,12).
25. If there is a long silence, rephrase the question (Ref. 8).
26. Highlight important points through repetition and through phrases such as "this is important" (Refs. 3,8,16).
27. Vary pitch and volume (Refs. 3,7,8).
28. Give verbal acknowledgement when a comment is made (for instance, "I see" or "Fine") (Refs. 3,7,8).
29. Sit at a round table or in a semicircle to encourage participation (Ref. 3).
30. Focus on main topics first while attention is peaked (Ref. 3).
31. Be aware of a false impression that concurrence has taken place in decision-making (Ref. 3).
32. If a conflict arises, suggest it be handled after the meeting (Refs. 3,4).
33. Use humorous devices (Ref. 16).
34. Use humor with care--avoid misunderstanding in the absence of nonverbal cues (Ref. 8).
35. Use imagery when explaining concepts (Ref. 16).
36. Don't overload people; keep the format simple (Refs. 6,16).
37. Pause to allow for comments (Refs. 3,8, 11,12,14,16).

Facilitator--After the Meeting

After ending the meeting, the participants should be allowed a few minutes for informal talk as they were at the beginning (Refs. 3,14,16), and the main points should be summarized in written form and sent out to participants (Refs. 3,8,16). The facilitator should ask for feedback on how the meeting went (Refs. 8,12,16) and should get feedback by listening to a tape of the meeting, if possible (Ref. 8).

Facilitator--Throughout the Meeting

The quality of a medium such as teleconferencing that affects the nature of interpersonal interaction is termed "social presence" (Ref. 13). "Quality is affected not simply by the transmission of single non-verbal cues, but by whole constellations of cues that affect the 'apparent distance'

of the other" (p. 157). The facilitator functions primarily to reduce the 'apparent distance' between participants. "A sense of social presence can perhaps be cultured by teleconference leaders or encouraged by initial learning sessions" (Ref. 6, p. 19).

The skills mentioned above function to reduce the distance, to "humanize" the interaction. Monson defines humanizing in the following manner: "creating an atmosphere which focuses on the importance of the individual and overcomes distance by creating group rapport" (Ref. 8, p. 2).

It is important to consider this more global perspective. While listing skills helps to identify behaviors targeted for change, recognizing this overall goal is important. Facilitators will find ways of decreasing distance through their own creative processes--introducing humanizing allows room for such creativity. Training programs should seek to expand creative processes while concurrently offering concrete suggestions.

The literature is peppered with references to humanizing. The facilitator must "exaggerate vocal participation to make up for the fact that conferees can't see each other" (Ref. 14) and he/she must also "exaggerate social niceties and give equal attention to remote participants" (Ref. 14).

Preventing "polarization" of the meeting or a "we versus they" attitude is part of humanizing (Refs. 14,16). The participants should feel as though they are in a shared space as much as possible. However, simply telling someone to "humanize" is not enough. It is important to set forth ways of accomplishing the goal while still allowing imagination. Both the facilitator and participants contribute to this process. The next section identifies participant skills.

Participant Skills

The following list of skills are focused on participants, although they tend to apply to both participants and facilitators:

1. Be concise (Refs. 3,14).
2. Do not leave the meeting unannounced (Ref. 4).
3. Prepare ahead of time (Ref. 3,15).
4. Speak at the correct distance from the microphone (Ref. 3).
5. Turn away from the microphone to cough or sneeze (Ref. 3).
6. Don't make excess noise like chewing gum, rustling papers, and so on (Refs. 3,15).
7. Avoid side conversations (Refs. 3,15).
8. Speak naturally (Refs. 3,15).
9. Vary the pitch and volume of your voice (Refs. 7,17).
10. Identify yourself by name when you speak (Refs. 3,4,11,14,15).
11. Give verbal acknowledgement when you are addressed ("I see," "o.k.") (Refs. 3,8,15).

12. State the number of the visual you are using and describe it if everyone doesn't have one (Refs. 3,15).

13. Pause occasionally to allow others to comment (Refs. 3,15).

14. Mentally picture the people you're talking with (Ref. 3).

15. Address people by name (Refs. 3,15).

16. Keep a record of people's names, their comments, and what you want to ask them (Ref. 3).

17. Stress important points verbally (that is, "This is important") (Ref. 3).

18. Spell out unusual terms (Ref. 3,15).

19. Notify the facilitator about your audio-visual needs (Ref. 15).

20. Deactivate the microphones during a lengthy presentation (Refs. 3,15).

Examination of this list reveals the numerous skills which overlap for both facilitator and participant. The next section describes a research effort to provide verification for identifying important skills.

What the Experts Say about Training: Validating the Literature Review

Because of the relatively small number of studies in the literature, we conducted research to identify user skills by seeking expert opinion. The method employed was a modification of the DELPHI technique. Seven experts were selected from across the country and they provided a 100% response rate to two rounds of questionnaires.

The first questionnaire was an open-ended instrument. The experts were simply asked to list what, in their opinion, were the top ten skills necessary for competent performance as an audio-conferencing user, and they were also asked to list the five most common user mistakes.

The total number of original written responses was 95. Many of these responses were similar and were subsequently grouped together, for example, "Ability to speak concisely" and "Be concise and to the point." The skills and mistakes were also collapsed together because they overlapped. What one respondent had listed as a skill, another had listed as a mistake--for example, skill: "Introduces self as necessary" and mistake: "Failure to identify self." This reduction process generated a list of 36 general statements. The purpose of this survey was to compile a list of trainable skills. Therefore, the 36 responses were translated into 36 behavioral statements and listed with a Likert-type scale beneath each one for ranking. For example:

The purpose of the second round was to determine whether or not the experts agreed with the translations and to determine how they would

rank the statements in order of importance. The following table indicates the mean ranking for each of the identified skills:

1. Keeps group on topic (6.57).

1. Numbers and organizes visual materials (6.57).

2. Verbally introduces documents to be discussed (6.42).

3. Designates a chairperson to control the flow of conversation (6.28).

3. The agenda should be well-timed--the most important topics should be discussed first and the participant shouldn't try to accomplish too much in one meeting (6.28).

4. Allows time for personal comments and remarks (6.14).

4. Selects visual material that directly corresponds to verbal presentation (6.14).

4. Is concise when making contributions to discussion (6.14).

4. Is prompt (6.14).

4. Remembers to dial in for meet-me conferences (6.14).

4. Concludes conference with clearly defined next steps (6.14).

5. Waits 15 seconds after asking a question (6).

5. Frequently summarizes and synthesizes meeting process (6).

5. Does not leave meeting without announcing departure (6).

6. Uses names when directing questions (5.85).

6. Waits for a pause before speaking--does not interrupt others (5.85).

6. Varies pitch and tone of voice (5.85).

7. Remembers to use quiet button when talking at one location (5.8).

8. Tries to incorporate print materials instead of relying solely on the audio presentation (5.71).

9. Modulates loudness of voice--neither too loud nor too soft (5.71).

9. Modulates speed of voice--neither too fast nor too slow (5.71).

9. Asks for feedback by directly addressing specific participants--does not just ask "Any questions?" (5.71).

10. Remembers to tell participants what to write down--stresses salient points (5.57).

10. Paraphrases--restates in different words what another participant has said (5.57).

10. Uses name to identify self when speaking (5.57).

10. Allows for normal silences--doesn't try to fill them up with irrelevant comments (5.57).

10. Uses names to connect participants comments to each other (5.57).

11. Rarely makes references to documents not available to everyone (5.28).

12. Keeps track of participation and contributions on a participant list (5.14).

13. Laughs at own mistakes (5).

14. Introduces humorous statements (4.85).

15. Verbally follows other participants--uses an exact repetition of another

participant's last few words and inter-
jects short comments such as "I see"
or "uh-hum," and so on (4.71).

16. Laughs at others attempts at humor
(4.57).

(Note: Two skills were excluded because of
a very low number of responses and notes
from participants indicating that they were
confusing.)

These data were used primarily to add to
the literature review and should therefore not be
accepted as a definitive ranking of skills necessary
for audioconferencing. However, it can be safely
assumed that skills with a rating of six or higher
should be given careful consideration for inclusion
in an orientation program. Also, these expert
judgments provide support for the importance
of the skills identified in the literature. All of
the skills listed here were also listed (in some
form) in the literature review.

Implications of Findings

Some conclusions may be drawn and questions
raised regarding the results of the literature review
and the survey:

1. To date, the common wisdom of what
to do comes from personal opinions
more than systematic evaluations.

None of the articles mentioned a systematic method of determining necessary
skills. The skills in the literature and
in the survey are based on expert opinion.
We cannot be completely sure that this
covers all of the skills, nor can we be
sure that training is necessary to change
them. The opinions are useful because
they come from experts who have worked
with users and they have seen what works
and what doesn't. In a sense, they have
been doing an "informal" component
analysis over the years.

However, we need to discover what
skills remain deficient over time without
training. It could be the case that many
skills are acquired simply through use
of the system. We also need to know
if training actually can help users overcome skill deficiencies--by experimentally determining training effects.

2. The literature on training is scant and
does not encourage the development of
programs.

Much of the literature bemoans the
fact that organizations are not including
training programs as part of the implementation of their new systems. However, it is a small wonder that training
usually consists of a manual when there
is such a lack of literature in support
of the effectiveness of more expanded
training programs.

3. The survey tends to validate the skills
identified in the literature.

There are not many people who are
writing about audioconferencing skills;
therefore, in identifying a panel of experts some overlap was unavoidable.
However, the survey represents an attempt to: (1) identify skills that might
have been missed in the literature review,
(2) provide an expert ranking of those
skills, and (3) show how the skills could be
broken down into behavioral statements.

All three of these observations lead to one
overwhelming conclusion: We must conduct empirical
research regarding training effectiveness. This
research should be two-pronged, including behavioral
measures as well as global measures. In other
words, we should know at the end of a training
session if Susan is using her name to identify her-
self and we should have raters listen to tapes and
decide if she is more effective overall after the
training. Until we do this, the case for "Training
helps" will be flimsy.

The next section of this paper will focus
on research in the area of communication skills.
This information is included because it provides
examples of skills that have been empirically measured. This literature can provide a guide for
conducting similar research in audioconferencing
training.

Can these skills be taught?

At this point it seems valuable to draw upon
other literature which has assessed behavioral
change for various skills. Bassett and Boone (Ref. 2)
identified studies in which communication skills
improvement was empirically measured. We concluded that a variety of skills could be changed
in varying populations with varying methods. We
identified over 22 types of skills that had been
changed successfully. Among those skills that
would relate directly to audioconferencing are
the following:

1. Pronunciation and articulation
2. Fluency
3. Rate of speech
4. Inflection
5. Volume of speech
6. Pausing
7. Positive feedback statements
8. Conversational questions
9. Compliments and appreciation statements
10. Anecdotes
11. Latency of response time
12. Duration of oral statements.

Therefore, it is possible to conclude that
communication skills can be improved through
training. It should also be pointed out here that
many of the other trainable skills identified in
the Bassett and Boone review were nonverbal
skills (for example, facial expressions, gestures)--
thus pertaining to videoconferencing as well.

The next logical question to be asked is "How should these behaviors be taught?" Baird and Monson (Ref. 1) suggest using a variety of methods to teach users. The Bassett and Boone (Ref. 2) review indicates that many methods may be effective.

Selecting a Method

What has worked before? The Bassett and Boone (Ref. 2) review identified the following methods that produced successful results: written instructions, coaching, oral instructions, discussion of skills to be improved, modeling of desired behaviors (either live or videotape), overt rehearsal of skills, and finally covert rehearsal of skills. The final method, covert rehearsal, is quite interesting. By imagining themselves or others successfully performing the behavior to be changed, subjects experienced a concrete improvement. The most frequently used model consisted of a variety of methods employed simultaneously (a kind of "kitchen sink" approach). The most effective combination of methods seems to consist of: (1) instructions about skills to be learned, (2) modeling, (3) practice, and (4) feedback with reinforcement. This use of a combination of methods thus supports Baird and Monson's (Ref. 1) assertion that using a variety of methods is effective.

This paper has served to: (1) identify skills necessary for effective audioconferencing, (2) identify empirical studies that demonstrate that many of these behaviors can be taught, and (3) identify the methods that have been empirically proven successful. The next section will identify what methods organizations currently use to conduct training for audioconferencing, and will suggest factors for consideration in implementing training programs.

Implementation--Current Methods

Many training groups suggest hands-on training with the equipment. Pereyra (Ref. 10) states that "experience has shown that training sessions conducted in person or in some other medium are not as satisfactory because participants have to make the leap from theory to practice" (p. 240). She also suggests modeling as an effective method.

The University of Wisconsin-Extension conducts a three-hour hands-on workshop, and Baird and Monson (Ref. 1) state: "Training new teleconferencing users means more than writing a single how-to instruction booklet and more than an hour's introduction to your new teleconferencing system" (p. 281). Wisconsin uses simulations, games, modeling, hands-on experience, and post-simulation discussion and provides audio and video tapes and written materials.

Some user organizations have developed their own training programs. One organization stated that it used "various modes" to train users, specifying peer training, videotaped instructions, and practice on the system (Ref. 9). Another organization also required hands-on training and "stressed the participative nature of the medium";

they also provided assistance in developing agendas and gave suggestions for conducting a meeting over the system (Ref. 17).

Still another organization's training consisted of a "dry run" simulation one week prior to a first conference and also provided accompanying written materials in the form of a manual (Ref. 18).

Several references are made in the literature to determining the appropriate "level" of training for users (Refs. 5,9). This is obviously an important part of designing a program due to varying proficiency levels. Chute (Ref. 5) provides a helpful model in determining the "level" of a user through his adaptation of Hall's 7 stages for adoption of an innovation. These stages include awareness, informational, personal, management, consequence, collaboration, and refocusing. The staff (for the workshop described in the Chute article) assessed needs and concerns through interviews and questionnaires, and the workshop consisted of hands-on experience and instruction in how to design a program for an audioconference.

Simulated meetings are sometimes used as a method of training for new teleconferencing media. Two of these simulations include Spinoff (developed by the Institute for the Future) and SIMeeting (developed by SBS).

Spinoff is a useful simulation for various types of media, including audioconferencing. It consists of eight basic roles and can accommodate up to 25 people (Ref. 6). There is a basic structure for the meeting; however, participants create their own scripts. This tutorial allows people to experiment with the medium and to develop an understanding of what skills and protocols are necessary for conducting a successful teleconference. A full description of Spinoff is available in the book Electronic Meetings (Ref. 6).

SIMeeting was originally developed for videoconferencing; however, it could potentially be adapted to an audioconferencing format. It is similar to Spinoff in that it does not provide an exact script, but gives background and supporting materials for a simulated meeting. The purpose of the "meeting" is a planning session to produce a tentative program for an annual Board of Director's meeting. The materials for SIMeeting are available from SBS (Satellite Business Systems).

The next section will examine some of the considerations for implementing training programs for audioconferencing.

Suggestions for Implementation

Several questions should guide the efforts of anyone planning to implement a training program. The first of these is What attitudes toward training currently exist in the culture of this organization? Every organization is different. Some organizations have experienced great success with their training programs while others have not. If existing training programs within an organization have poor reputations, it may not be effective to use the term training or to turn over this instruction to an existing training department. On the other hand, if training is well-accepted in an organization, it may be appropriate to work with the existing trainers by having them receive instruction from a commercial facility. This brings us to a second question: Who should do the training?

If a commercial service is under consideration, it should be examined carefully. The success of a system might be influenced by the success of training. Whoever does the training should have a thorough knowledge of the constraints and advantages of the equipment. They should also know about how to design for an interactive format, and what skills are necessary for user effectiveness.

The next question to consider is Who should receive training? It would be difficult to give appropriate hands-on guidance to huge groups of people simultaneously. Criteria should be set for selecting trainees and unless a full-time facilitator is appointed, users should probably be trained as facilitators if the chances are good that they will have to assume that role from time to time.

This paper has already addressed the questions of What skills should be taught? (in the literature review and survey), and How should skills be taught? (in the section on selecting a method). The information found therein can serve as a guideline for answering these questions.

How can training effectiveness be evaluated? It is beyond the scope of this paper to list all of the possible methods of evaluation. Evaluation of teleconferencing, up to this point, has focused on evaluating the system, not the training. However, if money is to be invested in training and if training is considered to be an important factor in system success, then we need to find out if it is working. Taping conferences before and after training can provide feedback. Questionnaires and interviews can be used with trainees. However, a behavioral evaluation--where baseline measures are taken before and after training--is what is really needed to provide specific, meaningful insight into whether or not behavior change is actually occurring as a result of training.

How can the cost of training programs be reduced? It's a lot cheaper (on the surface) to simply put together a manual. However, ingenious designers of training programs can make use of opportunities to combine training programs. Meeting skills, listening skills, presentation skills, and communication skills that are required for face-to-face interaction need to be practiced from time to time. Many of the skills required for audio-conferencing are the same as those that are needed for face to face interactions (Refs. 1,14,17) and programs can be designed to accomodate both. By learning how to have more effective audio meetings, trainees can easily learn to apply this instruction to regular meetings.

The final, and perhaps most important question: Is training really worth the time, effort, and expense? The experiences with training identified in the literature lean toward a "yes" answer. For example, after implementing a training program, Threlkeld and Pease (Ref. 17) stated: "Experiencing a training session that was both technically and purposefully relevant, encouraged staff to begin thinking how each might apply the system to meet their communication needs. These sessions nurtured budding interest by future users, and an institutionalization of the electronic meeting, planning and execution process, as simulated by the training" (p. 11). Young and Burrell (Ref. 19) reported on implementation in an article appropriately entitled "Some Lessons Learned": "Time spent in a thorough introduction and education programme will be recouped by higher and broader utilisation in the organization" (p. 438).

"All of the new media require new skills for effective use . . . ease of learning new skills will also differ among users, and these will be important considerations if the system is not to become a white elephant" (Ref. 6, p. 133).

While these experiences provide valuable evidence that training may be helpful, the answer to the question "Is it worth it?" cannot truly be provided until there is empirical evidence. People "seem" to be more effective after training; however, no one has reported measuring whether or not they actually are better. Studies assessing training through measurable effects must be conducted before a definitive answer can be arrived at.

Conclusion

There are many considerations facing the person who is designing or purchasing a training program. Not every practical situation will allow for research activities. However, it is important that such activities take place.

This paper is only a beginning. Trainers, users, and vendors must work together to ensure the quality of training given its potentially vital role in the success of teleconferencing systems.

References

1. Baird, Marcia, and Monson, Mavis, "Training Teleconferencing Users: How to Tackle It." In L. Parker and C. Olgren (eds.), Teleconferencing and Interactive Media. University of Wisconsin-Extension, 1982.

2. Bassett, Ronald, and Boone, Mary, "Improving Speech Communication Skills: An Overview of the Literature." In R. Rubin (ed.), Improving Speaking and Listening Skills. San Francisco: Jossey-Bass, in press.

3. Bolsky, M.I., and Stockbridge, C., "The Audio Teleconference: What it is . . . Why do it . . . How to do it." Bell Labs.

4. Carey, John, "Interaction Patterns in Audio Teleconferencing," Telecommunications Policy, Dec. 1981, p. 309-314.

5. Chute, Alan, "Selecting Appropriate Strategies for Training Teleconference Presenters." In L. Parker and C. Olgren (eds.), 1982 op.cit.

6. Johansen, R., Vallee, J., and Spangler, K., Electronic Meetings: Technical Alternatives and Social Choices. Menlo Park, CA: Addison-Wesley Publishing Company, 1979.

7. Josephson, H. & Dvorin, D., The Tele-skills Manual: A Working Book. Mediasense, Inc., 1981.

8. Monson, Mavis, Bridging the Distance: An Instructional Guide to Teleconferencing. University of Wisconsin-Extension, 1980.

9. Niemiec, Anne, "CMITS: Communication and Craft." In L. Parker and C. Olgren (eds.), 1982 op.cit.

10. Pereyra, Susan, "Human Factors in Establishing an In-House, Meet-Me System." In L. Parker and C. Olgren (eds.), Teleconferencing and Interactive Media. University of Wisconsin Extension, 1982.

11. Pereyra, Susan, "Conducting Business Meetings by Telephone." In L. Parker and C. Olgren (eds.), Teleconferencing and Interactive Media. University of Wisconsin Extension, 1980.

12. Rowan, Paul, The 1981 Audioconferencing Handbook, Paper P-117. Menlo Park, CA: The Institute for the Future, 1981.

13. Short, John, Williams, Ederyn, and Christie, Bruce, The Social Psychology of Telecommunications. London: John Wiley & Sons, 1976.

14. Stockbridge, C., and Bateman, T., "Procedures for Conducting a Successful Telemeeting," Ninth International Symposium on Human Factors in Telecommunication. Holmdel, NJ: Bell Labs, 1980.

15. Teleconferencing Conferee Kit (Quorum Teleconferencing System), Bell Labs.

16. The Teleconferencing Manager's Guide (Quorum Teleconferencing System), Bell Labs.

17. Threlkeld, R., and Pease, P., "Audio Teleconferencing in Northern New England." In L. Parker and C. Olgren (eds), 1982, op.cit.

18. Waller, E., "The Use of Teleconferencing for Training Bank Personnel." In L. Parker and C. Olgren (eds.), 1982, op.cit.

19. Young, Ian, and Burrell, Jim, "The Implementation of Teleconferencing: Some Lessons Learned." In L. Parker and C. Olgren (eds.), 1982, op.cit.

20. Bikson, T., Gulek, B., and Mankin, D., "Implementation of Information Technology in Office Settings: A Review of Relevant Literature," The Rand Paper Series (P-6697), Nov. 1981.

The Teleconferencing Resource Book: A Guide to Applications and Planning
Lorne A. Parker and Christine H. Olgren (eds.)
Elsevier Science Publishers B.V. (North-Holland)
©Center for Interactive Programs, University of Wisconsin-Extension, 1984

INTEGRATION OF MEDIA COMPONENTS FOR SUCCESSFUL TELECONFERENCING*

Burton W. Hancock, Ph.D.
Research Associate

Alan G. Chute, Ph.D.
Director, Teleconferencing Network

Robert R. Raszkowski, M.D., Ph.D.
Associate Professor, Department of Internal Medicine

Kathy D. Austad, M.S.
Teleconferencing Program Coordinator

The University of South Dakota School of Medicine
Sioux Falls, South Dakota

For a teleconference to be successful, a teleconference presenter needs to make the teleconference presentation both interesting and educational. Audio teleconferencing is frequently regarded as a weak instructional medium by new users because it does not provide the learners with enough visual cues to gain and maintain attention, or to structure and organize the presentation. However, audio teleconferencing, when carefully integrated with slides, overhead transparencies, videotapes, or handouts can be a successful educational medium. The integration of such media components in an audio teleconference expands a single channel presentation into a multi-sensory channel presentation.

The selection and utilization of media in an educational teleconference should be based upon instructional principles derived from research on human perception and human learning. In this article, examples of media utilization techniques for teleconference presentations are described.

Audience Attention

One of a teleconference presenter's first challenges is to gain the attention of the audience, and thereafter to continue to hold that attention. Although a certain amount of audience attention can be expected, maintaining

---

*The South Dakota Medical Information Exchange has been developed under Exchange of Medical Information Grant No. 80-001-03 from The United States Veterans Administration to the University of South Dakota School of Medicine. The opinions stated in this article are those of the authors and do not necessarily represent those of The Veterans Administration.

that attention is far from guaranteed. Generalized attending is often insufficient, for attention must be directed narrowly and precisely to critical aspects of the subject matter.[1]

Gaining Attention

Among the first steps to be taken in designing a teleconference is to make provisions for gaining attention. It is generally recognized that there are two kinds of attention-gaining techniques. The first is a general alerting technique in which the audience assumes a posture of readiness or anticipation for the learning experience. This can be accomplished by: using alerting sounds, light changes, or other sudden changes in the environment. The second attention gaining technique is accomplished through selective attention. The instructional materials are designed to give emphasis to the important concepts being presented.[2] For words in print, this is generally accomplished by underlining or italicizing the important points. For pictures and diagrams, feature cues including color, geometric shape, size, angular orientation, and brightness highlight the concepts to be perceived.[3] For audio teleconferencing, selective attention can be accomplished by changes in volume, changes in voice, inflection, use of pauses, or use of audio embellishments such as sound effects.

Maintaining Attention

Maintaining the audience's attention is an important component in developing instruction for face-to-face presentations. Maintaining an audience's attention is equally as important, if not more important, in a teleconference presentation. Since presentations via teleconference

lack the attention getting potential of face-
to-face presentations, special considerations
must be made to maintain the audience's level
of attention. The use of visually stimulating
media and frequent changes in the visual materials are methods the authors have used to
maintain the audience's attention during a
teleconference.[4]

Perceptual Principles

It is frequently impossible to predict
with certainty what individuals at remote locations will attend to during a teleconference.
However, it is possible to describe some of the
major ways in which perceptions will vary and
some of the conditions under which they will
vary. With an understanding of the principles
of perception, a teleconference presenter can
design a teleconference to increase the probability that a message will be attended to and
encoded.

Perception is Selective

Perception is relative rather than absolute. Individuals attend to only a few of the
sights, sounds, and smells available in an
environment at any one time.[5,6,7,8] If the
teleconference presenter can predict what the
individuals in the audience will selectively
perceive, or attend to, he will be able to
accentuate the critical aspects of a message
and de-emphasize or delete distracting stimuli.
Selective perception is in part physical, and
in part dynamic. What an individual perceives
depends upon what the individual considers to
be the critical aspects of the message.[5,8,9,10]

There are several strategies which can be
employed to enhance selective perception. In
a teleconference presentation, limiting the
range of the topic presented, going through a
complex process in a step wise manner, or
emphasizing the relevant aspects of the material can facilitate selective perception.
Visual materials can be prepared which illustrate the key information presented or present
complex information in a series of visuals.
Visual cueing can be employed for highlighting
relevant parts of the visual. An obvious cue
such as an arrow can focus attention to the
point being emphasized. During the teleconference, the presenter can also use voice cues
to direct the audience's attention to the visual
cues designed into the media materials. This
insures that the program participants located at
remote sites are attending to the same content
material. It is important for a teleconference
presenter to remember that the problem of directing perceptual selectivity includes control-

ling attention within the audio channel, among
visual elements, and between the auditory and
visual elements.

Once a teleconference presentation has been
designed to gain the attention of the audience,
the problem arises of how to maintain the optimum level of stimulation. An individual's capacity for encoding information is limited.
This restricts the amount of information which
can be presented during a teleconference. The
teleconference presenter's goal should be to design instructional teleconferences which reduce
perceptual demands by increasing the codability
of the message. By increasing the codability of
a message, the amount of information processed
can be increased.[11,12]

Perception is Organized

The organization or patterns of a message
also play an important role in the processing of
information.[4,11] Perception should be organized. Organization of the message influences
the ease and accuracy of perception.[5,13] The
teleconference presenter should make the organization of messages apparent to the audience.
For example, simply numbering steps in a
series of events gives organization to what the
audience perceives. Perception is also strongly effected by what the audience expects to perceive.[9,13] Handouts such as a topical outline
can provide a framework for perceiving the
organization of the presentation.

By using structuring strategies such as the
appropriate spatial arrangement or temporal
ordering of message elements, the teleconference
presenter can facilitate perceptual groupings
and reduce processing demands. Increasing an
individual's encoding capacity can also be
accomplished by combining the auditory and
visual channels to deliver instruction.[11] This
is especially important for teleconference presentations. Instruction via teleconferencing
is primarily auditory and lacks face-to-face
contact between the presenter and the audience.
Therefore, visual materials are essential for
successful teleconferencing.

Learning Strategies

In addition to employing perceptual principles to gain and maintain attention, the teleconference presenter needs to be aware of those
learning strategies which affect retention of
the content. Content in a teleconference needs
to be relevant to the learner so that the learner feels comfortable interacting with the presenter. The content also needs to be presented
in a way that the learner understands the struc-

ture and organization of the presentation. Media materials, in addition to their attention directing functions, can be used to structure the overall teleconference and to facilitate the learner's retention of the content material.

Relevant Content

Learning that is meaningful to an individual is acquired more readily and is retained longer than that which appears meaningless or arbitrary.[14],[15] When designing media for instructional teleconferences, it is important to incorporate meaningful visual examples with which the learner is familiar. The extent to which the material is familiar to the learner or relevant to the individual's prior experiences or current interests strongly influences the degree to which the material is retained.

Participation

Presenters need to encourage the participants to take an active role in teleconference programs. Interactive techniques and learner involvement are essential elements in the design of a successful teleconference.[16],[17] Learning theory emphasizes that the learner should be active in the learning process.[18] It is important that adult learners participating in continuing education via teleconferencing be allowed to contribute to the program by sharing their experiences and discussing perceptions they hold. Presenters on the South Dakota Medical Information Exchange teleconferencing network typically employ "summary or transition visuals" during the teleconferences to indicate stopping points for discussion.[4] While a "summary or transition visual" is being projected, the individual participants are asked to share their reactions to the content which was just presented. This activity not only increases interest and generates participation, but also reinforces the content and thereby increases the probability of retention.

Content Organizers

The extent to which material is organized and the extent to which the learner is aware of that organization provides a meaningful context in which the material can be remembered. The relative difficulty or ease of meaningful learning is a function of the extent to which the individual learner possesses a relevant cognitive structure. Meaningful learning proceeds most easily when it moves from the abstract to the concrete.[19] For instance,

"advance organizers" are devices employed by Ausubel to present an outline of the program content at a higher level of abstraction, generality, and inclusiveness. The purpose of an "advance organizer" is to provide an "ideational scaffolding" into which subsequently presented material can be presented.[19] Providing the learner with strategies for classifying such things as events, objects, or diseases, is a frequently used advanced organizer.

Integrating Media for Structure

When designing their presentations for South Dakota Medical Information Exchange programs, presenters are asked to use media materials to "preview, present, and review" the program. By previewing the program at the start of the teleconference, the presenter in effect, tells the participant how the teleconference content material is related to his current knowledge base.[4] The presenter furnishes the participants with an overview in the form of a welcome letter, a written outline, or a visual outline to introduce the presentation. This overview provides an advanced framework for perceptual organization and sets forth the most important terms and concepts that are to be presented to the learner. This will help to facilitate learning and retention for participants. The presenter uses the outline to explain what will be covered in the program, how much time will be spent in each area, what types of interactions are planned, and what is expected of the participants.

During the present portion of the teleconference, presenters are encouraged to have the participants focus their attention on the visual materials. As the presenter discusses his topic, he directs the audience's attention to relevant visual data being presented. Typically the instructor presents the first fifteen to twenty minutes of the teleconference relying heavily on media materials to structure and organize the content.[4] After this first period of content material is presented, the instructor directs the audience's attention to the handouts for the next ten minutes during a problem solving exercise such as a case study, or a hypothetical scenario. During the discussion period, the presenter may redirect the audience's attention back to his slides, overhead transparencies, handouts or perhaps to a video cassette segment to clarify salient points.

Presenters of educational teleconferences on the South Dakota Medical Information Exchange teleconferencing network use a combination of visual materials to organize their presentations. Slides supplemented with a handout packet are most often used. Overhead transparencies are frequently used when only a few visuals are required. Video cassettes have been used particularly with topics which require a motion component in the learning activity. After the preview and present portions of the teleconference have ended, the presenter typically uses a visual outline to review the major points of the program and to restate the objectives. When the presenter previews, presents and reviews the content, he facilitates participant learning.

Summary

If a teleconference presenter effectively employs perceptual principles in the design of his media materials, he should be able to gain and maintain an audience's attention in a teleconference. If a presenter correctly utilizes learning strategies which are appropriate for a teleconference situation, the presenter will be able to facilitate participant learning. This article has described how a teleconference presenter can integrate a variety of media components into a teleconference presentation to gain and maintain audience attention, structure and organize the presentation, and enhance retention of the material presented.

References

[1] Fleming, M. and Levie, W.H. Instructional Message Design: Principles from the Behavioral Sciences. Englewood Cliffs, NJ: Education Technological Publications, 1978.

[2] Gagne, R.M. The Conditions of Learning (3rd ed.). New York: Holt, Rinehart and Winston, 1977.

[3] McVey, C.F. An analysis, synthesis, and application of selected research findings to visual design and presentation by the visual specialist. Unpublished doctoral dissertation, University of Wisconsin, 1969.

[4] Raszkowski, R.R. and Chute, A.G. Guidelines For Effective Teleconference Presentations in Continuing Medical Education. Issues in Higher Education, Vol. 6, 1982

[5] Berelson, B. and Steiner, A. Human Behavior: An Inventory of Scientific Findings. New York: Harcourt, Brace, and World Inc., 1964.

[6] Helson, H. Adaptation-Level Theory. New York: Harper and Row, 1964.

[7] Rock, I. An Introduction to Perception. New York: MacMillan Publishing Co., 1975.

[8] Treisman, A.M. Selective attention in man. In P.A. Fried (Ed.), Readings in Perception: Principles and Practice. Lexington, MA: D.C. Heath & Co., 1974.

[9] Forgus, R.H. Perception. New York: McGraw-Hill Book Company, 1966.

[10] McGinnies, E. Emotionality and perceptual defense. In P.A. Fried (Ed.), Readings in Perception: Principle and Practice. Lexington, MA: D.C. Heath and Co., 1974.

[11] Moray, N. Where is capacity limited? A survey and a model. In A.F. Sanders (Ed.), Attention and Performance. Amsterdam: North-Holland Publishing Co., 1977.

[12] Posner, M.I. Cognition: An Introduction. Glenview, IL: Scott, Foresman and Co., 1973.

[13] Murch, G.M. Visual and Auditory Perception. Indianapolis: The Bobbs-Merrill Co., 1973.

[14] McKeachie, W.J. Instructional psychology. In P.H. Mussen and M.R. Rosenzweig (Eds.), Annual Review of Psychology (Vol. 25). Palo Alto: Annual Reviews, Inc., 1974.

[15] Mauly, G.J. Psychology for Effective Teaching. New York: Holt, Rinehart, and Winston, 1973.

[16] Monson, M.K. Bridging the Distance: An Instructional Guide to Teleconferencing. Instructional Communication Systems, University of Wisconsin-Extension, Madison, 1978.

[17] Parker, L.A. and Monson, M.K. More Than Meets the Eye. Instructional Communications Systems, University of Wisconsin-Extension, Madison, 1980.

[18] Bruner, J.S. Toward a Theory of Instruction. Cambridge: Harvard University Press, 1966.

[19] Ausubel, D.P. Educational Psychology: A Cognitive View. New York: Holt, Rinehart and Winston, 1968.

The Teleconferencing Resource Book: A Guide to Applications and Planning
Lorne A. Parker and Christine H. Olgren (eds.)
Elsevier Science Publishers B.V. (North-Holland)
©Center for Interactive Programs, University of Wisconsin-Extension, 1984

A TELEWRITING SYSTEM FOR TEACHING AND TRAINING
Some Principles and Applications

Mavis K. Monson
Instructional Design Coordinator
Instructional Communications Systems
University of Wisconsin-Extension

INTRODUCTION

One picture is worth more than ten thousand words.
—Old Chinese Proverb—

Fortunately, the meaning of this ancient saying needn't be taken too literally when we talk about teleconferencing. In fact, the extent to which teleconferencing is being done today on a world-wide basis using voice-only audio channels is a direct contradiction. It belies the notion that one needs to have visual information to communicate effectively.* Yet, we do know that for certain types of information and in certain communication processes, the capability to produce real-time graphics (hand-drawn pictures, diagrams, letters and numbers) is an asset whose worth is incalculable.

It is for these applications, that a system which provides real-time graphics capability becomes a vital component of teleconferencing. This paper focuses on one of those systems; a system called "telewriting." Telewriting may be defined as the process of converting hand-drawn pictures or written symbols into electronic signals which are then transmitted instantaneously down telephone wires (or a similar transmission system) to be reproduced at a number of remote locations simultaneously.

The discussion of telewriting will center on the electronic blackboard. Within the area known as telewriting, there are a host of options other than the electronic blackboard such as electrowriters, graphics tablets, light pens, etc. And, there are, in addition, other visual graphics systems such as facsimile, slow-scan televideo, computer graphics and full-motion video. The limits of this paper preclude a discussion of these systems. However,

a number of the design principles introduced in a later section can be applied to the successful use of all of the so-called "visual graphics" systems in teleconferencing.

THE ELECTRONIC BLACKBOARD

Based on components familiar to all of us, the electronic blackboard was developed by Bell Laboratories and pilot-tested at the University of Illinois for engineering courses. It is now being marketed nation-wide.

We had the unique opportunity to work with the blackboard on an intensive basis: exploring its capabilities and discussing its use in a distant learning environment. The information in this paper is based upon this experimental work as well as our past experience with another type of telewriting system: the electrowriter.*

The electronic blackboard is a very simple-to-operate system. Thus, little or no "training" is required for its use. Optimally, however, we would use some forethought--simply because we are working within a teleconferencing frame of reference. This means that we need to consider how the transmitted material will appear at the remote locations as well as how the visual component will be integrated into the overall design of the teleconference. Accordingly, this paper will explore these two aspects: I. Transmitting Technique and II. Designing for a Teleconference.

For example, the Educational Telephone Network, an audio teleconferencing system at the University of Wisconsin-Extension, was in use over 1800 hours in 1978-79, with a total of more than a quarter million "contact hours" (number of program hours times number of participants). Programs included instruction and meetings.

The SEEN system combines audio teleconferencing with an electrowriter system. It reaches twenty-three locations within the state. Visusl material is received at locations via an overhead-type projection system so that images appear greatly enlarged on a standard projection screen.

I. TRANSMITTING TECHNIQUE

To understand the "whys" of good transmitting, we need to know how the equipment operates. The basic components --the blackboard, the chalk and the eraser--look very similar to those found in an ordinary classroom (indeed, in the case of the chalk and the eraser, they are.). These components differ in that as they are being used in the usual way, they are also performing the "behind the scenes" electronics necessary to produce images at the remote locations.

The blackboard itself consists of two layers of tightly stretched Mylar™ which is electrically conductive. A minute gap separates the two layers. As chalk is used to write on the surface, the two layers come into contact, registering electronically as the x-y coordinate of that point. The impulse formed, along with thousands of others, is carried by the telephone line to the remote location where a memory unit stores it until it is displayed on a video monitor. Each x-y "point" thus becomes a part of a larger graphic. All of this occurs so fast that we would call the writing or graphic produced "instantaneous."

To remove a portion or all of the writing, the eraser is lifted from the tray--activating the erase mode of the unit. The physical act of lifting the eraser and using it to remove the chalk markings from the board also erases the visual images on the remote monitors.

The blackboard thus has two operating modes: "write" and "erase." The main thing to remember, then, is that any pressure on the board, whether that be from a piece of chalk (as intended for use), a fingernail or other object, will signal the equipment electronically to either "write" or "erase" depending on its mode. It is important to note, also, that since the transmission depends on electrical contact, the best quality will be obtained when stokes are deliberate--flourishes should be saved for regular classroom writing.

Legibility

Making deliberate strokes is only one of the ways to insure legibility. Legiblility becomes a critical factor because of the large difference between the size of the writing surface on which the information is originally produced and the size of the video monitor on which it will be viewed. In this case, what you see on the blackboard is not exactly what you get on the monitor, so that an instructor needs to make a conscious effort to make letters on the blackboard large enough to read well when they appear on the monitors.

(Fig. A. shows proportional difference between writing surface and an average monitor.)

Fig. A. Proportional size difference between blackboard writing surface and video monitor (23 inch).

The work of G. F. McVey (1970) on legibility and television display suggests a number of guidelines appropriate for the blackboard. Legibility, according to McVey's studies, depends upon a number of factors: size, form, shape and dimensions, stroke width, brightness, contrast, color, illumination, and the resolution of the display system. Several of these factors--stroke width and color, for example-- are not relevant to this discussion. However, size, form, legibility, etc., affect the choices a user of the blackboard might well make.

Size. McVey suggests a formula for determining symbol size on artwork to be produced for television. We would suggest that based upon this formula as well as our experimental trials that instructors begin by making letters at least 2 inches in height. Depending on the skill of the user in writing, this size can be adjusted upward or downward, accordingly.

Form, shape. Certain alphabets and numerals are so similar in shape that they are frequently confused with each other. For example, G, C and O; 3 and 8; R and B; 2 and Z. Instructors can make a deliberate attempt to make sure these particular symbols are distinct from one another.

Spacing. Lines set too closely together can be difficult to follow and reduce readability. It is generally recommended that line spacing be between 100% and 150% of symbol height.

Upper and Lower Cases. Upper case letters are more legible, lower case letters more readable. Thus, if you are producing individual alphanumerics, short statements, captions, etc., use upper case. When a great quantity of alphanumerics are presented, lower case letters are easier to read. Thus, if you are going to write sentences or any quantity of letters, use lower case.

Illumination. The amount of illumination in the viewing sector and the area surrounding the monitor are factors that will affect viewing accuracy and comfort. This particular aspect is not under direct control of the instructor but is noted here as a condition to be considered when preparing the locations for the learning experience. The reader is again referred to the work of McVey for suggestions to follow in terms of background lighting and monitor brightness.

To summarize, legibility depends on a number of factors such as size, form, spacing, use of upper/lower case and illumination.

Tracing Techniques

In certain subject matter areas such as accounting, there may be a need to precisely position lines or columns on the screen. In order to assure as accurate a transmission as possible, two options for tracing from an original graph or chart are possible:

1) Use of an acetate overlay the same size as the blackboard which incorporates a grid of 3 inch squares. The grid would facilitate the drawing of straight lines and organizing space more efficiently.

2) Use of an overhead projector or slide projector to display the visual information on the surface of the blackboard. This would enable the instructor to trace (and thus transmit) the material by following the lines projected.

Prerecording

Both the visual and the verbal message can be prerecorded using a standard two-channel audio cassette recorder. The advantages of the audio cassette format are two-fold. First, the obvious cost savings. Video cassette format runs roughly 25:1 as compared to audio, in terms of tape costs alone. (Over and above these costs are studio cameras and setups for video taping.) The second advantage lies in the ease of recording procedures. Because the recording is made directly from the impulses generated within the system, an instructor can easily do his/her own recording by following a few simple steps.

Although the procedure is simple, it is important to remember that if a recording is being made incrementally (with many "stops" and "starts") that the very first stroke when beginning the next segment must be made very deliberately. This helps to make sure that "contact" has been made to reestablish the recording process.

Ground Rules

Ground rules might be thought of as the process of providing structure or developing common expectations regarding how communications (both visual and verbal) will "happen." People react positively to ground rules in teleconferencing-- somehow the mutual understanding as to when and how the message is coming, when and how to interact, etc., makes the experience more comfortable and satisfying. For example, the work of Korzenny and Bauer (1979) points out that contrary to what we might expect, in a teleconferencing setting, "... *the level of communication rules during the interaction had a positive impact on communication satisfaction.*"

The exact ground rules established will, of course, depend on the individuals participating and the goals of the teleconference. However, one example of why developing common expectations is important results from the fact that at a remote location, one of the centers of focus for students will be the visual material on the monitors. Thus, ground rules for the visuals, i.e., what to expect, may well be of consequence. For example, if nothing will be transmitted for a number of minutes, the instructor could note this--either verbally or visually.

Pairing the Transmission of Visual Material with Spoken Material

The work of Severin (1967) is relevant to the pairing of the information being given verbally and visually. His research suggests that it is important for us to make sure both of the "cues" match each other to the extent that they are related or relevant, but not to the extent that they are one and the same. In other words, to have the same word or words displayed on the screen as those being given verbally will not result in better communication. As he states:

"*Multi-channel communication which combines words with related or relevant illustrations will provide the greatest gain because of the summation of cues between the channels.*"

However, it is the corollary to this relationship which concerns us even more here. If communications in two channels contain unrelated cues, it will cause "*interference between channels and result in less information gain than if one channel were presented alone.*" This suggests that it is critical for the information which is displayed on the monitor to "match" in some related way, the information being transmitted verbally by the instructor. And this is why some pre-planning of the instructional sequence is a recommended policy.

Planner Concept

Through past experience, we have found that the most successful instructors using a telewriting system have a planning device of some type. The form is not the important ingredient--it is the concept of some kind of visualization process before the actual program begins--that is.

The idea of "story-boarding" is one that comes out of the instructional design process. A story-board is simply a *"series of sketches or pictures which visualize each topic or sequence in an audio-visual material to be produced."* (Kemp, 1975) A typical story-board shows the visualized portion of the message on one side and the verbal portion on the other.

For easy use, many of our telewriting instructors have developed their "planners" using 5 x 8 inch cards. (The type of form used, of course, will depend on the teaching/working style of the individual.) Cards are an easy format to work with as they can be arranged and rearranged in sequence during the lesson-planning stage. During the actual program, notes made on each card can help pinpoint problems in student understanding or visual presentation, as well as strategies that worked well. For future use, after evaluation, the cards can be rearranged, added to and/or discarded easily so that improvements in the program can be made directly from the cards.

(Fig. B. shows a suggested format for a planer card. Fig. C shows an easy way to sketch the proportional size of a video monitor.)

Planner cards help make the transition from traditional classroom teaching to teaching via telewriter proceed more smoothly. By thinking in terms of "frames," the instructor becomes aware that once an illustration has been developed, it will be "lost" in terms of review as the next illustration takes its place on the monitor. As one of the instructors using our electrowriter system put it: *"In an ordinary classroom, I just write all around the room, coming back to whatever I need. With the electrowriter, I have to think ahead. I need to remember when students may have to refer back to the start of a math problem."*

To summarize: the planner concept is a form of story-boarding which encourages instructors using the blackboard to plan and organize the visual portion of the content. The form of the planner is not important; even the direct use of the planner during instruction is not critical. What is important is the introduction to instructors of the concept of thinking about instruction in terms of the visual portion of the message--how the visual is to be presented, pacing, how much information each frame should carry, size of lettering, size ratio of the video monitor (3:4), simplification of diagrams if necessary and interrelating information on the board to other media (slides, videotapes, printed material, etc.).

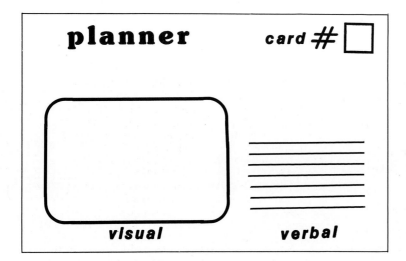

Fig. B. A suggested format for a "planner card" (actual size: 5" x 8")

Size ratio of monitor screen is 3:4. Draw a diagonal line through a 3 by 4 inch rectangle. Any rectangle formed by meeting this diagonal will be in the ratio of 3:4.

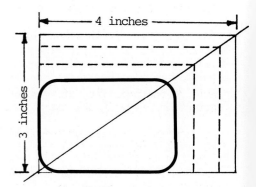

Fig. C. Determining proportional size for monitor "frame"

Summary of Transmitting Technique

To successfully produce the type of meaningful, clear visuals we want, we need to keep in mind a number of factors:

1) That although the blackboard looks like an ordinary blackboard, an instructor is actually generating electronic signals as well as chalk images. These signals will appear on a video monitor (ratio 3:4). Size of letters will appear at a smaller size on the remote monitors in comparison to the blackboard.

2) Legibility factors such as size, form, spacing, upper/lower cases and illumination are important considerations.

3) Tracing techniques, using either an acetate overlay or projecting the material via overhead or slide projector, can be used to transmit material that is precise in nature (such as accounting forms, music staffs, etc.).

4) Prerecording of both visual/audio message is easily accomplished. Entire programs might be prerecorded; short segments of complex material can be used as inserts. Format is standard audio cassette tape.

5) Ground Rules which establish when the visual (and verbal) messages will vary help provide participants with needed structure.

6) Pairing of the visual message which is on the monitor with the verbal message being given by the instructor is important to assure that "cues" are related and relevant to one another.

7) In general, most successful telewriting instructors use some form of "planner" which organizes their thinking in terms of visual and verbal messages.

(Chart A summarizes capabilities of the electronic blackboard as compared to the electrowriter [original equipment, Victor Comptometer].)

CHART A

CHARACTERISTIC AND/OR CAPABILITY	Electronic Blackboard	Electrowriter (Vict.Comp.)
Basic writing components	chalk on blackboard surface	ball-point like pen with mechanical linkage on paper platen
Writing surface Viewing surface	42" by 56" tv monitor	3 1/2" X 5" projection screen
Real-time: provides "building" of graphics, revisions, clarification, pacing	yes	yes
Prerecording option: prerecords can be edited, used as inserts, replayed for independent study	yes	yes
Tracing techniques to provide more exact visuals	yes	limited by writing area
Can be combined with other media such as slides	yes	yes, use with care because of high contrast between lighted screen and slide
Ability to underline, draw arrows, etc. for attention, repetition, clarification	yes	yes
Ability to erase all or portions of graphic for attention, emphasis	yes	no
Capability of having one site add notations to visual display on monitor	yes, w/ each site equipped with BB	very limited because of writing area
Hard copy (paper or acetate) available for review	yes, w/ special add-on equipment	yes, automatic paper and acetate copies

II. DESIGNING FOR A TELECONFERENCE

Now that we have explored the considerations important in the transmission of material,
let us turn to another fundamental area in the
success of a telewriting teaching/training session: designing a teleconference.

We would like to introduce in this section
a model for design--a model that suggests guidelines for adapting instruction to a distant
learning environment.

Although teleconferencing classes are in
many ways the same as those taught face-to-face,
they are, as one would intuitively guess, different in others. The unique characteristics of
the teleconferencing setting which relate to
the design of instruction are:

- individuals are physically distant from
  one another
- they are using telecommunications technology to interact with one another
- they may be initially uncomfortable in using the technology or may have different
  expectations of what the experience will
  be like
- they may be participating in a group or be
  all alone at a location
- individuals cannot see one another as they
  communicate.

These characteristics of the teleconferencing environment provide some basic guidelines
to use in thinking about designing instruction
so that it is effective and satisfying. For
example:

1) Listening "groups" aren't always a
   group.

   When students are dispersed geographically at many locations, it may be
   difficult to develop a "one group"
   rapport. Getting-to-know-someone is
   not as easily accomplished in teleconferencing as it is in a face-to-face
   setting.

2) Discussion doesn't "just happen."

   No matter how much students may want to
   make use of the opportunity to ask
   questions (and no matter how easy it
   may be to use the technical equipment),
   some students are reluctant to interact at first--they may feel they are
   infringing on program time, on other
   people's time, or that their questions
   or comments may not be worded correctly.

3) People don't listen very long

   It is hard work concentrating on a
   one-way message very long. For adults,
   that "very long" is not much more than
   12 to 15 minutes. For children, the
   time is even shorter.

4) Communication isn't complete until we
   get feedback.

   In any communication with others, we
   haven't completed the "loop" until we
   know that the message has been received and understood. In teleconferencing, non-verbal cues from individuals are not available as a feedback
   mechanism.

In many cases, an effective teacher in a
face-to-face setting will intuitively recognize the differences inherent in teleconferencing and quickly adapt by incorporating new
communication techniques into his/her teaching
style. However, to help new instructors make
the transition as smoothly and rapidly as possible, we have organized what we know from the
research and what we have learned from the
experiences of those instructors who program on
our network--into four design components:

HUMANIZING

PARTICIPATION

MESSAGE STYLE

FEEDBACK

To operationalize these components in a
teleconference, we provide instructors what
we call "hip pocket" techniques. These hip
pocket techniques can be incorporated into
instruction to adapt it to teleconferencing.
Later, after some initial success with the
medium, an instructor may wish to work more in
depth on polishing his/her teleconferencing
technique.

To demonstrate how the techniques might
be used in a teleconference using telewriting,
we will briefly define and discuss each design
component, and then suggest a number of ways
to use the blackboard in implementing some of
the strategies.*

---

*We will focus on the electronic blackboard in
this paper. For a more detailed description
of how to design for an audio teleconference
see: Bridging the Distance or TELETECHNIQUES,
An Instructional Model for Interactive Teleconferencing.

H U M A N I Z I N G

Humanizing is the creation of an environment which emphasizes the importance of each individual and overcomes any barriers of distance by generating a feeling of group rapport.

How important is humanizing? It may or may not directly affect communication effectiveness but it's very important when we talk about the "satisfaction quotient," that is, how well people like teleconferencing. However, an added benefit of humanizing is that in gaining information about the members of your group, you influence the effectiveness of your communications with them.

One of the main ways of beginning to humanize is to establish some type of personal information base about group members. This may help counteract the feeling expressed by some participants that teleconferencing is "less personal" than face-to-face settings.

Instructors can develop a roster list of locations with information about each individual at the sites. By using names from the roster when calling on individuals, the instructor can make a beginning step in the getting-to-know-each-other process.

If the instructor is not able to visit each location at some time during the series--a highly recommended practice--then providing a slide or picture of him/herself will help students associate that individual with the voice being heard. (Ongoing classes have sometimes even shared roster lists and pictures of all students participating, via a "round robin" circular.)

Another key to successful humanizing is the use of a "local facilitator." This individual's responsibility is to assure that students feel comfortable with this new learning experience, help orient them to the use of the equipment, see that any materials needed for the session are available--in short, see that "all is well" at the local site and serve as a link between the students at the distant location and the instructor.

The success with which humanizing strategies are realized has a direct influence on participation of the group members.

Some "hip pocket" techniques for

H U M A N I Z I N G

☐ Use the blackboard during an "informal roll call." To create an informal, friendly atmosphere and to give students a chance to become accustomed to using the microphones, ask for a few responses from selected locations. During the first session, use the blackboard to add interest. Sketch a map. As you talk with individuals at each site, note the name and approximate point geographically of that group.

☐ Introduce yourself. Don't be afraid to use the blackboard as an extension of your personality. Use it during the first few minutes of the program to create an atmosphere for learning and to "humanize." Use it to introduce yourself--your name, your course, perhaps some phone-in "office hours." If your personality allows--you might add a touch of whimsy by sketching a picture of yourself on the blackboard.

☐ Remember your remote locations. If you're like most instructors, you'll want to have students with you at your home location as you teach. If so, you'll want to make a special effort to treat both home and remote locations alike. Remind your face-to-face group to use mikes when speaking so that distant groups can hear the questions. And make sure your instructions and references are tailored to a teleconference by using the blackboard to note any special instructions or important points you wish to make.

☐ Continue to use names. Remind individuals to identify themselves by name when they speak. Continue to use names as you talk with individuals. People like to be recognized by name; it is an especially important courtesy in teleconferencing, as it links a personality with a voice.

PARTICIPATION

Participation is transcending technological factors to allow a more natural interaction to take place--by providing for and encouraging opportunities to communicate on the part of students.

We want active participation because we know it is important in any learning or communications process. Two guidelines that might be noted in thinking about teleconference participation are:

1) Participation opportunities and strategies should be carefully planned in advance of program time.

2) Participation involves not only encouraging group members to contribute to the program, but also the vital process of interacting with the content--what you are saying and the visual material you are providing via the blackboard or other media.

The first guideline tells us that participation doesn't always "just happen" in teleconferencing. Studies suggest that there is an automatic training over time in which individuals consciously and unconsciously adapt their ways of communicating as they use an interactive system. However, some individuals adapt more quickly to new situations; others may need encouragement to do so. Building participation formats into the program is one way to help assure that interaction will take place. Asking some individuals to be responsible for taking the lead is another.

Our second guideline concerns the fact that most of us think of participation as the kind in which we see an individual or a group engaged in an activity of some sort: asking a question, making a comment, jotting down notes. Seeing the activity gives us confidence that the person is in the process of actively hearing and remembering what is being said.

However, the unobserved participation such as active listening (anticipatory alertness) and mental work in thinking about a problem are important processes. By skillfully using questions, both written and oral, and by guiding group members in some mental exercises, you can stimulate this internal participation.

Some "hip pocket" techniques for

PARTICIPATION

☐ Get specific groups involved. Ask a group at one of the locations to be responsible for the answer to a specific question to be presented at the next session. This responsibility encourages students to sit down and talk with one another after program time. Cooperation is fostered--and there is a built-in leadoff for the next program's discussion. If there is a blackboard at each location, have the groups use it as they present their answers.

☐ Stimulate participation. Stimulate individuals to interact with what you are presenting. Leave some portion of the materials to be completed--or erase a portion of what you have written--so the group has a chance to do a little mental "filling in" of your presenta-. tion. Put a thought question on the board to get small groups discussing at their own locations.

☐ Use the blackboard as a participation tool. Put a list of key points on the board. By giving your students a "sneak preview" of what you are going to cover, you can create the kind of mental anticipation important to learning. You can use the "key points" technique as a discussion starter too. Jot down some points you think might need more clarification--ask group members to select one or two they'd like to see discussed.

☐ Present interactive formats via the blackboard. Use formats which encourage participation. Almost all types of participation formats have been used successfully in teleconferencing: case study, interview, panel discussion, role playing. Provide the background necessary to introduce these formats by writing it on the blackboard. Continue to use the board as you elicit participation from the group for illustration and key point summary.

MESSAGE STYLE

Message style is presenting what is to be communicated (both visually and aurally) in such a way that it will be received, understood and remembered by participants.

Basic to understanding how to design a message is the realization that at any given moment, we will have the full attention of only a fraction of our audience. The reason for this is simple--the differences between the rate at which our brains process information and the rates at which we can give that information-- either by speaking or writing. Because the brain processes at a much faster rate, there is a temptation for it to sidetrack and go off on a thought quite different from the subject at hand. Accordingly, pacing and repetition within a message become important considerations.

Research points out the strong link between interest in--and retention--of subject matter. Unfortunately, research doesn't give us many clues as to how to generate and maintain that interest! We do know that variety can be one way. If things are too orderly and logical, they become routine quickly. Thus, a point-by-point lesson, although "well-organized," may not have the excitement of one more

spontaneously presented. (An exception would be, of course, the necessary organizational structure inherent in learning a series of concepts which form a hierarchy.)

If learning is presented in a more "disorganized" fashion to provide interest and variety, we need to remember that an underlying structure is important. Ausubel's work with advance organizers has demonstrated the value of general, abstract overviews of the content. Research in listening shows that verbal cues which emphasize the important ideas the speaker wants remembered increase their retention dramatically.

Based on our discussion of attention span and the need to keep interest levels high, we recommend the concept of the "short learning segment," i.e., presenting the material in smaller bits of information, interspersed with question-and-answer or other activity. We would define "short learning segment" as approximately 10 to 15 minutes.

Message style, then, should be a combination of some structure (advance organizers and verbal cues--repetition and emphasis) as well as variety to add interest.

Some "hip pocket" techniques for

MESSAGE STYLE

☐ Plan ahead for visual communication. Use the "planner" concept to gear you for what you are going to "say" visually. It's important that what's on the blackboard complements the information you are giving verbally. Plan your topics for short learning segments--about 10-15 minutes. Note on your planner where questions might naturally arise in the minds of your students--or ask your participants to interrupt as you go along.

☐ Make creative use of the blackboard. Use an advance organizer--an overview of the material you are going to present by putting key phrases on the board. Use the blackboard to compare and contrast objects you draw, present sequential time segments relating to a single event, give meaning to an abstract idea, emphasize a fact or concept, or illustrate relationships such as parts to a whole.

☐ Use the blackboard for emphasis. Use repetition and summary to help individuals remember the message. Use the blackboard to provide some of this repetition by placing words, phrases, or diagrams on the board. Repeat any new words, concepts or phrases at least three times during an hour program. New words can be spelled out on the board for better understanding.

☐ Remember to use the prerecord capability. Do you have a drawing that is a bit on the complex side--one you would like to have come out "perfectly?" Use the prerecord capability. Tape the visual portion ahead of class time. When you're ready for that complex visual, you'll be all set. The prerecord technique lets you relax at program time--as well as for all those "next times" because you can use the tape you made over and over.

FEEDBACK

Feedback is the process of completing the communications loop which enables both speaker and listener to correct errors and omissions in the message.

How important is feedback in teleconferencing? It's an essential ingredient, just as in any other communications process. It is important to consider the mechanisms that might be incorporated into a program, because with members of the group physically separated from you, you won't have nonverbal cues to rely on.

The easiest way to incorporate feedback is, of course, to make use of the "short segment" concept and after each natural division in the material, ask questions. Was the message too fast? too complex? clearly explained? These questions may be asked spontaneously or by asking for feedback from some individuals.

Feedback's importance cannot be overemphasized. Whatever the teleconference's objective, people want and need to know "how they are doing." Channels other than verbal communication during the program can provide some of the feedback--a personal note, phone-in office hours, comments on homework, a "feedback form" (pre-stamped, preaddressed form provided with class

materials). All are ways to keep communication channels open.

The simplest form of feedback, so obvious it's often forgotten, is a brief verbal acknowledgment whenever a comment is made by a group member or whenever a question is answered. For example, "Yes, I see," or "Fine" (or whatever phrase is appropriate) is feedback in its most basic form, but it's important because it tells participants that the message was received--and helps the flow of communications.

Another type of feedback often forgotten is self-feedback. This can be an excellent tool to improve next week's or next year's program. Many instructors use the previous program tape to do some self-diagnosing. During the actual instructional process, the pacing usually doesn't allow an individual to observe what's happening. By listening to and watching the tape of the session, instructors can catch both strengths and weaknesses of the program. Did the message come across with enthusiasm? Naturalness? Were opportunities for participation provided? Did individuals take advantage of those opportunities? This type of feedback can be as important as that received from the program participants.

Some "hip pocket" techniques for

FEEDBACK

☐ **Ask for feedback.** Make use of the interactive capability--verbal (and if you have blackboards at each location--visual) to ask for feedback. Check on your pacing and the quality of the reception at the locations. Use a "test pattern" to check out the monitors at each location. This can be prerecorded and played back just before the program begins--or as part of your opening informal roll call.

☐ **Assign feedback.** Because individuals new to teleconferencing may be shy about giving feedback, preassign it. Call several students in advance of the program and ask them to be responsible for feedback--on content, process--and the technical quality. Another alternative is to prerecord an entire program, allowing you to visit one of the sites to observe. Use the audio system at that location to interact with other sites.

☐ **Use on-the-spot application of information.** Find out if the material being presented is relevant to participants' situations. Pause several times during the presentation to ask one or two participants to briefly comment on the value of the information. Would the person use it? If so, how? Although this technique would not be applicable in certain subject matter areas, it can be very meaningful in many continuing education courses. It stimulates others to think creatively about their own applications.

☐ **Listen and look at tapes.** Teleconferences tend to be concentrated action--with little time for reflection. You may miss an important ingredient in the program because of your involvement in it. Tape your program--then review what happened during the session at your leisure.

SUMMARY

In this paper we have explored a number of principles and applications regarding the use of a telewriting system for teleconferencing, specifically the electronic blackboard. Two major aspects were discussed.

Under "Transmitting Technique" we focused on the considerations critical to successful use of the blackboard in reproducing the telewriting itself. Such topics as basic operation, legibility factors, tracing techniques, prerecording, ground rules, pairing visual/verbal transmission and the "planner" concept were discussed.

In the section entitled "Designing for a Teleconference," we presented a model for adapting instruction to a distant learning environment. Basic design components of Humanizing, Participation, Message Style and Feedback were defined and illustrated by simple techniques incorporating the blackboard.

The electronic blackboard is one of the electronic tools which will enable an instructor to reach beyond the single classroom by providing the capability to present that proverbial "picture" that we talked about at the beginning. With good transmitting technique and good teleconferencing technique, that picture, then, indeed, may be worth more than "ten thousand words."

REFERENCES

Ausubel, David P. The Psychology of Meaningful Verbal Learning. New York: Grune and Stratton, 1963.

Fleming, Malcolm and Levie, W. Howard. Instructional Message Design (Principles from the Behavioral Sciences). Englewood Cliffs: Educational Technology Publications, 1978.

Kemp, Jerrold E. Planning and Producing Audiovisual Materials. New York: Thomas Y. Crowell Company, Inc., 1975.

Korzenny, Felipe and Bauer, Connie. "A Preliminary Test of the Theory of Electronic Propinquity: Organizational Teleconferencing." Mimeographed. Paper presented to the Information Systems Division of the International Communication Association at its annual conference in Philadelphia, May 1-5, 1979.

McVey, G. F. "Legibility and Television Display," Educational Television, November, 1970.

McVey, G. F. "Television: Some Viewer Display Considerations," AV Communication Review, Vol. 13, No. 3, Fall , 1970.

Monson, Mavis K. Bridging the Distance. University of Wisconsin-Extension, Madison, 1978.

Monson, Mavis K. "Designing for the Participants," Journal of Communication, Vol. 28, No. 3, Summer 1978.

Olgren, Christine H. "Visual Systems for Teleconferencing: Telewriting, Televideo and Facsimile," Technical Design for Audio Teleconferencing. University of Wisconsin-Extension, Madison, 1978.

Parker, Lorne A. and Monson, Mavis K. TELETECHNIQUES: An Instructional Model for Interactive Teleconferencing. (Vol. 38 of the Instructional Design Library). Englewood Cliffs: Educational Technology Publications, 1980.

Parker, Lorne, Monson, Mavis and Riccomini, Betsy, editors. A Design for Interactive Audio. University of Wisconsin-Extension, Madison, 1976.

Severin, W. "Another Look at Cue Summation," AV Communication Review, Vol. 15, 1975.

394

FUTURE TELECONFERENCE MEETINGS: PATTERN AND PREDICTION

Lesley A. Albertson
Senior Psychologist
Research Laboratories, Telecom Australia

Given the proliferation of behavioural studies of teleconferencing that has taken place in the past decade, a disinterested observer might well ask why, in 1980, psychologists should be proposing more studies of this kind. Those of us who have been involved in teleconferencing research, however, hardly need to be reminded of the complexity of researching this subject, and of the difficulty in reaching definitive answers. One of the earliest teleconference researchers, Alex Bavelas, commented:

Even a partial list of the behavioural variables that could reasonably be expected to affect the process and outcomes of a teleconferencing facility would virtually define the field of social psychology...Obviously, what is required...is the selection of a relatively small number of variables -hopefully those most immediately relevant to the questions that must be answered about teleconferencing (1963:1)

When behavioural research into teleconferencing began in earnest in the early 1970s, most researchers concentrated their efforts on empirical investigations of the effects of channel type (audio, audio-video or face-to-face) upon meeting outcomes and user attitudes. While channel type seems a logical choice for an independent variable, it has emerged from these studies as far less important than was initially expected. Channel type has been reported as having either no effects upon objectively-measured group outcomes (Chapanis, 1973) or quite subtle qualitative effects, depending upon the type of task being carried out (Short, Williams and Christie, 1976). And while statistical differences have been found in user attitudes towards the different media, these differences have been described as "relatively trivial compared with the task variable" (Short et al, 1976: 152). Unfortunately, however, by the time these researchers had arrived at this conclusion, channel type had become an inextricable part of their research edifice: neither the "Type Allocation Model" nor the

notion of the "Social Presence" of a medium can survive unless channel type accounts for a substantial part of the variance between teleconferencing media.

Despite its empirical base, then, the prevailing research framework cannot be retained if key variables and assumptions are to be changed. The evidence thus points to the conclusions that if this area of research is to progress, the selection of research variables and a theoretical model should begin anew.

Selecting Variables Relevant to Teleconferencing Research

Selection criteria

In tackling the question of what conditions are necessary if a small group is to perform at maximum capacity, it is important that the results of investigation are not only significantly different, but are also practically useful. In elaborating this basic principle, the following criteria for selecting research variables were developed:

(i)The selected variables should be capable of being reliably controlled by the provider of the facility.

This criterion leads to an emphasis on selecting variables that will lead to the specification of design changes in the facility. While recognizing that some user education will be necessary, complex behavioural changes cannot be controlled reliably by the system provider. Hence, the teleconference facility should be designed as far as possible to fit existing patterns of conference behaviour, rather than that the user should be required to adapt to the system.

(ii)Variables selected should be capable of

testing the upper limits of the facility.

It can be shown, for example, that for a group to perform extremely simple tasks, only minimal facilities are required: written messages handed to the chairman of a conference produces a solution as efficiently as a "real time", fully interactive communication medium. This information, however, is hardly useful for designing a teleconferencing facility. It is not practical to design a range of teleconferencing media and to expect users to match their meetings with the medium that minimally meets their needs. In designing a general-purpose facility, some features may well be redundant; but it is important to know at what point the facility is no longer adequate to the task, and for what reasons distortion or breakdown occurs - even if, in practice, these limits are rarely tested.

(iii)The selected variables should account for a substantial proportion of the variance between teleconference meetings.

This criterion suggests that new studies of teleconferencing should be based upon variables that have been shown to be important determinants of group behaviour. In particular, the variables should have sufficient theoretical backing that specific and falsifiable hypotheses can be made at the outset of the research. Valuable discoveries in both the natural and the social sciences have begun with empirical research, and there are no general methodological grounds for rejecting this approach. However, in the assessment of teleconference facilities, empirically-based studies have multiplied without adding greatly to our basic understanding of the psychology of teleconferencing.

With these three criteria in mind, the literature on small group behaviour was searched for a model capable of generating variables suitable for behavioural studies of teleconferencing.

Evaluating theoretical models of small group behaviour

In the study of small group behaviour, the three recognized attempts at theoretical model-building centre around group processes, group structure and leadership(see Lindzey and Aronson, 1969). Initially, neither process nor leadership models appeared capable of satisfying the criteria outlined above.

Process models centre around the concept of distribution: the distribution of information among the members, the distribution of responses required to achieve co-ordination and control of the group, and, in such cases as conflict resolution and mixed-motive games, the distribution of outcomes. While such models offer an understanding of group dynamics, but such an understanding seems mainly useful for a retrospective understanding of why certain outcomes occurred. Since applications of process models to teleconferencing would require changes in user behaviour, it was considered that variables derived from process models would not satisfy the criteria established above .

Leadership models have developed from the conventional wisdom that the personality and style of the leader affect group functioning. Since leadership style is also a user characteristic that can not be controlled reliably by the teleconference provider, variables generated by leadership models also failed to meet the criteria and were excluded.

Structural models consider the small group as a communication network. Laboratory studies which contributed to this model, such as those of Leavitt and Bavelas, analysed the effects of restricting group communication to such configurations as circles, chains and wheels (see Figure 1). A summary of the results of such studies is shown in Table 1 below.

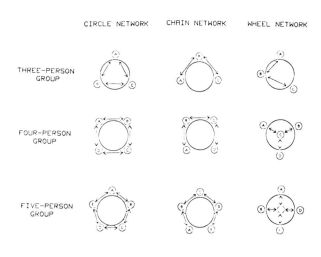

Figure 1: Communication flow in circle, chain and wheel networks

Because group structure is potentially controllable by the teleconference provider, structural models satisfy criterion (i). They also offer the possibility of satisfying criterion (ii). Generally speaking, a centralized network (such as Y, wheel) is the most efficient for simple tasks, whereas a decentralized network (all-channel, or "comcon") is the most efficient for complex tasks. The model thus indicates that for complex tasks, the configuration of the conferring group should not prevent decentralized interactions.

Criterion (iii), however, requires that the variables selected should account for a significant proportion of the variance between conferring groups; and this criterion is more difficult to satisfy. First, in a teleconference, interactions between participants are not physically prevented in the manner of the early laboratory experiments, in which handwritten messages were passed through slots in wooden partitions separating the subjects. More recent work, however, has shown that in groups within which all interactions are possible, group structure still accounts for an appreciable proportion of the variance (see Faucheux and Mackenzie, 1966). Nevertheless, structure is seen as a mediating variable between group task performance and group process, rather than itself determining group outcome. Network structure may assist in the formation of an efficient group organization, or it may hinder it. Furthermore, the most efficient network possible with a given network structure may not emerge in the course of one trial. Few laboratory studies have been able to examine the effect of repeated experience on the efficiency of network organization, but recent studies using computer simulation of network organization suggest that up to 200 trials may be required before the most efficient group organization emerges (Shane, 1979).

Table 1: Performance of circle, chain and wheel communication networks

---

	Circle	Chain	Wheel
SPEED	slow	fast	fast
ORGANIZATION	no stable form	slowly emerging but stable	almost immediate and stable
EMERGENCE OF LEADERSHIP	none	marked	very pronounced
MORALE	very good	poor	very poor

Other research indicates that the group leader contributes substantially to the emergent network organization. In the network studies conducted in the laboratory, the person occupying the topologically central location is usually appointed by the group as leader, and this appointment is usually also the most effective from the point of view of network organization.

In field conditions, however, the chairman will have been selected in advance, normally on the basis of status within an organization. One of the most influential theories of leadership, Contingency Theory (see Fiedler, 1967) indicates that leaders tend to maintain their leadership styles across a range of situations; and, as noted above, this style is not regarded as amenable to change by the teleconference provider. Nevertheless, what has traditionally been called leadership style can be viewed in behavioural terms that have direct implications for network organization. A democratic (or participative) leader attempts to decentralize interactions within the group, while an authoritarian (autocratic) leader attempts to centralize interactions upon himself. This behaviour can be expressed in network terms as follows:

> ...the effect of a participative leader is effectively to assign an equal weight to all members' resources, whereas the autocratic style is characterized at one extreme as one with a matrix of group members' weights equal to 0 and a leader weight equal to 1. (Shifflet, 1979:75).

The effect of these leader-induced variations in the information flow produced by a network depends upon the type of task being performed by the group. An authoritarian leader induces better group performance for structured tasks, while a democratic leader induces better group performance for unstructured tasks.

It would also be expected that the authoritarian chairman will prefer centralized network configurations, while the democratic chairman will prefer decentralized networks. Further, as shown in Table 1, laboratory studies have also found that network configuration affects group morale: should this aspect then be taken into account in designing a teleconference facility? Fiedler (1967) dismisses member satisfaction as "an interesting by-product"; but while this position may be tenable if a once-only performance is being considered, satisfaction of the members is related to group cohesion, and is therefore important if long-term maintenance of the group and repeated use of the facility are considered. Thus, the theoretical model from which experimental variables are to be derived, although based on a structural theory, has become more comprehensive, incorporating as well some concepts of group process and of leadership behaviour.

Observations of Teleconferencing Behaviour

With this emerging model in mind, observations were carried out of inter-member interactions over three teleconferencing media available in Australia:

(i)Confravision (an audiovisual facility operating between Sydney and Melbourne)

(ii)An experimental microprocessor-controlled group-to-group audio system

(iii)Multi-part telephone calls using a conference bridge

As noted above, with the three teleconferencing media under consideration no exchanges between the conference participants are actually prevented. Observations of teleconference meetings using these media, however, suggest that all inter-member communications are not equi-probable.

With the Australian Confravision facility, for example, videotapes show that a high proportion of the interactions are between-terminal; within-terminal communication is rare, particularly with inexperienced Confravision users. Part of the explanation for this interaction pattern may lie with the physical layout of the terminal (see Photograph 1).

A face-to-face conference configured in this way would almost certainly cause subjective discomfort: the proximity of participants violates the culturally acceptable distances for public utterances. It is difficult for the chairman to keep all participants at his terminal within his field of vision, and his location at one of the two central positions makes it somewhat awkward to address the participants at his own terminal, although it facilitates addressing members at the opposite terminal.

The group-to-group audio teleconference facility (see Photograph 2) constrains communication in a different way.

Within-terminal communication is encouraged both by the seating pattern and because there is an asymmetry in the availability of non-verbal cues to remote participants and those within the same terminal. In this type of conference, the leader's task is to make between-terminals interactions as probable as within-terminal interactions.

In the case of multi-part telephone calls, particularly with a large group, the only feasible organization is for the leader to sequentially poll the participants for inputs; any other communication pattern rapidly becomes chaotic. Functionally, then, although all interchanges are possible, the emergent network configuration is equivalent to a wheel network. Interviews conducted by John Short with users of the audio conferencing facility at the University of Quebec supports these observations. In this study, the respondents - mostly chairmen of meetings - felt that the system "subtly increased the power of the chairman and the importance of the chairman" while at the same time leading to a more "democratic" meeting, because the formal meeting procedure ensured that everyone's view was heard (Short, 1973:15-16).

Towards an Interactive Model of Teleconferencing

From the theoretical review and the observations of teleconferencing, the following proposition is put forward:

A teleconference group's capacity to achieve the most efficient organization is a function of the entropy of the medium, the behaviour of the chairman and the complexity of the communication task. "Entropy" here is defined as the medium's emergent tendency towards achieving decentralization [2].

From this general statement, the following specific predictions are made:

1.1 A Confravision meeting will be moderately difficult either to centralize or to decentralize.

1.2 The performance of both simple and complex task will be less effective at a Confravision meeting than at a face-to-face meeting.

1.3 Both autocratic and democratic chairmen will be less satisfied with a Confravision meeting than with a face-to-face meeting.

Photograph 1: Australian Confravision (Melbourne Terminal).

Photograph 2: Experimental Microprocessor-controlled Audio Teleconference System.

2.1 A group-to-group audio conference will be moderately difficult either to centralize or to decentralize.

2.2 Performance of both simple and complex tasks will be relatively less effective than in a face-to-face meeting.

2.3 Both autocratic and democratic chairmen will be less satisfied with a group-to-group audio conference than with a face-to-face meeting.

3.1 With a multi-part telephone call, it will be relatively easy to achieve a centralized group organization, and relatively difficult to achieve a decentralized group organization.

3.2 Simple tasks will be performed relatively well, and complex tasks relatively poorly.

3.3 Autocratic chairmen will be relatively satisfied with the facility, and democratic chairmen relatively dissatisfied.

Linking Micro Studies with Macro Studies

The above hypotheses are open to experimental tests that will show what happens within micro-networks under controlled conditions[3]. But given that the telecommunications network provides a substratum upon which innumerable micro-networks can potentially be based, the next problem is to predict the extent to which these potential micro-networks are actualized. Such predictions entail making assumptions that have no necessary relationship to those made in conducting laboratory studies. Macro analysis is an emergent level above micro analysis; each has its own peculiar characteristics.

Many teleconference researchers have employed the concept of travel substitution to bridge these two levels. Short et al (1976), for example, link their laboratory studies to the problem of macro prediction by arguing that if a teleconference facility is shown in behavioural experiments to be as effective as face-to-face communication, and it is cheaper than travel, then the teleconference meeting will substitute for the face-to-face meeting. Consequently, prediction of the future use of teleconferencing has been based upon analyses of the content of present face-to-face meetings, in order to determine what proportion are substitutable [4].

While estimates of demand can be made relatively easily in this way, a number of

questionable assumptions are contained in the argument. It is also necessary to assume that the individual who makes the decision not to travel (generally portrayed as a senior executive) is both aware of the results of behavioural experiments and actually makes trade-off decisions in the fashion of the "rational man" of market economics. In the light of recent research into organizational decision-making, this simplifying assumption is difficult to maintain. One management researcher, for example, characterized the decision-making of senior executives in the following terms:

> Senior managers will usually receive for review what amounts to a single option...rather than a set of fully developed choices. They usually face yes or no decisions rather than trade-offs. Rarely, moreover, do the proposals that they see include assessments of possible competitive responses or government constraints that will emerge over the longer term. (Peters, 1979).

It also seems peculiarly inappropriate, when one is dealing with the future of teleconferencing facilities, to consider only the mechanisms of individual choice, rather than the mechanisms of group choice.

Another problem in predicting the future use of teleconferencing is that the introduction of the facility itself is likely to bring about new modes of behaviour. While it can be legitimately argued that people are unlikely to make rapid changes in their behaviour, and that only present behaviour is available for study, the implicitly static view of needs underlying the substitution approach has demonstrable limitations. On the one hand, by failing to recognize that teleconferencing may serve needs that were previously unsatisfied, potential markets may be overlooked. On the other hand, the substitution approach may overestimate demand by assuming that the level of communication represented by present travel is fixed, and must be satisfied by one medium or another. A more systemic view would see present levels of communication and travel as means by which the organization functions within a given context. Levels of communication have not remained constant in the past, and no sound reason has been given for expecting that they will remain constant in the future . Thus, changes in the cost of travel may not mean that communication is automatically transferred to a telecommunications medium; rather, it may mean that the organization changes its mode of functioning - for example, by reducing the geographical range of its operations [5].

If demand is to be estimated from an analysis of existing face-to-face meetings, assumptions similar to those discussed above. are certainly necessary. However, there is increasing evidence that many of these assumptions are untenable. It seems reasonable to argue, then, that a necessary intermediate step to predicting the future demand for teleconferencing is to gain better insights into how demand for teleconferencing is generated. The results of research into social networks at the macro level suggest that network analysis may again be a useful way of gaining these insights.

Macro Studies of Communication Networks

Techniques for analysis of macro networks have in common the concept of an individual as a node connected by lines representing communication links; however, a variety of specific techniques for data collection and representation have been developed. The most appropriate technique in a particular case is likely to depend upon the population and the problem to be solved.

In business and industry, for example, needs for teleconference facilities are likely to be latent rather than manifest. In order to predict the organization's needs for teleconferencing, therefore, it may be necessary to conduct an audit of communication patterns, and, by examining the organization's goals, to recommend ways in which using teleconferencing could improve organizational effectiveness (see Albertson, 1975). Other research evidence suggests the importance of inter-organizational communication in the diffusion of technological innovations (Czepiel, 1975). An adequate network technique in such cases may therefore need to go beyond the single organization as the unit of analysis. One possible means of achieving this would be to explore a hierarchy of networks (see, for example, Conrath, 1973) that includes an inter-organizational level of analysis as well as intra-organizational levels of analysis.

In the area of community service, Kay and Kramer (1978) have shown that a network approach in which the population sample is derived using a "snowballing" technique provides a useful methodology for defining communication needs within a rural community. In urban communities, however, potential teleconference users may initially be quite unconnected [6]. In this context, the concept of "blockmodels", which is based upon

similarity of ties (rather than sociometric choices) and which identifies gaps in communication networks, may prove useful (see White, Boorman and Breiger, 1976).

New Directions in Teleconferencing Research

Attempts are currently being made to establish theoretical links between micro and macro social networks (see, for example, Clark, 1979). These developments open up exciting possibilities for developing a comprehensive framework for teleconferencing research. Meanwhile, the network approach promises to provide a workable alternative to the prevailing channel/substitution paradigm in teleconferencing research: a paradigm whose assumptions have become increasingly discredited.

Notes

1. This result in itself has implications for the conduct of teleconferences; few teleconferencing groups would have sufficient experience for the most efficient network organization to emerge.

2. An objective measure of the centrality of the emergent network was provided by Mackenzie's Total Expected Participation Index (TEP). The value of TEP ranges from 0 to 1, being greatest for a decentralized network that permits maximum information flow and least for a completely centralized network. A decentralized network, according to this scheme, is defined as having greater entropy than a centralized network.

3. Pilot tests of these hypotheses have just been completed at the Telecom Australia Research Laboratories, and results are currently being analyzed.

4. Champness (1972) estimated that up to 65% of existing face-to-face meetings could be replaced by teleconferences. Later estimates are more modest, although the basic approach remains unchanged.

5. The "generated" meetings discussed by Short et al (1976) are seen as necessarily increasing the estimates based on substituted traffic.

6. An example is the "Telelink" service operating in two Australian cities. This service uses multi-part telephone conference calls to provide socially isolated and housebound people with an opportunity for group communication (see Albertson and Bearlin, 1978).

References

Albertson, L. "Researching Business Needs for Teleconference Facilities", Unpublished Report, Telecom Australia, September, 1975

Albertson, L. & Bearlin, K. "Melbourne Telelink: A Case Study", Laboratories Report No 7421, August 1978

Bavelas, A. "Teleconferencing: Guidelines for Research". Institute for Defense Analyses, Research and Engineering Support Division. Research Paper P-107, November, 1963

Champness, B. G. "The effectiveness and impact of new telecommunications systems", University College, London. Unpublished Report (1972)

Chapanis, A. Parrish, R. N., Ochsman, R. B., & Weeks, G. D. "Studies in Interactive Communication:II. The Effects of Four Communication Modes on the Linguistic Performance of Teams during Cooperative Problem Solving", Human Factors, 1977, 19 (2) 101-126

Clark, A. W. "Interorganizational Network Analysis: A link between micro and macro social analysis". Unpublished Working Paper, La Trobe University, 1979

Conrath, D. "Communications Environment and Its Relationship to Organizational Structure". Management Science, 20, 4, Part II, December, 1973

Czepiel, J. A. "Patterns of Interorganizational Communications and the Diffusion of a Major Technological Innovation in a Competitive Industrial Community", Academy of Management Journal, Vol. 18. No. 1. March, 1975

Faucheux, C. & Mackenzie, K. D. "Task Dependency of Organizational Centrality: Its Behavioural Consequences". Journal of Experimental Social Psychology, 1966, 2, 361-375

Fiedler, F. E. A Theory of Leadership Effectiveness. N. Y. Wiley, 1967

Kay, P., and Kramer, J. F., "Planning for Rural Telecommunications Systems:Phase 1 - A methodological approach to community needs analysis". U.S. Office of Telecommunications Policy, January 1977

Lindzey, G. & Aronson, E. (eds) Handbook of Social Psychology, Second Edition, Addison-Wesley, Massachusetts, 1969

Mackenzie, K. D. "The Information Theoretic Entropy Function as a Total Expected Participation Index for Communication Network Experiments". Psychometrika. Vol 31, No 2, June, 1966

Peters, T. J. "Leadership: sad facts and silver linings", Harvard Business Review, November-December 1979

Shane, B. "Open and Rigid Communication Networks: A re-evaluation by simulation", Small Group Behaviour, Vol 10, No 2. May 1979, 242-262

Shifflet, S. "Toward a General Model of Small Group Productivity", Psychological Bulletin, Vol. 86, No. 1, 67-69

Short, J. "A Report on the use of the audio conferencing facility at the University of Quebec". CSG Report P/73161/SH

Short, J. Williams, E. & Christie, B. The Social Psychology of Telecommunications, Wiley, 1976

White, H. C., Boorman, S. A. & Breiger, R. L. "Social Structure from Multiple Networks I. Blockmodels of Roles and Positions. American Journal of Sociology, Vol 81, No 4, 730-780

The Teleconferencing Resource Book: A Guide to Applications and Planning
Lorne A. Parker and Christine H. Olgren (eds.)
Elsevier Science Publishers B.V. (North-Holland)
Center for Interactive Programs, University of Wisconsin-Extension, 1984

INTERNATIONAL POTENTIALS
OF
COMPUTERIZED CONFERENCING[1]

Murray Turoff and Starr Roxanne Hiltz
Computerized Conferencing & Communications Center
New Jersey Institute of Technology
323 High Street
Newark, N.J. 07102

ABSTRACT

Current experience with computer based
communication-information systems is projected
to more widespread international uses such as
the exchange of scientific and technical
problems and solutions, conflict resolution,
and the management of crises such as famine or
terrorism. Artificial price hikes and other
prohibitive regulations as well as a lack of
understanding of the dynamics of inter-cultural
communication in this medium are the major
barriers to international applications.

INTRODUCTION

The concept of the 'information society'
has perhaps become 'old hat'. So much has been
written about it that we may have convinced
ourselves that we understand what it means and
how it will shape the future. However, when
one leaves the realm of generalities and focuses
in on specifics, one's sense of certainty
becomes largely illusory. In no single area is
this perhaps truer than when we are dealing
with interactions between information societies.
In this paper we will disregard the scant
empirical evidence and bravely or foolheartedly
step out on the "ice" of conjectures about what
the international arena will be like in the
coming "information age" of computer mediated
telecommunications.

Our intention is to base these conjectures
on some substance, however nebulous. We
therefore will draw on a microcosm sample of an
information society to extrapolate the potential
opportunities, alternatives and consequences of
a future 'wired world'. Our basis of experience
includes four years of operational field trials
of an experimental computerized conferencing
system, tying together approximately seven
hundred individuals in the U.S., Canada and a
number of foreign countries. In January of

1980, EIES had about 30 non-U.S. members,
including seven from Switzerland and three each
in Sweden and Austria. There were none from
developing nations because they currently lack
TELENET ports to make the international
connection. Though there has been some international use, we will thus have to rely mainly
upon intra-national experiences which may be
generalizable to international applications.
In addition, we have some initial results from
controlled experiments in group communication
and problem solving in face to face versus
computerized modes.

The system we will use as a prototype is
called the Electronic Information Exchange
System (EIES). It represents a joining of a
communications space and an information space
in one coherent man-machine interface designed
to allow access to a vast computer resource
and network by individuals with no prior
computer experience or expertise. We will
assume basic familiarity with the technology
in this paper, and concentrate only on a
delineation of characteristics and applications
which are particularly relevant for international communications. (See Hiltz and Turoff,
1978, for a complete general overview of the
technology, its applications, and impacts).
We will make some initial generalizations, and
then turn to current case studies and empirical
observations on the impacts of the use of this
technology, and finally suggest some possible
or probable international applications in the
next decade.

Information is "the message of human
experience" (Becker 1978) which has a content,
a medium of communication, a sender and a
receiver. The medium can, in some cases, be
more important than the content in determining
the impact upon any decisions or actions of the
recipients. Equally important are the
perceptions the senders and receivers or
exchangers have about one another and the
degree to which they share or do not share a
common base of knowledge. It is quite clear

that the relationships between nations and the actions of nations through their leadership have historically been influenced by these factors. Whether it is the negotiation of a commercial agreement, the settlement of a conflict situation or a crisis management operation, the results often turn upon factors of psychology, sociology, culture and technology.

The introduction of the computer as a facilitator of a human group communication process means that all these concerns come into play. It further means that the communication process can be intentionally designed to control these factors to a far greater degree than possible in other communication media. Our purpose in this paper is to look at the differences between face to face meetings and human communications facilitated through a computer, and to infer from this some of the more crucial impacts on common objectives of international communications such as negotiation, persuasion, information exchange, consensus formulation, conflict resolution, problem solving and crisis management. We do not, of course, propose that this new medium might replace the dramatic summit meetings at the top level or treaties and fundamental agreements. We do suggest that it might supplement some meetings, allowing a wider channel of continuous communication for the preparation and follow-up on face to face negotiations and the daily administration of the details of international cooperative meetings. We also feel that it can make possible many more ongoing cooperative efforts by trans-national working groups than is now possible with other available communications media.

But first, we shall fall prey to the explorer's tendency to describe a whole continent on the basis of a limited expedition, and share our visions of an international information anarchy a decade hence. There will be a tremendous computer capacity available in all industrialized nations, and even in many that are not industrialized in the traditional sense. The capacity to transmit data and communications in vast quantities and at low cost will exist. Whether the low costs will be realized is more of a regulatory and political question than a technical one. Information and data will take on the characteristics of a commodity as basic as money or energy. The electronification of information will mean that its value or worth in currency can be easily pinpointed and marketed. Concerns about information flowing out of countries have already reached the point where regulatory barriers are being adopted in some countries. One can easily imagine the same controls and superstructure imposed on the flow of information that nations have imposed on money, though it is probable that this will be even

less successful than is the case for money. The realities of the technology are such that any individual having access to a phone line to an information system will be able to capture in electronic form anything to which he or she has access. Such concepts as copyrighting will suffer from the same difficulties that the citizen with a video recorder presents to the producer of TV programs. People and organizations with information and communication facilities to market will gravitate to whichever countries offer the least restrictions on their international operations. This of course has its analogy with the flags flown today on the majority of the tankers that ply the sea lanes. All things considered, we are likely to go through a period that can be characterized by 'information wars' as opposed to 'trade wars'.

Nations will be faced with the fundamental choice of capitulating to information technology or fighting it by trying to apply outmoded regulatory concepts and artificial controls. It is our belief that the latter approach would be comparable to cutting off one's nose to spite one's face.

Having made these sweeping introductory generalizations, we will attempt to summarize what is known and what has been observed in the utilization of computer technology to facilitate human communication. This will serve as a basis for extrapolation to specific international communication objectives.

FACTORS IN COMPUTERIZED CONFERENCING

TECHNOLOGICAL FACTORS

The concept of utilizing a computer for facilitation of human communications introduces two key technological elements that make this form of communication distinctly different from other forms. The electronic equivalent of a group memory allows people to be non-coincident in time as well as space. It also allows individuals to interact at their own rate in terms of reading at their own desired time and speed and deciding when they wish to respond and how fast they need to compose that response. Furthermore, it means individuals can join a communication process after it has started and not miss out on what has taken place.

Secondly, the logical and computational processing possible within a computer means that the actual protocols of the communication process can be tailored to the application, and that the historical record can be organized in whatever data retrieval structure makes sense for a particular objective. Perhaps most importantly, it eliminates for the people involved the bookeeping chores associated with any group communication process. The computer can keep track for each individual of what is new or of interest for him or her. The result is that the communication process can be tailored around a particular application just as a parliamentary or judicial set of protocols are utilized to tailor a coincident face-to-face group for a particular objective. No other form of remote communication that has been available to date has allowed this flexibility common to coincident groups.

Many otherwise impossible mechanisms for structuring communications occur with such a system. Communications can be addressed by content as well as by individual, allowing discussions to be self organizing both with respect to content and participants--the content can be the address. This feature alone holds tremendous promise for utilizing systems with large populations of users. It also implies a possible fluidity in the nature of human groupings that has never before been possible.

In the following sections we have chosen our examples to highlight the fact that a computerized conferencing system emphasizes structuring of both the communication and information process. Terms commonly used in the current literature like "electronic mail" are highly misleading in that they conjure up merely the process of automating a letter. To view any computerization as a mere automation of what has been done before is the height of naivete with respect to understanding the future of these systems. In essence, we can now do remotely most things that face to face groups have been able to do and there is every promise we may be able to do them somewhat better. It is the resulting changes in psychological and sociological processes that make the real difference.

SOCIAL AND PSYCHOLOGICAL FACTORS

In any negotiations involving such matters as arms control or a political crisis that threatens to escalate to military action, it is crucial that the parties be able to:

1. Exchange information accurately and comprehend it

2. Give and review orders

3. Gain trust in one another; the converse of this being to know if the parties are attempting to lie or mislead

4. Be willing to change their attitudes or

opinions as a result of the negotiations

Many people who have never used a written form of communication or an interactive computer system might intuitively dismiss computerized conferencing as totally inferior to face to face negotiations for any such crucial interactions, and feel that the non-verbal cues in face to face interaction are necessary in order to facilitate the resolution of the conflict. The fact is that these intuitions are wrong. The experimental evidence indicates that media which decrease or eliminate the non-verbal cues can be at least as effective. Briefly summarized, here is some of the evidence:

1. In our current research we have asked both face to face (FTF) groups and users of EIES (CC) to solve two problems and to rate the media on a 1 to 7 scale. We have found no differences in the quality of solution of a task that has a "correct" solution to serve as a criterion. In a qualitative human relations task, we found similar solutions, except that CC groups were less likely to decide on negative sanctions and more likely to decide on "face saving" solutions for both parties in the conflict situation. (See Hiltz, Johnson, Aronovitch and Turoff, 1980 for a complete account of the controlled experiment and the findings.).

2. Davies (1971) compared face to face and teletype for the communication of factual information. Teletype was found to be the more effective mode.

3. Comprehension is generally found to be better with the written word (Toussaint, 1960). This may be because the written channel allows the possibility of rereading or checking difficult passages. (Short, Williams, and Christie, 1976, p. 84). In our controlled experiments, problems were seen as significantly clearer in the computerized mode as compared to face to face groups.

4. Written media are capable of introducing MORE, rather than less, attitude changes. Wall and Boyd (1971) presented subjects with a face to face, a videotaped, and a written message. The written message produced more attitude change than the other two media. In our experiment, however, we found no difference between the media in amount of opinion change.

Short (1972) had 30 pairs of subjects discuss controversial issues face to face, on closed circuit t.v., or loudspeaking telephone. An initial questionnaire was used to select pairs of subjects who disagreed. Medium of communication had no effect on the ability of the disputants to reach agreement. However, as measured by a post-discussion questionnaire, there was "significantly more opinion change in the audio condition than in the face to face condition with the video condition intermediate... Removing the visual channel (making the situation more anonymous) could be taken to decrease the loss of face associated with yielding; in this way there might be less 'opinion change' in the face to face condition where perceived commitment of one's initial position is higher". (Short et. al, 1976, p. 104). This finding has been replicated in four subsequent experiments.

"Lying" and its detection are particularly interesting. Reid (1970) had a speaker alternately lie and tell the truth at randomly determined 30 second segments. A subject sat and received this face to face, and a second subject received audio only. There was no difference in the ability to perceive lying; however, the face to face listeners THOUGHT that they could tell lying better.

While there is no doubt that a written mode of interactive communication represents a loss of non-verbal cues, it could very well be that this is a good thing. Non-verbal cues frequently detract from a participant's ability to focus on the cognitive content of what is being sent and received. Furthermore, cross-cultural differences can lead to the misinterpretation of non-verbal cues.

Goffman (1955) has written extensively about the "face work" that goes on in face to face behavior. He defines this as "the positive social value a person effectively claims for himself, ... and image of self delineated in terms of approved social attributes... A person tends to experience an immediate emotional response to the face which a contact with others allows him; he cathects his face; his 'feelings' become attached to it." (pp 5-6).

In a computerized conference, participants seem able to "let down face" in terms of their overt behavior (see Hiltz, Johnson, and Agle, 1978); and conversely, to be able to consciously control the emission of the "cues and clues" given to others (see Kerr and Bezilla, 1979). Without the necessity to constantly worry about the impression being made by one's facial expressions or other physical cues being "given off", in other words, a participant is potentially freer to concentrate on the content of what is being communicated.

One example of non-verbal communication which interferes with the ability of people to communicate effectively when they are from different cultures is the preferred conversational distance of South vs. North Americans. In Latin America, the

conversational distance preferred is much closer than is comfortable for North Americans; what is felt to be comfortable by the Latin American is likely to be subconsciously interpreted as aggressive or sexual by the North American. (Hall, 1959). Likewise, the Japanese never overtly disagree in face to face meetings; it would be interesting to experiment and see if they would express disagreement in a written mode.

This is not to say that all of the evidence indicates that computerized conferencing might be more effective than face to face meetings for international interactions; but it certainly indicates that it warrants some trials and experimentation. It should also be emphasized that it takes at least several hours of learning and practice time for people to become comfortable and effective with communication via computer. When they first sit down at a terminal, they can barely communicate at all. Moreover, our field trials and experiments indicate that groups which are distrustful of one another and who do not wish to cooperate simply avoid using the system. A high degree of motivation is thus a pre-condition for successful applications.

SPECIFIC EXPERIENCES

What we are going to try to illustrate in the next few sections is that experience with EIES indicates that both the standard and "custom structured" communication-information features can be successfully used for many tasks that involve cooperation or conflict-negotiation among participants in several nations. These tasks include coordination of plans for cooperative ventures, writing of joint reports, the construction of a stored dictionary of accepted meanings for terms that are being used for negotiation, technology and knowledge-resource sharing, and crisis management. Some of the technical capabilities illustrated in these examples have not been explained in the very brief overview of EIES included in this paper; for further clarification, see Hiltz and Turoff, 1978 and 1980.

INFORMATION EXCHANGE: LEGITECH

There is an increasing need for a trend away from competitive inter-national strategies and towards cooperative strategies. Given the major resource interdependencies, currently typified by global dependence on oil-exporting countries, there is little alternative to cooper-

ative strategies. Policies that
lead to a widening of the economic
gap between developed and less
developed societies are likely to
lead to dangerous instabilities.
As the global economy changes to
an information economy, the
knowledge of each society could be
shared for mutual benefit, without
loss to the sharing societies. The
result of such an exchange would
permit continued improvement in the
rate than would otherwise be possible.
This conclusion follows from the
nature of information; the more of
it you share the more you have left
at the end of the transaction. This
is because information exchanges
lead to improvements in the quality
of information, rather than a loss
of information. The information
game can be played on a global scale
as an 'everybody wins' game, rather
than the 'winners and losers' game
that is played with material
resources. (Parker, 1976 pp 14-15).

A current model of such international
knowledge resource-sharing is provided by the
"legitech" group now on EIES. This group
consists of 20 state legislative science
advisors and approximately thirty resource
organizations.

The members of this exchange enter
"inquiries" which are a maximum of three lines
in length. Other members then decide whether
or not to "select" the inquiry. If they do
not choose it, they have filtered out receipt
of all further information or exchange on the
topic. However, they can search the
alphabetical list of inquiry-topics at any
time and select the topic. Thus, among the
fifty members of the exchange, only those
interested in a specific topic among the
hundreds that have been entered are
participating in the additions to and
retrieval from that specialized data base of
current information on the topic. This kind
of scientific information exchange would
probably be a good mechanism for the dissemination of knowledge to developing nations.
This figure shows an actual interaction in an
exchange, including an initial attempt to
retrieve a keyword not in the index. The
inquiry on "biomass" was answered by responses
from Hawaii, California, Oklahoma,
Massachusetts and several resource organizations (the American Society of Mechanical
Engineers and the National Bureau of Standards).
Note that an individual within the responding
organization usually signs the response.
While states are not exactly sovereign nations,
the fact that they have found it possible and
fruitful to exchange their science and

technology-related problems and knowledge via
EIES suggests that a comparable model on an
international level would also be possible.

When a LEGITECH member signs onto EIES, he
or she receives a special menu of choices
geared to this information exchange function.
Figure one shows a printout of some typical
items.

Figure 1
Interacting with Politechs-Legitech

Welcome to Politechs, whose current exchanges
include:

Publictech (pt) & Legitech (le); message
PSI (700) about others.

Welcome to the PUBLICTECH Exchange.

As of 2/26/80 11:15 PM there are 139 topics
and 508 responses in PUBLICTECH.

REVIEW HOW MANY ITEMS WAITING (Y/N/A)?n

PUBLICTECH EXCHANGE CHOICE?

ACCESS TO:
    TOPICS/INQUIRIES   (1)
    BACKGROUND/RESPONSES   (2)
    YOUR SELECTION OF TOPICS   (3)
    ALPHABETIC INDEX OF TOPICS   (4)
    TOPIC MEMBERS AND MARKERS   (5)
    MEMBERS OF THIS EXCHANGE   (6)
    GROUPS IN THIS EXCHANGE   (7)
    MONITOR OPTIONS   (8)
    OTHER EXCHANGES   (9)
    EIES   (0)

PUBLICTECH EXCHANGE CHOICE?4

INDEX CHOICE?

DO YOU WISH TO:
    GET FULL ALPHABETIC INDEX   (1)
    GET ABBREVIATED INDEX   (2)
    ADD KEYS TO TOPICS   (4)
    DELETE KEYS FROM TOPICS   (5)
    RETRIEVE TOPICS BY KEYWORD   (6)

INDEX CHOICE?6
KEY (KEY/KEY PART/AAA-ZZZ)?biomass
TOPICS ARE INDEXED UNDER THE FOLLOWING KEYS:
  BIOMASS. SEE UNDER AGRICULTURE:BIOMASS
KEY?agriculture:biomass
Topics indexed under AGRICULTURE:BIOMASS to
which you have access: 52
To get these topics, enter:
Y (topics only), H (topic headings only), R
(topics + responses), or N (no).
GET (Y/H/R/N)?y

PUBLICTECH T52 DAVID L. JONES (HIPED, 755)

10/10/79   2:01 AM
KEYS: /AGRICULTURE?BIOMASS/
TOPIC: BIOMASS AS ENERGY SOURCE

To what degree is your state/country/
community involved in developing biomass energy,
either from waste materials or from crops grown
for the purpose?

Some Sample Responses to an Inquiry

PUBLICTECH EXCHANGE CHOICE?2

RESPONSE CHOICE?

DO YOU WISH TO:
    GET FULL TEXT OF RESPONSES   (1)
    DISPLAY TITLES OF RESPONSES   (2)
    COMPOSE/ADD RESPONSES   (4)
    MODIFY/DELETE RESPONSES   (5)
    REVIEW RESPONSES   (6)

RESPONSE CHOICE?1
TOPIC/INQUIRY (#)?52
RESPONSE (#/#-#)?1-5

PUBLICTECH T52R1 DAVID L. JONES (HIPED,755)
10/10/79   2:10 AM
RESPONSE: BIOMASS FOR ENERGY IN HAWAII

Hawaii has strong interest in developing
biomass as an energy source. The City and
County of Honolulu (embracing the Island of
Oahu) has begun actively evaluating bids/
proposals for installation of facilities for
conversion of solid wastes from garbage/trash
collections to energy, probably methane/
methanol.

The State Department of Planning and
Economic Development, the State Department of
Agriculture, and other agencies are also
studying various crops which might be grown
for biomass energy conversion. Under
consideration are sugarcane, corn, forest
products (including varieties of Eucalyptus
and Leucaena), kelp, and others. Different
conversion methods, processes, and products
are also being investigated for the best
combination of maximum energy output, minimal
energy input, minimal effects on the
environment, etc. End products may be
ethanol, methane, methanol, hydrogen, fuel
pellets, direct heat by burning, etc.

PUBLICTECH T5R2 VERNER R. EKSTROM (OKLEG,715)
10/11/59   7:20 PM
RESPONSE: OK AND BIOMASS

At the present time the OK Legislature
has formed a special subcommittee on
alternate energy sources to look into the
potential of biomass energy. For further

information contact Lisa Barsumian, State
Legislative Council, 305 State Capitol,
Oklahoma City, OK 73105. 405/521-3201
    Tom Clapper - Legislative Researcher

PUBLICTECH T52R3 VERNER R. EKSTROM (OKLEG,715)
10/12/79 11:34 AM
RESPONSE: /ENERGY-BIOMASS/

Oklahoma City is currently attempting to
assemble package to use municipal solid waste
in combination with the liquid waste from
agribusiness (meatpackers) plus whatever makeup
liquid is required from the municipal sewer
system for co-digestion in bacteriological
digester to produce methane as one of the output
products. This project is rather unstable as
the number of participants is large, the dollars
spent to date are small, and the only local
methane generator has been subject to criticism
for several years on the basis of odor.

This same system can work with any other
sort of biomass that is postulated, and the
same sort of problems will come out of the
woodwork. In any of these systems be sure to
establish who is responsible for the residue.
Chimo, Jim/OKC

PUBLICTECH T52R4 J.M.WYCKOFF/D.CUNNINGHAM
(NBS,744) 11/19/79 2:06 PM
RESPONSE: BIOMASS CONTACT

".TEXT" THE NATIONAL BUREAU OF STANDARDS IS NOT
DOING WORK INVOLVING BIOMASS AS AN ENERGY
SOURCE. HOWEVER A GOOD CONTACT FOR THIS
INFORMATION IS SOLAR ENERGY RESEARCH INSTITUTE,
DEPARTMENT OF ENERGY, 1536 COLE BLVD., GOLDEN,
COLORADO 80401 (303) 234-7171. THE SOLAR ENERGY
RESEARCH INSTITUTE HAS RECENTLY COMPLETED A
SURVEY ON BIOMASS GASIFICATION, SO IT MAY BE
GOOD TO CONTACT THEM FOR FURTHER DETAILS.

PUBLICTECH T52R5 RICHARD BRANDSMA (CAASY,714)
12/6/79 1:46 PM
RESPONSE: BIOMASS

The Calif State Energy Comm, Development Div
(916-920-6033), is developing a program whereby it funds private energy producers to convert
to burning ag and muni wastes. The funds will
be recovered by payback over a period of time.
(I will forward some material in the mail.)
The Calif State Solid Waste Mgmt Board is
involved, to a large degree, in Bio-Mass energy
application. It has one major facilcity in
Eureka, CA which, in 1983, will be processing
1500 tons of wood waste and 300 tons of muni
waste per day... *****Tom Buckles*****

The Legitech operation is a good example
of a tailored subsystem within the EIES system
designed for a group sharing information.
While each state legislature is independent,

there is a recognition that they have meaningful information to exchange both of a factual
and of a lore nature. In fact, one of the
revolutionary aspects of integrating
communication and information functions into
one system (Information Exchange) is the
unique ability to handle transitory lore and
subjective information in the same manner
computerized information systems now handle
factual and quantitative information. However,
the optimal analysis techniques for this type
of information are still a very open and
researchable question. (Legitech was created
by Peter and Trudy Johnson-Lenz, Harry Stevens,
and Jim Williams, using Interact, the EIES
programming language. See Johnson-Lenz 1969
for more details on this subsystem).

PLANNING AND COORDINATION: WHCLIS AND INTERNATIONAL MEETINGS

The most basic of the features of a
computerized conferencing system such as EIES
are the unstructured conference and the
private message system. One of the most
active groups on the system during the last
two years was the "World Symposium on Humanity".
The Symposium itself was a week-long event held
in April 1979...simultaneously in three
locations-- London, Toronto and Los Angeles.
They eschewed a single headquarters where all
decisions would be made by one person or a
small group and then dictated to the other
locations. EIES made possible, through a
joint conference participated in by several
people at each of the three locations, a
decentralized decision making process and
daily sharing of information, problems, and
issues. During the first six months, the
main conference alone for that group generated
about 800 entries. In summarizing the
experience of using computerized conferencing
as the primary means of communication to
coordinate this effort, one of the leaders
wrote:

Figure 2: A Testimonial
Use of a Computerized Conferencing System to
Coordinate an International Organization

Frank Catanzaro (1978)

Due to the highly innovative and
decentralized nature of the World Symposium
planning effort, it was felt that access to
EIES would provide both a highly synergetic
relationship between the Humanity Foundation
and other EIES users, and a unique experiment
in the effect of computer conferencing on the
evolution, management and organization of such
a large transnational event....an event
dedicated to peace and the future. Presently,

at any major Humanity Foundation office you
will find, somewhere, usually under a mound of
computer paper, a small desk-top acoustic
coupler terminal. These devices have a simple
keyboard and a roll of paper, and can print out
messages at a spunky 30 characters per second.
They are lightweight and portable, usually
coming in travelling cases...indeed, when
Symposium staff travel, they generally take
their terminals with them, so they can plug
in through the phone system wherever they go.
The terminals are used daily by our staff
members in Los Angeles, Toronto, Vancouver,
Boston, New York and London, to maintain a
constant flow of communication between offices.
The range of communications which this medium
is able to transmit is quite remarkable. In
the course of our use of EIES, these have
ranged from budget formulation and debate, to
the posing of motions and voting upon them,
to a key-worded random accessible production
notebook, to the sometimes soul-wrenching
torment of cutting through the obstacles on
the way to the World Symposium on Humanity.
The system is used as a confessional, a message
relay to others in our major cities, and a
scratchpad for articles and letters that find
their way into our publications and files. We
have even used EIES to link major conferences
together...for example, in August we had
Symposium staff members attend the Canadian
Futures Fair in Ottawa, the Annual Convention
of the Association for Humanistic Psychology
in Toronto, and the Cable Television Conference
in San Diego. Users accessed EIES through
telephone booths on each of the three sites,
and exchanged reports of the proceedings, with
participants in the conferences adding their
own personal notes. When one of us sits down
at a terminal and dials the local Telenet
computer network telephone number to connect
to the EIES computer in New Jersey, we have
come to expect, regardless of the content of
the latest messages, an immediate sense of
electronic camaraderie. A sense of the
disembodied presence of the group as a whole
seems to leap from the printed page with the
first few keystrokes. The record of all our
conversations, disagreements, fun, poetry and
even affection is maintained in a continuum
which becomes suffused with the vital spirit
of our effort. EIES has become a living element
in the ongoing organization of the World
Symposium.

A computerized conferencing system allows
decentralized control because it provides the
ability to coordinate actions and to establish
the accountability for actions necessary to
support decentralization. One would have to
view the transcript of the Humanities
Foundations's conference to really comprehend
that we have an example of a group of people
spread out half way around the world acting
as a single project team, sometimes on an

hourly basis. We know of no other way that a dispersed project team could have worked together with the same coordination of effort that can usually only be exhibited by a co-located team. Another example of meeting coordination was the White House Conference on Libraries and Information Services. During the six months prior to the large meeting in Washington D.C. in November 1979, approximately 35 members of the national advisory committee and central staff planned and coordinated the conference largely through the use of EIES. At the conference itself, approximately eight terminals were used to input and index the hundreds of resolutions produced by the working groups, according to an indexing sceme developed ahead of time (See Kerr, 1980 for a complete account of this application of EIES). EIES was also used for international coordination of the World Futures Studies Conference in 1979, as well as for remote participation in the Berlin meetings by U.S. members and onlookers. The executive secretary described his organizations use of and projections for computerized conferencing as follows (Menke-Gluckert, 1979, p.3.):

The Program Committee of WFSF tried to organize the Berlin gathering as a self organizing exercise or part of an ongoing learning process. To underline the concept of self structuring and self generating open conference style, the Berlin Conference has started already three weeks ago as part of a world-wide computer conferencing exercise, which will accompany this conference during the next three days. Every participant can exchange his thoughts on all conference items with more than 600 other computer conference participants all over the world. In five years time, Computer Conferencing will be a normal communication medium used everywhere, linking most easily and practically problem generators and problem finders to experts within and outside the scientific community. Information exchange and planning decisions are made reliable through computer-assisted conferencing.

CONSENSUS FORMULATION & PROBLEM SOLVING: JEDEC

One of the preliminary tasks in many negotiations is agreeing on the precise meaning of the terms which will be used. This may be especially time consuming in multi-lingual, multi national negotiations, where comparable translations must be agreed upon, to make sure everyone is talking about the same thing when a word or phrase is used.

A prototype for a system which can be used in the preliminary stages of negotiations has been designed for the EIES system by Peter and Trudy Johnson-Lenz. It is being used by the Joint Electronic Devices Engineering Council, which engages in the difficult mission of getting competing companies to agree on industry-wide standards for products. As they explain the TERMS software in the on-line documentation which they created (Johnson-Lenz, 1975):

+TERMS has been designed to increase the ease and efficiency by which glossaries of terms and definitions may be developed by a geographically dispersed group of people working together over EIES. This special software has the following features of particular interest:

o   any glossary member may add a term or an alternative proposed definition for a term.

o   any glossary member may enter a written comment about a proposed definition.

o   participants may vote on the alternative definitions; immediate feedback of tabulations of their preferences is available using a single command.

o   all information related to a specific term may be retrieved, including all definitions, comments, and results of voting, using brief, concise commands.

The process of continuously building and documenting a consensus introduces a new dimension to all sorts of groups that have to continually reach or negotiate agreements. It means that the process can be stabilized and made insensitive to changes in the individuals participating in the process.

Some of the associated properties we seem to be observing with respect to human problem solving in this medium are the consideration of larger numbers of options or alternatives and fewer psychological pressures towards forcing a concensus within a short period of time.

In other cases, such functions as contract negotiation and the designing of contracts have been accomplished using computerized conferencing systems. In all cases there seems to be agreement that the process is much faster than other alternatives when the individuals are not co-located.

INTERNATIONAL POTENTIALS

We believe that computerized conferencing can increase the quantity and effectiveness of international cooperative efforts to prevent the escalation of conflicts and crises to the point where war occurs. Among the kinds of functions which can be performed for international organizations or networks of individuals and groups are the exchange of information which promotes and builds peaceful cooperation; monitoring and detection of incidents or threats to international relations; negotiation of agreements; resolution of conflicts; and crisis management. In this section we will move beyond current data and applications to suggest developments and applications of these types which might occur in the next decade.

Among those who have seen the international potentials of computerized conferencing is Anthony Judge, who argues that "computer conferencing could really constitute the much needed breakthrough in international and transnational action.... In a special sense computer conferencing gives form and structure (however subtle and dynamic) for the first time, to the Sixth Continent-- The transnational non-territorial world-- over which so much activity takes place" (Judge, 1979, P. 402).

INFORMATION EXCHANGE

The concept of using a computerized conferencing system for information exchange on an internatinal basis opens the door to a host of decision aids that could greatly facilitate the process by overcoming both language and cultural differences. The first example of this would be a system for aiding in the use of on-line translators to support the information exchange process. Transcripts of the discussions could be kept in as many languages as needed. When a person writes an item, he or she would designate it as ready for translation, and the computer would find the next available translator qualified to do the translation. The writer could designate various options such as reviewing the translation if they can read in the language or having it re-translated from the translation back to the original language by another translator. Given enough translators in relation to the people engaged in the information exchange process, the dialogue could move quite rapidly.

The next concept is to introduce an extended version of the +TERMS system. This would allow the group involved to build up over time a data base of accepted translations for

various words and phrases that are specialized to the technical area they may be dealing with. Such a data base could be incorporated into the system given to the translators to get rough draft translations done. In the long run, this means the group would capture the 'lore' that governs their discussion and the process would become less sensitive to changes in who is actually participating. In other words, there would be less repetition of mistakes that are made by new individuals because of the language barriers and cultural differences.

The next phase of development is the incorporation of other types of decision and analysis aids that would facilitate the pinpointing of disagreements. The simplest of these are voting and estimation processes that already exist on EIES. Using such aids a group can see from the desirability or feasibility of an alternative or the distribution of cost estimates for carrying out the option as provided by the members of the group. This concept extends further to the incorporation of analytical models that can check consistency of viewpoints among a large set of data items.

Probably of greatest potential significance in this area of aids is the extension to inference type systems such as are now used in the medical diagnostic area. In this process individuals in a group can contribute English like statements representing premises about a given topic and the computer can pinpoint inconsistencies in the premises as they arise from different members of the group as well as making predictions about possible new premises. Early unclassified work of this sort in the international area is represented by North et. al (1963).

MONITORING AND DETECTION

One of the more recent public examples of shortcomings in international affairs is the African drought and resulting famine. It has been hypothesized that the national leadership in many of the affected countries were reluctant to admit, in the early stages, the seriousness of their problem because of an impression that it might be considered a local problem reflecting on their leadership. As a result, it took far longer than it should have to realize the extent of the developing situation. We would hypothesize that in a more open computerized conferencing environment with the proper analysis aids to correlate factual information contained in various conferences, it is very likely that situations such as the above would have been highlighted and more fully understood very early in the process.

Monitoring and detection is very often a function of imposing human interpretation by 'experts' on the collection of basic facts and data. Many situations require a broad range of experts unlikely to be located in any one place at any one time. Good examples of this are to be found in the environmental, pollution and energy areas. There are a multitude of regional and international bodies concerned with such problems within such contexts as a single river to the sea shared by a number of countries. One would expect computerized conferencing to greatly increase the operational efficiency of such organizations and allow a wider tapping of expert knowledge.

In many situations having to do with economic, political and technical change, the rate of change today seems faster than the ability to detect, plan and react in any anticipatory mode. It would have been interesting to see how a multi-national network of data gatherers and experts would have evaluated in real time the revolution occurring in Iran. Current communications and organizational structures impose a situation where information and analysis from small groups or individual experts are passed up a hierarchy with significant filtering at each stage. Lateral international networks usually only exist at the higher leadership levels and can suffer from the pre-elimination of minority views and data that supports such views as well as the natural bias of humans to emphasize positive aspects of any situation.

Another unique feature of computerized conferencing is the ability to hold anonymous discussions or utilize pen-names. This means that the views undergoing open discussion on an international net do not have to be perceived as an official position. It is a well known phenomenon that the view of an individual may be considerably influenced by who uttered something as opposed to what the utterance was. This could be quite significant in a negotiation situation or in evaluating an action. However, in the analysis and synthesis phases associated with any monitoring and detection operation, this could be a very dangerous bias. It could very well be that monitoring and detection networks could operate more effectively if anonymity were imposed upon the members of the network. In any case, the issue is a researchable one.

NEGOTIATION

It is doubtful that computerized conferencing or any form of teleconferencing is going to make inroads into the standard face to face negotiation process for some time. Those engaged as negotiators, whether on a commercial or political basis, have developed a rather fine art that has its own rules and protocols. However, any negotiator is backed up, in most instances, by a rather large team of support individuals who must review and assess alternatives brought up in the negotiation process and provide information. In addition, decision makers who must approve or endorse changes are often remote from the negotiation scene. Because of this, the first area of potential utility for this technology is the speed up of information exchange and analysis among the individuals involved on one side of the negotiation process.

The second likely phase for use of this medium is agenda setting prior to face to face meetings. Reports from EIES groups which have used the medium for this purpose indicate that this can greatly speed up the progress made in face to face meetings. However, there is a problem in projecting this application to international negotiations. Some analysts have indicated that the Russians, in their opinion, use the agenda setting phase as part of the crucial assessment of the intentions of the other side. Negotiation over the separate points of the agenda signals to the other side how the opponent may react during the actual negotiations. Softness during the agenda setting process may be interpreted by them as weakness. On the other hand, Middle Easterners often want to meet without ever setting an agenda or coming to the point, because they simply are not yet ready to resolve the situation. To insist on setting an agenda ahead of time and sticking to it rigidly will be likely to alienate them. (Hall, 1959) Thus, as in other areas, there is need for field trials which explore how outcomes observed within a single cultural tradition may alter when an international application is attempted.

CONFLICT RESOLUTION

The concept of Conflict Resolution still remains an exercise for academics. While promising work has taken place in this area, there currently seems to be some unwillingness by those in leadership roles to utilize either the conflict resolution processes that have been developed or the modeling aids that have been investigated. At the very minimum, a meaningful conflict resolution process forces people to make the implicit explicit. Apparent "facts" are re-examined to determine the range of belief among the participants; underlying assumptions must be exposed and perceptions of strengths and weakenesses of self as well as other parties must be put on a comparative basis. It is not clear that people holding power are willing to subject themselves to this form of "encounter". The process of conflict involves high states of emotion and/or commitment to opposing views. Most attempts to deal with such situations involve the structuring of communication protocols and process models to eliminate confusions and biases that result from the psychological factors involved. Furthermore, the face to face environment seems the least desirable of communication media to promote such processes in an effective manner.

However, the technology of computerized conferencing provides a much "cooler" atmosphere in which individuals can engage in communication on an entirely "self activating" basis. The cooler medium with less social presence tends to promote a reflective atmosphere that is desirable for attempts at conflict resolution.

An example of the type of conflict resolution work that lends itself to the computerized conferencing environment is that by Levi & Benjamin (1977, 409-410). They have developed a process model for the resolution of conflicts which they have tested in situations such as workshops including Arabs and Israelis trying to resolve conflicts in their national goals. Step one in their process is to clearly identify alternative solutions to the conflict; then to have each of the participants rate each solution on a -10 to +10 scale. Besides measuring the degree of conflict, these initial ratings provide a point of reference against which the participants can measure progress as the negotiations continue, and new solutions are proposed and voted on. The method uses a fairly structured approach to regulating the conflict and its negotiation. Based on their preliminary experiments, they conclude that the "approach can be a useful addition to the tools of international conflict resolution" (ibid, p. 422).

However, there are difficulties for the outside consultant/facilitator of such a model of conflict resolution. "Those using the model may not be able to use it consistently, either because of lack of skill or temporary factors such as fatigue. Secondly, the need to concentrate on the model can lead to missing other important developments which require attention and intervention (Benjamin and Levi, 1978). These developments have to do with interpersonal relationships between the participants, as opposed to the task of solving the underlying conflict between them (Levi, 1978, pp 2-4). Levi proposes that a computerized conferencing system be used to implement the model. "Computerizing the model should solve these difficulties. By taking over the human intervenor, the computer can assure reliable application of the model,

regardless of the skill or mental state of the consultant or participants. More attention can then be paid to interpersonal issues as they arise. The computer can even be programmed to remind the participants of such issues" (ibid, p. 4).

We believe, with Levi, that significant efforts are needed to integrate the efforts in conflict resolution with modern information and communication technology. Given that basis of development, the next and more crucial step is to get individuals in leadership positions willing to undergo these structured exercises. However, there is another possibility that may be an easier objective and have the same sort of long term impact. Given that modern computer based systems for conflict resolution are readily available it becomes possible to consider such conflict resolution systems as a new form of survey and polling. Provided large enough samples of the populations concerned with the issues can be put through these electronic workshops the resulting outcomes can potentially have the same influence on the involved leaderships that polls and surveys can have today. Because of the highly limited and simplistic nature of polls and surveys their application to conflictual issues are often more confounding than useful in terms of the influence they can have on decision processes and decision makers.

CRISIS MANAGEMENT

In the keynote speech to the 1978 ICCC, H.P. Gausmann pointed out that computers and communication technology are:

"The basis for new information infrastructures which will permit the transition from the present, essentially STATIC, information (such as represented in books, journals, printed matter) to DYNAMIC information in electronic form... selective, up-to-date, precise, relevant information... which is increasingly needed to master the intricacies of the modern world and the rapid changes with which we are confronted today." (Gausman, 1978)

In no application are these requirements more essential than crisis management. As Kupperman and Goldman pointed out in 1974: "We are living in an era of permanent crisis. Science and technology have accelerated the march of history while progress in astute and calculated political decision has not grown apace." (p. 71) Worldwide inflation, extensive famine, international terrorism, massive displacement of refugees as in

Southeast Asia and worldwide resource shortages
are all examples of international crises,
current or potential, as well as the constant
threat of a nuclear confrontation. Crises have
in common such factors as a paucity of shared,
accurate information, communications
difficulties, uncertainty, and an urgency
brought on by the sense that if nothing is
decided, things will get much worse. Information is the key element in successful crisis
management. If it is limited and inaccurate,
crisis managers will spend most of their time
just trying to find out what is happening, let
alone being able to control the sequence of
events. Who is to become involved, what
irrational acts could occur, who will have to
make a decision, who has the information needed
at the moment, what previous "lore" applies—
are all major questions and uncertainties
facing those involved in dealing with the
situation. The lack of critical information or,
even worse, the use of misinformation are the
fundamental sins of the crisis situation:
"Insufficient or misconstrued information not
only tends to aggravate crises but may also
tend to provoke them". (ibid., p. 74)

As a matter of record, one of the first
computerized conferencing systems, EMISARI, was
designed for domestic crisis management in the
Office of Emergency Preparedness (Executive
Office of the President of the U.S.) (See
Hiltz and Turoff, 1978 for a review of the
evolution of this system; McKendree, 1977 for
details about the components of this and
subsequent systems).

Those involved in EMISARI realized that
one of the most crucial aspects of computerized
conferencing applicable to crisis management is
that it can incorporate access to any existing
computerized data base and the use of models.
As Kupperman, Wilcox, and Smith put it in the
article in Science in 1975:

"The embedding of a dynamic model in
the conference could assist the
participants materially in understanding the real import of each other's
statements... Because of the power and
speed of the computer, a participant
could illustrate his proposals for
change by demonstrating those changes
in the model dynamically and producing
results which all could witness
identically. Similarly, 'what if'
questions could be explored with the
model by individuals or the group,
without problems of their misunderstanding either the options tried or
the effects obtained."

A crisis management system embedded within
a computerized conferencing system can provide
exactly the kind of dynamic, selectively

retrievable information and discussion space
needed to effectively manage a crisis. Crisis
situations usually impose deadlines whereby
actions and decisions must be made and,
furthermore, make clear to all that inaction
is actually decision as well. Perhaps at the
crux of the crisis management process is the
lack of a clear cut decision process pre-
defined by those forced to respond. The degree
of decision responsibility and the degree of
decentralization in the decision process must
be established, very often "on the fly" among
the parties involved, and as a result represent
a consensus process among the parties
representing otherwise autonomous bodies.
Computerized Conferencing technology is one of
the few pragmatic ways in which the decision
process can be adjusted by the participants in
response to the minute by minute developments
and at the same time allow opportunities for
inputs that might impact on a decentralized
decision.

A system for crisis management would
incorporate many of the other functional areas
we have been discussing. For any regular
international crisis management system to exist
we must have a cadre of international crisis
managers who utilize the underlying processes
and technology for information exchange and
monitoring and detection on a regular basis.
There is not the time, once the crisis has
started, to train the team in the technology
and its use.

THE FLY IN THE OINTMENT

While one can paint a very rosy picture of
the potential international applications of
this technology, there are ever present
political and regulatory problems associated
with the gray area of the merger of computers
and communications. We already have some
indicators that are symptomatic of the
problems yet to be faced. One problem is that
prices for international packet switching
networks are set artificially high by foreign
governments to protect their TELEX operations.
European charges are in the $40 - $80 per hour
range for instance.

Currently (at this writing), if a person
in England seeks a Telenet account from the
English PTT and indicates that he is planning
to use a message or conference system, he will
be denied an account. Whereas, if he or she
indicates plans to use a data base system, the
account is granted. It is as if the telephone
company in the U.S. asked who you wished to
talk to on the phone before determining if you
have a right to have a phone.

It seems a "fait accompli" that most

414 M. Turoff and S.R. Hiltz

European PTT's assume that they will be the only one offering computerized conferencing or message system technology. Also, it seems that most current thinking suggests that the 'system' to be offered will appear the same to everyone, so that the advantage of the computer being able to tailor communication structures to the needs of individual groups and applications will be unrealized. Basically, the monopoly maintained by the PTT's will prevent the emergence of the entrepreneurs willing to tailor small systems to select groups (such as duplicate bridge fanatics).

We have on EIES a multi-government sponsored research laboratory in Europe, with research associates in many different European countries. It was determined that it would be a cost saving to them to install an EIES system for less than $200,000 to tie these people into an ongoing network. However, it was also observed that it might take them ten years to ever get approval independently from each of the European governments to do this. Hence they have "netted" some of their key people by using our system in New Jersey, though they are all based in Europe.

Recently, a Swedish government agency applied for a license to operate a computer conferencing system they had developed. The Swedish privacy board came back with assorted regulations that included: 1) If a private message between two parties referred to a third individual, then the third party must be notified of the contents of the message; 2) Messages and comments could not talk about religion and politics. If these regulations were to be enforced, the operators of the system would have to read all private material. The issues, we believe, are still in some state of arbitration.

Without a doubt, the rates charged by European PTT's for telenet services and the rates planned for their own TELETEXT type services are protectionist in nature towards TWX and associated services, and do not show any flow through to the customer of the efficiencies of current technology. It is doubtful, in the long run, that these policies will hold back the "tide" of this technology. It is still possible for organizations that are large enough to put in their own private networks offering their own tailored services. Over the long term, new organizations will form to supply the users left out. The penalties will be a longer time lag in many countries and fewer potential benefits realized by the less affluent segments of the social structures in these countries. Perhaps the underdeveloped countries will suffer lost opportunities the most, since they are more inclined to follow the European directions these days than the U.S. efforts.

The existing regulations reflect fears of foreign dominance through "teleinformatics". Walsh (1978) summarized the situation with particular reference to France:

> Europeans share a conviction that information technology is profoundly changing the workings of industrial society and are convinced that a given country will have little chance to control its own destiny unless it develops a measure of self-sufficiency in the new technology. Evidence of the European view is found in rather pure form in a recent report to French President Valery Giscard d'Estaing on the impact of data processing on French society and on possible ways to develop and control it.

> The report puts the problem in the context of the long-term societal crisis caused by industrialization and urbanization. France is portrayed as facing a series of challenges which could deprive the country of the ability to determine its destiny. The "computerization of society" is identified as a key issue which could either worsen or help to solve the crisis. The stakes riding on information technology are described as nothing less than the country's "economic balance, the 'social consensus' and national independence." The report goes on to say that "changes in economic and social structures can only be brought about if France can escape from the pressures of foreign governments or groups whose objectives may stand in the way of her own ambitions.

INCLUSION OF THE DEVELOPING NATIONS

Thus far, new communication-information technologies have not served to close the gap between the industrialized and the developed nations. As one prominent spokesperson for the "Third World" put it, the information and communications system is characterized by the same imbalances that affect other aspects of the international system.

> The present structures and patterns of the telecommunications networks between developing countries are based solely on criteria of profitability and volume of traffic, and so constitute a serious handicap to the development of information

and communication. This handicap
affects both the infrastructure
and the tariff system.

With regard to the infrastructure,
in addition to the absence of direct
links between developing countries,
a concentration of communications
networks is to be observed in the
developed countries. The planning
of the infrastructure devised by the
former colonial powers precludes,
for certain developing countries,
all possibility of transmitting
information beyond their frontiers
(earth stations allowing only
reception of television programs
produced in the industrialized
countries, with no possibility of
broadcasting towards these
countries).

With regard to tariffs the
situation is even more striking and
in certain respects quite
irrational. Designed so as to
disadvantage small outputs, the
present tariff system perpetuates
the stranglehold of the rich
countries on the information flow.
It is strange, to say the least,
that, over the same distance,
communications should cost more
between two points within developing countries than between two
other situated in developed
countries. (Masmoudi, 1979, p. 177)

We believe that computerized conferencing
can be used as a tool to redress the balance,
rather than perpetuate and increase it. Though
this view has yet to be publicly embraced by
the developing nations, it does find support
among others familiar with the technology, as
for instance in the following statement by
futurist Robert Theobald, sent to the
participants in the World Futures conference in
Berlin via EIES:

I suppose the most dramatic result
of EIES for those who have not used
it before is that the person living
in a rural area of the state of
Arizona is able to do as well in
communication terms as anyone living
in a large city. I was born in
Madras, India and I see EIES as being
compatible with the communications
traditions of this part of the world
far more than it is with those of
the rich countries. This is not an
accepted view among many of my
colleagues on this system and we
continue to argue about it in various
places on this system.

I believe that EIES is part of what
I call the shift to the communications
era. I am convinced that this shift
will be as profound as that which
took place when the countries now rich
moved from the agricultural to the
industrial era. I am also convinced
that it is possible for the countries
new poor to skip the industrial era
and benefit from the increases in
understanding and knowledge to avoid
many of the costs of the industrial-
ization process. (Theobald, 1979)

CONCLUSION: KNOWLEDGE & WISDOM

The definition on "information" by Becker
which was quoted at the beginning of this paper
distinguishes it from "knowledge" and "wisdom".

KNOWLEDGE derives from the process of
understanding and analyzing information, so
that it can lead to intelligent decisions or
actions.

WISDOM, finally, is the ability to
successfully utilize knowledge to carry out a
decision.

It is our belief that the combination
communication-information systems which we call
"computerized conferencing" can be used by
multi-national organizations or groups to
create and share more information on a more
timely basis for dealing with priority or
crisis issues, and more easily make knowledgeable decisions, using the computer as a tool.
Or, to put it another way, experimentation
with and application of computer based
communication-information systems for international problems may help us to move from
poorly utilized information to knowledge and
wisdom in the conduct of international affairs.

[1]The EIES system and part of the work
reported here is supported by a contract from
the Division of Information Science and
Technology, National Science Foundation
(DSI-77-21008). The views in this paper are
also partially influenced by work under grants
from the Division of Mathematical and Computer
Sciences (NSF MCS 78-00519 and MCS-77-27813)
for the exploration of computerized
conferencing technology and its social impacts.
The opinions and views expressed here are
solely those of the authors and do not
represent those of any organization with which
the authors may be associated. An initial
version of this paper was presented at the
International Studies Association, Toronto,
Canada, March 1978 in a Panel on "Foresight,
Conscience and Strategy in World Control
Options for Nuclear Peacekeeping". (Earlier

Title: One Critical Factor: Information,
Both Factual and Speculative)

REFERENCES

Becker, Joseph
    1978 "Information as a National Resource".
         In Eight Key Issued for the White
         House Conference on Library and
         Information Services. Pamphlet
         published by John Wiley and Sons for
         the American Society for Information
         Science.

Catanzaro, Frank
    1978 EIES Conference 822, comments 549 and
         554, December 19, 1978.

Davies, M.A.
    1971 Communication Effectiveness as a
         Function of Mode. Unpublished M.A.
         Thesis, University of Waterloo.

Gaussman, H.P.
    1978 "Data Networks as New Information
         Infrastructures: A Challenge to
         International Co-operation. Keynote
         Speech, International Conference on
         Computer Communication, Kyoto, Japan.

Goffman, Erving
    1955 "On Face Work: An Analysis of Ritual
         Elements in Social Interaction.";
         "An Interaction Ritual: Essays on
         Face-to-Face Behavior", Garden City,
         Doubleday and Company, 1967.

Hall, Edward
    1959 The Silent Language. New York:
         Doubleday.

Hiltz, Starr Roxanne, Johnson, Kenneth and
Agle, Gail
    1978 Replication of Bales Problem Solving
         Experiments on a Computerized
         Conference: A Pilot Study. Newark,
         N.J., Computerized Conferencing and
         Communications Center, NJIT Research
         Report No. 8.

Hiltz, Starr Roxanne, Kenneth Johnson, Charles
Aronovitch, and Murray Turoff
    1980 "The Effects of Computerized
         Conferencing vs. Face to Face Modes
         of Communication on the Process and
         Outcome of Group Decision Making: A
         Controlled Experiment", Newark, N.J.,
         Computerized Conferencing and
         Communications Center, NJIT, Research
         Report No. 11.

Hiltz, Starr Roxanne and Turoff, Murray
    1978 The Network Nation: Human
         Communication via Computer. Reading,
         Mass.: Addison Wesley Advanced Book
         Program.

Hiltz, Starr Roxanne and Turoff, Murray
    1980 "The Evolution of User Behavior in a
         Computerized Conferencing System",
         Paper to be presented at the
         International Communications
         Association, Acapulco, Mexico, May.

Johnson-Lenz, Peter and Trudy
    1978 "+TERMS: Software System for
         Collective Development of Glossaries.
         EIES, November.

Judge, Anthony
    1977 "Enhancing Transnational Network
         Action". Transnational Associations,
         10, 401-402.

Kerr, Elaine and Bezilla, Robert
    1979 "Cues and Clues: The Presentation of
         Self in Computerized Conferencing".
         Paper prepared for the National
         Computer Conference, June 1979.

Kerr, Elaine B.
    1980 "Conferencing Through Computer:
         Evaluation of Computer-Assisted
         Planning and Management for the White
         House Conference on Library and
         Information Services". Unpublished
         Report.

Kupperman, Robert H. and Goldman, Steven C.
    1975 "Toward a Viable International
         System: Crisis Management and
         Computer Conferencing". In
         Nathaniel Macon, ed., Computer
         Communication: Views from ICCC '74.
         Washington, D.C., ICCC.

Kupperman, Robert H., Wilcox, Ricard H. and
Smith, Harvey A.
    1975 "Crisis Management: Some Opportu-
         nities". Science, 187, 404 - 410.

Levi, A.M.
    1978 "Problem Solving and Conflict
         Resolution: Developing and
         Evaluating Computerized Communication."
         Proposal submitted to the United
         States-Israel Binational Science
         Foundation.

Levi, A.M. and Benjamin, A.
    1977 "Focus and Flexibility in a Model of
         Conflict Resolution". Journal of
         Conflict Resolution, 21:405-425.

Mencke-Glucker, Peter
1979 "Opening Address to the World Futures
Society Conference on Science,
Technology and the Future".

North, Robert C., Holsti, George Z and Zinnes,
Dina
1963 Content Analysis. Northwestern
University Press.

Parker, Edwin B.
1976 "Social Implications of Computer/
Telecoms Systems". Telecommunications
Policy (December): 3-20.

Reid, A.A.L.
1970 Electronic Person-to Person
Communications. Communications
Studies Group paper no. P/70244/RD.

Short, J.A.
1972 Conflicts of Opinion and Medium of
Communication. Unpublished
Communications Studies Group Paper no
E/722001/Sh. Reported in Short,
Williams and Christie, 1976, 102-4.

Short, J. Williams, E., and Christie, B.
1976 The Social Psychology of Telecommunications. London, New York,
Sydney and Toronto: John Wiley and
Sons.

Theobald, Robert
1979 Message 1801, April 26, 1979 to the
participants in the Working Group on
Implications of Computer Conferencing
for Developed and Developing
Countries; World Future Studies
Conference, Berlin, May.

Toussaint, J.H.
1960 "A Classified Summary of Listening
1950-1959", J. Communication, 10,
125-134.

Wall, V.D. and Boyd, J.A.
1971 "Channel Variation and Attitude
Change", J. Communication, 21,
363-7.

Walsh, John
1978 "There's Trouble in the Air Over
Transborder Data Flow". Science,
202,6 (Oct.), 29-32

The Teleconferencing Resource Book: A Guide to Applications and Planning
Lorne A. Parker and Christine H. Olgren (eds.)
Elsevier Science Publishers B.V. (North-Holland)
© Center for Interactive Programs, University of Wisconsin-Extension, 1984

POLICY-ORIENTED TEAMS IN COMPUTER CONFERENCE

Manfred Kochen
University of Michigan

1. Introduction

Exchanges of information and coordination of activities requiring rapid flow of information and control signals have been accomplished through computerized store-and-forward messages over telephone networks for about a decade (Turoff, 1970;[19] Hall, 1971;[6] Umpleby, 1971;[20] Engelbart, 1970[5]). The idea had been under development already during the sixties (e.g. Bohnert and Kochen, 1965;[3] Bush, 1945[4]). It is now relatively easy for anyone with a terminal to become a participant in a computer conference. The number of computer conferencing systems and the number of computer conferences within each has recently begun to accelerate dramatically, for several reasons. One is decreasing cost of reliable packet-switched data over existing telephone networks. Another is awareness and know-how about computer conferencing and its benefits. Yet another is the availability of terminals by microcomputers usable as such in offices and homes and the existence or pending existence of several commercial services (EIES: M. Turoff at New Jersey Institute of Technology; PLANET: J. Vallee, Institute for the Future; CONFER: R. Parnes, Advertel Co.) to mention but three.

Those of us who have used existing computer conferencing systems -- CONFER and EIES in my case -- have encountered several problems. We regard these as challenges and believe that they will be met very soon, making computer conferencing one of the most important means of group communication. This is likely to provide us with a powerful new tool for more enlightened policy-making. The formulation of policies on a complex issue requires the collective brainpower of the best team that can be mobilized for that issue, when a single well-connected and knowledgeable policy-maker with an expert staff and his braintrust of consultants might have sufficed for a simpler issue. The most important problem or barrier to the use of computer conferencing (or any other technology) to aid in policy-formulation on a complex issue is the recognition and mobilization of an appropriate team.

If the mix of people who participate via computer conference is insufficiently coherent, representative, motivated, large, diverse and organized by new procedural rules, these key participants are likely to drop out or participate very marginally. They are likely to find the density of information too low, the opportunity cost of the time they spend in participation too high, the pace of moving toward closure too slow, the diversions too distractive, the incentives too feeble, etc.

The aim of this paper is to illustrate and clarify this central problem. I will argue that lack of incentives, incompatibility of world views, of values and personality traits are the main barriers to a set of participants acting as an effective team. I will analyze this claim, suggest how the barriers can be overcome in a computer conference and how to test such hypotheses experimentally.

2. Four Cases

To illustrate how lack of incentives and incompatibilities are chief barriers to success in policy-oriented teams, we describe four experiences. Two used computer conferencing and two did not. Two succeeded and two did not. We begin with an ordinary group process that failed, a very common occurrence.

Case 1. I was one of a group of researchers who participated in writing a proposal that led to a grant. The system was developed as a policy-oriented information system to be shared by a county mental health clinic and a university-based mental health service. Its aim, through shared record-keeping and research effort, was to improve utilization and planning of mental health facilities. This included the policies for governing such shared use. Success would have meant that the two institutions would continue to share the use of this information system. This was not the case. No policies emerged that are now in effect. A rather dismal conclusion, corresponding to a belief held by the principal investigator at the outset, was reached: the barriers to an information system for mental health planners are predominantly political; they inhibit the production of adequate-quality inputs as well as imaginative use of the possible outputs.

As in every failure of a group process, the cause is partly in the group and partly in its management. Outstanding management can get almost any group (within limits) to produce results. A truly great mix of people can produce outstanding results (within limits) despite the grossest mismanagement. In this case, the initial group comprised some experienced and creative systems analysts. The PI's wish to have that group act as a team -- in deciding on a system design, research designs, equipment, conceptualizations -- was undermined by his unwillingness to delegate any part of the research (and budget) to any participant (a classic management error). Constant lack of support for ideas generated by the more experienced participants provided enough disincentives and raised opportunity costs to effect their withdrawal, resulting in a much smaller group of inexperienced assistants to the PI who adopted perspectives and values that coincided rather completely with his.

Tightly knit research teams consisting of a senior scientist assisted by compatible junior assistants and technicians have been highly successful. This is less frequently the case for policy-oriented groups, especially when two or more potentially conflicting interests are involved, as was the case here. The managerial style and the team that ultimately emerged (it no longer included members from the county mental health center) was consistent with the conclusion it reached (and started with). The process had elements of a self-fulfilling prophecy and the group built a web that entrapped them by the very barriers they were trying to identify and overcome.

While incompatibilities among team members accounted for some of the causes of failure, the lack of incentive or disincentives -- project- generated barriers -- seemed to be the major cause.

Case 2. This is another group process that failed, this time with the help of a computer conference. Several of us involved in computer conferencing were invited to a session of the International World Future Congress devoted to this technology. We used EIES to prepare for the meeting several weeks in advance. Enthusiasm and expectations were high by all concerned. The conference accumulated well over 100 lengthy items in a few weeks. I found it prohibitive to find enough time to read the huge backlog of items the two times a week I signed on, not to mention breaking into the discussion with worthwhile contributions or sorting out which of several randomly interwoven themes to latch on to. It was fascinating to watch increases in the heat of a debate between a key participant in Brussels (who had invited me and who seemed to many of us in the U.S. the initial organizer) and an enthusiastic participant in Berlin (who was to be our host) about the agenda of the session, whether some of us from the U.S. could stop for a prior conference in Brussels on the way, etc.

None of us realized it prior to the meeting in Berlin, but we were a policy-oriented group. At the meeting we formulated a statement toward a proposal for linking scientists in developing regions with their colleagues in other regions so that they would feel less isolated, enabling them to participate in mainstream science as if they were in the major centers. The proposal (entered on-line over EIES in Berlin) was circulated and improved by contributions from the U.S. and Hawaii before the meeting ended. There was even considerable follow-up activity but it gradually flickered and has not yet resulted in action toward the desired goal. While a success in many ways (also true for Case 1), the group did not actualize what it was probably capable of accomplishing.

This might be attributed to the lack of committed leadership (though that was present to some extent); although computer conferences are supposed to be less dependent on charismatic leadership with strong organizational ability. The existence of this activity could have been known to over 600 participants in EIES, who could have joined the team that finally emerged. It would have been most helpful if key people in potential funding agencies with the potential interest and power to commit resources were active participants. That would have greatly improved the mix of people.

The sets of persons who participated most actively before, during and after the face-to-face meeting were quite different, with an overlap of about a half dozen at most. The values, goals, perspectives of the individuals in any one set were quite diverse. At one time, all the participants were nearly "addicted" to the conferencing activity, despite the relatively low density of information of direct value to them. Conferencing generates incentives of its own -- beyond the toy appeal for the novice -- that may not necessarily contribute to group productivity.

Case 3. This illustrates successful teamwork without computer conferencing. I was asked to formulate a position or policy for research in artificial intelligence at the IBM Research Center in 1960. To assess the state of the art, develop ideas and formulate a strategy, I enlisted the help of two younger colleagues. We started our discussions with an examination of the major claims about the progress and promise of artificial intelligence that were most popular at the time. We concurred in the conclusion that many of the claims and promises were irresponsible and not of uniformly high quality, and began to analyze the ideas, methods, problems and issues. We conceptualized these and structured the area into: 1. storing; 2. searching; and 3. mapping mechanisms of cognition and information processing. We divided work of surveying the literature among us into these three aspects. Each of us taught the other two what he learned and jointly we composed an overall assessment. We then considered plausible criteria on which to base recommendations for IBM to launch activity in this area and checked our assessment of the field and its potential against these.

The resulting report (Kochen, 1960)[7] succeeded in its aim of helping the IBM Research Center make an informed decision about whether or not to build a program in this field. (We advised against it, and that advice was followed. We developed ideas that eventually led to research in complexity theory, representation theory and models for the organization of relational data bases.) The incentives for each of us were to learn about a very interesting aspect of computer science, active involvement and intellectual responsibility for a major decision, recognition within and outside of IBM (our report was used as a text in an advanced psychology seminar at Harvard), and experience that proved to be valuable in our subsequent careers. Though we started with widely differing views, we quickly reached consensus about the essential values and ways of looking at this field, partly because we wanted to reach consensus quickly (and were receptive to one another's ideas) and partly because each of us stressed only the few most important issues and justified his beliefs about these issues sufficiently well to persuade the others.

Case 4. This is a partially successful computer conference related to a proposal for research on information deficits in medical judgments. The proposal did not result in a grant. This was not unexpected. The granting agency wanted a different form of project organization than the one we proposed; they advised us to resubmit a revised version of the proposal to a different program in the agency, and at the time of this writing is still underway. The computer conference was used not only to prepare the proposal but also to work out the first task of the proposed research. That involved having a team, consisting of a physician and a cognitive psychologist, interview a physician-subject just before and just after he sees patients to determine what information he uses and does not use in forming judgments. Ideas for formulating the questions to be used in the interview and for structuring the interview were exchanged and tried over the conference. The results of a pilot interview were reported and stimulated further revisions in the interview design. They also led to more refined hypotheses, methods for testing them and for better exploratory studies.

The participants were an extraordinary group. The two physician participants, despite interest and intent, never got around to keying in their entries. A research assistant did so on their behalf. They tended to rely and depend exclusively on face-to-face or telephone communications. (It suggested to us that until computer conferencing could accommodate voice input and output, such participants required an intermediary, which was less than satisfactory.) Their commitment was also less than what was required for such a project to succeed, with or without computer conferencing. According to their own testimony, it takes an enormous incentive (malpractice suits, deaths) for most physicians to change their patterns of practice; hence, knowing about, revealing, using information that might call for such change is resisted or at least not embraced.

The other participants were key researchers in this field in other parts of the country. A consensus of views and values seemed to emerge among all the participants. The incentive for most of them to participate was their respect for all the others and their interest in exchanging ideas with them on what they all felt was an important and interesting task.

3. Team Formation

Despite the impressive literature on small group research, we know little about the conditions under which a highly effective mix of, say, 20 persons out of a total of 10,000 "subscribers" to something like a computer conferencing system, seek out one another and form themselves into an effective team to work toward resolving a specific issue over six weeks or so. Small group research is the most relevant body of knowledge, and it has a long history. In the first three decades of this century, the study of gangs led to the beginning of social work. There were studies of children by Terman, Allport, Sorokin, etc. Investigation of small groups in industry by Taylor and of togetherness/apartness led to the theoretical formulations of Burgess, Sheffield, Piaget, Newcomb and Murphy. The study of group problem solving, started by Munsterberg as early as 1914, followed a course of its own.

By the middle of this century, the influence of mathematical ideas and methods, especially as applied to communication theory, was evident in Bales', Strodtbeck's and Stephans' analyses of participation content and rates by participants in group discussion. Graph and probability theoretic models based on binary relations among members were fruitful. Another promising start was the formalization of such hypotheses as Homans' about the relations between the amounts of interaction, I, friendliness, F, activity generated by (A) and imposed by the environment on (E) the group members. Thus, Simon (1952)[18] assumed:

$$I = f(A,F) \text{ with } \partial f/\partial A > 0, \partial f/\partial F > 0,$$
e.g. $f(A,F) = a_1 F + a_2 A$

$$dF/dt = g(I,F) \text{ with } \partial g/\partial I > 0, \partial g/\partial F < 0,$$
e.g. $g = b_1(I - b_2 F)$

$$dA/dt = h(A,F;E), \text{ with } \partial h/\partial A < 0, \partial h/\partial F > 0,$$
$\partial h/\partial E > 0$, e.g. $h = c_1(F - kA) + c_2(E - A)$

to deduce that there is at most one stable equilibrium solution. Experimentally it was found that the average conference has dual leaders, an "idea-person" and a "best-liked" person (Bales, 1957).[2] A practical instrument for assessing compatibility was developed (Schutz, 1958).[17] It was based on a three-dimensional view of personality aspects relevant to interpersonal behavior (Inclusion -- I join or like to be invited, include myself or like to be included; Control -- I influence others or like to be influenced, control others or like to be controlled; Affection -- I act friendly or like others to be friendly to me, act close or like people to get close to me). Two or more people were found to be compatible to the extent that they

have the same orientation in the affection dimension, i.e. when they agree on the degree of closeness, complement one another in giving and receiving affection, etc. If two people disagree on the desired amount of interdependence -- about how much to participate, conform, and show closeness -- conflicts will surface. Disagreement about who should initiate the selection of others, a plan of what to do, a close relation also leads to conflict and incompatibility.

In the past three decades, the most relevant research included experiments with DELPHI Linstone and Turoff, 1975),[11] a variety of studies of consensus (e.g. Reed, 1979),[16] the use of n-person game theory (Marschak and Radner, 1972),[12] "intervention theory" of group processes (Argyris, 1970)[1] and general 1970)[1] and general theory (Miller, 1975).[13] It is quite possible, however, that entirely new conceptualizations, not even based on existing lines of thought, may be created to meet the need for understanding the processes by which teams form themselves to deal effectively with the complex issues facing us. To the best of my knowledge, the present state of understanding and know-how in group dynamics does not meet this need; nor is there evidence that progress in that direction is likely.

The first consideration in modelling team dynamics is to specify performance variables. For a policy-oriented team, this should be an assessment of progress toward a satisfactory policy on an issue that requires teamwork. Such an assessment is likely to require several variables, such as:

R: the extent to which a situation is seen as presenting an issue, after it has become one that requires management if it is not to develop into a situation inconsistent with the community's values and before it has progressed so far that intervention can no longer help.

S: the quality of ideas for appropriate and timely management.

A high rating on R requires that the team grasp the dynamics of the situation as a whole. It is unlikely that any single team-member could grasp in its entirety a complex task-situation such as inflation, decline in productivity rate, a medical emergency in a remote region, docking two spacecraft, a natural or manmade disaster, etc. Each person has a partial picture, as if he were one of the proverbial blind men feeling part of an elephant. For them to function as a team, they must share a minimal conceptual repertoire. The image of an elephant must be common to the minds of the blind men, or those that have it must be able to impart it to those who do not. More importantly, the potential teammates must each give a small number of shared concepts top priority during the critical moments of team-formation, which is concurrent with "situation-grasping." The blind men would not all be touching parts of the same elephant if they did not simultaneously share a common interest in identifying that animal just then. Somehow, the situation may "reveal itself" just as the potential team with its potential for grasping the situation dynamics "unfolds," i.e. may be actualized. The elephant, through the information it provides to the blind men via tactile modalities, and through the resulting synthesis of that information with what they already know and through communication "reveals" itself to the team, and its potential to "see" an elephant is realized.

A high rating on R also requires that the team members share the values by which they judge situations to be acceptable. In addition, they must all have sufficient incentive to continue to participate in teamwork. Thus, the blind men must all consider it valuable to them to discover the elephant and consider the benefits to exceed the costs to them to continue their joint inquiry until they succeed. In Marschak's[12] formulation, each member has a total preference ordering on the "states of the world" in which the state of "seeing the elephant is preferred to all others, and this is the group preference ordering as well. In Schutz's[17] terms, the team members should all have similar feelings about the most desirable degree of closeness, about who should initiate communication, etc., about the optimal degree of interdependence.

It is highly improbable that among 10,000 people, even selected people, that more than a dozen potential teams with all these properties could exist. (The number of groupings of, say, 10 people out of 10,000 is about 10^{29}, and if one group after another tried, at 1 day per trial, it would take many times the age of the universe to get through a small fraction of the possibilities.) How then could and do appropriate teams ever form? The first group that forms is probably already a fair approximation to the team that finally emerges -- and with surprising speed.

Somehow, an appropriate dozen rise, simultaneously, spontaneously and independently to the situation, perhaps responding to the same cues. Then, they quickly sort themselves out, with a few who are incompatible with the rest replaced by others who fit better.

A good team is, almost by definition, one that rates high on S, provided that the situation in response to which the team emerged is one that required the team to produce ideas for appropriate and timely management. To describe the process by which such ideas are generated, analyzed, realized, discussed, utilized and phased out, we once again need new concepts and methods of observations. The main concept may resemble the one used to explain team formation: the appropriate mix of concepts that comprise a good idea arise simultaneously, spontaneously and independently to the situation, perhaps in response to the same stimuli. Then, the mix rapidly refines itself, with the few that do not fit well replaced by those that fit better.

4. Toward Teamwork by Computer Conference

Once enough of a conceptualization to formulate testable hypotheses has been produced, experiments to observe the formation of policy-oriented teams can be conducted. A system such as CONFER could announce to a mailing list of persons at various levels of government, industry, academia, think tanks and citizens who have expressed their concern with certain issues of urban or regional development that a computer conference will start at a certain date, and how to join it. The issue might deal with solid waste disposal, traffic control, housing, pollution control, nutrition, health care, etc. It could also be announced in local newspapers. Citizens without terminals or accounts of their own could use a public terminal at the nearest library. The debate might be partially financed by industry and government as long as it is an experiment, and then by the participants themselves (perhaps with government matching funds to subsidize would-be participants who cannot afford it). New procedural rules are to encourage the spontaneous formation of a variety of potential teams.

A number of commercial services of this kind are already underway. PLANET has been used for specific conferences in the service of a particular industrial or governmental client. LEGITECH and POLYTECH are being used successfully for networks of state legislatures. The much larger number of informal networks used by government and industry to perform this function are likely to adopt computer conferencing as soon as the users and financiers are persuaded that this is a more cost-effective way to manage.

While we cannot measure R and S except by asking for the judgment of participants and observers, we can try to observe the light-to-heat ratio and information density during the discussion. Verbal exchanges can be content analyzed for words pertaining to the issue and the ratio of such words to all words used can estimate information density. Using a thesaurus, such as Stone's, certain words indicative of emotion -- e.g. anger, anxiety, impatience, etc. could be screened for. The intensity of such emotions could be monitored by providing participants with a means for expressing them more directly. Discovery of relevant ideas can stimulate heated feelings of a positive nature as readily as frustration or incompatibility can stimulate heated negative feelings. The former are likely to reinforce further participation positively and the latter negatively. Low information density is likely to evoke negative feelings and lower participation. Some mix of light and strong feelings of a positive nature ("good heat") may be optimal, though it may require a certain level of "bad heat" (creative tension) on the way.

We cannot objectively rate S, the quality of ideas toward an effective policy to manage the situation except by hindsight. Perhaps the best we can do is to rate the quality of justifications given for various beliefs and for proposed steps toward a policy. This requires a more sophisticated content analysis of communications than the search for words in a checklist. The fraction of the content in what is communicated that denotes well justified beliefs or suggested steps toward a policy is the light ratio.

A variety of hypotheses can be tested: 1. the light ratio increases with information density; 2. the light-to-bad-heat ratio increases with the compatibility of the team, averaged over all pairs of members; 3. team compositions shift toward a team of greater mean compatibility as the light ratio increases; 4. the rate at which justified ideas are generated increases as team compatibility increases. We have already mentioned the hypothesized relation between participation and density as well as heat-ratios and incentives. Testing of such hypotheses is not a straightforward experimental design problem, because it is difficult to control all the relevant variables as required for rigorous statistical inferences. Empirical studies should therefore be regarded as exploratory rather than confirmatory research. Nonetheless, the designs (selection of topics, rules of

procedure) and methods of observation should be much more systematic and carefully planned and implemented than has been the case so far for studies in computer conferencing.

A great deal of additional conceptualization and pilot experimental will, of course, have to be done. The notion of "closeness" for example in Schutz's[17] FIRO test takes on a different operational meaning in a computer conference. It is not a well-defined concept in face-to-face groups. It has a physical aspect, measurable by the distance between them when they converse. It has a psycho-linguistic aspect, perhaps measurable by the extent to which each person makes reference to intimately personal concerns of the other(s). It has a visual aspect, measurable perhaps by frequency of eye contact, pupil dilatation, etc., an olfactory component, possibly a tactile and auditory one as well. Only the psycholinguistic aspects might -- with some distortion -- be in effect over a computer conference. But wholly new components of "liking" and "closeness" could perhaps be expressed through the Eugrams (Lederberg, 1980)[10] of computer conferencing than was possible via face-to-face. Expressions of like or dislike that persons would not have communicated, thought of communicating or had opportunities, time or resources to compose could now be possible.

It is easy to see the kind of model that would emerge for team dynamics. It would extend the Simon-Homans[17] model (without assuming linearity, however), combined with relational or network approaches (Pool and Kochen, 1978)[14] and introduce the methods of analyzing dissipative systems to account for the formation of novel teams and the ideas they generate (Prigogine, 1976;[15] Kochen, 1979).[8,9]

If such models are enriched by a cumulative body of empirical findings and developed to a higher level, then we may begin to understand the process of team-formation and function and perhaps by appropriate management of computer conferences. This should contribute to the art of team-formation, which is increasingly inportant for coping with the complex situations we are likely to face.

References

1. Argyris, C.   Intervention   Theory   and
    Method,   Reading,   Mass:
    Addison-Wesley, 1970.

2. Bales, R.F.   In Conference, Harvard
    Business Review, March-April, 1954,
    44-50.

3. Bohnert, H.   and Kochen, M.   The
    Automated Multilevel Encyclopedia as
    a   New   Mode   of   Scientific
    Communication.   In M. Kochen, ed.,
    Some Problems in Information Science,
    Metuchen:Scarecrow, 1965, 156-160.

4. Bush, V.   As We May Think, Atlantic
    Monthly, 176, 1, July 1945, 101-108;
    reprinted in The Growth of Knowledge,
    M. Kochen ed., New York:Wiley, 1967,
    23-35.

5. Engelbart,   D.C.   Intellectual
    Implications of Multi-Access Computer
    Networks,   Proceedings   of
    Interdisciplinary   Conference   on
    Multi-Access   Computer   Networks,
    Austin, Texas, April 1970.

6. Hall, T.W.   Implementation of   an
    Interactive Conferencin System, AFIPS
    Conference Proceedings, 38, 1971,
    Spring, Joint Computer Conference,
    217-229.

7. Kochen, M.   Cognitive   Mechanisms.
    RAP-16,   Yorktown   Heights,   New
    York:IBM Research Center, April,
    1960.

8. Kochen, M. Peer Cooperatives to Improve
    Research: Potenntial   of   Human
    Networks for Bringing Out Better
    Ideas and for Improving Published
    Literature.   In Proceedings of the
    3rd National Research Forum in
    Information Science, Oslo, Norway,
    August 1979.

9. Kochen, M. Social Know-How and Its Role
    in Invention and Innovation. To be
    published in Proceedings of a
    Workshop on Innovation in the Public
    Sector held in December 1979, Berlin:
    Science Center of Berlin, K.W.
    Deutsch and R. Merritt, eds.

10. Lederberg, J. Digital Communications and
    the Conduct of Science.   The New
    Literacy, Human Systems Management,
    1, 1, February 1980, 29-37;
    originally published in Proceedings
    IEEE, 66(11), 1978, 1314-1319.

11. Linstone, H.A.   and Turoff, M.   The
    Delphi Method, Reading, Mass:
    Addison-Wesley, 1975.

12. Marschak, J. and Radner, R.   Economic
    Theory of Teams, New Haven: Yale
    University Press, 1972.

13. Miller, J.G.   Living   Systems,   New
    York:McGraw-Hill, 1978.

14. Pool, I. de Sola and Kochen, M.
    Contacts   and   Influence,   Social
    Networks, 1, 1978-1979, 5-51.

15. Prigogine, I. Order Through Fluctuation:
    Self-organization and Social System.
    In Evolution and Consciousness, E.
    Jantsch and C.H.   Waddington, eds.,
    Reading, Mass: Addison-Wesley, 1976.

16. Reed, R.W.   Unanimous Decisions and
    Consensual   Plans.   Ph.D.
    Dissertation,   1979,   Ann
    Arbor:University of Michigan.

17. Schutz, W.C. FIRO: A Three Dimensional
    Theory of Interpersonal Behavior, New
    York: Rinehart, 1958.

18. Simon, H.A.   A Formal Theory of
    Interaction in Social Groups, Am.
    Soc. Rev., 17, 1952.

19. Turoff, M.   The Design for a Policy
    Delphi, Technol. Forecasting and
    Social Change, 2, No. 2, 1970,
    149-172.

20. Umpleby, S. Structuring Information for
    a   Computer-Based   Communications
    Medium, AFIPS Conference Proceedings,
    39, 1971, Fall, Joint Computer
    Conference, 337-350.

The Teleconferencing Resource Book: A Guide to Applications and Planning
Lorne A. Parker and Christine H. Olgren (eds.)
Elsevier Science Publishers B.V. (North-Holland)
©Center for Interactive Programs, University of Wisconsin-Extension, 1984

STRATEGIC PLANNING VIA INTERNATIONAL COMPUTER CONFERENCING

G. M. Krembs
International Business Machines Corporation
Kingston, New York 12401

Introduction

International corporations have product research and development resources distributed throughout the world. Traditionally, their strategic planning process is accomplished by travel to group meetings, telephone conversations, and the exchange via mailing of written materials. This paper is a case study which describes an on-going effort at IBM to prepare a technical strategy via international computer conferencing.

The initial objectives were: 1. to improve the productivity and efficiency of key management plus senior technical personnel; 2. to minimize their personal hardships; 3. to assess the associated cost benefits.

There were several initial start-up problems, that have gradually been overcome, including instructive training, compatibility and availability of dispersed equipment, and involvement of untrained personnel who need to participate on an occasional basis. Begun in March, 1981, the tangible, real benefits of this international computer conferencing procedure for strategic planning are now known, while some areas that need further improvements have been identified, and are described in detail.

One of the positive experiences has been the administrative ease for introducing new members into the conferencing procedure, and another is the significant decrease in the typing/copying workload required by this project compared to the traditional administrative methods of planning in the multinational corporations. Specific requirements for future systems are desired in regard to human factors and software psychology, for 1. data base management; 2. telecommunications from locations remote from the office environment.

IBM VM/370 Networking (VNET)

As used in this paper, the term VNET denotes a single purpose operating system for a virtual machine dedicated to processing files spooled to it, and transmitting these files via communication lines to remote nodes or work-stations.[1] The telecommunication I/O facilities operate in binary synchronous mode and are attached, temporarily or permanently, to the

VNET virtual machine running under the Control Program of an IBM VM/370. No operator intervention is required. VNET automatically stores and forwards files, spools and sequences to the next destination link on VNET, performs indirect or alternate routing automatically, and other significant features that simplify system operation for the end user and provides compatibility within the networks. See Figure 1 for a representative VM/370 installation using VNET.

This paper presents an application of the VNET approach to computer conferencing. This case study had as its goal the creation of a technical strategy by an international consortium of professional employees, each possessing an expertise in some facet of the contributory disciplines. The following headings outline the contents of this paper: 1. Initial Objectives. 2. Start-up Problems. 3. Real, Tangible Benefits. 4. Areas for Further Improvements.

Initial Objectives

Complex technological products involve contributions from many professionals of different technical expertise; often these professional skills are so unique that different locations within international corporations must establish a task force mode of operation, or schedule Spring and Fall planning cycles that concatenate. Traditionally, these strategic planning processes are accomplished by travel to group meetings, a long sequence of telephone conversations, and the constant exchange of written material, such as correspondence, "viewgraphs", and position papers. Interactive telecommunications offers the potential to improve the productivity and efficiency of these key management plus senior technical personnel, to minimize their personal hardships, and to achieve a simultaneous cost benefit to the international corporation.

Within IBM, systems exist for applying audio teleconferencing, enhancing that audio with facsimile, and in some locations adding two freeze-frame video cameras for document and people pictures. However, there also exists a display-oriented network of workstations, called VNET, which seemed to offer additional tangible

benefits for the preparation of a technical
strategy. These added values were:

1. Participants may attend the computer
conference whenever they want; they can enter
contributions to the conference, selectively
read those of other contributors, plus comment
on the strategy content in a one-on-one, one-on-
many, or broadcast manner. With IBM geographic
locations in various time zones, computer conferencing permits top-quality technical people to
participate in a perpetual manner without leaving their home laboratories.

2. Each contribution would be automatically recorded and can be dated to show the changes
and corrections made since the prior release by
each responsible contributor; this series of
contributions could be centrally controlled and
edited into a full record of the conferences by
topic and subject matter.

3. The same central administrator may then
organize the record into the proper sequence,
according to an agree-to outline, and the resultant file can be debated via the computer conferencing network to achieve a consensus for a
major portion of the technical strategy. The
audio teleconferencing methods could then be
introduced to enhance further the group dynamics;
business travel would be restricted to resolving
the remaining residue of issues.

4. Summary presentations (overhead projection, slides, and hardcopy) would be produced conveniently from the same data base.

In the above scenario, the display workstation provides the interactive blackboard and
"viewgraph" equivalent; no two-way full-motion
video would be assumed to exist, and all data
links are shared with other data transmission
applications on the VNET. Any single production
execution is spontaneous by each contributor
based on linked traffic; response has been
rapid, since each file is usually transmitted
in a few seconds of time; no one person need
always be present to lead the computer conference (as in audio teleconferencing) because the
record maintains continuity of the interaction;
new people can be easily brought into the conference by simply requesting them to read the
appropriate sections of the full electronic
document.

Start-up Problems

In any mode of interactive teleconferencing,
procedures must be established to get and maintain participation. Because the process involves
people of wide dispersion in training and language familiarity, human factors such as ease-
of-use are real concerns in implementing the
procedures for the group. The time involvement
of the participants was anticipated to range
from occasional to daily. The hardware/soft-
ware interface to all interested people had to
be uniform in order to integrate the individual
contributions into a published document, and
achieve a positive initial experience for those
directly involved in the process.

When this concept was begun in March, 1981
by IBM, the "founding fathers" traveled to one
location to work out the initial outline for
the document and agree on protocol for the central administrator. The most difficult task
at the initial meeting was the educational
training of the participants, given that most
of them had ready access to a VNET display workstation in IBM. The span of experience ranged
from none to "expert" on the VNET and related
document control facilities. Once the "founding
fathers" returned to their different geographic
locations, experts started contributing, while
others were obstructed by the steep learning
curve to use the system. Hence the first problem encountered in computer conferencing was
the span of educational preparation. Audio
teleconferencing does not need to overcome this
initial obstacle.

Faced with the important realization that
participation meant that contributors had to
learn a set of functions, we adopted support
software for the conference that was generally
available for text processing and electronic
mail on IBM systems. With over 500 VNET nodes
installed, it was impossible to suggest new
functions for computer conferencing until electronic mail systems had been applied to handle
basic computer conferencing. Thereafter, we
could design a set of unique computer conferencing functions as extensions of message
switching. That way, users learned from released IBM publications, and could attend IBM
training classes offered for general education
of its professional employees. As it turned
out, there were some "nice-to-have" functions
discussed later in this paper, but the selected
software support served the group very well.
Thus, the second start-up problem, training of
conferees, was overcome by adopting the general-
purpose software system installed on the network
for text processing and electronic mail.

For this VNET case study, we adopted the
Generalized Markup Language (GML) and the Document Composition Facility (DCF), which is available on VM/370.[2] The DCF/GML programs will format the input into conventional hardcopy. Lists
figures, footnotes, etc., are processed by the
program along with text. Because editors are a
personal preference, no restriction was necessar
editorial commands to move, delete, and change
contents on the display screen are instantaneous
and do not get stored on the file which is maile
to conferees or processed by the document for-
matter. Setting up headings, adjusting line

lengths, pagination, and the tables of contents, figures and footnotes are all generated automatically by DCF/GML. In this manner, the quality of the software and service support were maintained at a high level by the computer center personnel at each node. Conferees referred their operational problems to the site EDP organization first, and not to the central administrator of the computer conference.

Because the participants have display workstations of differing hardware features, such as full-screen capacity, the files are mailed electronically unformatted and subsequently processed at each node of the VNET if required. After initial start-up, GML is understood enough to visualize the final appearance of the document, and conferees send and receive files without formatting them to read and react. The presentation on a user's screen is managed by the node, because it knows the hardware features of each workstation on that node. Figure 2 indicates the worldwide distribution of computer conferees as of December 1981.

Real, Tangible Benefits

After the conference began with those familiar with their workstation, VNET and DCF/GML, the momentum induced others to learn and join the computer conference, partly to avoid being left out of the decision-making process. The only essential ingredient was dedicated central administration; first, a comprehensive outline of the document was published so conferees knew where they fit into the "big picture", and who was the starting contributor for each section of the document; second, the central administrator had to provide a clearing house function so that each section conformed to the outline in scope of content, and so that differing opinions of contributions were overcome. From May 1981 when the initial outline was published, until September 1981 when the first comprehensive management review occurred, this technical strategy has blossomed into 80 pages of printed material involving 45 people. The September publication was a varied and complicated structure of a long document, organized by central administration. All participants only had to learn the functions necessary for their purpose; some entered the computer conference only for a brief time, so that a wide range of interest coul be handled. People were added at various times, and brought into the computer conference by sending them the necessary portions of the on-going draft for them to read and add to.

The following discussion provides further examples of the cost benefits of this approach to computer conferencing:

1. Participation was governed by the willingness of the people to learn to use the computer conferencing system, if they had no prior experience. They had no reason to be excluded, because some professionals simply wrote out a contribution in conventional long-hand, and then gave it over to one of their colleagues to send and receive on the system. We arranged for the "founding fathers" to sign-on to one node (Kingston) and avoid the delay of message switching. However this was found not to be important, as group meetings could be held via a separate audio teleconference and the files transmitted as ordinary electronic mail.

2. User activity is fundamentally related to the presence of the central administrator. Because that person maintains the document outline, and assigns technical writers, he or she is the managing editor of the document, just as in the commercial printing and publishing industry. The central administrator must carry out these tasks:

- Modify the outline, which is really a hierarchial data base of headings, by inserting and deleting as experience occurs and consensus is gained.

- Solicit contributions from known experts, assign them to responsibilities within the outline, and publish their bylines.

- Organize the contributions so that terminology is uniform among participants, and the "layout" is balanced. A few writers are very enthused about certain topics, and will prepare a dissertation on the subject rather than a strategy.

- Set deadlines as in any press release. In computer conferencing, the lack of travel pressure also provides an opportunity to avoid completion of the responsibility.

- Arrange audio teleconferences to resolve conflicts in opinion. When the participants did meet face-to-face in IBM Kingston at the end of September 1981 for the first time, the week long meeting consisted of final review of the published document, and assignments to further enlarge the computer conference into more depth of subject matter and additional conferees.

If the central administrator is not continuously available to perform these above tasks, a brisk startup will quickly die out and the impact will be lost. With consistent central administration, this method of telecommunications will receive a positive response, spread throughout the international company, garner commitments for future action, and, the concept will become accepted by senior management. From a cost benefit viewpoint, one or two full-time key personnel are necessary to staff the central administration function of a computer conference as organized and described in this paper.

3. In conventional office procedures for developing a corporate strategy, the administrative support costs include salaries, benefits and workspace for these personnel. There is a constant stream of travel arrangements, preparation of overhead foils, and phone messages or physical mail passing back and forth among the participants. Once computer conferencing has begun, the administrative support workload actually decreases. Instead, more support is required from system programmers to explain, debug, and improve the functions of the electronic system. Because the interactive process by its nature requires some keyboard skills, the typing workload decreases, less time is spent on travel arrangements, and the number of phone messages also decreases. It has been observed that once participants learn these electronic mail techniques, they more often send messages on the VNET rather than pick up a telephone. Messages are worded precisely, pleasantries are reduced, and returned calls are avoided.

4. Foreign nationals involved in the computer conference generally will have difficulty with the English language keyboard. "Hunt and Peck" was a decided handicap compared to speed typing by some conferees in their native language. For an analogy, imagine that you work for an international corporation with the headquarters in Japan, and the keyboard will be Kanji. At least the Japanese in this case study know English and practice it; in fact as computer conferees they seemed to comprehend the displayed document quicker and with less misunderstanding than would occur in audio teleconferencing with foreign nationals. Again they had the opportunity to seek assistance, because the recorded message could be read over several times.

Areas for Further Improvements

When any form of interactive teleconferencing has been implemented and used, there are bound to be suggestions for improvement to achieve a greater degree of success. Some ideas are indeed a personal preference and "nice-to-have"; others are clearly essential to a broad spectrum of users. Below is a list that addresses the latter category of need, followed by a detailed explanation of each item in the list:

1. Emulators or translators to integrate incompatible displays into the network.

2. Work-at-home and portable ancillary displays.

3. On-line prompts for assisting the occasional user.

4. Program products specially designed to feature computer conferencing.

5. Interactive projection displays for group meetings.

The list is not all-inclusive; as we gain more practical computer conferencing experience, more human factors will be recognized.

1. Emulators or translators to integrate incompatible displays into the network. Given the initial success of this VNET computer conferencing tool, we have decided to expand further into all the laboratories of the IBM Corporation. Herein comes a fundamental concern. Some departments in IBM have installed small systems loosely coupled to the network, or not on the VNET. Because these systems cannot process and display the data stream architecture associated with VNET protocols, the end user is forced over to another display workstation for computer conferencing. More and more it will be required that display workstations be able to change personality under program control to avoid several terminals in the office, each dedicated to a particular system architecture. A like statement can be made about dissimilar printers.

2. Work-at-home and portable ancillary displays. Now that so much effort has been committed to VNET computer conferencing by this IBM group of professionals, access to the centralized data base is essential at off-site locations. Unless there is an equivalent electronic attache case, taking one's work home for the night or on weekends is impossible, and no one wants to conduct all job-related activities at the office. An IBM 3101 was installed at the home of the central administrator, and attached to the VNET via the VM/370 program product called "VM/Pass-Through Facility".[3] Together with another program called "3101 Pass-Through VM/370 Support",[4] the program can upon dial-up from the 3101, via a modem, simulate the 3270-type full-screen support used by the central administrator in the office. At 1200 baud half-duplex, the response time is tolerable, but both the keyboard command and program function keys are different in logical implementation.

3. On-line prompts for assisting the occasional user. If a diligent participant asks for all user manuals for VM, VNET, RSCS, CMS, GML, DCF, and text editors, the stack is over a foot high. Together with related hardware operator references for the display workstation, hardcopy devices and other system components that can become visible to the end-user, this represents a formidable start-up learning curve. Most computer conferees are not the least bit interested in the inner mechanism for data processing. They rarely use any of the options, and sometimes a terse message like "ERROR (00045); RESTART" flashes

on the screen. To realize a positive result in a short time, untrained conferees were advised to "turn on the ignition" and search through the publications when a problem arises; then seek further help from more experienced colleagues when necessary.

4. Program products specially designed to feature computer conferencing. While this area of improvement was not essential to this IBM group of professionals, a group removed from the data processing industry would have less tolerance to the long sequence of simple commands sometimes necessary to accomplish a task in computer conferencing on the VNET. Here is just a partial tabulation of convenience features for future consideration (all can be accomplished currently with several key entry steps by the user on the file record):

. annotating   . controlling security

. checking syntax . tagging file instructions

While the user may have less flexibility with such a higher-level language, there is also less opportunity for error.

5. Interactive projection displays for group meetings. Whether the group that is meeting are the conferees themselves face-to-face, or a summary presentation for others, the ability to exhibit the data base on a projection display is essential. When the conferees met in September 1981, the conference room contained a projection display that created the appearance of the 3270 full-screen. With the related keyboard, the central administrator was able to change wording, insert new material, and delete any portion of the file. As agreements were reached, they were converted into the record at that point in time; not only did all participants see the minutes of the meeting, they helped to write them as a group. Later when presentation was given to non-conferees, the 35mm-slides and overhead transparencies were prepared on MVS/TSO to gain color emphasis. Unfortunately this software support is currently available on a separate network, and these files are not visible in monocolor on all the differing display workstations used by the VNET conferees. Also projection displays are not readily available as are overhead and slide projectors.

Conclusion

Many organizations are seeking better office procedures to improve productivity and efficiency of their personnel. In the past, the major progress has occurred in the secretarial and clerical support operations or in computer aids to finance, design, and manu-

facturing. This paper described a case study of professional managers and technical planners creating a technical strategy. In many corporate environments, as IBM, this important application of computer conferencing holds promise for significant cost benefits.

Footnotes

[1] IBM publication SH20-1977-0 "Virtual Machine Facility/370 Networking: Program Reference and Operations Manual".

[2] IBM publications G320-5776-01 "General Markup Language (GML) Beginner's Guide", and SH20-9160-0 "Document Composition Facility: Generalized Markup Language (GML) User's Guide".

[3] IBM publications GC24-5206-2 "IBM Virtual Machine/System Product: VM/Pass-Through Facility General Information", and SC24-5208-0 "IBM Virtual Machine Facility/370: VM/Pass-Through Facility Guide and Reference".

[4] IBM publication SH20-2608-0 "3101 Pass-Through VM/370 Support Program Description/Operations Manual".

G.M. Krembs

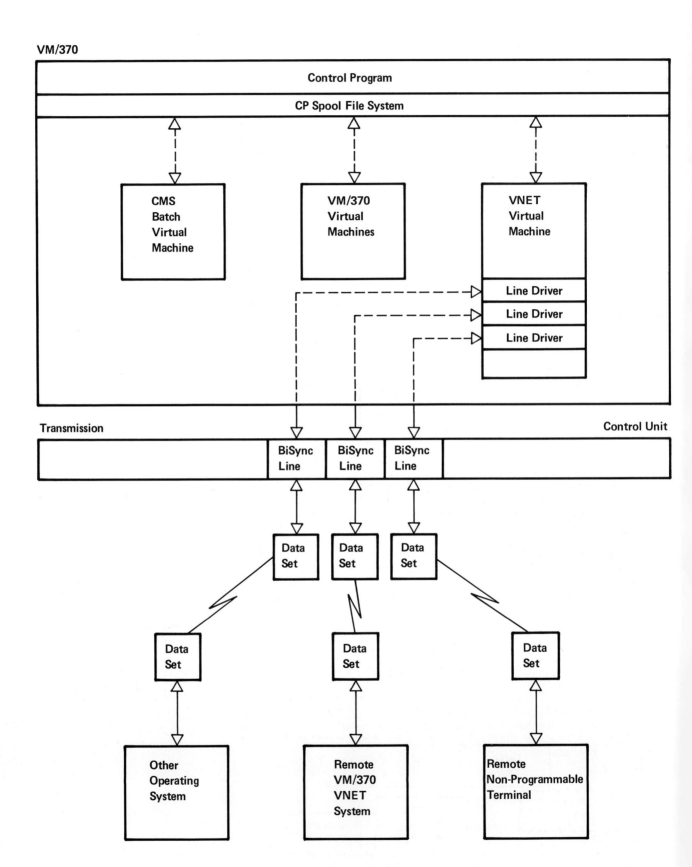

Figure 1   Representative VNET Installation

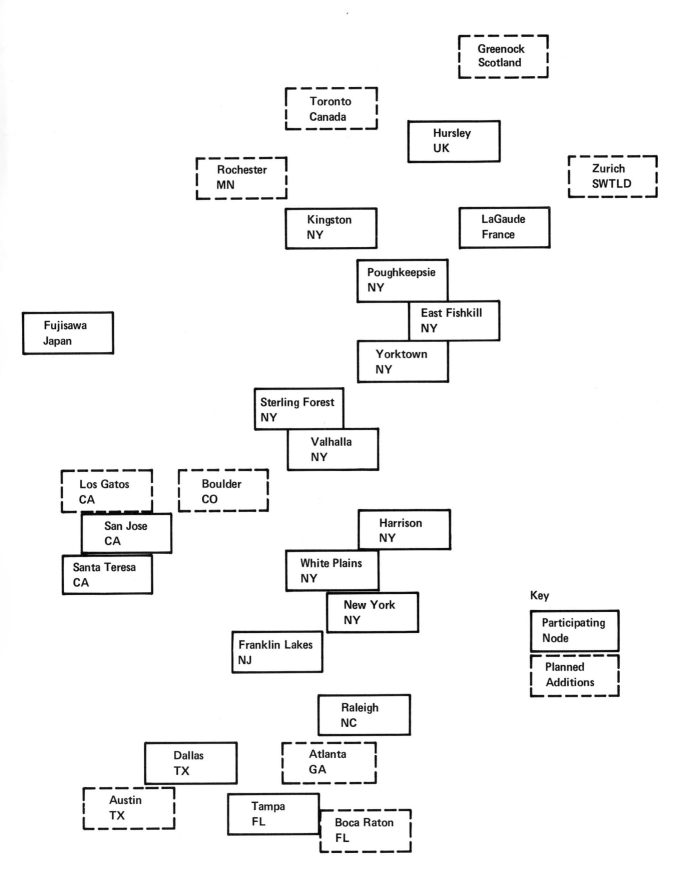

Figure 2   IBM VNET For Case Study

The Teleconferencing Resource Book: A Guide to Applications and Planning
Lorne A. Parker and Christine H. Olgren (eds.)
Elsevier Science Publishers B.V. (North-Holland)
© Center for Interactive Programs, University of Wisconsin-Extension, 1984

432

LEARNING WITHOUT GOING THERE

EDUCATION VIA COMPUTER TELE/CONFERENCING

Thomas B. Cross
Cross Information Company
934 Pearl - Suite B
Boulder, Colorado 80302

Several years ago, Torsten Husen, an eminent Swedish educator stated that by the 1980s, "it should be the rule rather than the exception for a working career to be dramatically affected by technology which translates human talent into machinery, information systems, computer programs, and "precooked" knowledge which is immediately retrievable for use." The 1980s are here, and computer tele/conferencing meets the challenge of the "knowledge explosion." Computer tele/conferencing makes possible what was potential only a few years ago. As stated by Husen:

"That which looms on the horizon is the possibility of communicating both visual and verbal information to individual terminals installed in classrooms or in homes. Learning transmitted in this way can be controlled by information-producing programs which are stored in computer memory units or otherwise kept in some central facility where it is accessible in the form which suits the individual."

Computer tele/conferencing is a system which enables two or more individuals at two or more locations to communicate. Without having to interrupt their work schedules and without having to pay for costly travel, these individuals can exchange information and learning aids. Through keyboard terminals, printers, and telephone lines, participants access a common computer for extremely efficient direct communication. The advantages of computer tele/conferencing are best stated in terms of its extraordinary flexibility: it overcomes geographical constraints (the necessity for travel

either of the educator or the student) and time constraints (conflicting schedules, time zones). Computer tele/conferencing also brings expertise to people where it is not otherwise available. Computer tele/conferencing in short, brings the expert to the student, wherever that student may be located and whenever that student has time to study.

REASONS FOR COMPUTER TELE/CONFERENCING

● NO TIME RESTRICTIONS - "NEVER LATE FOR CLASS AND NO MORE TELEPHONE TAG"

● NO GEOGRAPHICAL RESTRICTIONS - "ALWAYS THERE"

● LOWEST COST OF TELE/CONFERENCING TECHNOLOGIES

● SELF-DOCUMENTING AND FILING - "ELECTRONIC DESK"

● NO ACTORS OR PERFORMING SKILLS REQUIRED

● ALLOWS TIME FOR 'THOUGHT' IN WORK AND MANAGEMENT

● SELF-PACING AND TRAINING

● MULTIPLE PARTICIPATION IN MANY LECTURES IN A DAY

The possibilities of computer tele/conferencing in the field of education are numerous. It has the potential to close the gap between the educational institution and the society it serves. It can rationalize the educational administrative process and facilitate educational planning. Most importantly, it can make available a superior education that utilizes top experts in each field for those who do not have the time or opportunity to travel to their local institution.

CLOSING THE GAP

A continuously changing society with an ever-changing job market challenges educational institutions to develop new eudcational systems. Further education and reorientation should be made available by educational institutions if the needs of today's job market are to be met. Or, when viewed from the perspective of the individual rather than that of society, a person should be allowed to benefit from more education and a higher quality education. Computer tele/conferencing enables a person to reenter the system any time he or she feels like it, despite work schedules, home schedules, and other commitments. In this manner, computer tele/conferencing closes the gap between learning and life, between the educational institution and the job market.

Today, the pattern of overeducated and underemployed (and, increasingly unemployed) is clear. Why has this state of affairs evolved? College and universities, despite the fact that they are major social institutions, often operate in a social vacuum. In some areas of the country, people with master's degrees in English cannot find jobs. While in other areas of the country illiteracy remains a deep problem. There are also large gaps in some professional fields, not enough job applicants with required skills. From another perspective, many corporations are choosing those states with a strong skilled labor force and the educational institutions to back that up.

Often the conclusion reached by both educators and students is that there is no place in industry for someone with a liberal arts degree. Therefore, funds are redirected into MBA programs. The irony, however, is that business is finding that an MBA doesn't necessarily get the job done. What is desired in a job candidate is vocational training interwoven with a solid liberal arts education. Business increasingly places a premium on verbal and written skills and ability to identify and solve problems, grapple with abstract concepts, plan independently, and carry out the plan to its implementation.

What is missing in education today is what has been termed education for work, training that is not simply the assimilation of a technical or business skill per se, but education that cultivates in the student the old-fashioned and time-honored ability to think. Towards this end the ideal courses could be, for example, an engineering course that is taught by an engineer and an English composition teacher, or a management course taught with a corporate manager and a systems theory expert in organizational sociology.

Computer tele/conferencing makes possible such multidisciplinary courses and programs. Diverse expertise on a given topic is available to the student. The computer tele/conferencing network becomes a 'electronic environment' in the thought processes of those connected to each other via the network. Teaching and learning become independent of time and travel restrictions, because busy experts can make effective contributions and give feedback to their student's course of study. Thus, with computer tele/conferencing, a course can be organized through the network rather than in person and taught by the instructor in an asynchronous fashion, i.e., none of the teachers or students need be in the same place at the same time, yet they can be in touch at all times.

Additionally, with computer tele/conferencing, assessment of job market demands in each field can be continually updated and made available to administrators, department chairpersons, faculty, and students so that both teaching and learning can be tailored to meet specific vocational needs as well as particular skills that transcend various professions. Educators can take advantage of computer tele/conferencing to update their skills and become better informed about current ideas and issues in their field.

CONTINUING EDUCATION

Faced with the "knowledge explosion" of an "information-rich" society, education is increasingly and of necessity being viewed as a life-long process rather than a stage or period in an individual's life. Higher and higher costs of education both to the state and to the student are compelling a radical change in the educational system, one that allows for higher teacher-student ratios on one hand, and one that makes possible periods of schooling interspersed and interwoven with gainful employment on the other. Computer tele/conferencing makes expertise available to greater numbers of people who found cost a factor yet are interested in broadening their skills and perspectives, or starting over again in another field.

For those individuals bound by the demands and schedule of their current profession or family life, as is often the case of women with small children, computer tele/conferencing affords the opportunity to take advantage of leisure or nonbusy hours for continuing education.

One of the most innovative applications of computer tele/conferencing is being carried out at the Western Behavioral Sciences Institute in La Jolla, California. The WBSI School of Management and Strategic Studies is a two-year program that incorporates a multi-media approach to the education of executives. Twice a year the participants gather in La Jolla for an intensive 8-day seminar, and the rest of the time they remain 'connected' to the program through print, audio tapes, telephone, and, as well as, computer tele/conferencing.

According to Mr. Darrell Icenogle, Director of Educational Resources at WBSI, "only about 40% of all business telephone calls are completed, and the percentage of completions for the kind of high-level executives we have in this (WBSI) program is of course much lower. With computer tele/conferencing, on the other hand, one has the confidence that a message sent is a message received. Also, communication of this kind tends to be less redundant, more precise, and less pressured. People communicate at their best because they have the time to say exactly what they mean to say."

"Writing is the more personal form of communication, the one which permits the most natural expression of feeling. The message, once detached can cross time and space, acquiring objectivity, permanence and mobility."

Andrew Feenberg
Western Behavioral Sciences
    Institute
Paper on computer tele/conferencing

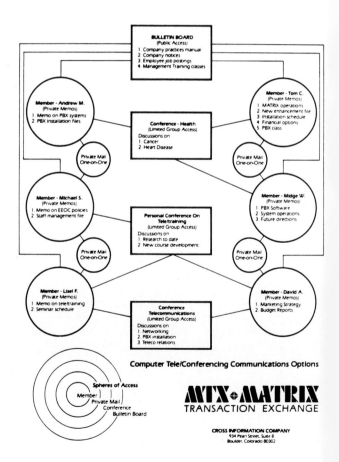

Computer Tele/Conferencing Communications Options

Industries which utilize computer tele/conferencing for meetings, electronic mail, and other ongoing business needs can also make available to employees educational classes that will afford them the opportunity to advance their skills. Industry's need to reeducate employees and train new workers and staff can be accomplished easily with computer tele/conferencing without adding inhouse personnel or taking too much time away from ongoing duties of their training personnel. With computer tele/conferencing, no matter how many individuals are taking a course, basically the material needs to be taught only once. The company can use a combination of in-house managers and university instructors to teach the classes. Such ongoing education and reeducation imparts to employees a sense of functional participation in a company, a broader perspective of the company's overall objectives, and the potential opportunities for advancement within it.

TEACHING/LEARNING METHODS IN THE 1980'S

In the introduction of On Knowing, Jerome Bruner of Harvard University made a challenging declaration: "there is no subject matter," he stated, "which a pupil cannot learn by intellectually honorable means." Too often what prohibits effective learning, educators have discovered, are teaching methods which are standardized and aimed at the average student working at an average speed in an average classroom situation. However, it is felt that the classroom often fosters harmful and unnecessary competition, teaching methods that ignore the above- and below-average student, and a milieu where it is assumed that the student doesn't want to learn and must be forced to learn. As a result, apathy on the part of both teachers and students--is often the result. Swinehart's Rule becomes the norm, rather than the exception "the lecture is that procedure whereby material in the notes of the professor

is transferred to the notes of the student without passing through the mind of either!" Such conditions have led Torsten Husen to conclude "everything must be done to organize school that work along lines that permit learning at individually set speeds and let the methods of class instruction die a slow death."

NOT COMPUTER-PROGRAM EDUCATION

For many years, computer-based education has failed to provide the personal and intimate instruction required by students. Computer tele/conferencing meets the challenge of individualized instruction. It provides an opportunity for self-motivated learning and makes it possible for each student to discern the best learning strategy. Computer-aided instruction (CAI) should not be confused with other computer-driven instructional systems. A clear distinction should be drawn, therefore, between computer tele/conferencing and CAI. CAI is programmed and packaged course material. The computer is programmed to respond to the student's actions. Computer tele/conferencing, on the other hand, allows for genuine communication between instructor and the pupil. With computer tele/conferencing, there is a constant back-and-forth dialogue with the teacher as well as other students, both online.

According to reports from some adults involved in vocational retraining, the distance from the instructor afforded by the computer makes it easier to be told when they're wrong. Along these same lines, computer tele/conferencing eliminates needless comparison of one student to another which, as educators all seem to agree, contributes to a negative self-image that causes feelings of frustration and fear of failure, that can inhibit the learning process considerably. For effective learning to take place students need first and foremost to be compared themselves. That is, the most stimulating comparison and the one that is most appropriately motivating is the one that lets students see their own problem areas and progress. With computer tele/conferencing, students can be totally anonymous to everyone, even to the teacher.

Computer tele/conferencing offers the student not only invdividualized instruction but also the possibility of a <u>healthy</u> cooperation among students. The following list capsulates some of the advantages of computer tele/conferencing in teaching and learning:

● Instruction is individualized in terms of schedule, time methodology, teaching aids, texts, and tests. With individualized instruction, there arises the greatest opportunity to learn by <u>insight</u> rather than by rote, as there is less emphasis on or need for memorization.

● Working at one's own pace means there is improved ability to organize work for the most logical presentation. The discipline of putting thoughts into writing before communicating them improves the quality of communication.

● Instructions are clearer when written rather than spoken. Often two students can have different impressions altogether of what is being taught or asked by the instructor. Computer tele/conferencing eliminates such unnecessary confusion, thereby making both the teacher's job, such as grading the the student's task or writing a report, clearer and therefore easier and more efficient.

● Learning independently increases the possibility for relaxed learning. Instead of the usual competition for grades, custom learning creates a situation of less stress, and <u>higher quality</u> learning is the result. Exit the tyranny of the bell curve! Evaluation can be an ongoing process and can be carried out by peers as well as teachers. Each student has the means readily available for evaluating him/herself.

● Cooperation among students is greater. Students can <u>discuss</u> topics with other students taking the course and even with those who have already completed the course, i.e., there can be an ongoing file of previous students' work, contributions, insights, etc. This translates to mean that all points of view can be exposed for comparative and critical analysis, resulting in richer feedback to the student. This also means that we can stop '<u>reinventing the wheel</u>' and push on to solving more difficult and unknown problems. Team work can take place among students involved in a research project on the same topic.

● Improved project tracking becomes possible for both students and teachers. This allows everyone involved in the project to be informed from the beginning to the end. People can enter the process at any point and have full documentation to evaluate the process todate.

● Far from discouraging dialogue, a computer tele/conferencing system encourages users to put down ideas as they occur to them, and gives them time for reflection which might not be possible in a traditional classroom situation. With computer tele/conferencing, all students have <u>equal time</u> and <u>equal access</u> to speak before the group. Moreover, shy or inhibited students can feel easier about speaking before the audience.

● One of the greatest aspects of this technology is the ability for <u>fully contemplated thought</u>. In most real-time group activities, individuals must respond immediately with the answer or complete a task. In computer tele/conferencing environments, each person has the time to review the materials and requirements with the time to develop a more thought-out answer.

EDUCATIONAL PLANNING AND ADMINISTRATION

Planning for education means taking into consideration the shape of future society. The economy's increasing demand for a combination of general education plus vocational training plus continuing education has already been noted. Creating such programs within the university and linking them to other professional institutions demands a new approach. The educational system must be viewed in its entirety, which means planning for all levels within the university and society at the same time. The development of a worldwide educational network is the goal of EDUNET, a service of EDUCOM. One of the critical tools for this network will be computer tele/conferencing and electronic mail.

A computer tele/conferencing system readily lends itself to faculty activities, work groups, and university management. It creates a forum in which strategies of educational planning for the future can be worked through. Such multilayered strategies would of necessity have to take into account and synthesize everything from future curricula, for example, new teaching methods and materials and administration of new continuing education programs. Expertise of education specialists, administrators, business professionals, and faculty would be called upon. The advantage of computer tele/conferencing is that it lets each of these individuals <u>meet</u> at their convenience independent of others' schedules to review all previous input, respond to it, and add additional thoughts and insights. Planning can thus become an ongoing process that takes into consideration all the expertise and variables necessary for a viable future education program.

DAY-TO-DAY

Computer tele/conferencing facilitates the day-to-day administration of the university, as well as long-term educational planning. A whole myriad of administrative tasks can be brought online, updated, and simplifed enormously with the use of a computer tele/conferencing system. Some of the areas which can be kept online with access by students as well as faculty are:

● Course administration

   Course descriptions
   Course procedures, guidelines, protocols
   Research reports
   Tests/handouts
   Test reports/grade distribution reports
   Reports of technical aids
   Operating manuals
   Bibliographies
   Research aids and guides

● Administering the department

   Faculty load composites
   Faculty review reports
   Minutes of and "online" faculty meetings

● Administering the university

   Admissions information
   Directories
   Notices
   Annual reports
   Updates on alumni information

There are a myriad of other university administration activities which can be enhanced by a computer tele/conferencing system. Electronic mail distribution throughout the university not only speeds the information across the campus, but can be retrieved remotely, for example, by faculty. Maintenance and police department can use the information to keep updated building and facility information. Even the food service can use the system for menu planning.

CONCLUSION

The aim of all education is to train students to assimilate their own new knowledge. Computer tele/conferencing achieves this goal better than the traditional teaching environment because it creates a new environment of motivation. It affords the opportunity for people to get more out of a course, take advantage of leisure time, continue their education, and reeducate themselves. Computer tele/conferencing thus provides the means for students to close the gap between their skills and the demands of the job market. With computer tele/conferencing the university moves out of the realm of ivory tower isolationism and into the society of which it should play a key role.

Likewise the university can through computer tele/conferencing can carry out educational planning and management to better meet the needs of those people who want to be on the cutting edge of the business and professional world. The university can also reach out for people to take courses from anywhere in the whole world. This will create new financial opportunities for universities with declining enrollments as they encourage students who want to <u>be</u> <u>there without going there</u>.

COSTS OF TELE/CONFERENCING

<u>APPLES TO APPLES COMPARISON</u>

<u>TELE/CONFERENCING SYSTEM $ PER HOUR</u>

FULL-MOTION VIDEO - AT&T PICTUREPHONE MEETING SERVICE (PMS) $2400.00*

FULL-MOTION VIDEO SPECIAL EVENT STYLE
$ 600.00**

AUDIO CONFERENCE - Audio bridge
$ 110.00

SLOW-SCAN VIDEO $ 110.00

ELECTRONIC BULLETIN BOARD
AT&T GEMINI $ 66.00

COMPUTER CONFERENCING
MATRIX $ 25.00

This comparison is to demonstrate <u>relative</u> comparisons, actual costs may differ. This comparison does include the necessary hardware where applicable amortized over two years. *Coast-to-coast only - shorter distances are less. **Average estimate.

<u>COMPUTER TELE/CONFERENCING FEATURES</u>

● Conference and Discussion Areas - Have you ever gone to a meeting with a colleague after which you left with entirely different impressions of what took place? Are meetings continually put off because one of the key players is out of town? A conference is the principal area which serves much like a book title. Within the conference there are 'dicussion areas' like the chapters in a book or lessons in a class. Writers use computer terminals to access text (data) files simultaneously by different persons located in many locations. This allows the group to share information in a central area, thus the concept of many-to-many communications with special areas for electronic meetings and discussions is created. These areas are accessed with approval of the conference manager much like a normal meeting. Discussions may be held on various topics, specific interests, and work activities.

● Personal 'yellow pad' notepads and files. Each writer has private secure (optional password protection) files or memo areas which are kept online for ease of use. These files perform the function of an electronic desk - file folders containing memos, plans, and correspondence. These notes or memos can be sent either to other conferees or to the discussion file.

● Online bulletin board - This is the electronic version of the hallway bulletin board. Ads, notes, notices can be posted to this area for system-wide access. Many electronic bulletin boards are used for employment postings, meeting announcements - job notices, company activities, policy and procedures, and a wide range of other bulletins.

● Status and tracking functions - Have you ever missed a staff meeting and wanted to know 'whats going on'? This function provides the conference member with information about what new comments have been sent to the conference, new mail, and information about 'keeping track of' project activities.

● Management reports and directories - Lets the conference member know about how long they have been working at an activity. This is useful to know for people who may be working at distant locations, for charge-back systems, or management reports. The directory is a useful system for finding out what conferences are on the system, conference members, or who else is interested in a conference on current topics such as quality control.

● Online Search Operations - Have you ever lost anything on your desk? Probably everyday. The capability of the computer system to perform functions far more rapidly than by hand helps find information quickly. Text searching of conference members, discussions, and personal notepad areas can save large amounts of time as well as allowing the conference members to be far better organized than in manual systems.

In addition, after the search is completed the results can be dispatched instantanously next door or around the world. Searching is generally performed by allowing the person to, for example, find comments on Venus and Mars but not Jupiter or Saturn. The computer will then retrieve the information requested and present the conference member with a listing of the occurences of these events. These can then be organized into a report or for personal filing.

● Online Voting/Polling/Testing - Reaching a consensus is never easy. Sometimes the group needs to make decisions before proceeding, then again at some point during the project, and then later evaluate what actually took place.

Managers may need to determine proficiency for promotion or allowing use of certain machinery. Marketing research departments are increasingly examining employees to help determine corporate policy toward benefits, hiring practicies, or promotional requirements. This feature allows for secret, roll call (where the voter's are known), true-false or multiple choice votes, polls, or tests.

● Gathering - Once you have a lot of information in separate file folders an important function is to reorganize it into a summary report. The 'gathering' function collects personal memos, mail, conference discussion comments and organizes them in any way you want them.

● Online 'Real-time' Meetings - while most computer conferences are actually performed in a non-real time environment, this features provides for an online conversation mode. The system knows who is presently 'on' the system and can help organize a conversational session like the telephone or face-to-face meetings except that the statements are made in text form. This feature is critically important in emergency situations, decision-making processes, or rapid information exchange as well as a verbatim transcript of the event.

AUTHOR

Thomas B. Cross is Managing Director of Cross Information Company of Boulder, Colorado. He teaches Trends in Information Technology in the graduate program at the University of Denver and is writing a book on tele/conferencing for Prentice-Hall. CIC developed MTX - Matrix Transaction Exchange an advanced computer tele/conferencing and electronic mail system - Winner of the Associated Information Managers - Outstanding Information Technology Award.

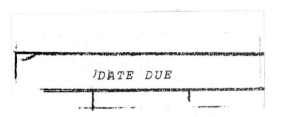
DATE DUE

DATE DUE